Edited by
Aránzazu del Campo and
Eduard Arzt

**Generating Micro- and
Nanopatterns on
Polymeric Materials**

Related Titles

Samori, P., Cacialli, F. (eds.)

Functional Supramolecular Architectures

for Organic Electronics and Nanotechnology

2010

ISBN: 978-3-527-32611-2

Mittal, V. (ed.)

Optimization of Polymer Nanocomposite Properties

2010

ISBN: 978-3-527-32521-4

Ohtsu, M. (ed.)

Nanophotonics and Nanofabrication

2009

ISBN: 978-3-527-32121-6

Harrison, P.

Quantum Wells, Wires and Dots

Theoretical and Computational Physics of Semiconductor Nanostructures

2009

ISBN: 978-0-470-77097-9

Zehetbauer, M. J., Zhu, Y. T. (eds.)

Bulk Nanostructured Materials

2009

ISBN: 978-3-527-31524-6

Rogers, J. A., Lee, H. H. (eds.)

Unconventional Nanopatterning Techniques and Applications

2009

ISBN: 978-0-470-09957-5

Mack, C.

Fundamental Principles of Optical Lithography

The Science of Microfabrication

2007

ISBN: 978-0-470-72730-0

Heinzel, T.

Mesoscopic Electronics in Solid State Nanostructures

2007

ISBN: 978-3-527-40638-8

Edited by
Aránzazu del Campo and Eduard Arzt

Generating Micro- and Nanopatterns on Polymeric Materials

WILEY-VCH Verlag GmbH & Co. KGaA

The Editors

Dr. Aránzazu del Campo
Max-Planck-Institut
für Polymerforshung
Ackermannweg 10
55128 Mainz
Germany

Prof. Eduard Arzt
Leibniz-Institut für
Neue Materialien GmbH
Campus D2 2
66123 Saarbrücken
Germany

All books published by **Wiley-VCH** are carefully produced. Nevertheless, authors, editors, and publisher do not warrant the information contained in these books, including this book, to be free of errors. Readers are advised to keep in mind that statements, data, illustrations, procedural details or other items may inadvertently be inaccurate.

Library of Congress Card No.: applied for

British Library Cataloguing-in-Publication Data
A catalogue record for this book is available from the British Library.

Bibliographic information published by the Deutsche Nationalbibliothek
The Deutsche Nationalbibliothek lists this publication in the Deutsche Nationalbibliografie; detailed bibliographic data are available on the Internet at http://dnb.d-nb.de.

© 2011 WILEY-VCH Verlag & Co. KGaA, Boschstr. 12, 69469 Weinheim, Germany

All rights reserved (including those of translation into other languages). No part of this book may be reproduced in any form – by photoprinting, microfilm, or any other means – nor transmitted or translated into a machine language without written permission from the publishers. Registered names, trademarks, etc. used in this book, even when not specifically marked as such, are not to be considered unprotected by law.

Composition   Thomson Digital, Noida, India
Printing and Binding   Fabulous Printers Pte Ltd, Singapore
Cover Design   Schulz Grafik-Design, Fußgönheim

Printed in Singapore
Printed on acid-free paper

ISBN: 978-3-527-32508-5

Contents

Preface *XIII*
List of Contributors *XV*

Part One   Molding *1*

1   Materials and Processes in UV-Assisted Nanoimprint Lithography *3*
         Marc Zelsmann and Jumana Boussey
1.1   Introduction *3*
1.2   UV-Assisted Nanoimprint Lithography   *5*
1.2.1   Process Details and Variants   *5*
1.2.2   Resist Flow in Thin Layers   *7*
1.2.3   Imprinting Examples and Resolution   *8*
1.2.4   Tools and Industrialization Issues   *9*
1.3   Imprinting Materials   *11*
1.3.1   State-of-the-Art   *11*
1.3.2   Shrinkage   *13*
1.3.3   Plasma Etching Processes and Resist Stripping   *14*
1.4   Mold Fabrication and Anti-Sticking Strategies   *16*
1.4.1   Mold Fabrication and Characterization   *16*
1.4.2   Anti-Sticking Strategies   *17*
1.4.3   Mold Treatment   *18*
1.4.4   Fluorinated Mold Treatment-Resist Interactions   *20*
1.4.5   Resist-Oriented Anti-Sticking Solutions   *21*
1.4.6   Polymeric Mold Materials and Stamp Copies   *21*
1.5   Conclusion   *22*
         References   *23*

2   Roll-to-Roll Nanoimprint Lithography and Dynamic Nano-Inscription *27*
         Se Hyun Ahn and L. Jay Guo
2.1   Introduction   *27*

Generating Micro- and Nanopatterns on Polymeric Materials. Edited by A. del Campo and E. Arzt
Copyright © 2011 WILEY-VCH Verlag GmbH & Co. KGaA, Weinheim
ISBN: 978-3-527-32508-5

2.2	Roll-to-Roll Nanoimprint Lithography 29
2.3	Dynamic Nano-Inscription 36
2.4	Summary 40
	References 40

3	**Solvent-Assisted Molding** 43
	Ho-Sup Jung and Kahp-Yang Suh
3.1	The Principle of Solvent-Assisted Molding 44
3.2	Solvent-Assisted Molding with a Good Solvent 47
3.3	Solvent-Assisted Molding with a Poor Solvent 49
3.4	Other Techniques 52
3.5	Applications of Solvent-Assisted Molding 53
3.6	Conclusions 55
	References 55

4	**Soft Lithography and Variants** 57
	Elena Martínez and Josep Samitier
4.1	Introduction 57
4.2	Key Features of Soft Lithography 58
4.3	Microcontact Printing of Self-Assembled Monolayers 59
4.4	Soft Molding Techniques 61
4.4.1	Nanoimprinting with Soft Stamps 62
4.4.2	Micromolding in Capillaries and Microtransfer Molding 63
4.4.3	Solvent-Assisted Molding 64
4.4.4	UV Molding 64
4.4.5	Forced Soft Lithography 65
4.5	Summary 66
	References 66

Part Two Writing and Printing 69

5	**Transfer Printing Processes** 71
	Luciano F. Boesel
5.1	Introduction 71
5.2	Techniques 72
5.2.1	Microtransfer Molding 72
5.2.2	Reversal Nanoimprinting 72
5.2.2.1	Reversal Nanoimprinting with Soft Inkpads 75
5.2.3	Decal Transfer Microlithography 76
5.2.4	Duo-Mold Imprinting 77
5.2.5	Printing on Topographies (Multilayer Printing) and on Flexible Substrates 79
5.3	Key Issues in Transfer Printing Methods 82
5.3.1	Surface Treatments of Mold and Substrate 82
5.3.2	Residual Layer 83

5.4	Advantages and Disadvantages	85
5.5	Applications	88
	References	89

6	**Direct-Write Assembly of 3D Polymeric Structures**	**93**
	Sara T. Parker and Jennifer A. Lewis	
6.1	Introduction	93
6.2	Polyelectrolyte Inks	95
6.2.1	Ink Design and Rheology	95
6.2.2	Polyelectrolyte Structures as Templates for Photonic and Biomimetic Applications	97
6.3	Silk Fibroin Inks	98
6.3.1	Ink Design and Rheology	98
6.3.2	Silk Scaffolds for Tissue Engineering Applications	99
6.4	Hydrogel Inks	101
6.4.1	Ink Design and Rheology	101
6.4.2	Direct-Write Assembly of 3D Hydrogel Scaffolds	102
6.5	Opportunities and Challenges	103
	References	104

Part Three Laser Scanning 107

7	**Three-Dimensional Microfabrication by Two-Photon Polymerization**	**109**
	Tommaso Baldacchini	
7.1	Introduction	109
7.2	Fundamentals	110
7.3	Materials	113
7.3.1	Photo-initiators	113
7.3.2	Mixtures of Monomers and Oligomers	114
7.4	Experimental Setup	116
7.5	Resolution	119
7.6	Microstructures: Properties and Characterization	123
7.7	Applications	128
7.8	Limitations and Future Directions	131
	References	133

8	**Laser Micromachining of Polymers**	**141**
	Chantal G. Khan Malek, Wilhelm Pfleging, and Stephan Roth	
8.1	Introduction	141
8.2	Principles of Beam-Matter Interaction in Ablation Processes	142
8.3	Laser Ablation of Polymers	147
8.3.1	Ultraviolet Laser	147
8.3.2	CO_2 Laser	151
8.3.3	Femtosecond Laser	153

8.4	Laser-Induced Roughening	*155*
8.5	Generative Laser Processes	*158*
8.5.1	Microstereolithography	*158*
8.5.2	Selective Laser Sintering	*160*
8.6	Conclusion	*161*
	References	*162*

Part Four Self-Organization *169*

9   Colloidal Polymer Patterning *171*
Eoin Murray, Philip Born, and Tobias Kraus

9.1	Introduction	*171*
9.2	Emulsion Polymerization	*172*
9.3	Forces and Mechanisms in Polymer Dispersions	*173*
9.3.1	Brownian Motion	*173*
9.3.2	Sedimentation and Viscous Drag	*174*
9.3.3	Particle–Particle and Particle–Surface Interactions	*174*
9.3.4	Mechanisms and Techniques of Structure Formation	*176*
9.3.5	Setups	*178*
9.3.5.1	Fabrication of Three-Dimensional Arrays	*178*
9.3.5.2	Fabrication of Two-Dimensional Arrays	*179*
9.4	Polymer Patterns from Colloidal Suspensions	*181*
9.4.1	Polymer Particle Assembly	*182*
9.4.2	Colloidal Lithography	*185*
9.4.3	Polymer Opals	*187*
9.4.4	Applications	*188*
9.5	Summary and Outlook	*191*
	References	*191*

10   Directed Self-Assembly of Block Copolymer Films *199*
Gordon S.W. Craig and Paul F. Nealey

10.1	Introduction	*199*
10.2	Energetics of the Basic Directed Assembly System	*200*
10.3	Examples of Directed Assembly	*207*
10.3.1	Non-Bulk Morphologies	*207*
10.3.2	Lithographic Applications	*210*
10.4	Conclusion	*214*
	References	*214*

11   Surface Instability and Pattern Formation in Thin Polymer Films *217*
Rabibrata Mukherjee, Ashutosh Sharma, and Ullirich Steiner

11.1	Introduction	*217*
11.2	Origin of Surface Instability	*220*
11.3	Polymer Thin Film Dewetting	*221*

11.3.1	Spontaneous Instability or Spinodal Dewetting: Theoretical Aspects *221*	
11.3.2	Dewetting: Experiments *225*	
11.3.3	Heterogeneous Dewetting *229*	
11.3.4	Influence of Residual Stresses on Film Rupture and Dewetting *231*	
11.3.5	Pattern Formation in Polymer Thin Film Dewetting *232*	
11.4	Dewetting on Patterned Substrates *233*	
11.4.1	Dewetting on Chemically Patterned Substrates *234*	
11.4.2	Dewetting on Physically Patterned Substrates *238*	
11.4.3	Pattern Directed Dewetting: Theory and Simulation *241*	
11.5	Instability due to Externally Imposed Fields *246*	
11.5.1	Electric Field-Induced Patterning: Theory *250*	
11.5.2	Electric Field-Induced Patterning of Polymer Bilayers *252*	
11.5.3	Thermal Gradient-Induced Patterning *255*	
11.6	Conclusion *257*	
	References *258*	

Part Five   Applications *267*

12   Cells on Patterns *269*
           Aldo Ferrari and Marco Cecchini
12.1   Introduction *269*
12.2   Physicochemical Properties of the Substrate Read by Cells *270*
12.2.1   Density of Adhesion Points *272*
12.2.1.1   Topography *275*
12.2.2   Rigidity of the Substrate *280*
12.3   Conclusions *285*
           References *286*

13   Polymer Patterns and Scaffolds for Biomedical Applications and Tissue Engineering *291*
           Natália M. Alves, Iva Pashkuleva, Rui L. Reis, and João F. Mano
13.1   Introduction *291*
13.2   Cell Response to 2D Patterns *292*
13.3   Cells onto 3D Objects and Scaffolds *296*
13.4   Concluding Remarks *300*
           References *301*

14   Nano- and Micro-Structured Polymer Surfaces for the Control of Marine Biofouling *303*
           James A. Callow and Maureen E. Callow
14.1   Introduction *303*
14.1.1   The Fouling Interface *303*
14.1.2   Surface and Coating Designs to Investigate and Control Biofouling *305*

14.2	Replica Molding in PDMS and Other Polymers	*306*
14.3	Stretched Topographies in PDMS	*309*
14.4	Structured Surfaces by Self-Assembly	*309*
14.4.1	Block Copolymers with Amphiphilic, Fluorinated and PEG-ylated Side Chains	*309*
14.4.2	Polystyrene-Based Diblock Copolymers	*311*
14.4.3	Hyperbranched Amphiphilic Networks	*311*
14.4.4	Phase-Segregating Quaternized Siloxanes and Siloxane-Urethane Nanohybrids	*311*
14.5	Nanocomposites	*312*
14.5.1	Carbon Nanotube-Filled Polysiloxanes	*313*
14.5.2	Nanostructured Superhydrophobic Surfaces	*313*
14.6	Nanostructured Polymer Surfaces by Vapor Deposition Methods	*314*
14.7	Conclusions	*315*
	References	*316*
15	**Bioinspired Patterned Adhesives**	*319*
	Marleen Kamperman, Eduard Arzt, and Aránzazu del Campo	
15.1	Introduction	*319*
15.2	Vertical Structures	*320*
15.2.1	E-Beam Lithography	*320*
15.2.2	Filling Porous Membranes	*321*
15.2.3	Photolithographic Templating	*321*
15.2.4	Hot Embossing	*323*
15.3	Tilted Structures	*324*
15.3.1	Filling Nanoporous or Photolithographic Templates	*324*
15.3.2	Drawing Polymer Fibers	*325*
15.3.3	Electron-Beam Irradiation	*326*
15.4	Coated Structures	*326*
15.5	Hierarchical Structures	*327*
15.5.1	Filling Stacked Membranes	*327*
15.5.2	Multistep Exposure in Photolithography	*327*
15.5.3	Microfabrication Technologies	*329*
15.5.4	Two-Step Embossing	*330*
15.6	3D Structures	*330*
15.6.1	Reactive Ion Etching	*330*
15.6.2	Post-Molding Inking	*331*
15.7	Switchable Adhesion	*332*
15.8	Outlook	*332*
	References	*333*
16	**Patterned Materials and Surfaces for Optical Applications**	*337*
	Peter W. de Oliveira, P. Rogin, M. Quilitz, and Eduard Arzt	
16.1	Introduction	*337*
16.1.1	Optical Materials, Light and Structures	*337*

16.1.2	Interaction of Light and Matter – Basic Considerations	*338*
16.1.3	Optical Microstructures in Nature	*342*
16.2	Optical Micro- and Nanostructures for Applications	*343*
16.2.1	Effects of Reflection and Refraction	*344*
16.2.1.1	Anti-Glare	*344*
16.2.1.2	Dielectric Antireflective Coatings	*345*
16.2.1.3	Moth Eyes	*346*
16.2.1.4	Holograms	*347*
16.2.2	Waveguides	*353*
16.2.2.1	Fabrication of Waveguides	*354*
16.2.2.2	Waveguide Devices	*354*
16.3	Conclusion	*357*
	References	*357*

Index *361*

Preface

Recent innovations in the area of micro- and nanofabrication have created a unique opportunity for patterning surfaces with features with lateral dimensions spanning from the nano- to millimeter range. The microelectronics industry and the need for smaller and faster computing systems have pushed this development during the past two decades, mainly focused on obtaining patterns with the smallest possible lateral dimensions via optical lithography in its multiple variants. In parallel, new application fields for miniaturized devices (i.e., lab-on-a-chip) and interesting properties for structured coatings (see Part Five of this book) have emerged. These applications have pushed the development of alternative patterning technologies more suited for plastic manufacturing.

Classical processing techniques in the polymer industry such as molding and printing have now been adapted to the micro- and nanoscale to produce such devices at laboratory and manufacture scales with submicrometric resolution using different materials. The most illustrative examples (UV, thermal, solvent, and soft molding) are collected in Part One (Chapters 1–4) of this book. In its simplest conception, molding is limited to 2D surface designs. However, molding can be combined with printing and transfer steps to create more complex geometries by stacking layers with different pattern and material designs in a multistep process (Chapter 5). In this way, surface structures with several hierarchy levels, tilted, reentrant, or suspended geometries become realizable. Patterning curved substrata is also a difficult task that has been achieved by direct writing with small quantities of material using micronozzles (Chapter 6). This strategy allows generation of almost any 3D pattern on any material provided the material solution has the appropriate rheological properties.

Laser scanning can be used for microfabrication of complex, 3D geometries via ablation (Chapter 8) or via two-photon polymerization (Chapter 9). Both are serial techniques that offer great flexibility in the geometrical design but cannot be applied to any material of choice. Finally, colloidal assemblies (Chapter 9), phase-separated block copolymers (Chapter 10), and surface instabilities (Chapter 11) exploit soft-matter inherent ordering phenomena to build periodic patterns. Feature size can be tuned by the molecular architecture and/or an external field and nanodomains with

Generating Micro- and Nanopatterns on Polymeric Materials. Edited by A. del Campo and E. Arzt
Copyright © 2011 WILEY-VCH Verlag GmbH & Co. KGaA, Weinheim
ISBN: 978-3-527-32508-5

a variety of motifs, chemistries, and tailored size, and periodicity might be created without the need of a mold, a great advantage against techniques described in Part One. This book offers a detailed description of all these techniques and highlights some applications.

December 2010

Aránzazu del Campo, Mainz, Germany
Eduard Arzt, Saarbrücken, Germany

List of Contributors

Se Hyun Ahn
The University of Michigan
Department of Mechanical Engineering
1301 Beal Ave.
Ann Arbor, MI 48109-2122
USA

Natália M. Alves
University of Minho
Department of Polymer Engineering
3B's Research Group - Biomaterials,
Biodegradables and Biomimetics
Campus de Gualtar
4710-057 Braga
Portugal

Eduard Arzt
INM Leibniz-Institut für Neue
Materialien gGmbH
Campus D2 2
66123 Saarbrücken
Germany

Tommaso Baldacchini
Newport Corporation
Technology & Applications Center
1791 Deere Avenue
Irvine, CA 92606
USA

Luciano F. Boesel
Max-Planck-Institut für
Polymerforschung
Ackermannweg 10
55128 Mainz
Germany

and

Swiss Federal Laboratories for Materials
Science and Technology
EMPA
Lerchenfeldstrasse 5
9014 St. Gallen
Switzerland

Philip Born
INM Leibniz-Institut für Neue
Materialien gGmbH
Campus D2 2
66123 Saarbrücken
Germany

Jumana Boussey
Laboratoire des Technologies de la
Microélectronique (LTM)
CNRS, CEA-LETI-MINATEC
17 rue des Martyrs
38054 Grenoble
France

James A. Callow
University of Birmingham
School of Biosciences
Birmingham B15 2TT
UK

Maureen E. Callow
University of Birmingham
School of Biosciences
Birmingham B15 2TT
UK

Marco Cecchini
NEST
CNR-INFM and Scuola Normale Superiore
Piazza San Silvestro 12
56126 Pisa
Italy

Gordon S.W. Craig
University of Wisconsin-Madison
Department of Chemical and Biological Engineering
Madison, WI 53706
USA

Peter W. de Oliveira
INM Leibniz-Institut für Neue Materialien gGmbH
Campus D2 2
66123 Saarbrücken
Germany

Aránzazu del Campo
Max-Planck-Institut für Polymerforschung
Ackermannweg 10
55128 Mainz
Germany

Aldo Ferrari
ETH Zurich
Laboratory of Thermodynamics in Emerging Technologies
Sonneggstrasse 3
8092 Zurich
Switzerland

L. Jay Guo
The University of Michigan
Department of Electrical Engineering and Computer Science
1301 Beal Ave.
Ann Arbor, MI 48109-2122
USA

Ho-Sup Jung
Seoul National University
School of Mechanical and Aerospace Engineering
Seoul 151-742
Korea

Marleen Kamperman
INM Leibniz-Institut für Neue Materialien gGmbH
Campus D2 2
66123 Saarbrücken
Germany

and

University of Wageningen
Physical Chemistry and Colloid Science
Dreyen plein 6
6703 HB Wageningen
Netherlands

Chantal G. Khan Malek
FEMTO-ST Institute - UMR CNRS 6174
Department of Micro Nano Sciences & Systems
32 Av. de l'Observatoire
25044 Besançon
France

Tobias Kraus
INM Leibniz-Institut für Neue
Materialien gGmbH
Campus D2 2
66123 Saarbrücken
Germany

Jennifer A. Lewis
University of Illinois at Urbana-
Champaign
Beckman Institute for Advanced
Science and Technology
Department of Materials Science and
Engineering and Frederick Seitz
Materials Research Laboratory
Urbana, IL 61801
USA

João F. Mano
University of Minho
Department of Polymer Engineering
3B's Research Group - Biomaterials,
Biodegradables and Biomimetics
Campus de Gualtar
4710-057 Braga
Portugal

Elena Martínez
Institute for Bioengineering of
Catalonia (IBEC)
Nanobioengineering Group
Baldiri Reixac 10-12
08028 Barcelona
Spain

and

Centro de Investigación Biomédica en
Red en Bioingeniería
Biomateriales y Nanomedicina
(CIBER-BBN)
50018 Zaragoza
Spain

Rabibrata Mukherjee
Indian Institute of Technology
Department of Chemical Engineering
Kharagpur 721 302
India

Eoin Murray
INM Leibniz-Institut für Neue
Materialien gGmbH
Campus D2 2
66123 Saarbrücken
Germany

Paul F. Nealey
University of Wisconsin-Madison
Department of Chemical and Biological
Engineering
Madison, WI 53706
USA

Sara T. Parker
University of Illinois at Urbana-
Champaign
Beckman Institute for Advanced
Science and Technology
Department of Materials Science and
Engineering and Frederick Seitz
Materials Research Laboratory
Urbana, IL 61801
USA

Iva Pashkuleva
University of Minho
Department of Polymer Engineering
3B's Research Group - Biomaterials
Biodegradables and Biomimetics
Campus de Gualtar
4710-057 Braga
Portugal

Wilhelm Pfleging
Karlsruhe Institute of Technology (KIT)
Institute for Materials Research I
Hermann-von-Helmholtz-Platz 1
76344 Eggenstein-Leopoldshafen
Germany

M. Quilitz
INM - Leibniz-Institut für Neue
Materialien GmbH
Campus D2 2
66123 Saarbrücken
Germany

Rui L. Reis
University of Minho
Department of Polymer Engineering
3B's Research Group - Biomaterials,
Biodegradables and Biomimetics
Campus de Gualtar
4710-057 Braga
Portugal

P. Rogin
INM - Leibniz-Institut für Neue
Materialien GmbH
Campus D2 2
66123 Saarbrücken
Germany

Stephan Roth
Bayerisches Laserzentrum GmbH
Konrad-Zuse-Straße 2-6
91052 Erlangen
Germany

Josep Samitier
Institute for Bioengineering of
Catalonia (IBEC)
Nanobioengineering Group
Baldiri Reixac 10-12
08028 Barcelona
Spain

and

University of Barcelona
Department of Electronics
C/ Martí i Franquès 1
08028 Barcelona
Spain

and

Centro de Investigación Biomédica en
Red en Bioingeniería
Biomateriales y Nanomedicina
(CIBER-BBN)
50018 Zaragoza
Spain

Ashutosh Sharma
Indian Institute of Technology
Department of Chemical Engineering
and DST Unit on Nanosciences
Kanpur 208 016
India

Ullrich Steiner
University of Cambridge
Department of Physics
Cavendish Laboratory
J.J. Thomson Avenue
Cambridge CB3 0HE
UK

Kahp-Yang Suh
Seoul National University
School of Mechanical and Aerospace
Engineering
151-742 Seoul
Korea

Marc Zelsmann
Laboratoire des Technologies de la
Microélectronique (LTM)
CNRS, CEA-LETI-MINATEC
17 rue des Martyrs
38054 Grenoble
France

Part One
Molding

1
Materials and Processes in UV-Assisted Nanoimprint Lithography
Marc Zelsmann and Jumana Boussey

1.1
Introduction

Nanoimprint lithography (NIL), first proposed by S. Chou in 1995 [1], is a high resolution and high throughput lithography technique based on the mechanical deformation of a resist layer with a stamp (or mold) presenting a surface topography (including eventually three-dimensional (3D) features). A schematic of this technique is shown in Figure 1.1. After the pattern formation, the polymer layer may be used as a resist mask for additional processing steps (transfer etching in the substrate, ion implantation, material deposition, lift-off . . .), or this layer may be used as it is, as a functional material. A residual resist layer is always observed under the mold protrusions after imprinting. This layer can be removed with an anisotropic "breakthrough" plasma etching step to obtain a conventional lithography resist mask.

Molds are in general fabricated with high resolution techniques, mainly electron-beam lithography [2], allowing NIL to achieve resolutions beyond the limitations set by light diffraction in optical projection lithography. Furthermore, due to parallel fabrication of features over large areas, high throughput production is possible. NIL is considered as a next generation lithography (NGL) technique in the microelectronics industry for the fabrication of integrated circuits (IC) [3]; in addition it is also being developed as a fabrication technique for applications where electron-beam lithography or state-of-the-art photolithography cannot achieve sufficiently high resolution at reasonable cost or where the capability of 3D imprinting or imprinting in a functional material is needed.

Two main process families can be distinguished: thermal NIL [1] and ultraviolet-assisted NIL (UV-NIL) [4]. The first one is described in Chapter 2 of this book. In this case, the mold is usually pressed into a thin thermoplastic polymer film heated above its glass transition temperature where the polymer can flow under quite high pressure. The viscosity of the thin heated polymer layer remains a few orders of magnitude higher than that of a monomer layer [5, 6]. The mold is generally made of silicon, using advanced and established processes from the microelectronics industry. Furthermore, the nonflatness of the mold and the substrate is in general compensated by quite a high imprinting pressure (with the help of mold/substrate

Generating Micro- and Nanopatterns on Polymeric Materials. Edited by A. del Campo and E. Arzt
Copyright © 2011 WILEY-VCH Verlag GmbH & Co. KGaA, Weinheim
ISBN: 978-3-527-32508-5

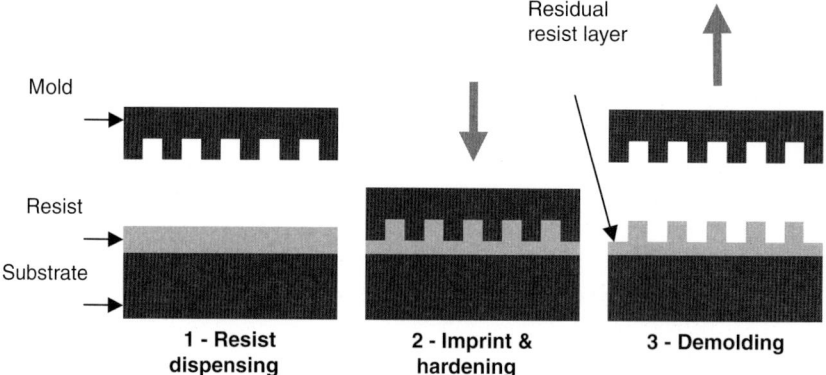

Figure 1.1 Schematic of the nanoimprint process.

bending and compliant layers). Also, resists are more readily available as most of the resists used in electron-beam or photolithography are thermoplastic polymers.

In UV-NIL, a low viscosity monomer resist is pressed at room temperature with a transparent mold at a limited pressure and later polymerized by UV light to form solid structures. In this case, a rigid and UV-transparent mold, usually from fused silica, is more difficult to fabricate. In addition, only few dedicated resists are commercially available and flatness issues are more critical due to the low viscosity resist used. Nevertheless, key advantages of UV-NIL make it a very attractive process for industrial applications:

- UV-NIL can be performed at room temperature, there is no need to heat and cool the mold/wafer stack as in thermal NIL. This leads to higher throughput, as UV curing takes only a few seconds, and improved fidelity of patterns. Also the process does not suffer from thermal expansion mismatch, as in thermal NIL, which might induce distortions of the printed patterns. (For this reason, it is mandatory to use the same material for mold and substrate in thermal NIL).
- With UV-transparent stamps, high alignment accuracy between different lithography levels is easier to implement in UV-NIL [7].
- Low viscosity resists allow an improved resist flow and redistribution leading to better printing uniformity, especially when the mold design includes micro- and nanometer features at the same time or areas with different densities. Additionally, with these low-viscosity resists, UV-NIL can be performed at low pressure (<1 bar). This is essential for molding films onto delicate substrates and releases constraints on the mechanical properties of the mold.
- A step-and-repeat process allows the fabrication of smaller and cheaper stamps and a better control of the placement accuracy [8]. Such a process is easier to implement with a UV system than with a thermal system.
- Finally, the properties of the cured polymer after imprinting can be adapted to dedicated applications by changing the resist formulation or the irradiation time.

In this chapter, we will deal with some issues involved in UV-NIL, in particular concerning the molding process itself, the resists used, the mold fabrication and its anti-sticking treatment. Some of the questions treated may apply to thermal NIL as well. It should be noted that we will not consider embossing or molding of thick materials in this chapter, but of thin layers of resists (i.e., channels in which the resist flows in very narrow spaces). This affects many issues such as printing uniformity and resist flow, among others. Nevertheless, for lithography, it is absolutely necessary to use a resist layer as thin as possible to be able to transfer the features to the underlying substrate. Also, a reduced resist thickness might be an important point in many applications using as-printed polymer layers as functional materials. This chapter will not cover specific nanoimprinting processes such as solvent-assisted NIL [9, 10] (see Chapter 3), reversal NIL [11–13], which involves resist transfer from the mold to the substrate, or hybrid processes (combination of NIL and photolithography for example [14]) that may use UV light and room-temperature imprinting as well.

1.2
UV-Assisted Nanoimprint Lithography

1.2.1
Process Details and Variants

In UV-NIL, a liquid material is coated onto a substrate and pressed, at room temperature, against a rigid UV-transparent mold. The displacement of the low viscosity resist is not only due to the imprinting pressure applied (squeeze flow), but also due to the capillary forces in the system. Consequently, only a small pressure (<1 bar) is sufficient to fill the mold cavities. Furthermore, this low force ensures a uniform imprint. Indeed, mold and substrate can be assumed to be completely rigid so that they approach in a perfectly parallel manner, insuring a uniform residual layer. Nevertheless, this ideal case works only with perfectly flat surfaces, which are difficult to obtain in real experiments. To ease the process and to be able to imprint over topography, planarization layers might be necessary [15, 16]. Also, imprinting at low force using two rigid surfaces is facilitated when the contact area is limited; this requires, in general, working with small molds. This is also an advantage for mold fabrication, as it is easier and cheaper to write a limited area for example by electron-beam lithography. Also, defect inspection and repair are also facilitated on smaller stamps. This is the reason why UV-NIL is often developed in a "step-and-repeat" process, where the mold, with dimensions in the range of 25×25 mm^2, is stepped to pattern the whole wafer area as in a stepper lithography tool.

To apply resists onto wafers, two techniques are used. The first one is spin-coating, already used in standard lithography. Its main advantage is its excellent thickness uniformity over large areas. As liquid low-viscosity resists are used in UV-NIL, the formation of a stable thin liquid layer by spin-coating is not straightforward. The resist has to wet the substrate surface, which depends on the resist used and might require a special wafer treatment. Additionally, low-viscosity resists with a limited vapor pressure must be chosen to ensure that all dies will have the same initial resist layer thickness ant to allow imprinting in vacuum. In practice, this means that the viscosity cannot be smaller than about 30 mPa.s.

On the other hand, the resist can be drop-dispensed on the wafers. This can be done for only one imprinting area just before contact with the mold, limiting the possible evaporation of resist and allowing working with resist viscosities as low as a few mPa.s [17]. Furthermore, drops as small as 1 pL can be used [18], which corresponds to individual imprinted area of about $10 \times 10\,\mu m^2$ on the wafer; depending on structure size and density, this corresponds to about 60 000 drops per die in standard step-and-repeat conditions. This large number of drops gives the possibility to adapt the resist quantity to the mold design in the die itself, which leads to an improved printing uniformity, and shorter imprinting times, due to the limited resist flow. The combination of step-and-repeat processing with tunable drop-dispensing (drop-on-demand™) has given rise to "step and flash™" imprint lithography (S-FIL) and lately to "jet and flash™" imprint lithography (J-FIL), trademarks of Molecular Imprint Inc. [19, 20]. Also, the multidroplet geometry seems to favor filling of the mold cavities [21]. When the mold approaches the coated substrate, the fluid droplets spread out and fill the cavities under capillary action, and the capillary force around each drop attracts the mold, enhancing the effective imprinting force. It was shown that smaller drops (~pL) induce an improved capillary action [21] but are more difficult to produce at high speed with good placement accuracy, generating issues on the drop dispensing unit. On the other hand, larger drops (~100 pL) are more difficult to displace (longer flow time). They induce constraints on the mold design and will obviously generate a larger quantity of expelled resist at the mold edges. Nevertheless, not all resists can be drop-dispensed, especially in small drops. This depends on their visco-elastic properties and can be a limitation, in particular in applications when especially dedicated resists are used.

Another development of S-FIL is S-FIL reverse [20, 22] (S-FIL/R, see Figure 1.2). In this case, a purely organic resist is first imprinted as in S-FIL. Then, a more plasma-etch-resistant resist (silicon-containing) is spin-coated onto the imprinted features

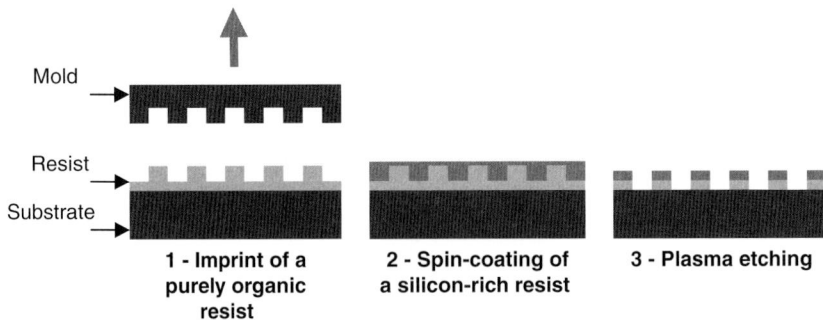

Figure 1.2 Schematic of the S-FIL/R nanoimprint process.

acting as a planarization layer. Finally, this stack is etched in a plasma reactor to produce the patterned resist mask. This technique is able to generate higher aspect ratio features and is less sensitive to non-uniformities in the imprinted residual resist layer. Also, it allows the fabrication of patterns with the same polarity as on the mold and, due to the two-layer resist, a lift-off process is possible. Additionally, there is no need to imprint the silicon-containing etch-resistant resist. Also, resist contaminations on the mold are easier to remove since the mold is in contact only with a purely organic resist. Finally, the silicon-rich resist does not have to exhibit low viscosity and can be applied with high uniformity by spin-coating.

Alternatively to the step-and-repeat process, UV-NIL is also used with larger stamps [23, 24]. In this case, the imprinting pressures applied are in general much larger than in the step-and-repeat process to ensure a conformal contact between mold and substrate. In order to avoid imprinting non-uniformities, it is then very important that the tool used is able to apply a uniform (isotropic) imprint pressure even on nonflat surfaces. This is possible using soft pistons or membranes. Another alternative for ensuring conformal contact, even with quite low imprint forces, is to use a soft stamp [25]. Nevertheless, with these techniques, it becomes more difficult to realize high accuracy alignments of successive lithography levels due to the tool architecture or to the fact that the mold is not stiff enough.

1.2.2
Resist Flow in Thin Layers

As explained above, in UV-NIL, resist displacement is promoted both by the applied imprinting force and by the capillary forces. The balance between the two phenomena is not clear and depends strongly on process conditions (mold treatment, wafer treatment, resist viscosity and surface energy and resist coating type); in UV-NIL the effect of capillarity is invariably increased. Nevertheless, the squeeze flow of a supposedly perfectly viscous resist can be described quite simply to a first approximation by Stefan's law [26]. For a line, the imprinting time can be written as [27]:

$$t_{IMPRINT} = \frac{\eta_0 s^2}{2p}\left(\frac{1}{h_f^2} - \frac{1}{h_0^2}\right)$$

where η_0 is the zero shear viscosity of the resist, s the width of the line, h_0 the initial resist layer thickness, h_f the final resist layer thickness (residual layer) and p the effective imprinting pressure on the line (see also Figure 1.3).

We note that a shorter imprinting time can be achieved by a higher imprinting pressure. Furthermore, the linewidth s as well as the residual resist thickness h_f plays a large role. Indeed, for a mold containing isodense 100 nm lines and spaces, the time needed to press a resist, with 10 mPa.s viscosity, from 100 to 15 nm thickness applying a pressure of 1 bar is about 1 μs. (In comparison, in the case of thermal NIL, where the viscosity is higher than 1000 Pa.s, the imprinting time under the same conditions is larger than 0.1 s.) This time is extremely short, but if the linewidth is thousand times larger (100 μm), the viscosity ten times higher (100 mPa.s, classical

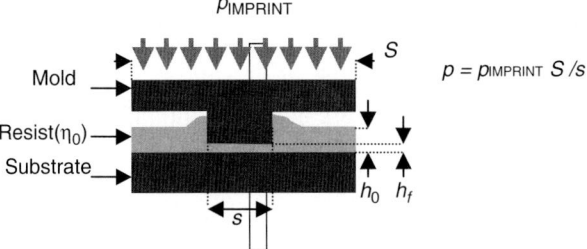

Figure 1.3 Illustration of quantities in Stefan's law for a one-dimensional line [27].

case of spin-coated resists) and if a residual layer of 10 nm is targeted, then the imprinting time becomes 0.2 s (2000s in the case of thermal NIL). In the case of UV-NIL, the process is still very fast. Nevertheless, as the mold and substrate are assumed to be perfectly rigid due to the low imprinting pressure, the sinking rate of the mold will be governed by the largest protrusions. Thus, large protrusions (>100 μm) must be avoided on the total surface of the mold, or fake cavities must be included in the mold design, in order to reduce the maximal linewidth.

As an example, with a 1000 mPa.s viscosity resist, a 13 bar pressure and an optimized initial resist thickness (final residual layer <20 nm), the mean free path (or flowing distance) of a resist molecule was experimentally estimated to be about 1 mm [28]. This value is important for design of the mold, which has to exhibit homogeneous protrusion density (protrusion to cavity area ratio) on any 1 mm^2 areas on its surface in order to favor a uniform residual layer. In this example, the resist redistribution area was quite small but, when using lower viscosity resists ($\eta_0 <$ 50 mPa.s), this area should reach almost the stamp size in the case of small stamps.

1.2.3
Imprinting Examples and Resolution

In Figure 1.4, some examples of imprinted resist layers, made by our group on an EVG 770 step-and-repeat system, are illustrated. With an optimized initial resist thickness and a known mold design, an excellent contrast between the feature height and the residual layer thickness can be obtained; this is a very important point in the case of additional process steps.

Even in the first NIL experiment by S. Chou [1], a very high resolution of 25 nm was demonstrated using thermal NIL. The fabricated polymer mask could be used successfully in a lift-off process. In UV-NIL, a resolution of 5 nm linewidth and 14 nm pitch using NIL and lift-off was demonstrated in 2004, also by the group of S. Chou [29]. Here, the mold used was a cleaved facet of a sample containing molecular-beam epitaxy-grown superlattices. In the same year, Hua et al. demonstrated that it is even possible to reproduce the shape of a single-wall carbon nanotube in a polymer (Figure 1.5) [30]. Indeed, the surface roughness of the mold is in general very well

Figure 1.4 Examples of UV-NIL imprinted patterns: (a) 100 nm lines and spaces; (b) 100 nm contact holes.

reproduced in the polymeric material during the molding process. The resolution of the molding technique itself does not limit the overall resolution of the technique, but much more the high resolution mold availability and the pattern transfer (as will be detailed in Section 1.3.3 below).

1.2.4
Tools and Industrialization Issues

As described in more detail in the review by H. Schift [6], many different tool architectures exist in NIL, depending on the process used or on the targeted application. In UV-NIL, two main types of tools can be identified. The first one is the step-and-repeat tool. The main commercial players are Molecular Imprints [31],

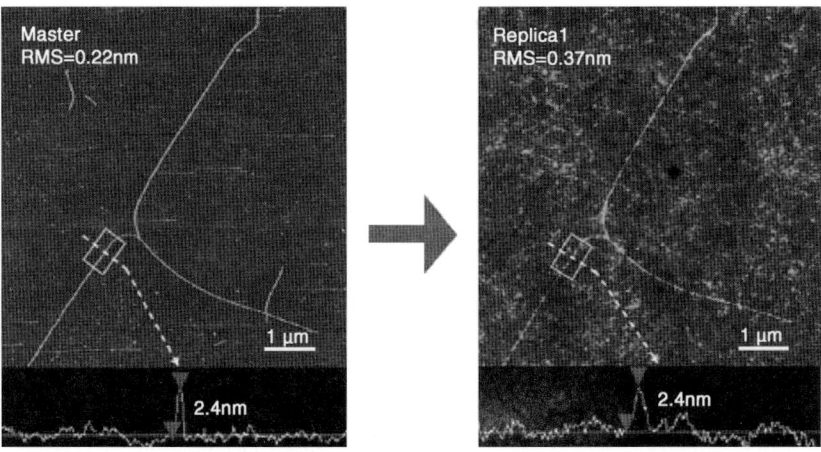

Figure 1.5 Demonstration of single-wall carbon nanotube reproduction using UV-NIL. (Reproduced with permission from [30]. Copyright © (2004) American Chemical Society.)

EVG [32], and SET/Karl Süss [33]. In this case, hard stamps and substrates with excellent flatness are brought into contact at low pressure. The advantages of these tools are the possibility of fabricating stamps with smaller active areas (in general around $25 \times 25\,\text{mm}^2$) and of more easily implementing a high accuracy alignment system. Indeed, the high accuracy alignment (<30 nm) requires stiff molds as it is performed when the mold is already in contact with the liquid resist before UV exposure. The second type of tool is composed of equipment using soft pistons and full wafer molds. Here, flexible membranes are used on one or both sides of the wafer/mold stack in order to apply an isotropic imprinting pressure. The higher imprinting pressure used (up to more than 100 bars) can bend the mold and substrates, thereby ensuring conformal contact. Known companies in this field are Nanonex [34], Obducat [35], Molecular Imprints [31], EVG [32], and SET/Karl Süss [33]. The main advantages of these tools are a lower flatness requirement for substrate and mold and very high throughput. However, alignment is more problematic due to the larger molds and the soft piston tool architecture.

Finally, some other companies (Toshiba, Hitachi) have reported the fabrication of UV-NIL tools, but for internal use and especially developed for microelectronic or bit-patterned media applications [36, 37].

In a research laboratory, nanoimprint lithography works very well. If high throughput and high yield are required in an industrial process, specific issues has to be solved or tolerated, for example:

- **Air entrapment, evacuation and dissolution**. This problem depends on the tool and process conditions used. At high aspect ratio features, the flow front can arrest and create bubbles. Also, the geometries of the resist drop placement and of the contact play an important role: for example, an inclined template or mold bending may enhance bubble evacuation, especially for larger molds. Air bubbles are able to dissolve in the liquid resist, but this might limit the speed of the global imprinting process. The bubble dissolution rate in the resist can be enhanced in special gas environments (small helium molecules [38] or carbon dioxide [39] dissolve well in organic materials). The use of pentafluoropropane, in which condensation starts when the gas pressure exceeds 0.15 MPa, was also demonstrated [40]. Another possibility is to work in vacuum [41]. Then, one has to be aware of the resist evaporation.
- **Flatness issues**. At low imprinting force, the imprint quality depends on the flatness of mold and substrate. In the step-and-repeat process, the common 6.3 mm thick quartz mold can be prepared with sufficient flatness. As substrates, double-side polished wafers are preferable to single-side polished ones due to their better flatness and reduced total thickness variation. The wafer chuck flatness also has to be controlled.
- **Local geometry of the mold features**. The geometry of the mold features might induce some problems. In particular, some applications in microelectronics are very sensitive to the line edge roughness (LER). The shape of the features (vertical or inclined sidewalls, undercuts) may also induce demolding problems (possible rip-off of the structures, increased demolding force) [6].

- **Mold inspection and repair**. Fabrication of defect-free molds is extremely difficult. Some applications, microelectronics for example, require molds with almost no defects. Therefore, automatic mold inspection and repair tools are being developed. This is not a simple task because very high resolution is required (much smaller than in photolithography mask inspection as no reduction coefficient exists in NIL) and the mold material is, in general, insulating; the use of scanning electron microscopy is therefore more problematic due to charging effects.
- **Mold design issues**. Uniform residual layers are obtained when large protrusions are avoided in the mold design and when the protrusion density is uniform to avoid resist displacement over longer distances. To avoid long resist displacements and to compensate for large density variations in the mold design, one may use a reduced initial resist thickness (much less than the mold depth). Then, larger cavities will be only partially filled. Nevertheless, and especially in UV-NIL, where capillary phenomena are very important, this is not a good solution. Indeed, the resist will not remain flat in the incompletely filled cavities, but will create capillary bridges between the mold and the substrate, leading to dewetted areas [42].
- **Mold and mold treatment, mold contamination**. This will be detailed in Section 1.4 below.

All the aforementioned issues may lead to defects in the imprinted resist [43, 44]. Additionally, defects may also come from atmospheric particle contamination. A simple particle trapped between mold and substrate will produce a large noncontact area. Hence it is absolutely necessary to work in a very high quality clean environment. Another issue concerns the distance between imprinted dies in the step-and-repeat process. Due to the expelled resist at the stamp border, a distance of about 100 µm or more is set in general between the dies. This resist excess at the die border is a problem for the resist stripping step (removal of the resist mask or residues after use) because of its increased thickness. Additionally, imprinting of an incomplete die at the border of a wafer is difficult without damaging or contaminating the mold. Finally, real time measurement and simulation tools (including all interfacial aspects) would be of great help to understand more precisely some of the phenomena [45].

1.3
Imprinting Materials

1.3.1
State-of-the-Art

Resists developed for UV-NIL must have the following characteristics:
- low viscosity
- high curing speed
- high etch-resistance

- high adherence to the substrate
- low adherence to the mold
- good film formation (spin-coated resists)
- low contraction during photopolymerization
- adapted mechanical properties and, finally,
- low evaporation rate (reduced vapor pressure).

In general, they are composed of monomers and a UV-sensitive photo-initiator. Additionally, they may contain sensitizers [46] (to improve UV light absorption), surfactants [47] (for example fluorinated surfactants to reduce adherence to the mold), more than one photo-initiator [48] (to improve the speed of conversion and the conversion ratio), inhibitors (to reduce the effect of free radical annealing by scavengers like oxygen) and solvents (to adjust the thickness of the spin-coated layer). Solvents are not used in the case of drop-dispensing in order to maintain initial spherical drop shapes, facilitating resist flow and limiting bubble entrapment.

The monomers used may present one or more polymerization sites. In the case of one site only, the cured material will be composed of linear chains and remain soluble in common solvents. This may lead to lower mechanical stiffness and lower plasma etching resistance, with the advantage of easy removal from the mold (resist contamination on the mold can be cleaned more easily) or wafer (resist stripping). In the case of more than one polymerization site, the material will consist of a 3D crosslinked network.

Three main monomer families are used in UV-NIL. The first one is epoxy monomers [46, 49]. In general, they have the advantages of insensitivity to oxygen during the polymerization (cationic mechanism) and higher mechanical stiffness; but the polymerization does not proceed rapidly as compared to acrylates or vinyl ether formulations. Acrylates are most often used, mainly because of their extensive commercial availability, low viscosity, and capability for rapid polymerization via radical propagation. However, oxygen is a strong radical inhibitor. Dissolved oxygen in the resist may slow down the polymerization mechanism, or some uncured resist contamination may appear at the border of the template where resist is in contact with air. Vinyl ethers are the third category of monomers used in UV-NIL [50]. Their polymerization also proceeds via a cationic mechanism which is insensitive to oxygen and very rapid. In addition, vinyl ether monomers have even lower viscosities than acrylates at otherwise comparable molecule composition [51]. Nevertheless, there is not a large choice of commercial raw materials and vinyl ether resists adhere more strongly to molds [51]. This latter limitation is partly overcome by the higher tensile strength of vinyl ether formulations [52]. Finally, the UV sensitivity is not a limiting point of UV-NIL since exposure times of 1 s or less are reported using acrylates or vinyl ether formulations [53].

Concerning the mechanical properties, high crosslinking of the material after UV exposure may induce improved plasma etch resistance [54] as well as improved mechanical properties. This last point is important for the demolding process. However, a long UV exposure time may cause excessive shrinkage and brittleness of the resist, increasing the possibility of features breaking and mold contamination. On the other hand, insufficient UV curing will lead to low cohesive strength of the polymer, increasing the probability of pattern distortion and collapse [55]. This implies that the irradiation dose will have to be optimized for each material and pattern type in order to achieve appropriate mechanical properties.

In addition, some resists have been developed for special applications and with special properties. This is the case for example in microelectronics, where special hybrid organic-inorganic materials with low dielectric constants and high thermal and mechanical stability are being developed to simplify the dual damascene electrical interconnection process [56]. Also, resists loaded with functional nanoparticles can be patterned using NIL [57].

Commercially, NIL products are available from Nanonex [34], Obducat [35], micro resist technology [58], Molecular Imprints [31], AMO [59], Toyo Gosei [60], or AGC [61], for example. Some manufacturers sell their resists only with a proprietary process and equipment. It is then difficult to compare the properties of these commercial products. Adhesion of low viscosity thin films of monomer solutions to the silicon (or fused silica) substrates can be enhanced by using an intermediate adhesion promoter applied prior to resist dispensing. This can be a thin polymeric film with high adhesion to the substrate and high affinity to the monomers used. Also, this thin additional layer can play the role of substrate planarization layer. However, its thickness might be as high as a few tens of nm and will increase the residual resist thickness, but it may help the pattern transfer (for lift-off processes, for example [62]). An alternative approach consists in functionalizing the silicon surface with a self assembled monolayer (SAM) whose molecules are designed to fulfill two requirements: one head group bonds covalently to the silicon surface via a silanization reaction with hydroxyl groups while the other head group copolymerizes with the UV-NIL resist during UV exposure. This solution was adopted by Hewlett Packard's research group in 2005 and has allowed 30 nm half pitch dense lines to be imprinted and transferred using a methacrylate based resist [63].

1.3.2
Shrinkage

During UV curing, the resist volume shrinks. Due to a ring-opening polymerization, epoxy resist have a very limited shrinkage rate, in general around 3%. On the other hand, acrylates and vinyl ether resists show shrinkage rates of around 10% in most cases [64]. For high aspect ratio features embossed in a thick UV-curable material, resist shrinkage was demonstrated to facilitate demolding [65]. It seems that an optimum level of shrinkage exists at which the stress experienced by the polymer during demolding is minimized [66]. Such phenomena have not yet been demonstrated in UV-NIL and large resist shrinkages must be avoided. Indeed, due to the thin residual resist thickness and the mechanical rigidity of the substrate, stresses at the bottom of the fabricated structures cannot relax. Consequently, the resist shrinkage will induce a change in the top lateral dimensions and height of the features, reducing their lateral slope. This is a problem for the control of critical dimensions, especially when an additional plasma etching transfer step is required.

In order to limit shrinkage, resist formulations have been made with special monomers. For example, steric hindrance might be used. Another possibility is to include an oligomer or a polymer in the resist formulation as a binder [67], but one has to be careful not to prohibitively increase the resist viscosity.

1.3.3
Plasma Etching Processes and Resist Stripping

The formed polymer layer can be used as a mask in a plasma etching equipment.

First, a "breakthrough" etch is performed in order to remove the residual resist layer. In general, a pure oxygen plasma is used here which is very reactive with organic materials but will not etch substrates like silicon or silica. Addition of ammonia (NH_3) or larger atoms (Cl or Ar) in the plasma was shown to improve the anisotropy of the process and to reduce the etching speed of the resist, leading to improved process control [68]. Even with an anisotropic etching process, it is important to have a uniform residual layer. In the case of an over-etch intended to compensate for the non-uniformity of the residual layer, reactive species at locations were the resist layer is already removed will react isotropically with the surrounding features and might change their lateral dimensions.

Subsequently, the "opened" resist mask obtained can be used to etch the underlying substrate. In the case of silicon or silica, plasmas with fluorine (mainly CF_4 and SF_6 gases), bromine (HBr) or chlorine (Cl_2, HCl) can be used. Usable aspect ratios are in general larger than 1, that is, the feature height exceeds the line width. To achieve this, imprinted resists have to exhibit the correct selectivity to the underlying material; this means that the etch rate of the resist in the plasma has to be sufficiently small compared with the etch rate of the substrate.

In general, in UV-NIL, the material is crosslinked due to the use of monomers with multiple photopolymerizable groups. This is an advantage compared with the photolithography case (in particular positive tone resists). Indeed, correct etching resistance can be obtained when choosing hybrid organic–inorganic monomers, for example molecules containing siloxane bonds, characterized by their weight percentage of silicon atoms. Perfluorinated monomers or monomers containing aromatic cycles (benzenic or norbornene cycles, for example) can also be used. Figure 1.6 shows 40 nm wide lines which were etched 120 nm deep in silicon using a 80 nm thick layer of the commercial product AMONIL, from AMO, which is a silicon-containing resist.

Figure 1.7 shows examples of silicon structures etched with three different resists under exactly the same conditions (same imprinting conditions, mold, etching plasmas, resist thickness and resist removal process). We observed that for the purely organic product, the plasma etching resistance was not sufficient, leading to a "faceting" of the silicon lines (left). In the case of silicon-containing or perfluorinated resists, nonfaceted line profiles were observed (center and right).

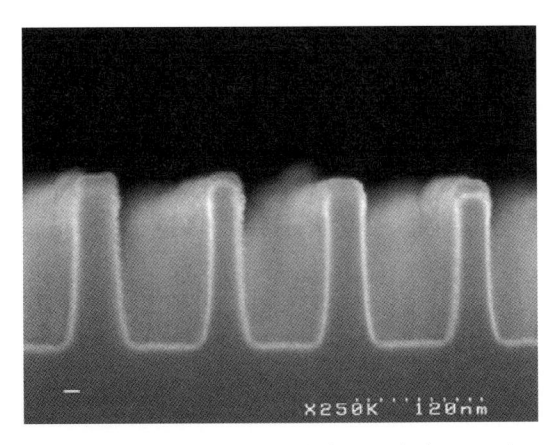

Figure 1.6 Example of 40 nm wide lines etched 120 nm deep in silicon using a silicon-containing resist, before removing the resist residue.

Indeed, incorporation of fluorine-rich monomers in the UV curable resist may enhance the etch properties. Kim *et al.* [69], where an acrylate-based formulation was doped with a fluorine-substituted acrylate chain, reported that the incorporation of 20% of fluorinated monomers increased the oxygen plasma etching rate of the photopolymerized material by more than 15%. But another example reported by AGC [70] highlighted a weakening of the adhesion to the substrate when the amount of fluorinated monomer increased, necessitating the application of a suitable adhesion promoter to the substrate.

The resist removal process, or resist stripping, is generally done with pure oxygen plasma or using an acid solution (mixture of hydrogen peroxide and sulfuric acid, for example). Additionally, a dip in hydrofluoric acid (HF), removing silica-like compounds, may help to remove the passivation layers and resist residues that appear during silicon etching. In the case of Figure 1.7, resist stripping was done with an oxygen plasma and HF dip. We observed that a large number of residues still exist on the wafer surface for the silicon-containing resist. Indeed, removing resist residues of crosslinked materials, especially those with a high etch resistance, is difficult;

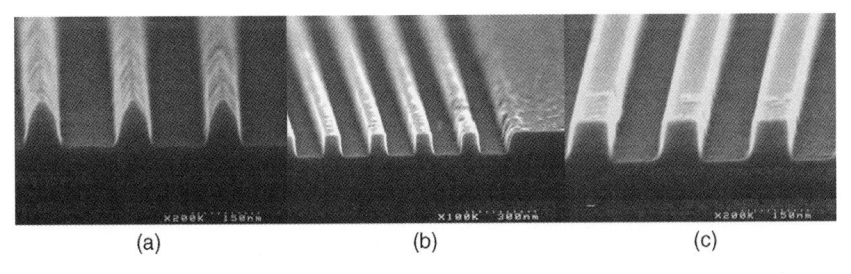

Figure 1.7 Examples of silicon etched with different UV-NIL resists in exactly the same conditions: (a) purely organic resist; (b) silicon-containing resist; (c) perfluorinated resist.

Perfluorinated materials seem to present both a sufficient etch resistance and ease of stripping. Alternatively, a very elegant method allowing an efficient resist stripping process for crosslinked materials is the development of "degradable crosslinkers" [71].

1.4
Mold Fabrication and Anti-Sticking Strategies

1.4.1
Mold Fabrication and Characterization

In UV-NIL, molds can be made of organic materials, using for example polymers or elastomeric materials (to be described in more details in Section 1.4.6 below) or inorganic materials: glass, fused silica, transparent conductive oxide such as indium tin oxide [72] (ITO, to overcome charge problems), sapphire [73], fluorinated diamond-like carbon [74], boron nitride [75] or silica-like electron-beam inorganic resist such as hydrogen silsesquioxane [76] (HSQ). Using HSQ, one can fabricate directly the features without etching as the resist has almost the composition and density of silica after thermal treatment.

Among all these transparent materials, 6.35 mm thick, 150 mm square fused silica plates (so-called 6025 fused silica blanks in the microelectronics mask industry) are mostly used as base material for UV-NIL stamp fabrication thanks to their commercial availability, their high degree of purity, their excellent flatness and their inert behavior with respect to photocurable monomers solutions. In addition, mask shops have acquired, for the purpose of phase shift masks fabrication, accurate and reliable electron-beam writing as well as plasma etching processes on such type of plates [77–79]. Moreover, quartz stiffness is high enough (9.3 GPa) to avoid the deformation of nano-features during repetitive imprinting sequences, which may induce unacceptable misalignment in mix and match lithographies. However, the resolution needed here is much smaller than required for optical masks, as NIL is a 1X lithography technique. Also, it is more difficult to etch an insulating material like silica to very small dimensions in a plasma (as compared to silicon for example) due to surface charge issues.

More details on fused silica mold fabrication can be found elsewhere [80]. In general, a resist is patterned using electron-beam lithography and then transferred to a thin (between 8 and 20 nm) chromium layer with a Cl_2/O_2 plasma. Afterwards, fused silica is etched in a fluorocarbon plasma using the chromium layer as a hard mask. Finally, this hard mask is removed in a wet or dry process.

As explained before, mold inspection [81] and repair [82] is mandatory in applications like microelectronics, requiring a very low defect rate. Some specific commercial tools, derived from wafer and mask inspection equipments, are under development [83, 84]. Also, the LER is an important characteristic of fabricated templates and becomes a critical issue when the dimensions shrink. One powerful method, used in our lab, to characterize the roughness of fabricated UV-NIL molds is

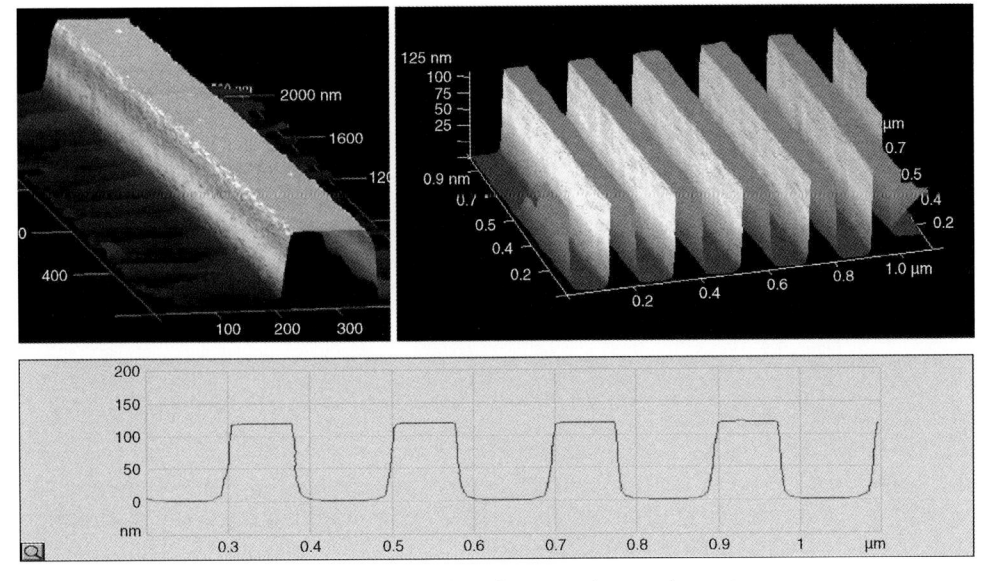

Figure 1.8 Example of mold characterization by three-dimensional atomic force microscopy (AFM3D) (100 nm wide lines).

three-dimensional atomic force microscopy (AFM3D). Such a tool is able to completely reconstruct a 3D surface (see example in Figure 1.8) and to measure directly the LER value. Also, AFM3D can be used to follow the roughness of NIL fabricated features at the different steps of the process, that is, after imprint, residual layer etching, transfer and finally after stripping.

1.4.2
Anti-Sticking Strategies

A major challenge of the nanoimprint technique is to perform a correct detachment of the mold from the cured resist. Its difficulty is inherent to the high density of nanoscale protrusions patterned on the mold surface, which effectively increases the total area in contact with the UV-cured resist. Therefore, sticking between imprinted polymeric structures and the mold surface can have the three following detrimental consequences:

- Improper release of the polymer film from the template creates defects in the imprinted layer.
- Any residual photopolymer that remains on the template from a previous incomplete release creates defects in subsequent imprinted patterns. (Some work has shown a "self-cleaning" effect of the mold [85], removing contaminants after a few imprints, but this effect is not very clear and seems to depend on the contaminant size and on the process used.)

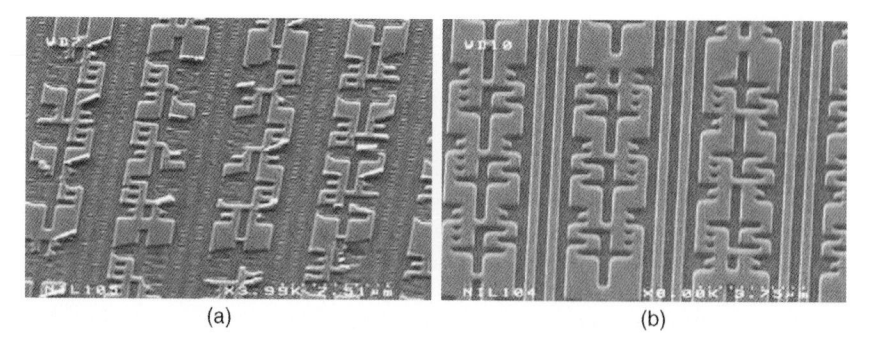

Figure 1.9 Resist imprinted with an untreated (a) and a treated (b) mold under otherwise the same conditions.

- Attempts to systematically remove photopolymer residues from the mold surface by wet or dry chemistry are not only time consuming processes but also harmful to the template integrity and lifetime.

An example of imprinted features, with and without mold treatment (Optool DSX, from Daikin Chemicals [86]), using otherwise the same imprinting conditions, is illustrated in Figure 1.9. In the case of the untreated mold, the imprinted resist layer is clearly not usable. The origin of this problem can be manifold: weak adhesion of the resist film to the substrate, strong adhesion of the cured materials to the mold, poor mechanical strength of the material or, more likely, a combination of all. However, improving adhesion of the resist film to the substrate and tuning its mechanical properties cannot totally prevent sticking. Therefore much work has been done to reduce mold-resist sticking by minimizing the interfacial energy as detailed in the next section.

1.4.3
Mold Treatment

Quartz is a hydrophilic material exhibiting a rather high surface free energy ($\sim$50 mJ m^{-2}) facilitating the adhesion of cured resist to its surface. Yet, according to industrial criteria for process throughput and cost of ownership, a quartz template should be able to undergo thousands of imprinting sequences before being cleaned or re-treated. To meet such severe requirements, quartz templates must be coated with an anti-adhesion layer that effectively lowers its adherence to the cured resist. In addition to this criterion, the anti-adhesion layer must have good adhesion to the template surface, should be deposited in a conformal way onto the mold features, should have good mechanical properties (high stiffness) and should not reduce excessively the UV transmittance properties of the mold. Moreover, in the case of sub-100 nm features patterned on the mold surface, the thickness of this layer should not exceed a few nanometers, its surface roughness has to be as low as possible and should not increase the initial roughness of the mold.

Diamond-like carbon (DLC) coatings have been considered as a good choice for this application due to the combination of relative hydrophobicity with outstanding mechanical properties [74]. They can be easily obtained by plasma enhanced chemical vapor deposition (PECVD) of hydrocarbon gases (methane) and exhibit a lower surface energy than silica ($\sim$40 mJ m^{-2}) and high stiffness (20 GPa). However, the deposition rate of those films is high in standard deposition conditions (few nm per second), which makes it difficult to obtain uniform and conformal layers thinner than 20 nm, making them unsuitable for sub-100 nm features. Besides, the transmittance of such DLC films is low at typical wavelengths used in the UV-NIL process (13% transmittance for a 100 nm thick layer, 50% for a 10 nm one) [87].

Other attempts have been made in order to improve the optical properties of DLC coatings via partial doping in the vapor phase. N_2 or Si doping has been proven to enhance the UV transmittance of ion beam synthesized DLC coatings [88], but the stiffness was noticeably reduced. Teflon-like thin films deposited by plasma show a very good hydrophobic performance but suffer from poor adhesion to the template surface [89]. More generally, the trade-off between the tribological properties, chemical composition, surface energy and optical band gap is critical in defining the performance of these coating materials for UV-NIL applications.

Another approach is the use of fluorinated silane molecules able to covalently bond to the mold surface. The principle consists in first preparing the mold surface to generate the required terminal hydroxyl groups. When fluorinated silanes are then adsorbed on the template surface, their polar head groups undergo a hydrolysis reaction forming silanol terminations (Si–OH). Finally, thermal annealing is performed in order to form covalent siloxane bonds (Si–O–Si) between the mold and fluorinated molecules [90]. Hydrophobic properties are provided by the CF_2 or CF_3 groups in the molecule [91], as predicted by molecular dynamics considerations [89].

Several fluorinated molecules have been reported as being release agents for NIL templates. Among them are tridecafluoro-1,1,2,2-tetrahydrooctyltrichlorosilane [CF_3–$(CF_2)_5$–$(CH_2)_2$–$SiCl_3$, F_{13}-TCS], 1,1,2,2-perfluorodecyltrichlorosilane [CF_3–$(CF_2)_7$–$(CH_2)_2$–$SiCl_3$, F_{17}-TCS], 1H,1H,2H,2H-perfluorooctyltrimethoxysilane [CF_3–$(CF_2)_5$–$(CH_2)_2$–$SiO(CH_3)_3$, F_{13}-TMS], and a perfluoropolyether molecule (Optool DSX from Daikin Chemical [86]). Almost all these molecules lead, when properly deposited, to water contact angles higher than 100° and a free surface energy on the order of 11 mJ m^{-2}, which is able to reduce adhesion between the treated mold surface and the cured resist.

Fluorinated self-assembled monolayers (F-SAM) can be deposited either in liquid phase, by dipping the mold directly into a diluted solution of anti-sticking molecules [92], or in a vapor phase process. The latter can be done either by thermal evaporation of the liquid precursor at atmospheric pressure [93] or by vacuum evaporation at room temperature [94]. Chlorosilane molecules are very reactive and able to polymerize, producing particles that can precipitate onto the mold. This is why it is preferable to use the vapor phase process with this type of molecules, leading to smoother surfaces [95].

Compared with inorganic deposited films, F-SAMs have several advantages:

- Their thickness is very low. Experimental values reported in literature vary from 0.2 nm for a F13-TCS based F-SAM up to 3 nm for an Optool DSX-based one [96]. Although the thickness measurement techniques (ellipsometry, X-ray reflectometry) are not always well calibrated at the sub-nanometer scale, it is now widely admitted that the deposition of such SAMs does not significantly change the critical dimensions of nanostructures fabricated on the mold surface.
- Their adhesion to the mold is excellent due to covalent bonding of the molecules.
- Deposition parameters like dip time, concentration and temperature of the solution or evaporating time and pressure can be monitored to optimize the uniformity, roughness and density of the deposited F-SAM.
- F-SAM deposition can easily be implemented in an industrial environment.
- F-SAM treatments are cleanable and can be entirely removed from silica surfaces by a suitable combination of wet and dry etching processes without altering the surface roughness of the mold [96]. This property enhances the ability of F-SAM treatments to be implemented in a whole repetitive and reliable patterning process and increases the mold lifetime.

Despite these interesting properties, even the best F-SAM has been widely reported, by several research teams from academia, to be not very durable. Indeed, sticking problems between a treated mold and a UV-cured resist may appear after ten imprints only [96]. This releasing default is accompanied by an increase of the surface energy of the template [90] and a loss of fluorine on the mold surface [97]. With optimized conditions and using the Molecular Imprint Inc. process, a mean lifetime of the release layer of about 800 cycles was reported [6] (i.e., 6 wafers in the used conditions). This is a poor figure of merit with respect to Sematech's board specifications [98] and can seriously impede the spread of UV NIL as a large volume nanopatterning technique.

1.4.4
Fluorinated Mold Treatment-Resist Interactions

Over the last few years, several studies have aimed at determining the accurate mechanisms that are responsible for the premature degradation of molds, but their conclusions do not systematically converge [99, 100]. However, outlines of numerous studies confirm a chemical reactivity between the cured resist and the F-SAM with acrylate or vinyl ether formulations [99]. The species that are presumed to attack fluorinated molecules are free radicals or cationic charges generated during photopolymerization. Recently, resist free radicals were precisely identified, for the first time, by an electron spin resonance analysis, as having a chemical affinity with Optool DSX fluorinated molecules [101]. Other studies have also highlighted the impact of the resist formulation on the F-SAM degradation rate and mechanisms. For instance, the incorporation of crosslinkers or silicon containing components has a clear impact on the adhesion between mold and cured resists [102].

Even if some experimental trends are not fully understood (in particular the balance between chemical and mechanical degradation), it is found that fluorinated release layers have a limited lifetime when imprinting conventional acrylate or vinyl ether resists. The only criterion of low surface energy is not sufficient to guarantee durable and effective anti-sticking behavior and there is a need to develop resist systems having a low reactivity with fluorinated molecules. Alternatively, a recent study by Houle *et al.* [103] showed that thin metal oxide compounds coatings, with water contact angles lower than 50°, can be good candidates for anti-sticking layers with well controlled behavior when used with free radical or cationic resists.

1.4.5
Resist-Oriented Anti-Sticking Solutions

Studies of the degradation of F-SAM anti-sticking layers have highlighted the fact that the free radicals in resists are, in some case, reactive species responsible for fluorine atom removal from the template surface. To minimize the interaction between the resist components and the template surface, two resist oriented solutions are reported.

By adding fluorine-containing monomers to the UV curable mixture, it was demonstrated that hydrophobic properties of the UV cured resist are excellent and allow easy and repetitive demolding [70, 104]. Besides, the viscosity of the UV curable mixture is not degraded because fluorine-containing monomers are available with viscosities as low as 10 mPa.s. Also, when associated with F-SAM template anti-sticking treatments, the use of fluorine rich resist seems to slow (but not eliminate) its degradation [104].

Another possibility is the use of fluorinated surfactants in the resist composition. Surfactants are small molecules that will not participate in the polymer network and can move in the resist before curing [105]. According to this explorative work, it seems that, to fully take advantage of their incorporation, surfactant migration from the resist to the resist/template interface has to be effective. This is possible only with low surface energy templates. Indeed, fluorinated surfactants are segregated at the resist/template interface only if there is an affinity at the considered interface [47]. For the same reason, and when used in relatively small quantities, surfactants will not degrade the adhesion of the resist to the substrate. Additionally, an elegant method being developed is the use of "reactive" surfactants able to regenerate the mold treatment in-situ during imprinting [106].

1.4.6
Polymeric Mold Materials and Stamp Copies

A different solution to fabricate molds and control contamination consists in copying an initial master mold into a daughter mold using a polymeric transparent material with good intrinsic release characteristics. In general, polymeric materials have low surface energies. Furthermore, master molds can be fabricated in silicon using standard tools and processes from the microelectronics industry. Also, this solution would solve the anti-sticking treatment aging problem as well as lower the cost of working templates.

AGC has produced fluorinated high transparency polymers (F-template) that can be embossed by thermal NIL to replicate a master template, made of either silicon or quartz [70], into a working stamp with a water contact angle higher than 100°. Other research teams have also synthesized thermally [107] and UV [108, 109] curable formulations for the need of template replication. Nevertheless, to be definitely adopted, this alternative solution has to be well characterized in terms of mold degradation and mechanical properties. Also, some replication materials, with very high stiffness, contain a high inorganic part and might need a surface treatment to be perfectly effective [110].

When no high accuracy alignment is needed and low pressure is used in UV-NIL, the mechanical properties of the mold can be relaxed. Then, mold copies can also be obtained by replicating elastomeric materials like Poly(dimethylsiloxane) (PDMS), which is a soft UV-transparent material in the 340–600 nm wavelength region [111, 112]. Finally, the ultimate stamp copy process was presented by Obducat with the Intermediate Polymer Stamp [113]. Here, the polymer stamp is used only once, avoiding any mold contamination or erosion issues.

1.5
Conclusion

On a laboratory scale, UV nanoimprint lithography is able to produce very high resolution features at high throughput and low cost potentially on any surface. It is essential to consider the mold design, resist quantity and formulation, resist flow, environmental contamination, and flatness issues. To proceed a step further in development and industrialization, some critical issues must be considered such as air inclusions, fast and uniform curing, mold inspection and repair, distance between imprinted dies, and border imprinted dies in the step-and-repeat process. Some solutions exist but will need further development to be really effective. One of the major problems is the sticking between the mold and the cured resist, which may lead to additional defects, mold contamination, mold abrasion and may require time-consuming mold cleaning and re-treatment. Most likely, a combination of resists with low reactivity towards fluorinated treatments, resist surfactants and mold copies will be needed to overcome this issue. The direct measurement of demolding forces, implemented in some of the prototype step-and-repeat systems, should reveal the stamp degradation and predict the need for change or re-treatment, almost certainly in this last case using automated mask cleaning sequences [114].

With regard to the future, nanoimprint lithography is still considered as a next generation lithography technique in microelectronics, but it will most probably remain a generic technique for specific applications with special processes and tools in the fields of optics, data storage [115], and applications requiring 3D fabrication or imprinting in a functional material. In the microelectronics industry, the cost of ownership (CoO) of NIL was calculated not to be systematically below the CoO of photolithography or extreme UV (EUV) [116]. Furthermore, even if the CoO of NIL is lower, industries will not change their technology completely unless the gain exceeds 30% (Personal communication, Jean Massin (STMicroelectronics) 2009). In any case, a major industrial laboratory claims that perfection in NIL will be too expensive and suggests the development of alternative integrated circuits technologies that are defect tolerant [117].

References

All web sites mentioned in the References were last accessed on 2 August 2010.

1. Chou, S.Y., Krauss, P.R., and Renstrom, P.J. (1995) *Appl. Phys. Lett.*, **67**, 3114.
2. Dauksher, W.J., Mancini, D., Nordquist, K., Resnick, D.J., Hudek, P., Beyer, D., Groves, T., and Fortagne, O. (2004) *Microelectron. Eng.*, **75**, 345.
3. International Technology Roadmap for Semiconductors (ITRS) (2009) http://www.itrs.net/.
4. Haisma, J., Verheijein, M., van den Heuvel, K., and van den Berg, J. (1996) *J. Vac. Sci. Technol. B*, **14**, 4124.
5. Masson, J.-L., and Green, P.F. (2002) *Phys. Rev. E*, **65**, 031806.
6. Shift, H. (2008) *J. Vac. Sci. Technol. B*, **26**, 458.
7. Fuchs, A., Vratzov, B., Wahlbrink, T., Georgiev, Y., and Kurz, H. (2004) *J. Vac. Sci. Technol. B*, **22**, 3242.
8. Schuetter, S.D., Dicks, G.A., Nellis, G.F., Engelstad, R.L., and Lovell, E.G. (2004) *J. Vac. Sci. Technol. B*, **22**, 3312.
9. Khang, D.-Y., and Lee, H.H. (2000) *Appl. Phys. Lett.*, **76**, 870.
10. Voicu, N.E., Ludwigs, S., Crossland, E.J.W., Andrew, P., and Steiner, U. (2005) *Nano Lett.*, **5**, 1915.
11. Borzenko, T., Tormen, M., Schmidt, G., Molenkamp, L.W., and Janssen, H. (2001) *Appl. Phys. Lett.*, **79**, 2248.
12. Huang, X.D., Bao, L.-R., Cheng, X., Guo, L.J., Pang, S.W., and Yee, A.F. (2002) *J. Vac. Sci. Technol. B*, **20**, 2872.
13. Kehagias, N., Reboud, V., Chansin, G., Zelsmann, M., Jeppesen, C., Schuster, C., Kubenz, M., Reuther, F., Gruetzner, G., and Sotomayor Torres, C.M. (2007) *Nanotechnology*, **18**, 175303.
14. Cheng, X. and Guo, L.J. (2004) *Microelectron. Eng.*, **71**, 277.
15. Sun, X., Zhuang, L., Zhang, W., and Chou, S.Y. (1998) *J. Vac. Sci. Technol. B*, **16**, 3922.
16. Hao, J., Lin, M.W., Palmieri, F., Nishimura, Y., Chao, H.-L., Stewart, M.D., Collins, A., Jen, K., and Willson, C.G. (2007) *Proc. SPIE*, **6517**, 651729.
17. Kim, E.K., Ekerdt, J.G., and Willson, C.G. (2005) *J. Vac. Sci. Technol. B*, **23**, 1515.
18. Sreenivasan, S.V., Lu, X., Cherala, A., Schumaker, P., Choi, J., and Mc-Mackin, I. (2007) Abstract Book EIPBN 2007 Conference, Denver, CO, 29 May–1 June.
19. Resnick, D.J., Sreenivasan, S.V., and Willson, C.G. (2005) *Mater. Today*, 8 (2) 34.
20. Jet and Flash™ Imprint Lithography (J-FIL™) process (2010) http://www.molecularimprints.com/Technology/technology2.html.
21. Colburn, M., Choi, B.J., Sreenivasan, S.V., Bonnecaze, R.T., and Willson, C.G. (2004) *Microelectron. Eng.*, **75**, 321.
22. Cheama, D.D., Kumar Karrea, P.S., Palardb, M., and Bergstrom, P.L. (2009) *Microelectron. Eng.*, **86**, 646.
23. Sim, Y.-S., Kim, K.-D., Jeong, J.-H., Sohn, H., Lee, E.-S., and Lee, S.-C. (2005) *Microelectron. Eng.*, **82**, 28.
24. Lentz, D., Doyle, G., Miller, M., Schmid, G., Ganapathisuramanian, M., Lu, X., Resnick, D., and LaBrake, D.L. (2007) *Proc. SPIE*, **6517**, 65172F.

25 Viheriälä, J., Tommilaa, J., Leinonena, T., Dumitrescua, M., Toikkanena, L., Niemia, T., and Pessa, M. (2009) *Microelectron. Eng.*, **86**, 321.

26 Stefan, M.J. (1874) *Akad. Wiss. Math. -Natur., Vienna*, **2**, 713.

27 Schift, H. and Heyderman, L.J. (2003) in *Alternative Lithography: Unleashing the Potential of Nanotechnology* (ed. D.J. Lockwood,), Kluwer Academic, New York, Chap. 4.

28 Perez Toralla, K., De Girolamo, J., Truffier-Boutry, D., Gourgon, C., and Zelsmann, M. (2009) *Microelectron. Eng.*, **86**, 779.

29 Austin, M.D., Ge, H., Wu, W., Li, M., Yu, Z., Wasserman, D., Lyon, S.A., and Chou, S.Y. (2004) *Appl. Phys. Lett.*, **84**, 5299.

30 Hua, F., Sun, Y., Gaur, A., Meitl, M.A., Bilhaut, L., Rotkina, L., Wang, J., Geil, P., Shim, M., and Rogers, J.A. (2004) *Nano Lett.*, **4**, 2467.

31 Molecular Imprint (2010) www.molecularimprints.com.

32 EV Group, (2010) www.evgroup.com.

33 SUSS Microtec (2010) www.suss.com.

34 Nanonex (2010) www.nanonex.com.

35 Obducat (2010) www.obducat.com.

36 Yoneda, I., Mikami, S., Ota, T., Koshiba, T., Ito, M., Nakasugi, T., and Higashiki, T. (2008) *Proc. SPIE*, **6921**, 692104.

37 Ando, T., Kuwabara, K., Haginoya, C., Ogino, M., Ohashi, K., and Miyauchi, A. (2005) *Proc. SPIE*, **5931**, 59310B.

38 Liang, X., Tan, H., Fu, Z., and Chou, S.Y. (2007) *Nanotechnology*, **18**, 025303.

39 Kim, K.-D., Jeong, J.-H., Sim, Y.-S., and Lee, E.-S. (2006) *Microelectron. Eng.*, **83**, 847.

40 Hiroshima, H., Komuro, M., Kasahara, N., Kurashima, Y., and Taniguchi, J. (2003) *Jpn. J. Appl. Phys.*, **42**, 3849.

41 Fuchs, A., Bender, M., Plachetka, U., Hermanns, U., and Kurz, H. (2005) *J. Vac. Sci. Technol. B*, **23**, 2925.

42 Landis, S., Chaix, N., Hermelin, D., Leveder, T., and Gourgon, C. (2007) *Microelectron. Eng.*, **84**, 940.

43 McMackin, I., Martin, W., Perez, J., Selinidis, K., Maltabes, J., Xu, F., Resnick, D., and Sreenivasan, S.V. (2008) *J. Vac. Sci. Technol. B*, **26**, 151.

44 DiBiase, T., Ahamdian, M., and Malik, I. (2007) *Microelectron. Eng.*, **84**, 989.

45 Yoneda, I., Nakagawa, Y., Mikami, S., Tokue, H., Ota, T., Koshiba, T., Ito, M., Hashimoto, K., Nakasugi, T., and Higashiki, T. (2009) *Proc. SPIE*, **7271**, 72712A.

46 De Girolamo, J., Chouiki, M., Tortai, J.-H., Sourd, C., Derrough, S., Zelsmann, M., and Boussey, J. (2008) *J. Vac. Sci. Technol. B*, **26**, 2271.

47 Wu, K., Wang, X., Kim, E.K., Willson, C.G., and Ekerdt, J.G. (2007) *Langmuir*, **23**, 1166.

48 Chan, E.P. and Crosby, A.J. (2006) *J. Vac. Sci. Technol. B*, **24**, 2716.

49 Cheng, X., Guo, L.J., and Fu, P.-F. (2005) *Adv. Mater.*, **17**, 1419.

50 Kim, E.K., Stacey, N.A., Smith, B.J., Dickey, M.D., Johnson, S.C., Trinque, B.C., and Willson, C.G. (2004) *J. Vac. Sci. Technol. B*, **22**, 131.

51 Long, B.K., Keitz, B.K., and Willson, C.G. (2007) *J. Mater. Chem.*, **17**, 3575.

52 Kim, E.K., Stewart, M.D., Wu, K., Palmieri, F.L., Dickey, M.D., Ekerdt, J.G., and Willson, C.G. (2005) *J. Vac. Sci. Technol. B*, **23**, 2967.

53 Jiang, W., Ding, Y., Liu, H., Lu, B., Shi, Y., Shao, J., and Yin, L. (2008) *Microelectron. Eng.*, **85**, 458.

54 Voisin, P., Zelsmann, M., Ridaoui, H., Chouiki, M., Gourgon, C., Boussey, J., and Zahouily, K. (2007) *J. Vac. Sci. Technol. B*, **25**, 2384.

55 Viallet, B., Gallo, P., and Daran, E. (2005) *J. Vac. Sci. Technol. B*, **23**, 72.

56 Stewart, M.D., Wetzel, J.T., Schmid, G.M., Palmieri, F., Thompson, E., Kim, E.K., Wang, D., Sotoodeh, K., Jen, K., Johnson, S.C., Hao, J., Dickey, M.D., Nishimura, Y., Laine, R.M., Resnick, D.J., and Grant Willson, C. (2005) *Proc. SPIE*, **5751**, 210.

57 Reboud, V., Kehagias, N., Sotomayor Torres, C.M., Zelsmann, M., Striccoli, M., Curri, M.L., Agostiano, A., Tamborra, M., Fink, M., Reuther, F., and Gruetzner, G. (2007) *Appl. Phys. Lett.*, **90**, 011115.

58 Micro resist technology GmbH (2010) www.microresist.de.

59 AMO (2010) www.amo.de.

60 Toyo Gosei (2010) www.toyogosei.co.jp/eng/index.html.

61 Asahi Glass company (AGC) (2010) www.agc.co.jp.

62 Li, M., Tan, H., Kong, L., and Koecher, L. (2004) *Proc. SPIE*, **5374**, 209.
63 Jung, G.Y., Wu, W., Li, Z., Wang, S.Y., Tong, W.M., and Williams, R.S. (2005) *Proc. SPIE*, **5751**, 952.
64 Voisin, P., Zelsmann, M., Cluzel, R., Pargon, E., Gourgon, C., and Boussey, J. (2007) *Microelectron. Eng.*, **84**, 967.
65 Chan-Park, M.B., Yan, Y.H., Neo, W.K., Zhou, W.X., Zhang, J., and Yue, C.Y. (2003) *Langmuir*, **19**, 4371.
66 Chan-Park, M.B., Lam, Y.C., Laulia, P., and Joshi, S.C. (2005) *Langmuir*, **21**, 2000.
67 Lee, H. and Jung, G.-Y. (2005) *Microelectron. Eng.*, **77**, 168.
68 Le, N.V., Dauksher, W.J., Gehoski, K.A., Resnick, D.J., Hooper, A.E., Johnson, S., and Willson, C.G. (2005) *Microelectron. Eng.*, **78–79**, 464.
69 Kim, J.Y., Choi, D.-G., Jeong, J.-H., and Lee, E.-S. (2008) *Appl. Surf. Sci.*, **254**, 4793.
70 Kawaguchi, Y., Nonaka, F., and Sanada, Y. (2007) *Microelectron. Eng.*, **84**, 973.
71 Heath, W.H., Palmieri, F., Adams, J.R., Long, B.K., Chute, J., Holcombe, T.W., Zieren, S., Truitt, M.J., White, J.L., and Willson, C.G. (2008) *Macromolecules*, **41**, 719.
72 Dauksher, W.J., Nordquist, K.J., Mancini, D.P., Resnick, D.J., Baker, J.H., Hooper, A.E., Talin, A.A., Bailey, T.C., Lemonds, A.M., Sreenivasan, S.V., Ekerdt, J.G., and Willson, C.G. (2002) *J. Vac. Sci. Technol. B*, **20**, 2857.
73 Komuro, M., Tokano, Y., Taniguchi, J., Kawasaki, T., Miyamoto, I., and Hiroshima, H. (2002) *Jpn. J. Appl. Phys.*, **41**, 4182.
74 Nakamatsui, K.-I., Yamada, N., Kanda, K., Haruyama, Y., and Matsui, S. (2006) *Jpn. J. Appl. Phys.*, **45**, L954.
75 Altun, A.O., Jeong, J.-H., Rha, J.-J., Kim, K.-D., and Lee, E.-S. (2007) *Nanotechnology*, **18**, 465302.
76 Mancini, D.P., Gehoski, K.A., Ainley, E., Nordquist, K.J., Resnick, D.J., Bailey, T.C., Sreenivasan, S.V., Ekerdt, J.G., and Willson, C.G. (2002) *J. Vac. Sci. Technol. B*, **20**, 2896.
77 Resnick, D.J., Mancini, D., Dauksher, W.J., Nordquist, K.J., Bailey, T.C., Johnson, S., Sreenivasan, S.V., Ekerdt, J.G., and Willson, C.G. (2003) *Microelectron. Eng.*, **69**, 412.
78 Schmid, G.M., Thompson, E., Stacey, N., Resnick, D.J., Olynick, D.L., and Anderson, E.H. (2007) *Microelectron. Eng.*, **84**, 853.
79 Sasaki, S., Hiraka, T., Mizuochi, J., Nakanishi, Y., Yusa, S., Morikawa, Y., Mohri, H., and Hayashi, N. (2009) *Proc. SPIE*, **7271**, 72711M.
80 Voisin, P., Zelsmann, M., Gourgon, C., and Boussey, J. (2007) *Microelectron. Eng.*, **84**, 916.
81 Hess, H.F., Pettibone, D., Adler, D., Bertsche, K., Nordquist, K.J., Mancini, D.P., Dauksher, W.J., and Resnick, D.J. (2004) *J. Vac. Sci. Technol. B*, **22**, 3300.
82 Dauksher, W.J., Nordquist, K.J., Le, N.V., Gehoski, K.A., Mancini, D.P., Resnick, D.J., Casoose, L., Bozak, R., White, R., Csuy, J., and Lee, D. (2004) *J. Vac. Sci. Technol. B*, **22**, 3306.
83 Pritschow, M., Butschke, J., Irmscher, M., Parisoli, L., Oba, T., Iwai, T., and Nakamura, T. (2009) *SPIE*, **7271**, 72711U.
84 Selinidis, K., Thompson, E., McMackin, I., Sreenivasan, S.V., and Resnick, D.J. (2009) *Proc. SPIE*, **7271**, 72711W.
85 Dauksher, W.J., Le, N.V., Gehoski, K.A., Ainley, E.S., Nordquist, K.J., and Joshi, N. (2007) *Proc. SPIE*, **6517**, 651714.
86 Daikin Chemical Europe GmbH, (2010) Optool DSX www.daikinchem.de.
87 Ramachandran, S., Tao, L., Lee, T.H., Sant, S., Overzet, L.J., Goeckner, M.J., Kim, M.J., Lee, G.S., and Hu, W. (2006) *J. Vac. Sci. Technol. B*, **24**, 2993.
88 Altun, A.O., Jeong, J.H., Rha, J.J., Choi, D.G., Kim, K.D., and Lee, E.S. (2006) *Nanotechnology*, **17**, 4659.
89 Sun, H., Liu, J., Gu, P., and Chen, D. (2008) *Appl. Surf. Sci.*, **254**, 2955.
90 Garidel, S., Zelsmann, M., Chaix, N., Voisin, P., Boussey, J., Beaurain, A., and Pelissier, B. (2007) *J. Vac. Sci. Technol. B*, **25**, 1179.
91 Nishino, T., Meguro, M., Nakamae, K., Matsushita, M., and Ueda, Y. (1999) *Langmuir*, **15**, 4321.
92 Srinivasan, U., Houston, M.R., Howe, R.T., and Maboudian, R. (1998) *J. Micro Electro Mech. Syst.*, **7**, 252.

93. Beck, M., Graczyk, M., Maximov, I., Sarwe, E.-L., Ling, T.G.I., Keil, M., and Montelius, L. (2002) *Microelectron. Eng.*, **61–62**, 441.
94. Schift, H., Saxer, S., Park, S., Padeste, C., Pieles, U., and Gobrecht, J. (2005) *Nanotechnology*, **16**, S171.
95. Jung, G.-Y., Li, Z., Wu, W., Chen, Y., Olynick, D.L., Wang, S.-Y., Tong, W.M., and Williams, R.S. (2005) *Langmuir*, **21**, 1158.
96. Truffier-Boutry, D., Galand, R., Beaurain, A., Francone, A., Pelissier, B., Zelsmann, M., and Boussey, J. (2009) *Microelectron. Eng.*, **86**, 669.
97. Truffier-Boutry, D., Beaurain, A., Galand, R., Pelissier, B., Boussey, J., and Zelsmann, M. (2010) *Microelectron. Eng.*, **87**, 122.
98. Lloyd, L.C. and Malloy, M. (2009) *Proc. SPIE*, **7271**, 72711Q.
99. Houle, F.A., Rettner, C.T., Miller, D.C., and Sooriyakumaran, R. (2007) *Appl. Phys. Lett.*, **90**, 213103.
100. Tada, Y., Yoshida, H., and Miyauchi, A. (2007) *J. Photopolym. Sci. Technol.*, **20**, 545.
101. Truffier-Boutry, D., Zelsmann, M., De Girolamo, J., Boussey, J., Lombard, C., and Pépin-Donat, B. (2009) *Appl. Phys. Lett.*, **94**, 044110.
102. Houle, F.A., Guyer, E., Miller, D.C., and Dauskardt, R. (2007) *J. Vac. Sci. Technol. B*, **25**, 1179.
103. Houle, F.A., Raoux, S., Miller, D.C., Jahnes, C., and Rossnagel, S. (2008) *J. Vac. Sci. Technol B*, **26**, 1301.
104. Schmitt, H., Zeidler, M., Rommel, M., Bauer, A.J., and Ryssel, H. (2008) *Microelectron. Eng.*, **85**, 897.
105. Lin, M.W., Hellebusch, D.J., Wu, K., Kim, E.K., Lu, K.H., Liechti, K.M., Ekerdt, J.G., Ho, P.S., and Willson, C.G. (2008) *J. Micro/Nanolith. MEMS MOEMS*, **7**, 033005.
106. Zelsmann, M., Alleaume, C., Truffier-Boutry, D., Beaurain, A., Pelissier, B., and Boussey, J. (2010) *Microelectron. Eng.* **87** (5–8), 1029–1032
107. Barbero, D.R., Saifullah, M.S.M., Hoffmann, P., Mathieu, H.J., Anderson, D., Jones, G.A.C., Welland, M.E., and Steiner, U. (2007) *Adv. Func. Mat.*, **17**, 2419.
108. Kim, W.-S., Choi, D.-G., and Bae, B.-S. (2006) *Nanotechnology*, **17**, 3319.
109. Kehagias, N., Reboud, V., De Girolamo, J., Chouiki, M., Zelsmann, M., Boussey, J., and Sotomayor Torres, C.M. (2009) *Microelectron. Eng.*, **86**, 776.
110. Micro resist technology GmbH (2010), Ormocer products http://www.microresist.de.
111. Bender, M., Plachetka, U., Ran, J., Fuchs, A., Vratzov, B., Kurz, H., Glinsner, T., and Lindner, F. (2004) *J. Vac. Sci. Technol. B*, **22**, 3229.
112. Viheriälä, J., Viljanen, M.-R., Kontio, J., Leinonen, T., Tommila, J., Dumitrescu, M., Niemi, T., and Pessa, M. (2009) *Proc. SPIE*, **7271**, 72711O.
113. Beck, M. and Heidari, B. (2006) OnBoard Technology p. 52 www.Onboard-Technology.com.
114. Singh, S., Chen, S., Selinidis, K., Fletcher, B., McMackin, I., Thompson, E., Resnick, D.J., Dress, P., and Dietze, U. (2009) *Proc. SPIE*, **7271**, 72712H.
115. Brooks, C., Schmid, G.M., Miller, M., Johnson, S., Khusnatdinov, N., LaBrake, D., Resnick, D.J., and Sreenivasan, S.V. (2009) *Proc. SPIE*, **7271**, 72711L.7280
116. Ina, H., Kasumi, K., Kawakami, E., and Uda, K. (2007) *Proc. SPIE*, **6517**, 65170M.
117. Xia, Q., Tong, W.M., Wu, W., Yang, J.J., Li, X., Robinett, W., Cardinali, T., Cumbie, M., Ellenson, J.E., Kuekes, P., and Williams, R.S. (2009) *Proc. SPIE*, **7271**, 727106.

2
Roll-to-Roll Nanoimprint Lithography and Dynamic Nano-Inscription

Se Hyun Ahn and L. Jay Guo

2.1
Introduction

Among many unconventional nanopatterning techniques that have been developed in the past decade [1], nanoimprint lithography (NIL) clearly stands out as a promising technology for high-throughput and high-resolution nanoscale patterning [2, 3]; it can achieve resolutions beyond the limitations set by light diffraction or beam scattering that are encountered in other traditional techniques. NIL has received great attention in nanoscale patterning [3–8] due to these characteristics [9–11].

Figure 2.1 shows the schematic of the originally proposed NIL [10] process. A hard mold that contains nanoscale surface relief features is pressed into a polymer material cast on a substrate at controlled temperature and pressure, which creates thickness contrast in the polymer material. A thin residual polymer layer is left underneath the mold protrusions acting as a soft cushion layer that prevents direct impact of the hard mold onto the substrate; this effectively protects the delicate nanoscale features on the mold surface. In most applications this residual layer needs to be removed by an anisotropic plasma reactive-ion-etching (RIE) process to complete the pattern definition. More detailed discussions of the NIL process, mold and resist material requirement have been provided in two recent review articles [5, 7]. Fixation of the polymer structure may be simply due to cooling below the glass transition temperature (thermal NIL) or be effected by curing of a resist in UV light (UV-NIL). UV-NIL, for example, that in step-and-flash imprint lithography (S-FIL) [12], allows fast processing speeds under ambient conditions because low-viscosity liquid resist quickly fills into mold cavities and no heating/cooling steps are required. However, for UV nanoimprint, at least one of the two parts, the mold or substrate, should be UV transparent, which is not required in thermal NIL.

Depending on the specific application, the imprinted polymeric material can be removed eventually or remain as a permanent component in the final device structure. When used as a resist layer, the patterned polymer (or polymer precursor)

Generating Micro- and Nanopatterns on Polymeric Materials. Edited by A. del Campo and E. Arzt
Copyright © 2011 WILEY-VCH Verlag GmbH & Co. KGaA, Weinheim
ISBN: 978-3-527-32508-5

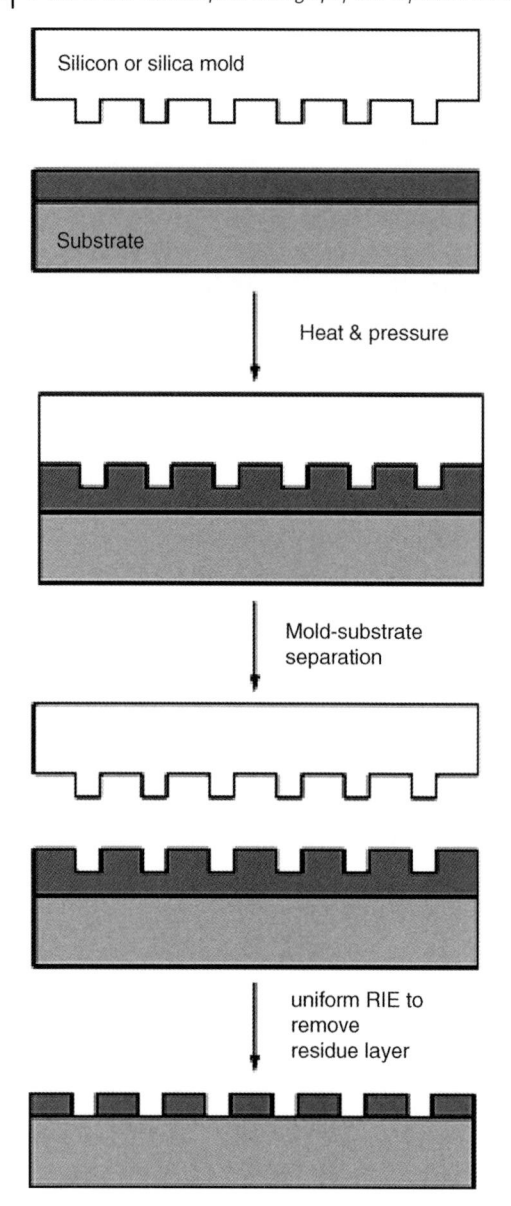

Figure 2.1 Schematic of the nanoimprint lithography process (Reprinted with permission from [10]. Copyright © (1997) American Institute of Physics.)

will be used to transfer the pattern into another material by RIE or a lift-off technique. This is the case for most current nanoimprint applications.

There is growing interest in direct-patterning of soft materials by NIL, particularly functional polymers. The patterned soft materials remain as a permanent structure or an active component in the final device structure. One such example is to nanoimprint the protein exchange polymer (Nafion®) for micro fuel cells with improved power efficiency by raising electrochemical active surface area (EAS), which therefore yields much enhanced catalyst utilization compared with conventional fuel cell membrane electrode assemblies (MEA) [13]. As another example of patterning functional polymers, NIL was recently used to create dense and ordered nanostructures in conjugated polymers to enhance their surface areas, which led to increased efficiency of organic solar cell devices [14]. Nanoimprint lithography is not only a successful patterning technique for these functional polymers, but it also improves the polymer properties through flow-induced polymer chain orientation during nanoimprinting. An example of improving organic electronics performance is presented to demonstrate the potential of manipulating chain orientation in polymer nanostructures. The origin and the factors that affect chain orientation in nanoimprinted polymer micro- and nanostructures are discussed in Ref. [15].

Since its inception, NIL has enjoyed great momentum in the past decade, and numerous applications such as in Si electronics [5, 16], organic electronic and photonics [14, 17], magnetics [18, 19], and biology [20–23] have been developed.

2.2
Roll-to-Roll Nanoimprint Lithography

The current process times in NIL amount to a few minutes or longer per wafer. Hence NIL is still far from meeting the throughput demands of many practical applications, especially in the areas of flat panel display, photonics, biotechnology and organic optoelectronics. To meet these demands, a faster and more economical process is needed. In this regard, a continuous roll-to-roll nanoimprint technique can provide a solution for high-speed large-area nanoscale patterning with greatly improved throughput; furthermore, it can overcome the challenges faced by conventional NIL in requiring large force, maintaining pressure uniformity and successful demolding in large area printing. In the original NIL process, huge contact area between the mold surface and the imprinted nanostructures can produce significant adhesion forces, making the mold–sample separation step difficult or even impossible to achieve without damaging the substrate. Moreover, in the thermal NIL process, with different thermal expansion coefficients of the Si mold and the polymer substrate, stresses can build up during a thermal cycle that even destroy the Si mold during mold separation.

Roll-to-roll nanoimprint lithography provides a unique solution to these challenges encountered in the conventional wafer-level NIL process. Imprinting in the roll-to-roll process takes place in a narrow region transverse to the moving direction of the substrate and thus requires much smaller force for pattern replication. Also since the mold is in the form of a roller, the mold-sample separation proceeds in a "peeling" fashion, which requires much less force, and reduces the probability of defect generation.

In an effort to realize the roller-based nanoimprint process, Tan *et al.* have used a solid rod to apply pressure to a piece of Si mold in a thermal nanoimprint process [24]. Lee *et al.* proposed a bilayer transfer process from a "rigiflex" mold to a Si wafer and demonstrated continuous thermal nanoimprinting process [25]. We developed a continuous roll-to-roll process capable of imprinting nanoscale structures on a flexible plastic substrate [26, 27], using a flexible fluoropolymer mold with a fast thermally curable polydimethylsiloxane (PDMS) [28] and a liquid UV-curable epoxysilicone resist [29]. Since in many applications such as flat panel displays, rigid substrates such as glass plate are preferred over flexible plastic substrates, we recently demonstrated large-area (10 cm wide) continuous imprinting of nanoscale structures by using a newly developed 15 cm-capable roll-to-plate/roll-to-roll apparatus, which can potentially be used in many practical applications [30].

Figure 2.2 shows the overall configuration of a typical continuous roll-to-roll NIL process [26, 27], which consists of three separate processing steps: (i) coating process, (ii) imprinting and separating process, and (iii) any of the subsequent processes. As an example, the last step in this schematic represents a continuous metal deposition process for making structures such as metal wire-grid polarizers [27] or transparent wire-grid electrode [31, 32] The fabrication and evaluation of metal wire-grid polarizers as an application of roll-to-roll NIL are described in the final section of this chapter.

True roll-to-roll nanoimprinting has been a challenge to the community because it requires a complete set of solutions to a number of interrelated material issues. Firstly, a special roller mold is required for continuous roll-to-roll imprinting of nanostructures. One approach is to use a NIL mold that is sufficiently flexible and can be wrapped on the roller surface. In addition, the mold should have sufficient modulus and strength to be able to imprint other materials. Secondly, liquid resists should have good coating properties and low viscosity to ensure fast imprinting; and

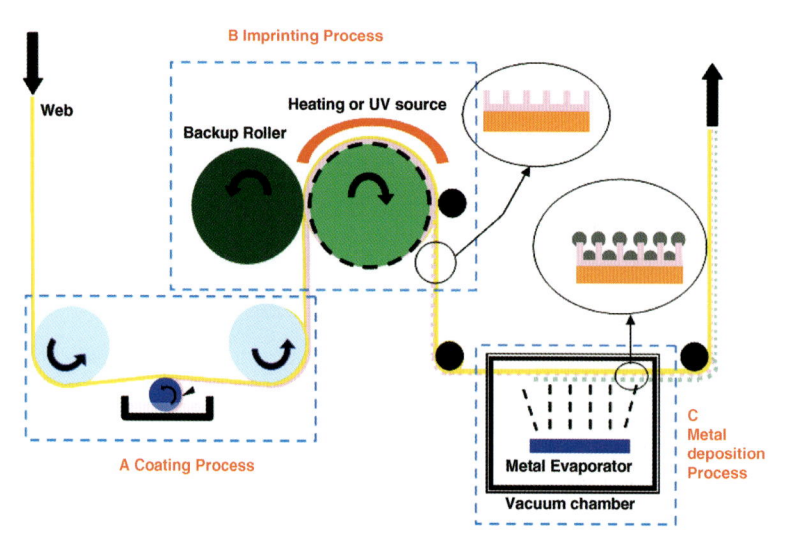

Figure 2.2 Schematic of a continuous roll-to-roll nanoimprinting setup, consisting of a coating (A) and an imprinting module (B), followed by a metal deposition process (C). (Reprinted with permission from [26]. Copyright © (2008) Wiley-VCH Verlag GmbH & Co. KGaA.)

they should be cured rapidly to maintain high-throughput and also should have minimal shrinkage. Therefore, conventional resist materials that are dissolved in solvents and require an additional baking process could make the roll-to-roll NIL process more difficult to control and thus more prone to defect generation.

The flexible mold can be made from metal or polymer materials. For example, a thin, electroplated Ni shim can be a durable mold for high volume manufacturing. Polymer molds, on the other hand, are much easier to replicate from original NIL masters. We introduced two types of flexible polymer molds that can be used for roll-to-roll NIL application. The first material is a commercially available fluoropolymer, ethylene tetrafluoroethylene (ETFE) [26, 30]. ETFE has high modulus ($\sim$1.2 GPa) at room temperature but can soften at elevated temperature. Therefore an ETFE mold can be easily replicated from an original Si mold by a thermal NIL process at 200 °C. Moreover, the exceptional anti-sticking property of ETFE (surface energy of 156 μN cm^{-1}, cf. PDMS $\sim$196 μN cm^{-1}) make demolding easy, even without any mold surface treatment and resulting deterioration in surface properties over many imprinting cycles. The second material is a new fluorinated photocurable silsesquioxane (SSQ) resin [33], which possesses outstanding properties for nanoscale patterning. With an appropriate viscosity, this resin can be easily imprinted by an original Si NIL master using a low pressure nanoimprinting process. The resin has a sufficient modulus in its cured state, which makes it suitable for nanoimprinting other polymeric materials. Due to the high thermal stability and UV transparency of SSQ materials, such a stamp can be used for both UV and thermal nanoimprinting. Furthermore, the fluoroalkyl groups contained in the silsesquioxane resin provide the low surface energy necessary for easy demolding. Details on the material composition and the SSQ mold fabrication can be found in Ref. [33].

In order to be able to imprint on both rigid substrate and flexible substrate, we constructed a prototype roller imprinter [30]. As shown in Figure 2.3a and b, the liquid phase UV-curable resist material is continuously coated on either a flexible poly(ethylene terephthalate) (PET) substrate in roll-to-roll NIL or a glass substrate in role-to-plate NIL by a 3-step roller coating system. The coating system is synchronized with the main imprinting roller to guarantee uniform coating thickness regardless of the web speed. For fast patterning speed, the system uses dual imprinting rollers and a tensioned belt supported by the two rollers to allow large curing area and to maintain constant pressure during the curing process. The pressure between the imprint roller and the back-up roller is adjusted by a clamping device and a force sensor. The system was designed to be able to handle up to 15 cm wide substrates.

In our experiment, several pieces of ETFE molds of proper size were replicated, wrapped and fixed onto a tensioned belt. For the roll-to-roll NIL process, the resist material is imprinted by the EFTE molds with pressure applied from the dual rollers and the tensioned belt, which is then cured by a high-power UV source under the web tension provided by the second roller. On the other hand, in the roll-to-plate NIL process, UV curing of the imprinted resist material takes place between the two rollers under the pressure provided by the belt tension. Even though the tangential component of the belt tension between rollers exerted on resist is smaller than that of the roller pressure, it is sufficient to maintain the resist film thickness after passing

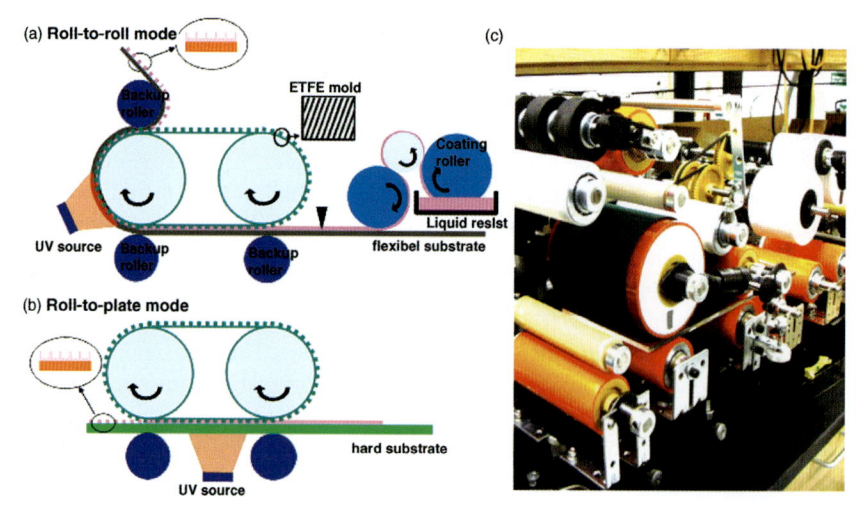

Figure 2.3 Schematics of (a) roll-to-roll NIL; (b) roll-to-plate NIL process; and (c) photograph of 15 cm-capable roll-to-roll/role-to-plate NIL apparatus. (Reprinted with permission from [30]. Copyright © (2009) American Chemical Society.)

through the first roller. Through this process, nano grating patterns were continuously created either on PET or glass substrate as the ETFE mold continuously detaches from the imprinted resist on the substrate [30].

For a fast roll-to-roll process, we used a UV-curable low viscosity liquid epoxysilicone [29] as the imprint resist material. Unlike acrylate-based resists often used in the UV-assisted NIL process such as step-and flash imprint lithography (S-FIL) [34], epoxysilicone is cured via a cationic curing mechanism, thereby free from the oxygen inhibition issue when exposed in air. Thus no special vacuum environment is required, which is convenient for the roll-to-roll process. Furthermore, its very low shrinkage after curing (only a fraction of the acrylate system) allows a faithful pattern replication. Owing to its low viscosity, the resist precursor can be imprinted at low pressures and cured within a second by focused UV light. The low pressure and the room temperature imprinting characteristics are advantageous for roll-to-roll NIL.

Figure 2.4a represents large area (10 cm by 30 cm) nanograting patterns produced on a flexible PET plastic substrate by the roll-to-roll NIL process. The inset is a SEM image of 700 nm period, 300 nm linewidth pattern from (a). Figure 2.4b shows the same imprinted nanograting patterns on a glass plate by roll-to-plate NIL. The strong light diffraction from the 700 nm period grating structures can be clearly observed. As described earlier, several pieces of ETFE molds were attached to the tensioned belt and the separation between them produced the seams between the square-shape patterned areas. The SEM image (inset in Figure 2.4a) shows that 300 nm linewidth, 600 nm height gratings are faithfully replicated onto the substrate. The fast curing of the epoxysilicone resist resulted in a web speed of 1 m per min, a 20-fold increase in imprinting speed as compared with that reported in Ref. [25], and can be increased further by using a more powerful UV curing light source. A requirement for

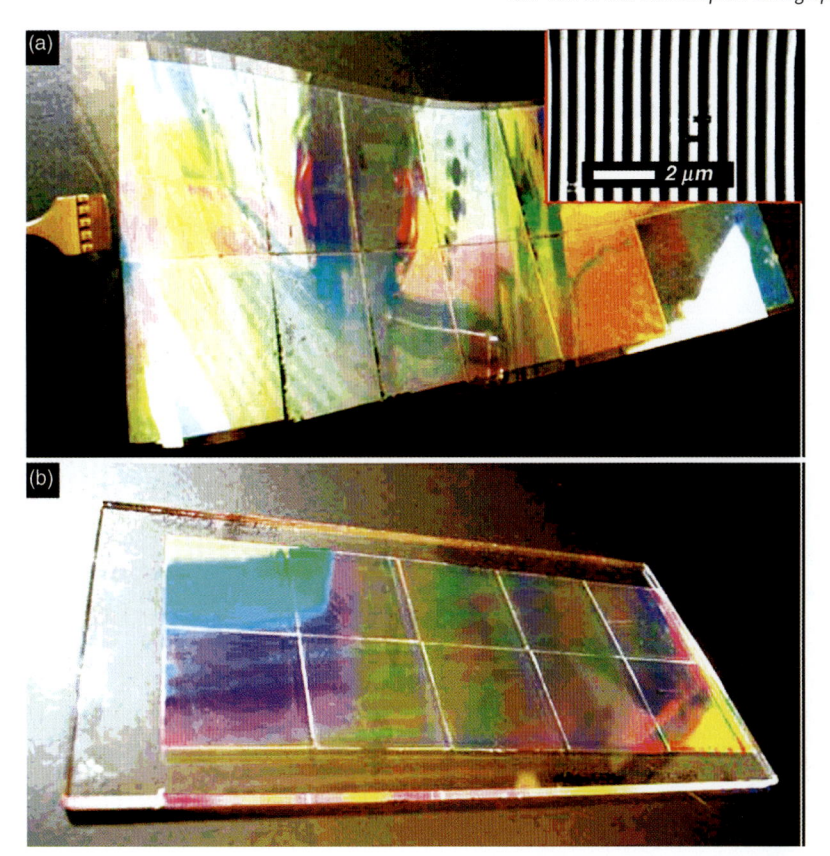

Figure 2.4 (a) A 10 cm wide, 30 cm long 700 nm period epoxysilicone patterns on a flexible PET substrate by roll-to-roll NIL process. The inset is a scanning electron microscope image of 700 nm period pattern from (a). (b) The same 700 nm period patterns on a glass substrate by roll-to-plate NIL process. (Reprinted with permission from [30]. Copyright © (2009) American Chemical Society.)

achieving such a large-area nanopattern is that the resist material has low adhesion to the mold and high adhesion to the substrate. For roll-to-plate NIL, we found that the epoxysilicone resist had relatively lower adhesion to the bare glass substrate and hence tended to peel off during the demolding process despite the low surface energy of the ETFE mold. To overcome this problem, the glass substrate was pretreated by oxygen plasma followed by the thermal deposition of an adhesion promoter (Silquest A187, GE Advance Materials).

High aspect ratio (AR = 5.4) grating structures with very sharp pattern definition can be fabricated by roll-to-roll NIL as is shown in Figure 2.5b and c [26]. Faithfully replicated epoxysilicone pattern should have the same geometry as in the original Si mold (Figure 2.5a) because the ETFE mold, replicated from the Si mold, has the exact inverse patterns of the Si mold. Comparing the grating structure of the original Si mold (Figure 2.5a) and the imprinted epoxysilicone pattern (Figure 2.5c), we observe

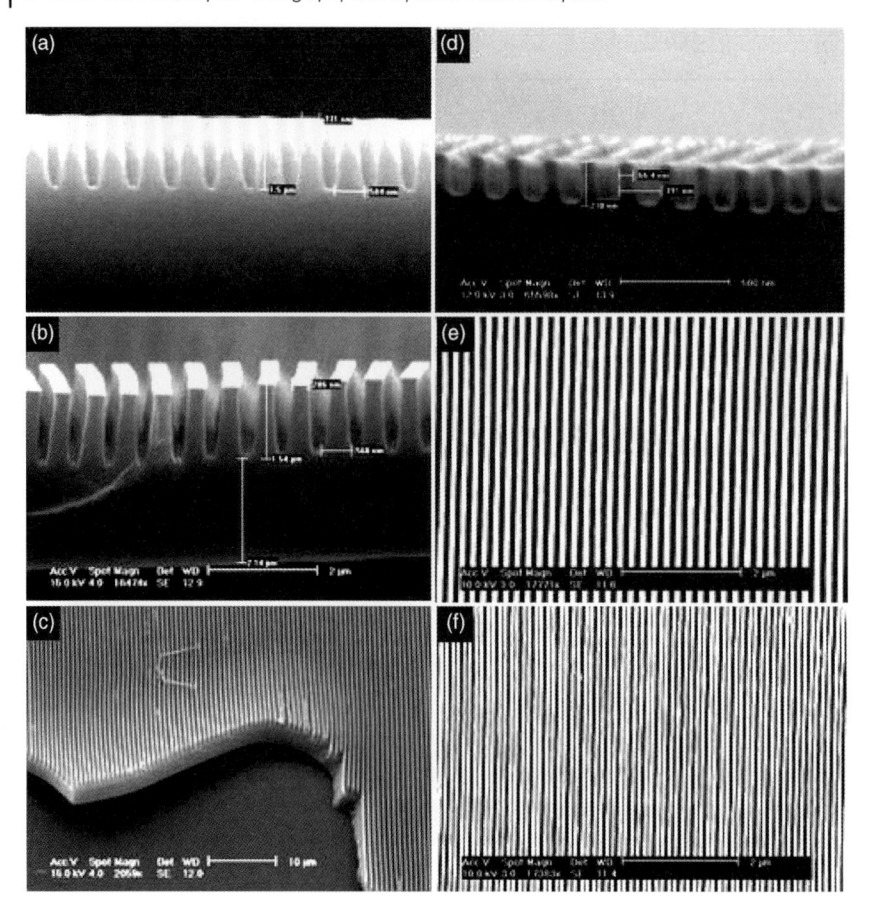

Figure 2.5 SEM images of epoxysilicone nano gratings fabricated by roll-to-roll NIL process: (a) the original Si mold; (b), (c) the epoxysilicone gratings replicated from the ETFE mold; (d), (e) SEM pictures of 200 nm period, 70 nm line width epoxysilicone pattern; (f) a 100 nm period, 70 nm line width epoxysilicone pattern. (Reprinted with permission from [26]. Copyright © (2008) Wiley-VCH Verlag GmbH & Co. KGaA.)

excellent pattern replication even for the very fine details at the bottom of the grating trenches. Even though the ETFE mold has good anti-sticking property, an imprinted structure with such a high AR tends to significantly stick to the ETFE mold due to the much larger interfacial area with the resist compared to shallow pattern. To achieve successful pattern transfer, we performed oxygen plasma treatment on the PET substrate before imprinting to improve adhesion of the resist pattern.

Continuous roll-to-roll imprinting of thinner and denser grating structure is more challenging because such patterns are mechanically fragile, and tend to collapse during demolding if the trench is very narrow. This requires the cured resist to have sufficient modulus and yield strength. Good adhesion of the resist to the substrate is also very important for such denser structures, which was achieved by using the aforementioned adhesion promoter. Figure 2.5d and e show 200 nm period, 70 nm linewidth epoxysilicone patterns by the UV roll-to-roll NIL process. The SEM photograph of 100 nm period grating structure in Figure 2.5f confirms successful replication.

To demonstrate an application of the roll-to-roll NIL process, a metal wire-grid polarizer was fabricated [26, 27]. The polarizer is an important optical element used in a variety of applications. Wire-grid polarizers in the form of subwavelength metallic gratings are an attractive alternative to conventional polarizers, because they provide high extinction ratio between the transmitted transverse magnetic (TM) – polarized light and the reflected transverse electric (TE) – polarized light over a wide wavelength range and incident angle with long-term stability [35, 36]. In addition, they are thin and planar structures and can be easily integrated with other thin-film optical elements. By depositing a thin metal (Al) layer over the imprinted grating structures, a high-efficiency polarizer in the form of bilayer metal wire grating was achieved [27, 35]. In an initial experiment, 200 nm and 100 nm period grating patterns (Figure 2.5d–f) were prepared by the roll-to-roll NIL process and aluminum of various thickness was thermally deposited on top of the grating as well as at the bottom of the trench (Figure 2.6a and b). To quantify the polarization effect, spectral transmittance was measured using UV/Vis spectrometer. Figure 2.6c shows the transmittance of the TM and TE polarized light through the fabricated metal wire-grid polarizer. The best polarizer result was obtained from the 100 nm period grating with 50 nm Al layer, with transmittance of about 30% at 800 nm wavelength and extinction ratio (transmittance of TM/transmittance of TE) over 2000 at 700 to 800 nm wavelength and 2500 maximum extinction ratio at 770 nm.

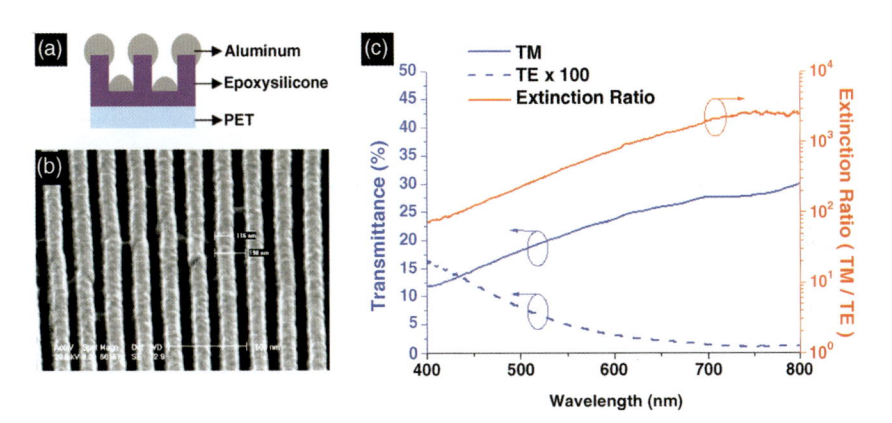

Figure 2.6 (a) Schematic of the metal wire-grid polarizer by depositing metal on top of the roller imprinted polymer grating; (b) SEM picture of a 200 nm period grating with 50 nm Al on top; (c) spectral transmittance (TM, TE mode) and extinction ratio (TM/TE) of metal wire-grid polarizer fabricated by roll-to-roll NIL. (Reprinted with permission from [26]. Copyright © (2008) Wiley-VCH Verlag GmbH & Co. KGaA.)

2.3
Dynamic Nano-Inscription

The roll-to-roll NIL process requires thermal or UV curing steps, which may not be compatible with many functional polymers; for example, bio- or sensitive organic semiconductor materials could be damaged by such process conditions. Moreover, creating large-area continuous pattern requires the same size original mold; also the imprinted patterns are not truly continuous due to the presence of a seam region where the two ends of the flexible mold meet on the roller.

Therefore, another nanofabrication technique, dynamic nano-inscription (DNI), was introduced by the authors [37]. DNI offers drastically increased throughput with nanoscale resolution, while providing continuous and seamless grating patterns at ambient conditions. Using DNI, we demonstrate high-speed nanopatterning of metals and several important functional polymers, and also illustrate its versatile use in producing free-form array patterns and patterning over curved surfaces.

Figure 2.7a illustrates the mechanism of the DNI process to create metallic nanopatterns on a polymer substrate. DNI uses the sharp edge of a slightly tilted cleaved Si mold to directly inscribe on a moving substrate, thereby creating seamless micro- and nanopatterns in a continuous fashion. DNI shares similarities with room-temperature nanoimprinting and relies on the plastic deformation of the inscribed

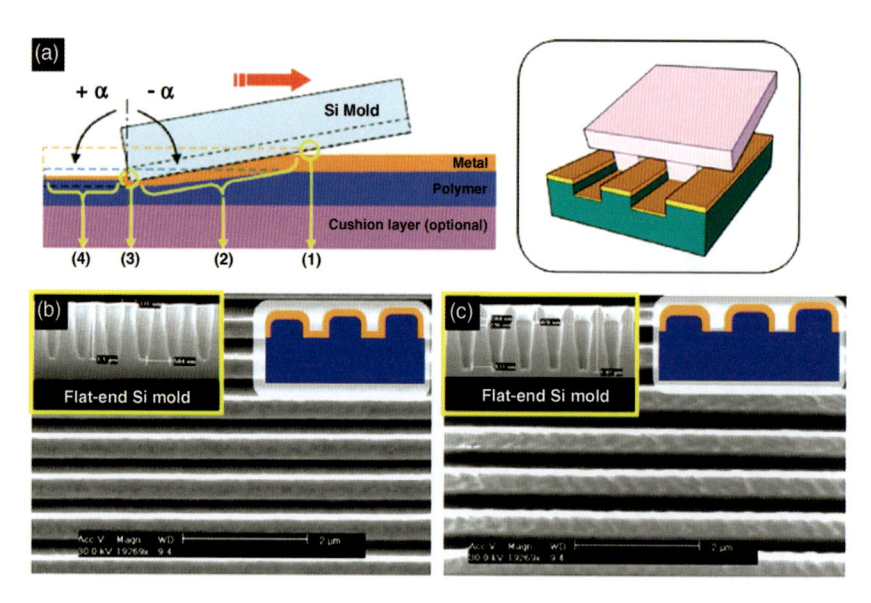

Figure 2.7 (a) Schematics of the DNI process for creating metal nanogratings: (1) Initial contacting point, (2) gradual imprinting region, (3) edge point responsible for plastic deformation, and (4) elastic recovery region; (b), (c) SEM images of the Si molds with two different tip geometries (flat-end, sharp tip) and the resulting silver nanogratings inscribed by each mold, respectively. (Reprinted with permission from [37]. Copyright © (2009) American Chemical Society.)

material. However, in contrast to nanoimprinting, the deformation in DNI takes place under gradually increased pressure over a very small contacting region where the sharp edge of tilted Si mold engages. As a result, continuous linear patterns with infinite length on various polymers, metals, or even hard materials such as indium tin oxide (ITO) can be successfully created by using very low applied forces (several newtons), and with a speed drastically faster than other nanopatterning techniques.

The DNI process can be separated into four sequential steps as depicted in Figure 2.7a. First, in region (1), the mold makes the initial contact with the substrate material to be inscribed. Second, in region (2), the polymer layer is imprinted with gradually increased pressure as the Si mold moves. Third, in region (3), where the sharp edge of the cleaved Si mold engages, most of the plastic deformation occurs since the pressure at this is the highest. Finally, in the elastic recovery region (4), the pressed polymer recovers by a certain amount and determines the final geometry. Note that, throughout the whole DNI process, the metal layer still remains on top of the imprinted polymer film without being removed, which supports the plastic deformation as the pattern formation mechanism rather than material removal.

There are several important properties a substrate material must possess for a successful DNI process. Since DNI relies on the plastic deformation of the polymer material, materials with lower modulus are preferred to achieve large deformation under a given applied pressure. Toughness is another very important property in DNI. Material having a low toughness can be easily fractured by the sharp edge of the Si mold and generates debris during the DNI process; and the accumulated debris further hampers the inscribing process. Therefore, materials with sufficiently high toughness are preferred for faithful and debris-free pattern formation. The optional elastomeric cushion layer underneath the inscribed polymer is to ensure conformal contact between the substrate and the edge of the Si mold that is not perfectly flat, but having terraces that result from the cleaving step. DNI can create debris-free nanoscale gratings without causing mold damage because it employs a large negative rake angle ($\alpha < 0$) in the process, which means the mold is almost parallel to the substrate surface, like in nanoimprinting. In general mechanics, a negative rake angle has low material removal capability but provides better mechanical stability for the brittle tool materials such as ceramic or diamond [38]. DNI uses both character-istics from a negative rake angle to create continuous nanogratings without gener-ating debris nor damaging mold.

The cross-sectional profile of the metal nanogratings created by the DNI process depends on the end shape of the mold grating structure. As shown in Figure 2.7b, when a Si grating mold with a flat-tip is used, the metal layer deforms and follows the shape of the inscribed polymer pattern and forms continuous metal gratings. On the other hand, when the metal layer is inscribed by a mold with a sharp-tip, the metal film breaks at the midpoint where it meets the tip of the mold tip. As a result, metal lines with discrete metal caps are formed on top of the inscribed polymer gratings (Figure 2.1c).

DNI can proceed in straight linear fashion or in free forms by changing the substrate moving direction with respect to the orientation of the mold grating. Figure 2.8a shows 70 nm trench, 200 nm period gold gratings on PET substrate

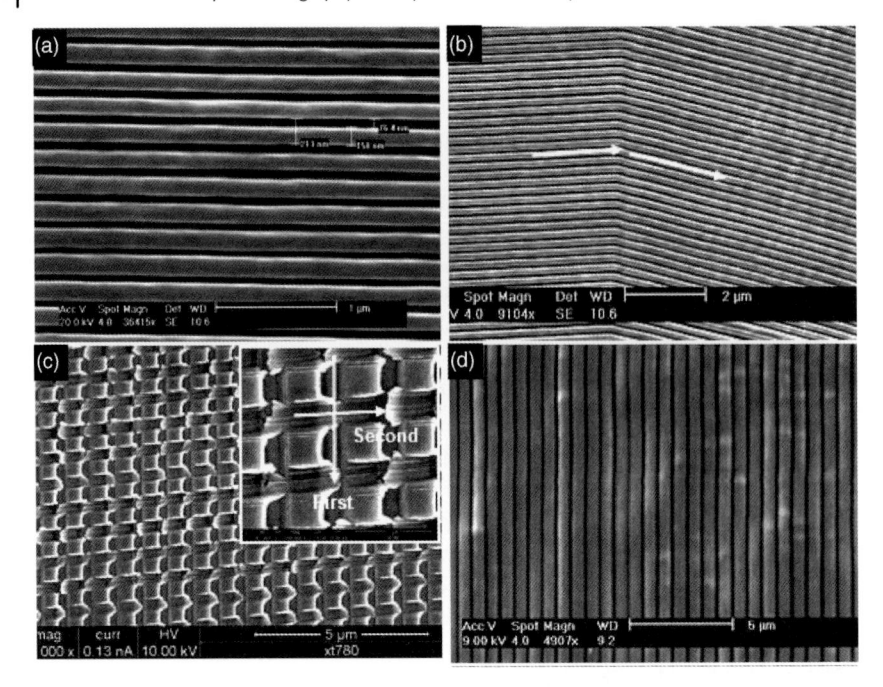

Figure 2.8 SEM micrographs of (a) 200 nm period, 70 nm trench gold nano gratings; (b) continuous 200 nm period gold gratings by DNI with sharp turns; (c) square-shaped gold nanopatterns fabricated by two sequential DNIs in orthogonal directions; (d) 700 nm period gratings directly fabricated on ITO surface. (Reprinted with permission from [37]. Copyright © (2009) American Chemical Society.)

fabricated by DNI at ambient temperature. Figure 2.8b shows angled but continuous 200 nm period gratings that were created by sharply turning the substrate during DNI process. Such a feature is impossible to achieve by the regular NIL process using a linear grating mold. Figure 2.8c represents square-shaped gold nanopatterns fabricated by two consecutive DNIs carried out in orthogonal directions. DNI is not limited to only polymeric and ductile metallic materials. Direct inscribing of much harder materials, such as conductive ITO, is also successfully demonstrated by RT-DNI. Figure 2.8d shows 700 nm period nanogratings directly created on an indium tin oxide (ITO) – coated PET substrate. ITO is a very hard material having a modulus (97 GPa) but slightly smaller than that of Si. Therefore, it is very difficult to directly pattern on an ITO surface by conventional fabrication techniques, however, DNI provides a unique solution for ITO nanopatterning.

Roll-to-roll printing technique can offer high-speed patterning in a continuous fashion. But roller-based printing technology generally requires a very large-area flexible mold to wrap around a roller surface, and at least one seam per revolution exists where the two ends of the flexible mold meet. However, DNI can generate an essentially infinitely long and seamless grating pattern by using just a small slice of a cleaved Si mold. This feature could find potential applications, for example, in large

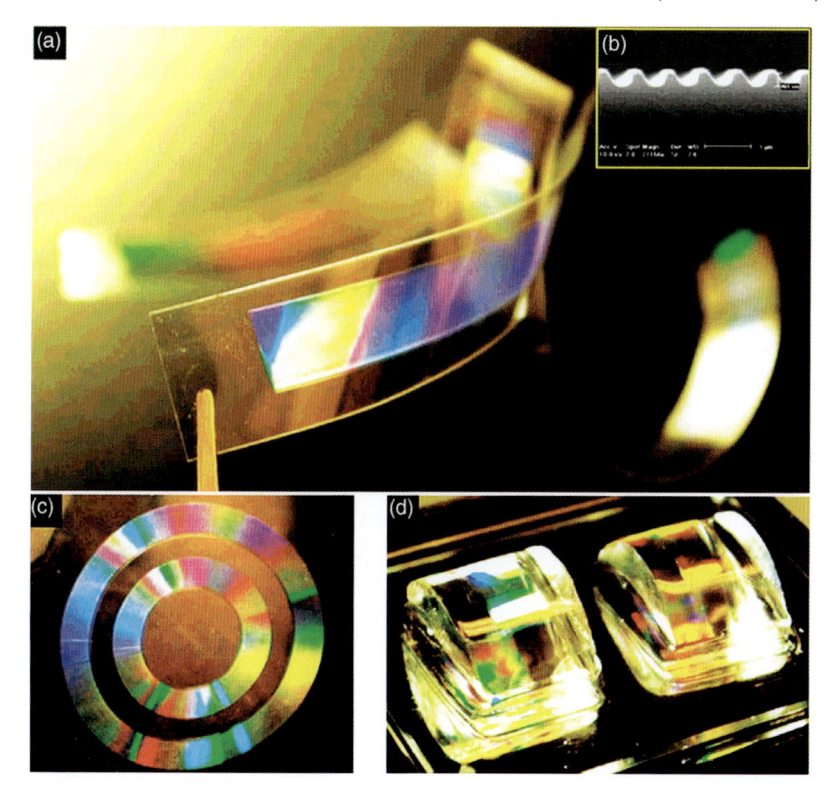

Figure 2.9 (a) A 56 cm long, 1.3 cm wide, 700 nm period, continuous nanograting pattern directly created on a polycarbonate strip by roll-to-roll DNI process and (b) its SEM image; (c) concentric nanogratings on ethylene-tetrafluoroethylene film fabricated by a rotating DNI process. Outer diameter is 5 cm; (d) 700 nm period gratings inscribed in cross-linked photoresist SU8 on curved surfaces. (Reprinted with permission from [37]. Copyright © (2009) American Chemical Society.)

format displays and for creating long polymer nanofibers. In this respect, a high speed roll-to-roll process can fully take the advantages offered by DNI. Figure 2.9a shows a 56 cm long, 1.3 cm wide, 700 nm period, seamless nanograting pattern created on a polycarbonate strip using roll-to-roll DNI. The inset (Figure 2.9b) shows the SEM image of the grating. In this case, the web speed is 10 cm s^{-1}, which is significantly faster than the NIL technique. We expect that the speed can be increased by at least an order of magnitude or higher in practical applications by proper tool design. Such high speeds are possible because DNI relies on the plastic deformation, which can occur on a time scale of microseconds [39]. In comparison, regular NIL relies on the filling of the cavity features on the mold by the polymer materials, which takes much longer time (seconds to minutes, depending on the material viscosity) [5]. In addition, continuous and curved nanograting patterns over a large area is another unique capability of DNI, which is very difficult to achieve by other nanopatterning techniques. In Figure 2.9c, we demonstrate concentric circular nanogratings on a fluoropolymer film, ethylene-tetrafluoroethylene, fabricated by a rotating DNI process. Such structures may find potential applications in magnetic data storage such as discrete track media [40]. Furthermore, since DNI only requires essentially line contact with the substrate, it offers a convenient method to pattern on curved surfaces. As an example, 700 nm period nanogratings created on a curved surface are shown in Figure 2.9d. In this result, the patterns are inscribed on a common photoresist, SU-8, which is coated on the hemicylindrical PDMS surface and subsequently cured by UV light. Nanoscale metal electrodes or patterned organic materials on the curved surface are potentially useful for light detection or energy-generating devices [41].

2.4
Summary

Nanoimprint lithography technology provides an effective solution to many challenges faced by conventional nanoscale patternings. To improve the nanoimprint throughput, a roll-to-roll NIL process has been developed for continuous imprinting of micro- and nanostructures on rigid or flexible substrates. This process will be particularly useful in the high-speed fabrication of nanostructures over very large areas. Because nanoimprinting creates surface relief structures by mechanical deformation rather than by relying on chemical etching, this technique is particularly suitable for polymer materials with special functionalities that might be compromised by the chemical process. In addition, we have introduced a new nanopatterning technique, dynamic nano-inscription, DNI, for directly creating true continuous nano grating patterns in a variety of metal or polymer materials at ambient temperature. It can be envisioned that DNI will be used in many practical applications such as in optics, display and bio-industries, all of which require nontoxic, room temperature environment processes to produce well-defined nanopatterns at high speed.

Acknowledgment

We are grateful to the National Science Foundation, the DOD STTR program and the University of Michigan Technology Transfer Office for the support of this work.

References

1 Rogers, J.A. and Lee, H.H. (2009) *Unconventional Nanopatterning Techniques and Applications*, John Wiley & Sons, Inc., Hoboken, USA.
2 Chou, S.Y., Krauss, P.R., and Renstrom, P.J. (1995) *Appl. Phys. Lett.*, **67**, 3114.
3 Chou, S.Y., Krauss, P.R., and Renstrom, P.J. (1996) *Science*, **272**, 85.
4 Cross, G.L.W. (2006) *J. Phys. D*, **39**, R363.
5 Guo, L.J. (2007) *Adv. Mater.*, **19**, 495.
6 Heckele, M. and Schomburg, W.K. (2004) *J. Micromech. Microeng.*, **14**, R1.

References

7. Schift, H. (2008) *J. Vac. Sci. Technol. B*, **26**, 458.
8. Stewart, M.D. and Willson, C.G. (2005) *MRS Bulletin*, **30**, 947.
9. Austin, M.D., Zhang, W., Ge, H., Wasserman, D., Lyon, S.A., and Chou, S.Y. (2005) *Nanotechnology*, **16**, 1058.
10. Chou, S.Y., Krauss, P.R., Zhang, W., Guo, L., and Zhuang, L. (1997) *J. Vac. Sci. Technol. B*, **15**, 2897.
11. Hua, F., Sun, Y., Gaur, A., Meitl, M.A., Bilhaut, L., Rotkina, L., Wang, J., Geil, P., Shim, M., Rogers, J.A., and Shim, A. (2004) *Nano Lett.*, **4**, 2467.
12. Ruchhoeft, P., Colburn, M., Choi, B., Nounu, H., Johnson, S., Bailey, T., Damle, S., Stewart, M., Ekerdt, J., Sreenivasan, S.V., Wolfe, J.C., and Willson, C.G. (1999) *J. Vac. Sci. Technol. B*, **17**, 2965–2969.
13. Taylor, A.D., Lucas, B.D., Guo, L.J., and Thompson, L.T. (2007) *J. Power Sources*, **171**, 218.
14. Kim, M.-S., Kim, J.-S., Cho, J., Stein, M., Guo, L.J., and Kim, J. (2007) *Appl. Phys. Lett.*, **90**, 123113.
15. Cui, D., Li, H., Park, H., and Cheng, X. (2008) *J. Vac. Sci. Technol. B*, **26**, 2404.
16. Zhang, W. and Chou, S.Y. (2003) *Appl. Phys. Lett.*, **83**, 1632.
17. Pisignano, D., Persano, L., Raganato, M.F., Visconti, P., Cingolani, R., Barbarella, G., Favaretto, L., and Gigli, G. (2004) *Adv. Mater.*, **16**, 525.
18. Wu, W., Cui, B., Sun, X., Zhang, W., Zhuang, L., Kong, L., and Chou, S.Y. (1998) *J. Vac. Sci. Technol. B*, **16**, 3825.
19. Martin, J.I., Nogues, J., Liu, K., Vicent, J.L., and Schuller, I.K. (2003) *J. Magn. Magn. Mater.*, **256**, 449.
20. Cao, H., Yu, Z.N., Wang, J., Tegenfeldt, J.O., Austin, R.H., Chen, E., Wu, W., and Chou, S.Y. (2002) *Appl. Phys. Lett.*, **81**, 174.
21. Guo, L.J., Cheng, X., and Chou, C.F. (2004) *Nano Lett.*, **4**, 69.
22. Falconnet, D., Pasqui, D., Park, S., Eckert, R., Schift, H., Gobrecht, J., Barbucci, R., and Textor, M. (2004) *Nano Lett.*, **4**, 1909.
23. Hoff, J.D., Cheng, L.J., Meyhofer, E., Guo, L.J., and Hunt, A.J. (2004) *Nano Lett.*, **4**, 853.
24. Tan, H., Gilbertson, A., and Chou, S.Y. (1998) *J. Vac. Sci. Technol. B*, **16**, 3926.
25. Seo, S.-M., Kim, T.-I., and Lee, H.H. (2007) *Microelectron. Eng.*, **84**, 567.
26. Ahn, S.H. and Guo, L.J. (2008) *Adv. Mater.*, **20**, 2044.
27. Ahn, S.H., Kim, J.-S., and Guo, L.J. (2007) *J. Vac. Sci. Technol. B*, **25**, 2388.
28. 28. Hernandez, C.P., Kim, J.-S., Guo, L.J., and Fu, P.-F. (2007) *Adv. Mater.*, **19**, 1222.
29. Cheng, X., Guo, L.J., and Fu, P.-F. (2005) *Adv. Mater.*, **17**, 1419.
30. Ahn, S.H. and Guo, L.J. (2009) *ACS Nano*, **3**, 2304.
31. Kang, M.G. and Guo, L.J. (2007) *Adv. Mater.*, **19**, 1391.
32. Kang, M.-G., and Guo, L.J. (2007) *J. Vac. Sci. Technol. B*, **25**, 2637.
33. Pina-Hernandez, C., Fu, P.-F., and Guo, L.J. (2008) *J. Vac. Sci. Technol. B*, **26**, 2426.
34. Colburn, M., Johnson, S.C., Stewart, M.D., Damle, S., Bailey, T.C., Choi, B., Wedlake, M., Michaelson, T.B., Sreenivasan, S.V., Ekerdt, J.G., and Willson, C.G. (1999) *Proc. SPIE*, **3676**, 379.
35. Ekinci, Y., Solak, H.H., David, C., and Sigg, H. (2006) *Opt. Express*, **14**, 2323.
36. Yu, X.J. and Kwok, H.S. (2003) *J. Appl. Phys.*, **93**, 4407.
37. Ahn, S.H. and Guo, L.J. (2009) *Nano Lett.*, **9**, 4392.
38. Kalpakjian, S. and Schmid, S.R. (2003) *Manufacturing Processes for Engineering Materials*, Prentice Hall.
39. Aretxabaleta, L., Aurrekoetxea, J., Urrutibeascoa, I., and Sánchez-Soto, M. (2005) *Polymer Testing*, **24**, 145.
40. Hattori, K., Ito, K., Soeno, Y., Takai, M., and Matsuzaki, M. (2004) *IEEE Trans. Magn.*, **40**, 2510.
41. Meitl, M., Zhu, Z.-T., Kumar, V., Lee, K.J., Feng, X., Huang, Y.Y., Adesida, I., Nuzzo, R.G., and Rogers, J.A. (2006) *Nature Mater.*, **5**, 33.

3
Solvent-Assisted Molding
Ho-Sup Jung and Kahp-Yang Suh

Micropatterning technology is vital to the development of various microsystems such as microelectronics, optical, mechanical, and bio-analysis systems. For bulk production, photolithography is the most commonly used patterning technique. Recently, photolithography has been challenged by the need to manufacture structures on the scale of less than 100 nm. At these small scales, conventional photolithography may not be an appropriate method considering difficulties and costs involved. For example, it is not an inexpensive technology [1–4]; it is poorly suited for patterning nonplanar surfaces [5]; it provides almost no control over the chemistry of the surface and hence is not very flexible in generating patterns of specific chemical functionalities on surfaces; it can generate only two-dimensional microstructures [6]; and it is directly applied only to a limited set of photosensitive materials (e.g., photoresists) [7].

As compared with conventional photolithography, nontraditional lithographic techniques provide a low expertise route to three-dimensional (3D) microscale patterning. The nontraditional lithographies typically involve the use of a mold. When the mold is soft such as silicone elastomers, the lithography method is termed soft lithography [8–11]. Soft lithography techniques are low in capital cost, easy to learn, straightforward to apply, and accessible to a wide range of users. Furthermore, this technique is a convenient route to fabricating templates or microstructures for biological applications due to its ability to control the molecular structure of surfaces and to fabricate channel structures appropriate for microfluids, and to pattern and manipulate cells [11]. In particular, biological applications often involve relatively large feature sizes ($>50 \mu m$), which is suitable for soft lithography. In soft lithography, an elastomeric stamp with patterned relief structures on its surface is used to generate patterns and structures with feature sizes ranging from 30 nm to 100 μm. Five major techniques have been developed: microcontact printing (μCP) [12], replica molding (REM) [13], microtransfer molding (μTM) [14], micromolding in capillaries (MIMIC) [15], and solvent-assisted microcontact molding (SAMIM) [9, 10].

When the mold is hard, for example, a silicon wafer, it is called imprint lithography [16]. Imprint lithography involves applying a very high pressure (3000 to 15 000 N cm^{-2}) for the imprinting of the mold pattern into an underlying polymer layer. Some apparent limitations [17] inherent in imprint lithography make it

Generating Micro- and Nanopatterns on Polymeric Materials. Edited by A. del Campo and E. Arzt
Copyright © 2011 WILEY-VCH Verlag GmbH & Co. KGaA, Weinheim
ISBN: 978-3-527-32508-5

somewhat unsuitable for the formation of 3D microstructures such as high pressure/ temperature, equipment complexity, and high cost. Moreover, the thermal cycle of heating and cooling processes may cause distortion of the mold and the imprinted pattern.

In contrast, solvent-assisted molding uses an elastomeric polydimethylsiloxane (PDMS) mold in combination with an appropriate solvent instead of a rigid mold and high temperature/pressure to emboss the polymer film. The solvent, rather than temperature, softens the polymer. Solvent is either briefly applied to the PDMS mold [9] or retained in the polymer film [18] before placing the two surfaces in contact. The polymer is drawn into the cavities of the mold as solvent is removed from the mold–polymer interface by transport and evaporation. The permeable mold prevents trapping of air pockets and non-uniform solvent evaporation. After evaporation of the solvent, the mold is removed, leaving behind a relief structure complementary to its topography.

In this chapter, we overview solvent-assisted molding techniques by describing the operation principle and several examples of methods. In order to assist molding, a good or a poor solvent can be used. First, a good solvent is used to soften polymer films at the time of contact, thereby forming a replica of the original mold by solvent absorption or capillarity. If a poor solvent is used, the process can be mediated by dewetting. Both molding procedures are generally carried out at low temperature and pressure, allowing for a simple way of pattern transfer to the substrate at mild conditions. Several techniques have been developed that can be referred to as solvent-assisted molding, which are described below. Also, several application areas of the techniques will be discussed.

3.1
The Principle of Solvent-Assisted Molding

The operational principle of solvent-assisted molding shares characteristics with both embossing and replica molding. The technique begins with an elastomeric mold (usually made from PDMS) [19–23]. Some years ago, Lee et al. reported an improved version of solvent-assisted micromolding (termed "soft molding") to form 3D structures [18, 23]. Some enabling conditions for solvent-assisted molding can be demonstrated as follows: the key element of this procedure is wetting of the PDMS mold by a solvent and conformal contact between the elastomeric mold and the substrate. Figure 3.1 illustrates the soft molding process. An elastomeric mold with the desired pattern on its surface is placed onto a polymer film, without baking, immediately after the film is spin coated onto a substrate and then pressed slightly at a pressure of less than 1 N cm^{-2}. After releasing the pressure, the mold and the substrate are allowed to remain undisturbed for a period of time, which is on the order of 10 min. During the molding procedure, the solvent contained in the polymer film diffuses toward the interface due to the concentration gradient and is absorbed into the mold. The solvent then permeates through the mold and finally evaporates into the air. After the evaporation is completed, the mold is removed, thus finishing

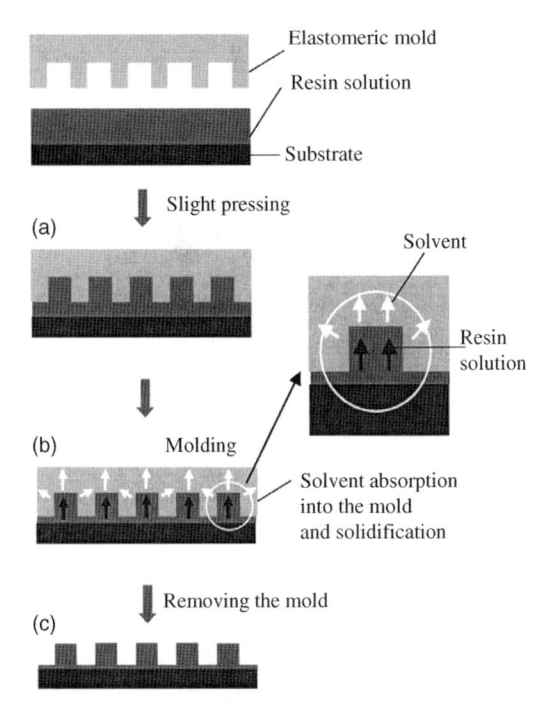

Figure 3.1 Illustration of the soft molding method. (a) An elastomeric mold is placed on a polymer film that is spin coated onto a substrate, and then slightly pressed down to form the pattern on the spin-coated polymer film. (b) After releasing the pressure, the whole structure is left undisturbed for a period of time for solidification, during which time the solvent in the molded structure is absorbed into the mold. (c) The mold is then removed. (Reproduced with permission from [23]. Copyright © (2001) American Institute of Physics.)

the soft molding process. In this molding, the solvent in the molded polymer is continuously removed by absorption into the PDMS mold at the interface between the mold and the molded polymer and replenished by solvent diffusion to the interface.

To arrive at a condition under which soft molding can be effective, solvent absorption into the mold and evaporation from the polymer surface are modeled for their rates. If it is assumed that the rate of solvent evaporation is proportional to the solvent concentration, then one has Equation 3.1:

$$\frac{\partial M}{\partial t} = -kM, \tag{3.1}$$

where M is the solvent concentration in terms of weight per unit area, t is time and k is a proportionality constant. Equation 3.2 gives

$$M = M_0 e^{-kt} \tag{3.2}$$

where M is determined experimentally by dividing the weight of solvent evaporated by the exposed surface area. In an experiment to determine the rate of absorption by the mold, the mold is floated in a bath filled with the solvent. Since the solvent concentration is constant, the rate of absorption per exposed surface area is independent of the concentration such that (Equation 3.3):

$$\frac{\partial M_a}{\partial t} = -\alpha, \qquad (3.3)$$

where M_a is the concentration of the solvent per unit area in the mold and α is a constant. For soft molding to be successful, the rate of absorption should be larger than the rate of evaporation. Since the evaporation rate is exponential, whereas the absorption rate is linear, the initial rates should be used for the condition that the absorption rate be greater than the evaporation rate.

It then follows from Equations 3.1 and 3.3 that

$$\alpha > kM_0 \qquad (3.4)$$

In soft molding, the solvent is absorbed into the mold at the interface between the polymer film and the mold. Therefore, the value of α in the soft molding would be smaller than that of α determined in a bath of solvent. Therefore, the condition should be modified to (Equation 3.5)

$$\alpha \gg skM_0. \qquad (3.5)$$

While 3D pattern formation in a film on a substrate has been reported, its transfer into the underlying substrate is a different issue [18]. There is always a residual layer remaining on the substrate when any method is used to form a pattern in a film, typically a polymer film. Therefore, a two-step etching is needed, first to remove the residual layer to expose the substrate surface, and then to transfer the pattern into the substrate with the remaining film as the etching mask. While this two-step etching may be much less of a problem in transferring a two-dimensional pattern formed in the polymer film into the substrate, it certainly can cause problems in faithfully transferring 3D structures.

To investigate the conditions under which single-step etching works, Lee et al. reported a modified soft molding technique and examined the behavior of solvent absorption into the mold and that of solvent removal from the polymer solution. In the original soft molding [23], the rate of absorption is much larger than the rate of evaporation such that the solvent reaching the interface between the film and the mold is immediately absorbed into the mold. In contrast, the solvent should be present in the polymer in the case of modified soft molding in an amount sufficient for continuous movement of polymer into the void space by capillarity. Therefore, the solvent and polymer pair should be chosen in such a way that the solvent would be removed slowly by absorption into the mold. In other words, the rate of absorption should be less than the rate of removal of solvent from the film or the rate of solvent evaporation from the polymer surface. It has been shown experimentally that the rate of evaporation can be described by first order kinetics and that of absorption by zeroth order kinetics. Therefore, the condition for the modified soft molding to work can be written as Equation 3.6:

$$\frac{\alpha}{kM_0} < 1. \tag{3.6}$$

where α is the rate constant for absorption, k is the first order rate constant for evaporation, and M_0 is the initial solvent weight in the polymer film per unit area. Given the pair of polymer and solvent, the rate constants can be determined experimentally.

3.2
Solvent-Assisted Molding with a Good Solvent

The original solvent-assisted molding technique is called solvent-assisted microcontact molding (SAMIM, Figure 3.1) developed by Whitesides' group [8–11]. The technique generates relief structures in the surface of a material using a good solvent that can dissolve (or soften) the material without affecting the PDMS mold. First, a PDMS mold absorbs a good solvent and it is brought into contact with the surface of the substrate (typically an organic polymer). The solvent dissolves (or swells) a thin layer of the substrate, and the resulting fluid or gel is molded against the relief structures in the mold. When the solvent dissipates and evaporates, the fluid solidifies and forms a patterned relief structure complementary to that in the surface of the mold. SAMIM shares an operational principle similar to that of embossing, but differs from the technique in that SAMIM uses a solvent instead of temperature to soften the material; furthermore, it uses an elastomeric PDMS mold rather than a rigid master to imprint patterns into the surface of the substrate.

SAMIM can be used with a wide variety of materials, although the initial demonstration has focused on organic polymers. The only requirement for SAMIM is a solvent that dissolves the substrate, and wets (but swells very little!) the surface of the PDMS mold. In general, the solvent should have a relatively high vapor pressure and a moderately high surface tension (e.g., methanol, ethanol, or acetone) to ensure rapid evaporation of the excess solvent and minimal swelling of the PDMS mold. Other materials can also be added to the solvent and will subsequently be incorporated into the resulting microstructures. Solvents with low vapor pressures (e.g., ethylene glycol or dimethyl sulfoxide) are not well suited for SAMIM. Hydrophilic elastomers or surface modifications of PDMS (for example, by plasma treatment) are required when solvents with high surface tensions (e.g., water) are used, because they only partially wet hydrophobic surfaces.

Figures 3.2 and 3.3 show SEM images of several 3D structures formed by soft molding or SAMIM. The circular cones shown in Figure 3.2a have a base diameter of approximately 1 μm. Rounded triangular channels with a period of approximately 3 μm are depicted in Figure 3.2b. A three-level structure is shown in Figure 3.2c. These results demonstrate the effectiveness of solvent-assisted molding in fabricating 3D structures. SAMIM can also replicate complex relief structures over large

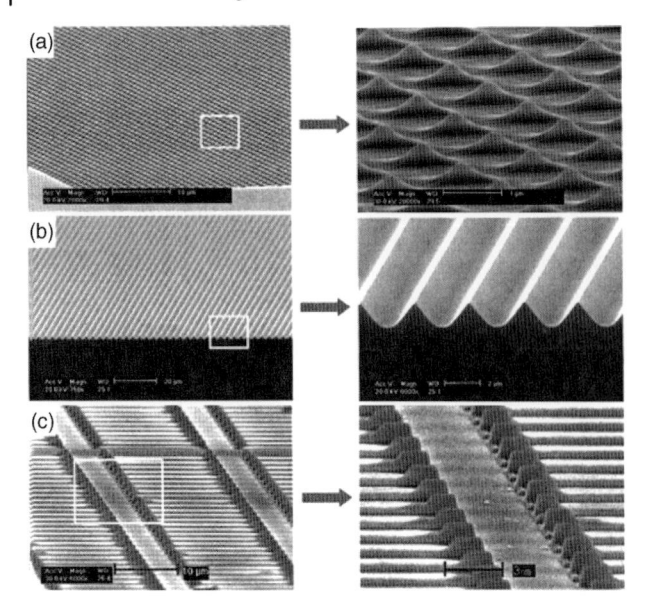

Figure 3.2 Tilted SEM images of several three-dimensional structures fabricated by soft molding: (a) circular cones, the enlarged part on the right showing that the base diameter of the cone is approximately 1 μm; (b) rounded triangular channels with a period of approximately 3 μm; and (c) a three-level structure. (Reproduced with permission from [23]. Copyright © (2001) American Institute of Physics.)

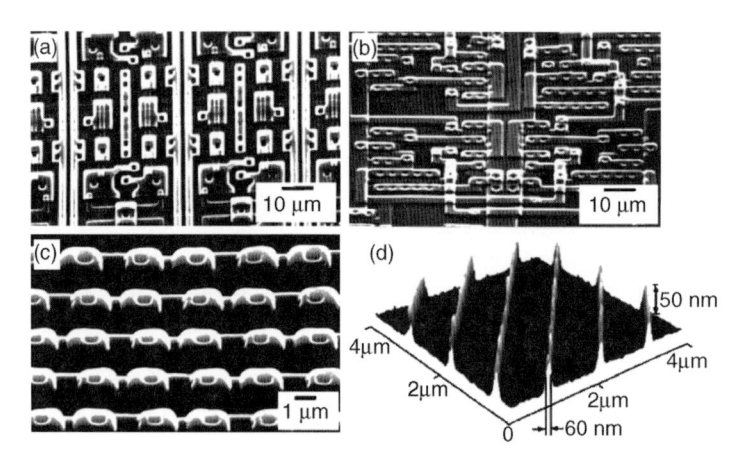

Figure 3.3 Images of polymeric microstructures fabricated by SAMIM. (a–c) SEM images of quasi three-dimensional structures in photoresist (Microposit 1805, Shipley; ∼1.6 μm thick) spin-coated on Si/SiO$_2$, polystyrene (PS, Goodfellow; 2.0 μm thick), and ABS (Goodfellow; 0.85 μm thick), respectively. (d) AFM image of nanostructures in a thin (∼0.4 mm thick) film of Microposit 1805 spin-coated on Si/SiO$_2$. The solvent was ethanol for the photoresist and acetone for PS and ABS. (Reproduced with permission from [15]. Copyright © (1995) Macmillan Publishers Ltd.)

areas in a single step (Figure 3.3). These quasi 3D structures are well defined and clearly resolved. Figure 3.3a shows an AFM image of the smallest features that were generated using SAMIM: parallel lines ∼60 nm wide and ∼50 nm high formed in a thin film of Shipley photoresist (Microposit 1805, the thickness of the film was ∼0.4 µm). A common characteristic of microstructures generated by SAMIM is that the resulting structures are joined by a thin underlying film of the polymer. This film can be removed by homogeneous thinning using O_2 RIE, and the resulting polymeric structures can be used as masks in the etching of underlying substrates.

3.3
Solvent-Assisted Molding with a Poor Solvent

To begin with solvent-assisted molding mediated by a poor solvent, it is necessary to briefly mention microcontact printing. Microcontact printing (µCP) is a useful technique for generating patterns of functional organic surfaces over large areas (>cm^2) [24–34]. The procedure uses a topographically patterned PDMS stamp, wetted with a solution of an alkanethiol or other molecules (phosphonic acids, alkylsiloxanes, isocyanides) that can form self-assembled monolayers (SAMs). The stamp is brought into contact with the surface of a metal, metal oxide, or semiconductor for a few seconds; an ordered monolayer forms rapidly at the points of contact. This procedure works best with alkanethiols and derivatives on gold, silver, palladium, and platinum. This methodology can tailor properties of the surface, such as wettability, biocompatibility, or reactivity, by transferring organic films that are only 1 to 2 nm thick [35, 36].

Recently, micromolding of a hydrophilic biopolymer such as HA or polyethylene glycol (PEG) [37–41] has been introduced as an alternative to µCP for the development of protein and cell arrays. Since dewetting often mediates the molding process, the technique can be termed solvent-assisted dewetting molding (SADEM). Suh *et al.* reported that hyaluronic acid (HA) is compatible with µCP and molding approaches (Figure 3.4). To pattern HA using µCP, PDMS stamps [42] were treated with oxygen plasma to improve adhesion of the HA layer to the substrate surface. The procedure is dependent on the wettability of the stamp at the time of contact, such that pattern transfer was successful only when the stamp was wet (i.e., prior to complete evaporation of solvent) [43].

Suh *et al.* also reported a direct method for molding of PEG polymers [40], which involves the placement of a patterned PDMS mold on top of a drop-dispensed or spin-coated PEG solution typically dissolved in water or ethanol, leaving behind a polymer replica after solvent evaporation followed by mold removal. In comparison with µCP, the technique generates features varying in height with precise control over cell migration and interconnection of cell arrays.

In the aforementioned techniques, water is typically used as the solvent. In this case, the wetting environment is very different from previous studies of micromolding of hydrophobic polymers such as polystyrene involving organic solvents [23]. For example, instead of filling into the mold, a water solution under the void space

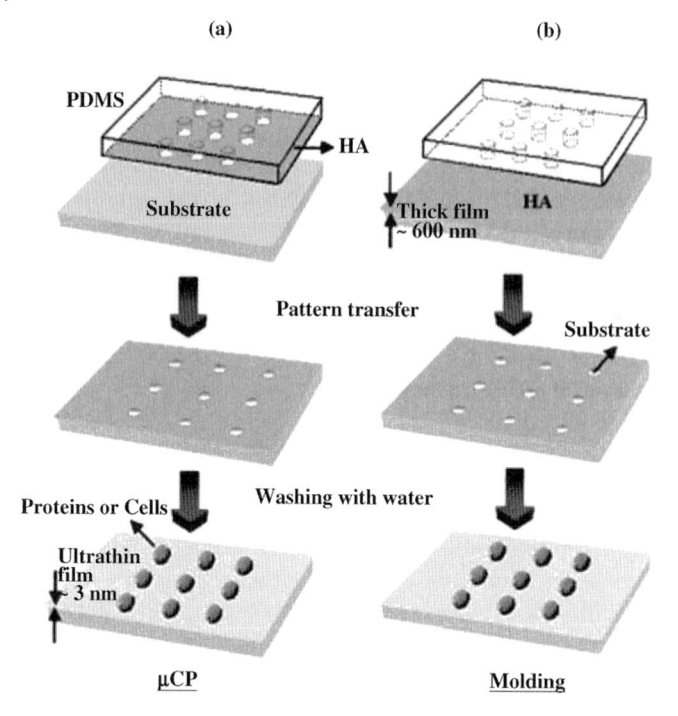

Figure 3.4 Schematic illustration of the two soft lithographic methods used for patterning a hyaluronic acid (HA) film: (a) microcontact printing (μCP) and (b) molding. μCP uses direct transfer from a stamp to a substrate whereas molding deals with pattern formation from a uniform polymer film into the features of the stamp. (Reproduced with permission from [42]. Copyright © (2004) Wiley-VCH Verlag GmbH & Co. KGaA.)

recedes downwards ("capillary depression") until the substrate surface becomes exposed ("dewetting") due to the obtuse contact angle of water on PDMS mold (~105°) [15]; by contrast, hydrophobic polymers rise into the void space ("capillary rise") by partially wetting the PDMS wall (acute contact angle). This phenomenon can be described by the well-known Young's equation [44].

As shown in Figure 3.5, dewetting takes place at the bottom of the substrate as evaporation proceeds, resulting in complete exposure of the substrate surface. This exposure is essential for subsequent selective deposition of proteins or cells. Based on a surface energy consideration, any capillary rise is very unlikely since the contact angle of water on PDMS is 105°. In the course of dewetting, the solution dewets from the substrate and appears to move to the edges of the circle to minimize the surface-to-volume ratio. Protruding features are present at the corner as a result of dewetting and mass pile-up. The height of the void space was quite high such that it is nearly impossible for the solution to touch the ceiling of the void space. The non-uniform distribution of film thickness is easily detected by the difference in contrast with an optical micrograph. It is noted in this regard that the microstructure is relatively flat when other organic solvents such as ethanol or methanol are used since these organic solvents are readily absorbed into the PDMS mold and evaporate into air by diffusion through the mold [45].

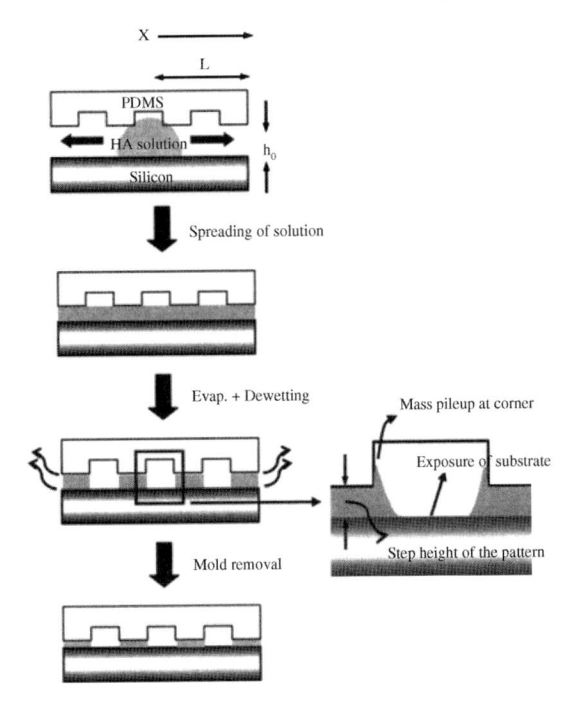

Figure 3.5 Schematic of the experimental procedure of solvent-assisted dewetting micromolding: water evaporates only from the edge of the mold due to the impermeable nature of the PDMS mold to water. The pile-up of hyaluronic acid (HA) during the dewetting process and the definition of the step height are shown on the right panel. (Reproduced with permission from [22]. Copyright © (2005) American Institute of Physics.)

Figure 3.6 shows AFM images of the resulting microstructures for several evaporation conditions. The numbers in the figure indicate the step height of the pattern and the four images were obtained at equal distance from center to edge for each evaporation condition. First, in terms of the step height the thickness profile becomes more non-uniform as the evaporation rate increases and vice versa. A qualitative explanation for this is that for high evaporation rates, the overall process becomes diffusion-limited, that is, water evaporates instantaneously when it reaches the edge of the mold. This process would then result in a non-uniform distribution. At low evaporation rates, when water evaporated at 27 °C while sealed in a petri dish, a uniform thickness profile was observed irrespective of the position (Figure 3.6, third column). In this case, evaporation would be slower than diffusion and thus the overall process becomes evaporation-limited. Interestingly, the average height of the pattern was dramatically reduced in comparison with other microstructures at the edge, except for protruding features around the circle. The height of the protruding features increased with decreasing evaporation rate, rendering a height as high as ∼620 nm for the slowest evaporation in the third column. One can hypothesize that the pile-up of HA at the corner is thermodynamically stable to minimize the surface-to-volume ratio and the slow evaporation might provide enough time to reach a thermodynamic equilibrium. At moderate or fast evaporation, there would be competition between

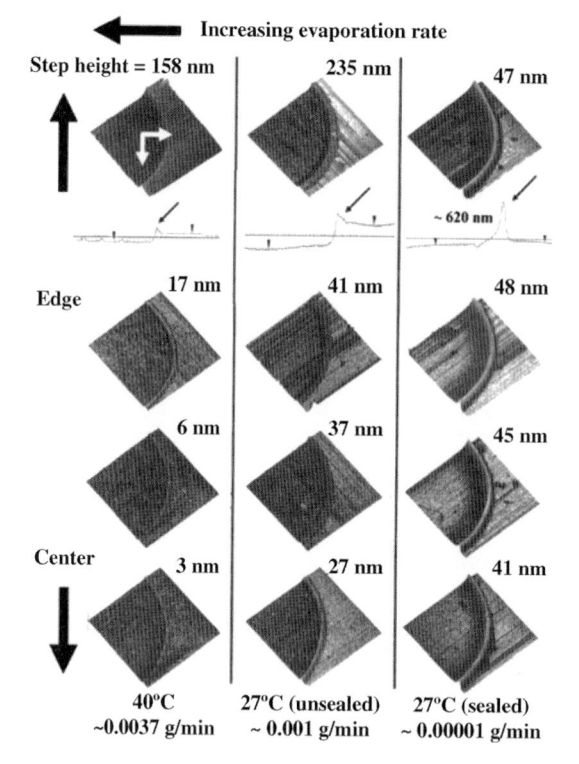

Figure 3.6 Three-dimensional AFM images of the hyaluronic acid (HA) microstructures, obtained under three evaporation conditions at four different locations from center to edge. Numbers indicate the step height of the pattern. Cross-sectional AFM images clearly demonstrate the presence of an HA pile-up. The thin black arrows under the first row indicate the pile-up of HA at the edge of the circle. (Reproduced with permission from [22]. Copyright © (2005) American Institute of Physics.)

mass pile-up along the PDMS wall and lateral diffusion towards the edge of the mold so that the protrusion would not be so significant. The protruding features are potentially useful for controlling cell migration or interconnection for cell arrays since they could be used as diffusion barriers. As a conclusion of this part, a uniform step height in the micromolding of a hydrophilic biopolymer could be achieved by using slow evaporation of water in a sealed environment [22]. Slow evaporation provides enough time for the solution to diffuse and construct a flat, uniform height profile.

3.4
Other Techniques

In solvent-assisted molding technique, capillarity may play an important role in pulling the solution into the void spaces of the mold. When a liquid wets a capillary tube, the surface free energy is lowered and thus capillary rise of the liquid occurs. This capillarity has been used in a method called micromolding in capillaries [15] and has been further exploited in a technique called capillary force lithography (CFL), which has proven to be quite fruitful [46–50]. CFL employs capillarity even in the absence of a solvent by increasing the temperature above the polymer glass transition. It combines the essential features of imprint lithography [51], that is, molding a polymer melt with the use of an elastomeric stamp. When the polymer film is thick enough, polymer melt can fill up the void space between the polymer and the mold completely, thereby generating a negative replica of the mold. Anisotropic dewetting occurs when the polymer film is relatively thin with respect to the penetration depth of the stamp. Separate lines with reduced dimensions located at the feature edges of the original stamp were also fabricated [48, 52, 53].

In CFL, the excellent elastomeric property of PDMS is a key factor. It allows a PDMS mold to be used widely in replicating patterns with high fidelity [27, 54]. But it also brings drawbacks such as deformation and swelling. To overcome the effect of deformation, composite stamps with a stiff layer supported by a flexible layer have been proposed, thus extending the capability of soft lithography to the generation of 50 to 100 nm features [55]. The deformation of the PDMS stamp has also been used beneficially: Xia *et al.* created patterns with size-reduced features by intentionally applying external forces to laterally deform the stamps during contact printing [56]. Guo *et al.* also gained size reduction and new patterns different from the features on the stamps by applying pressure perpendicular to the PDMS stamp to achieve vertical deformation [57].

3.5
Applications of Solvent-Assisted Molding

The solvent-assisted molding techniques presented here can serve as a simple, inexpensive way to fabricate micro/nano structures for applications in various research fields. Recently, biosensors of nanoscale dimension have attracted wide attention for sensitive, multi-targeting, and labeling-free biomolecular detection [58]. Various affinity-binding assisted analytical tools have been developed based on nanowires, nanoparticles, nanotubes, or microcantilevers. The widespread use of these tools, however, has been limited by the random configuration of binding molecules and the formation of protein–protein clusters on a solid surface, resulting in a loss or masking of binding sites. Since the binding event is highly specific and varies depending on the targeting strategy employed (e.g., chemical linking, physical binding, or bio-specific recognition), the biosensing platform needs to be constructed with many functional groups. Suh *et al.* reported a novel technique to achieve label free detection without nonspecific binding while maintaining the activity and stability of immobilized liposome on molded nanowell electrodes [58]. The nanowells were fabricated on a gold electrode using dewetting assisted nanomolding with a PEG copolymer. The nanomolding process involves only the contact of a nanopatterned polymeric mold with the PEG surface by drop dispensing;

it subsequently forms well-defined nanowell arrays with the gold substrate clearly exposed. Moreover, this fabrication process does not require any surface modification or external stimulus (e.g., temperature rise or mechanical pressure), offering several advantages over previous array methods for lipid membrane or liposome. The PEG nanowells thus formed were hydrophilic, offering less capillary resistances against deposition of liposomes, and strongly resisted nonspecific adsorption of liposomes on the PEG surface, resulting in distinctively separated, individual liposome nanoarrays.

Hammond et al. introduced a simple method to pattern electrostatic assemblies of viruses onto a polyelectrolyte multilayer (PEM) [59]. They were the first to apply solvent-assisted capillary molding to multilayer films; their study demonstrated micrometer-scaled dense patterns of viruses, where the accessible feature size could be correlated with the length scale of the virus and the swelling property of the underlying polyelectrolyte multilayer. Viruses were directly captured from solution onto the charged surface of a patterned polymeric template via electrostatic attraction and then spontaneously ordered into closely packed monolayer structures via the enhanced surface mobility provided by the underlying PEM patterns. This approach offers great advantages in that target surface molecules or ligands on the virus body can be readily concentrated into highly ordered and densely packed structures without any external physical forces or additional chemical functionality. The straightforward procedure of solvent-assisted capillary molding can also be potentially used as an alternative and more general tool for patterning layer-by-layer assembled polymeric films over large areas without the utilization of chemically pre-patterned surface chemistry or transfer stamping methods.

Rogers et al. demonstrated a method for fabrication of polymer light emitting devices (LEDs) that produced light patterns with sizes as small as $\sim 0.8\,\mu m$ [60]. Solvent-assisted micromolding of a film of a precursor to poly(p-phenylene vinylene) (PPV) produced variations in thickness that replicate the relief on the mold. Thermal conversion of this precursor formed a film of PPV with the same surface relief. LEDs formed with the structured PPV emitted preferentially in the thin regions of the film. Devices fabricated in this manner may become important for constructing plastic visual displays, and could eventually lead to new sub-wavelength sources of light suitable for applications in near field optics.

Samuel et al. reported polymer distributed feedback lasers fabricated using solvent-assisted microcontact molding [61]. The poly[2-methoxy-5-(3,7-dimethyloctyloxy) paraphenylenevinylene] film was patterned by placing it in conformal contact with an elastomeric mold inked with a suitable solvent. This approach does not involve the use of heat or pressure and so is a particularly fast and simple way of making photonic structures. The structures formed were characterized using atomic force microscopy and the effect of solvent choice and imprinting time were examined. To demonstrate the potential of applying this process to conjugated polymers, the authors also fabricated polymer distributed feedback (DFB) lasers and studied their operating characteristics. Thus, they demonstrated a simple, single step process in which a semiconducting polymer film was turned into a DFB laser in less than 2 min.

3.6
Conclusions

This chapter reviewed solvent-assisted molding techniques and discussed their principle and applications to various fields. Three major techniques can be distinguished: (a) solvent-assisted microcontact molding (SAMIM), (b) soft molding (SM), and (c) solvent-assisted dewetting micromolding (SADEM). The fabrication of complex functional nanostructures using theses techniques requires further understanding and control of several issues. These include (i) distortion or deformation of polymer micro/nanostructures, (ii) optimization of conditions for pattern transfer and for replication of nanoscale features, and (iii) theoretical understanding of mass transport and solvent interactions with the mold. It is envisioned that this simple method will continue to provide a versatile, economically viable route to the design and fabrication of functional surfaces and micro/nanoscale devices with many potential applications.

References

1 Brambley, D., Martin, B., and Prewett, P.D. (1994) *Adv. Mater. Opt. Electron.*, **4**, 55.
2 Rai-Choudhury, P. (ed.) (1997) *Handbook of Microlithography, Micromachining, and Microfabrication*, vol. 1, SPIE Opt. Engineer. Press, Bellingham, WA.
3 Levenson, M.D. (1995) *Solid State Technol.*, **38**, 57.
4 Geppert, L. (1996) *IEEE Spectrum*, **33**, 33.
5 Deninger, W.D. and Garner, C.E. (1988) *J. Vac. Sci. Technol. B*, **6**, 337.
6 Reichmanis, E. and Thompson, L.F. (1989) *Chem. Rev. (Washington, D. C.)*, **89**, 1273.
7 Miller, R.D. and Wallraff, G.M. (1994) *Adv. Mater. Opt. Electron.*, **4**, 95.
8 Xia, Y. and Whitesides, G.M. (1998) *Annu. Rev. Mater. Sci.*, **28**, 153.
9 Kim, E., Xia, Y., Zhao, X.-H., and Whitesides, G.M. (1997) *Adv. Mater.*, **9**, 651.
10 Paul, K.E., Breen, T.L., Aizenberg, J., and Whitesides, G.M. (1998) *Appl. Phys. Lett.*, **73**, 2893.
11 Marzolin, C., Smith, S.P., Prentiss, M., and Whitesides, G.M. (1998) *Adv. Mater.*, **10**, 571.
12 Kumar, A. and Whitesides, G.M. (1993) *Appl. Phys. Lett.*, **63**, 2002.
13 Xia, Y., Kim, E., Zhao, X.-M., Rogers, J.A., Prentiss, M., and Whitesides, G.M. (1996) *Science*, **273**, 347.
14 Zhao, X.-M., Xia, Y., and Whitesides, G.M. (1996) *Adv. Mater.*, **8**, 837.
15 Kim, E., Xia, Y., and Whitesides, G.M. (1995) *Nature*, **376**, 581.
16 Chou, S.Y., Krauss, P.R., and Renstrom, P.J. (1995) *Appl. Phys. Lett.*, **67**, 3114.
17 Scheer, H.-C., Schult, H., Hoffmann, T., and Sotomayor Torres, C.M. (1998) *J. Vac. Sci. Technol. B*, **16**, 3917.
18 Kim, Y.S., Park, J.H., and Lee, H.H. (2002) *Appl. Phys. Lett.*, **81**, 1011.
19 Kumar, A., Biebuyck, H.A., and Whitesides, G.M. (1994) *Langmuir*, **10**, 1498.
20 Xia, Y., Mrksich, M., Kim, E., and Whitesides, G.M. (1996) *Chem. Mater.*, **8**, 601.
21 Xia, Y., Mrksich, M., Kim, E., and Whitesides, G.M. (1995) *J. Am. Chem. Soc.*, **117**, 3274.
22 Jeong, H.E. and Suh, K.Y. (2005) *J. Appl. Phys.*, **97**, 114701.
23 Kim, Y.S., Suh, K.Y., and Lee, H.H. (2001) *Appl. Phys. Lett.*, **79**, 2285.
24 Xia, Y. and Whitesides, G.M. (1998) *Annu. Rev. Mater. Sci.*, **28**, 153.
25 Xia, Y. and Whitesides, G.M. (1998) *Angew. Chem. Int. Ed. Engl.*, **37**, 550.

26 Xia, Y., Rogers, J.A., Paul, K.E., and Whitesides, G.M. (1999) *Chem. Rev.*, **99**, 1823.
27 Xia, Y. and Whitesides, G.M. (1997) *Polym. Mater. Sci. Eng.*, **77**, 596.
28 Trimbach, D., Feldman, K., Spencer, N.D., Broer, D.J., and Bastiaansen, C.W.M. (2003) *Langmuir*, **19**, 10957.
29 Pavlovic, E., Quist, A.P., Nyholm, L., Pallin, A., and Gelius, U. (2003) *Langmuir*, **19**, 10267.
30 Delamarche, E., Donzel, C., Kamounah, F.S., Wolf, H., and Geissler, M. (2003) *Langmuir*, **19**, 8749.
31 Harada, Y., Girolami, G.S., and Nuzzo, R.G. (2003) *Langmuir*, **19**, 5104.
32 Martin, B.D., Brandow, S.L., Dressick, W.J., and Schull, T.L. (2000) *Langmuir*, **16**, 9944.
33 Delamarche, E., Geissler, M., Wolf, H., and Michel, B. (2002) *J. Am. Chem. Soc.*, **124**, 3834.
34 Koide, Y., Such, M.W., Basu, R., Evmenenko, G., and Cui, J. (2003) *Langmuir*, **19**, 86.
35 Whitesides, G.M., Ostuni, E., Takayama, S., Jiang, X., and Ingber, D.E. (2001) *Annu. Rev. Biomed. Eng.*, **3**, 335.
36 Kumar, A., Abbott, NL., Biebuyck, H.A., Kim, E., and Whitesides, G.M. (1995) *Acc. Chem. Res.*, **28**, 219.
37 Tae, G., Lammertink, R.G.H., Kornfield, J.A., and Hubbell, J.A. (2003) *Adv. Mater.*, **15**, 66.
38 Koh, W.G., Itle, L.J., and Pishko, M.V. (2003) *Anal. Chem.*, **75**, 5783.
39 Khademhosseini, A., Jon, S., Suh, K.Y., Tran, T.N.T., Eng, G., Yeh, J., Seong, J., and Langer, R. (2003) *Adv. Mater.*, **15**, 1995.
40 Suh, K.Y. and Langer, R. (2003) *Appl. Phys. Lett.*, **83**, 1668.
41 Suh, K.Y., Seong, J., Khademhosseini, A., Laibinis, P.E., and Langer, R. (2004) *Biomaterials*, **25**, 557.
42 Donzel, C., Geisssler, M., Bernard, A., Wolf, H., Michel, B., Hilborn, J., and Delamarche, E. (2001) *Adv. Mater.*, **13**, 1164.
43 Suh, K.Y., Khademhosseini, A., Yang, J.M., Eng, G., and Langer, R. (2004) *Adv. Mater.*, **16**, 584.
44 Adamson, A.W. and Gast, A.P. (1997) *Physical Chemistry of Surfaces*, 6th edn, John Wiley & Sons, Inc., New York, USA.
45 Lee, J.N., Park, C., and Whitesides, G.M. (2003) *Anal. Chem.*, **75**, 6544.
46 Suh, K.Y., Kim, Y.S., and Lee, H.H. (2001) *Adv. Mater.*, **13**, 1386.
47 Suh, K.Y. and Lee, H.H. (2002) *Adv. Funct. Mater.*, **12**, 405.
48 Suh, K.Y., Yoo, P.J., and Lee, H.H. (2002) *Macromolecules*, **35**, 4414.
49 Khang, D.Y. and Lee, H.H. (2004) *Adv. Mater.*, **16**, 176.
50 Zhan, L., Laird, D.W., and McCullough, R.D. (2002) *Langmuir*, **19**, 6492.
51 Chou, S.Y., Krauss, P.R., and Renstrom, P.J. (1996) *Science*, **272**, 85.
52 Zhang, H.L., Bucknall, D.G., and Dupuis, A. (2004) *Nano Lett.*, **4**, 1513.
53 Korczagin, I., Golze, S., Hempenius, M.A., and Vancso, G.J. (2003) *Chem. Mater.*, **15**, 3663.
54 Gates, B.D., Xu, Q., Love, J.C., Wolfe, D.B., and Whitesides, G.M. (2004) *Annu. Rev. Mater. Res.*, **34**, 339.
55 Odom, T.W., Love, J.C., Wolfe, D.B., Paul, K.E., and Whitesides, G.M. (2002) *Langmuir*, **18**, 5314.
56 Xia, Y. and Whitesides, G.M. (1997) *Langmuir*, **13**, 2059.
57 Guo, Q., Teng, X., and Yang, H. (2004) *Nano Lett.*, **4**, 1657.
58 Kim, P.N., Lee, B.K., Lee, H.Y., Kawai, T., and Suh, K.Y. (2007) *Adv. Mater.*, **20**, 31.
59 Yoo, P.J., Nam, K.T., Belcher, A.M., and Hammond, P.T. (2008) *Nano Lett.*, **8**, 1081.
60 Rogers, J.A., Bao, Z., and Dhar, L. (1998) *Appl. Phys. Lett.*, **73**, 294.
61 Lawrence, J.R., Turnbull, G.A., and Samuel, I.D.W. (2003) *Appl. Phys. Lett.*, **82**, 4023.

4
Soft Lithography and Variants
Elena Martínez and Josep Samitier

4.1
Introduction

Micro- and nanofabrication techniques have become, in the last decade, of extreme importance in life-science related disciplines [1]. However, the use of photolithography or other advanced lithographic techniques such as X-ray, electron beam or focused ion beam lithographies presents important drawbacks. These are largely expensive technologies because of the sophisticated instrumentation, the clean-room environment and the need of expert operators. This makes these technologies difficult to access for most chemists, biologists and materials scientists [2]. Moreover, conventional "hard" lithographic techniques are difficult to use with nonplanar surfaces; they are not optimized for the introduction of specific chemical functionalities and they are difficult to extend to materials other than silicon and other semiconductors such as glass or polymers [2].

A ground-breaking alternative method introduced by Kumar and Whitesides [3] has been proven to overcome most of these drawbacks and to make lithographic techniques available to life-science and scientists outside of microelectronics [3, 4]. It is named soft lithography and relies on using elastomeric polymers as soft molds to obtain patterned surfaces on other polymeric materials. Soft lithography overcomes many of the shortcomings of photolithography such as the use of trained technicians, expensive equipment and clean-room environment.

The soft elastomeric stamp can be used either as a vehicle for biomolecular surface patterning (an application called "microcontact printing" (µCP)) or to create three dimensional reliefs, particularly on polymer materials, as in micromolding in capillaries (MIMIC), microtransfer molding (µTM) or solvent-assisted molding. These techniques have been successfully applied in the fabrication of polymer patterns with dimensions down to the sub-100 nm scale and will be described in this chapter. These patterns have found many relevant applications in the life sciences, where scientists often need to spatially control topographical and chemical properties of surfaces at small scales [1, 5, 6].

Generating Micro- and Nanopatterns on Polymeric Materials. Edited by A. del Campo and E. Arzt
Copyright © 2011 WILEY-VCH Verlag GmbH & Co. KGaA, Weinheim
ISBN: 978-3-527-32508-5

In the last decade, another prominent application of soft lithography in biology and chemistry has emerged in the fabrication of microfluidic devices, in which control of very small sample volumes and their flow behavior is required [7, 8]. Devices made by this technique can be tightly sealed and active components such as valves or pumps can be easily manufactured. This chapter will, however, not cover microfluidic applications since they are outside the scope of this book.

A large number of review papers can already be found in literature about soft lithography techniques, [1, 2, 4, 6, 7, 9–13] and, more specifically, µCP technique [6, 9, 13]. This chapter focuses on the generation of topographies and is divided into two main parts: the first will deal with the key features defining the soft lithography technique and its main application, µCP; the second part will report on some of the attempts to pattern polymers by using "soft" molds.

4.2
Key Features of Soft Lithography

Soft lithography uses a fluid mixture as elastomeric precursor to fill a mold and generate an elastomeric, microstructured stamp, which can then be used to generate micro- and nanopatterns (Figure 4.1) [2]. This procedure is simple, inexpensive and feasible in a normal ambient laboratory environment.

The most commonly used elastomer for soft lithography is poly(dimethylsiloxane) (PDMS), which has many convenient properties: relatively high Young's modulus for an elastomer (1–4 MPa), nontoxic, commercially available and optically transparent down to a light wavelength of 300 nm. It is intrinsically very hydrophobic (water contact angle of 110°) but can be made hydrophilic by oxygen plasma treatment (water contact angle 10°) [14] and it is permeable to non polar gases such as O_2, N_2 and CO_2 [1]. Because of these properties, it can be bonded to glass and to itself and it can be used, for example, in channel systems for cell culture [1].

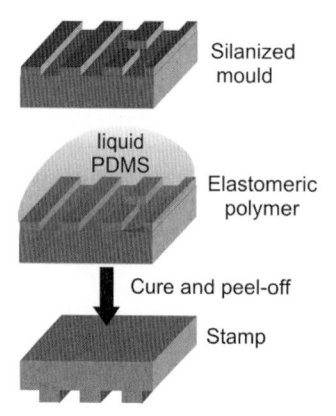

Figure 4.1 Schematic representation of the soft lithography process.

Moreover, its elastic properties allow for conformal surface contact with other materials over relatively large areas, and make operations, such as printing and demolding, easy [2].

Elastomeric stamps are the key element in soft lithography. They are usually prepared by casting the PDMS in liquid form (liquid pre-polymer mixed with curing agent) against a mold or master with topographical micro- or nanostructures (Figure 4.1), followed by a curing step at higher temperature. The mold is typically fabricated by lithographic methods. Although PDMS has a low surface energy [15], sometimes the mold needs to be pre-exposed to fluorinated silane vapor for 30 min to generate a perfluorinated coating and facilitate the subsequent demolding process. After demolding, the elastomeric stamp is ready to be used for applications such as μCP or as a soft mold in further replica processes.

PDMS has some limitations. These include its softness, which can produce bending or sagging of the stamp, leading to a loss in resolution [11]. Another drawback of PDMS is its propensity to swell in organic solvents, thus limiting its use in patterning chemicals dissolved in these solvents [15]. In order to improve the stamp properties and, thereby, the resolution of the technique, some alternatives to normal PDMS have recently been reported. One way of reducing the deformation of the stamp is to increase the PDMS stiffness by increasing the amount of cross-linker [11]. However, this leads to more brittle stamps that will impede conformal contact with rough or curved surfaces. A reported alternative is the use of composite stamps fabricated with a combination of harder and softer siloxane polymers [9, 16]. The harder thin layer then exhibits the patterned structures, whose size can be reduced while minimizing deformations. Another possibility is the use of a metal lamella covered with a thin structured PDMS [17]. Because of the stiffness of the metal, roof collapse between the structures is reduced; still, the metal is flexible enough to compensate macroscopic inhomogenities and to pattern curved surfaces [11].

4.3
Microcontact Printing of Self-Assembled Monolayers

Microcontact printing (μCP) in its original form uses PDMS stamps to transfer patterns of molecules on surfaces by simple contact of an inked PDMS stamp with the surface (Figure 4.2). The conformal contact between the stamp and the surface of the substrate is a key issue. Microcontact printing is a simple technique from the experimental point of view, and is rapid and can be parallelized. In fact, it can form patterned submicrometer features over an area of $50\,cm^2$ in a single impression within 30 s [2, 12, 18].

Initially, μCP was conceived and used as a method to transfer self-assembled thiol monolayers onto gold surfaces in a patterned fashion [5, 19]. The range of the ink molecules has been extended to alkylsiloxanes on silicon oxide [20], to particles and organic molecules with higher molecular weights, ranging from Langmuir–Blodgett films [21] to DNA [22, 23] and proteins [24].

4 Soft Lithography and Variants

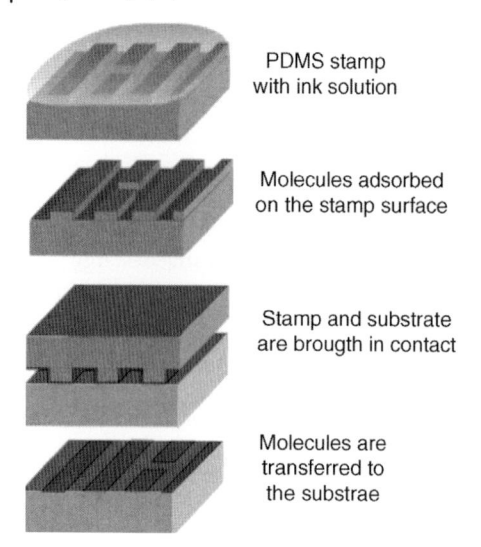

Figure 4.2 Schematic representation of μCP to transfer self-assembled monolayers of thiols on surfaces.

- PDMS stamp with ink solution
- Molecules adsorbed on the stamp surface
- Stamp and substrate are brougth in contact
- Molecules are transferred to the substrae

The main advantages of μCP are that is an easy, inexpensive, readily accessible processing method. Moreover, it is possible to use it on nonplanar surfaces and to generate three-dimensional structures; furthermore, it is a parallel method that allows the patterning of large areas even with nanometer size features.

However, μCP also suffers from some drawbacks. First, the stability of the PDMS stamps is a major concern [25]. PDMS shrinks by about 1% after curing and it can be swelled by nonpolar solvents such as toluene and hexane [2]. The use of a soft polymer also produces some deformations on the stamp structures such as pairing, buckling or roof collapse during contact with the surfaces (Figure 4.3), which result in distorted patterns [13, 26, 27]. Biebuyck and co-workers [28] have experimentally demonstrated that if the aspect ratio of the structures (depth/width) is too large, the

- pairing
- buckling
- roof collapse

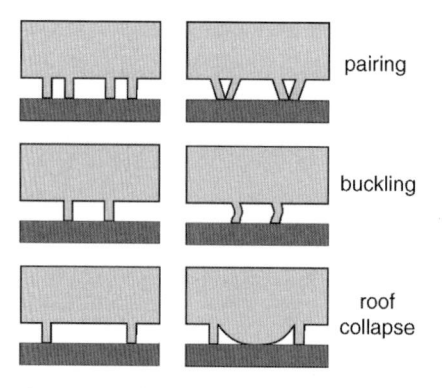

Figure 4.3 Schematic of the most frequent sources of loss of resolution in microcontact printing processes: pairing, buckling and roof collapse events.

structures can collapse when loaded or even under their own weight (buckling). These deformations become critical when the pattern has nanosize features [13]. Other problems come from the inking procedure. Swelling of the stamp during "inking" as well as an excess of ink result in enhanced diffusion of the imprinted molecules on the patterned surface. Diffusion of non-covalently-bound molecules occurs after the printing as well. Finally, the hydrophobicity of PDMS is a problem when working with polar inks [13].

Two variants of µCP have overcome some of the limitations stated above. Inverted microcontact printing (i-µCP), first applied by Dusseiller *et al.* [29], creates microwells that combine topographical structure with site-selective chemical modification. This has been applied to generate patterns for single cell culturing. The PDMS stamp has a flat surface, whereas the substrate to be printed (i.e., polystyrene) has a relief that would define where sites of ink transfer. In this way, chemical pattern can be added to the top of topographical structures. This structure seems to improve the mechanical stability of the stamp and eliminated the diffusion of ink molecules via the gas phase, thus improving the patterning resolution. Although it is not the most popular of the soft lithography techniques, the i-µCP method has also been successfully used in applications where the selective patterning of topographically structured surfaces was needed [30–32].

Microcontact printing in liquid media is another alternative to avoid stamp deformation and sample contamination. The incompressible liquid supports the stamp roof during printing and, at the same time, confines the inking molecules to the area of mutual contact. Xia and Whitesides [33, 34] have published results to show that, by performing µCP under water, thiol transfer improves and the dimensions of the features that are subsequently etched on gold substrates can be reduced. Bessueille *et al.* [35] have confirmed that liquid µCP enables the use of stamps with aspect ratios unsuitable for the conventional air-based approaches.

Another alternative to avoid structural collapse of the stamp, is a change in stamp design and stamp material. For instance, Renault and co-workers have shown that it is possible to make contact printing of proteins at nanoscale dimensions by using domed structural features in their stamps [36]. In this way, structural collapse is avoided, allowing the printing of proteins at the nanoscale. Other authors have used composite stamps, changing the crosslinking density and molecular weights of PDMS as well as using a variety of nanosized filler materials to improve its stiffness [37]. Another recent approach uses poly(methyl methacrylate) (PMMA) stamps (a rigid polymer) and, in order to produce a uniform pressure over the whole printing area, the µCP procedure was carried out in a nanoimprint lithography apparatus [38].

4.4
Soft Molding Techniques

Soft molding uses the elastomeric PDMS stamp instead of hard molds to obtain polymer replicas via molding processes. Using an elastomeric, flexible mold makes the demolding process easier and avoids damage of both the mold and the replica. Moreover, by taking advantage of the flexibility of PDMS, some authors have used this technique to create micro- and nanostructured replicas on curved surfaces [12]. Some of the techniques described in literature using PDMS or other elastomers as soft molds are described in the following sections.

4.4.1
Nanoimprinting with Soft Stamps

Soft molding has been used to obtain replicas of thermoplastic polymers by placing the soft mold on the top of the polymer surface and heating above the glass transition temperature of the polymer [12, 39]. In this way, the fluid polymer is forced to fill the void spaces in the PDMS mold (Figure 4.4). Then, the system is cooled and mold and replica are separated. The result of this process is different if applied to a thick polymer film (Figure 4.4a), which will result in a patterned polymer surface, or if applied to a thin polymer film (i.e., a spin-coated film, Figure 4.4b). In the latter case all the polymer will fill the mold cavities and no residual layer will be left behind [40]. It has been reported that the replicas obtained by this method exhibit high fidelity in pattern reproduction, without shrinkage or distortion that haunt other soft lithography techniques that employ solvents to allow flow of the polymer into the cavities [39–41].

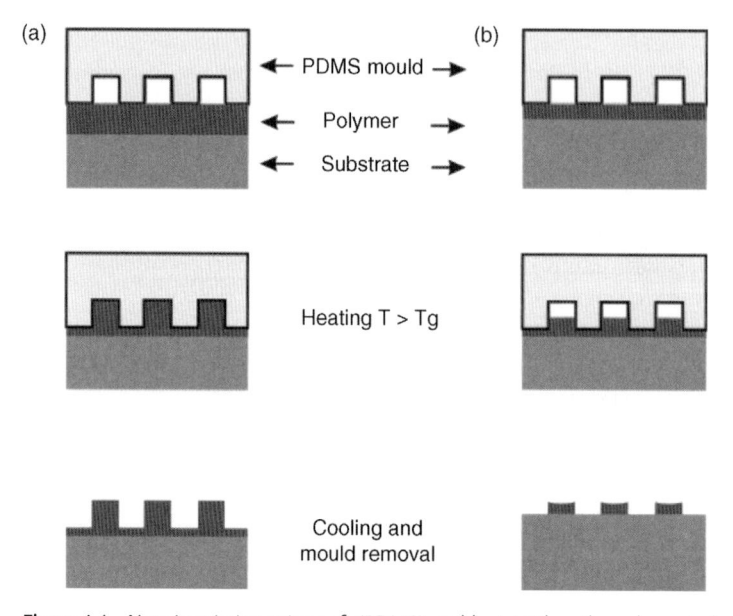

Figure 4.4 Nanoimprinting using soft (PDMS) molds. (a) When the polymer to be structured is a thick layer, the structure has a residual layer of polymer connecting the microstructures. (b) For thin polymer layers, the microstructures created can, eventually, be isolated on a supporting substrate.

4.4.2
Micromolding in Capillaries and Microtransfer Molding

Micromolding in capillaries (MIMIC) starts by bringing an elastomeric (PDMS) mold with channel or network features in contact with a substrate. Then, a prepolymer with low viscosity is added to fill the microchannels by capillary forces and is subsequently cured (by solvent evaporation or photolysis), leaving the structures formed after the removal of the PDMS mold (Figure 4.5a) [12]. This is an easy and cheap technique although it cannot be used to obtain features smaller than hundreds of micrometers [12]; the required processing time is too long for practical purposes because of the low filling velocity of small capillaries [42]. The filling velocity of capillaries is directly proportional to their radius and to the current fill length and inversely proportional to the viscosity of the pre-polymer filling medium. The filling procedure can also be assisted by vacuum.

Reported applications of MIMIC procedures are the patterning of ceramic and metal oxide materials in sol–gel processes [11]. Other authors used MIMIC to produce well-defined molecular imprinted polymer microstructures for applications in immunoassays, sensors and enzyme mimics [43].

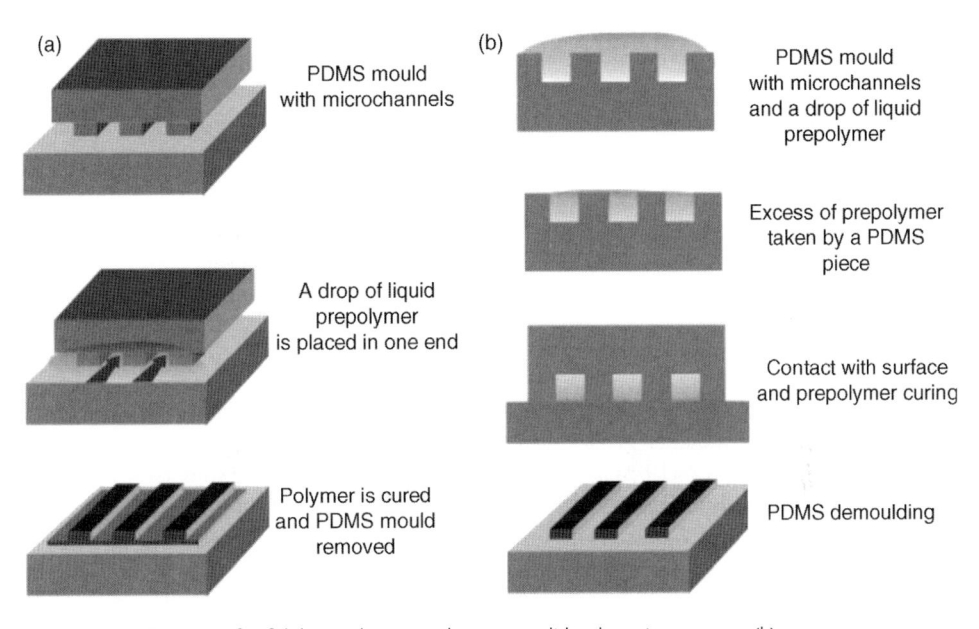

Figure 4.5 Schematic of soft lithography variations: (a) micromolding in capillaries: a PDMS mold with microchannels is brought in contact with a substrate and a drop of low-viscosity, liquid pre-polymer fills these channels by capillary action. The pre-polymer is then solidified *in situ* and the PDMS is removed, leaving a solid polymeric structure. (b) Microtransfer molding: a drop of pre-polymer is poured on the top of a PDMS mold and the excess is removed by another flat PDMS piece. The mold is then brought in contact with a substrate, the pre-polymer is cured and the mold is removed.

In order to overcome MIMIC limitations, some authors have proposed the use of a closely related technique that is called microtransfer molding (μTM) (Figure 4.5b). In this case, a drop of liquid pre-polymer is poured over the patterned surface of a PDMS mold, thus letting the liquid fill the mold cavities. Then, the excess of pre-polymer is removed and the filled PDMS is placed on top of a substrate, where the pre-polymer solidifies inside the mold. Zhao et al. [12] have shown that is possible to fabricate structures with 100 nm in lateral dimension by this method, although a thin polymer layer remained on the substrate. This method can be adapted for creating structures on curved surfaces and for the fabrication of multilayered structures.

4.4.3
Solvent-Assisted Molding

Solvent-assisted molding (SAMIM) uses a soft elastomeric mold (PDMS) that is wetted with a solvent that dissolves the target polymer to be patterned but does not damage the PDMS. The wetted PDMS mold is then brought in contact with the polymer to be structured in such a way that the solvent swells or dissolves the surface of the polymer, which will become softer and adapt to the shape of the mold (Figure 4.6a) [11]. Afterwards, the solvent evaporates and the polymer hardens again, allowing for the release of the PDMS mold. This is a useful approach for structures that do not form capillaries, as required by MIMIC. The reported problems deal with shrinkage of the features after solvent evaporation and difficulties when working with textured surfaces [11]. In an attempt to overcome these difficulties, which are believed to be related to a non-uniform solvent distribution on the mold surface, Kim et al. [41] proposed polymer solutions to be spun coated onto a substrate. The only requirement is that the solvent has to be absorbed by the elastomeric (PDMS) mold. The authors have reported the successful use of this technique with a commercial available novolac resin dissolved in propyleneglycol monomethylether acetate (PGMEA).

4.4.4
UV Molding

Another option of soft lithography takes advantage of the optical transparency of PDMS; UV radiation is used through the mold structure to cure a pre-polymer. This technique, as described by Choi et al. [44], is depicted in Figure 4.6b. First, a mixture of styrene monomer and UV initiator is prepared and spun onto a substrate. The monomer mixture is then partially polymerized by 10 min of UV exposure to increase its viscosity and avoid the PDMS being swollen by the monomer. Then the PDMS mold is placed on top of the substrate and full polymerization of the pre-polymer is achieved by further exposure to UV light. By this technique, Choi et al. have reported polymer patterns of submicron dimensions, that is, lines 150 nm wide and dots of 770 nm in diameter. The fact that this is a process carried out at room temperature avoids problems such as the tendency of the PDMS mold to separate from the substrate due to differential thermal expansion. Because of their rapid polymerization with minimal change in volume, the monomers isobornyl acrylate

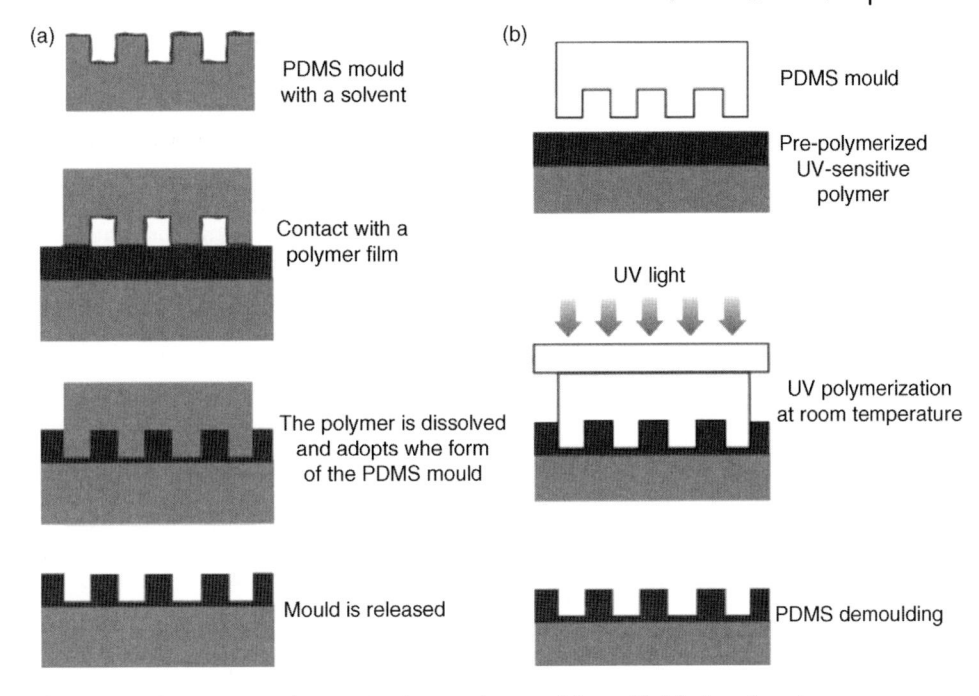

Figure 4.6 Schematics (a) solvent-assisted molding (SAMIM): a PDMS mold wetted with a solvent that dissolves the polymer to be patterned is brought into contact with the polymer surface; the solvent softens the polymer and the mold sinks into the polymer. (b) UV molding: UV light is radiated through a PDMS mold placed against a pre-polymerized polymer; the pre-polymer is cured and the mold can be removed.

and tetraethylene glycol dimethacrylate (crosslinker) also constitute an excellent choice. These materials can be polymerized into hard structures under UV irradiation in the presence of the photo-initiator Irgacure 651 [45]. More examples of the application of this technique can be found in the literature.

4.4.5
Forced Soft Lithography

In 2008, Fernandez et al. [46] have introduced a technique, called forced soft lithography, by which solution-processable polymers can be patterned without relying on capillary forces. In this process, the mold is placed inside a chamber with the structured surface upward. Then, a liquid pre-polymer (chitosan or PDMS) is poured into the chamber and the pressure is increased by a plunger at a determined rate. After pressure release, the chamber is opened to let the solvent evaporation and the curing of the pre-polymer. The pressure applied acts in the same direction as the capillary forces to displace the air inside the mold cavities. Researchers have used this method to successfully structure chitosan polymers with features in the nanometer range (400 nm) [46, 47].

4.5
Summary

Soft lithography is an unconventional method of micro- and nanofabrication with important applications, especially in the life sciences. The technique uses a mold with a topographical relief that is transferred to elastomeric polymer replicas. These replicas can be further used as stamps to transfer chemicals to a surface in a patterned fashion or as soft molds in soft lithography process variations. Advantages of soft lithography include: the process is parallel in nature, inexpensive, involves process conditions which are not detrimental to (bio)-molecules, requires no special equipment and can be undertaken by unskilled scientists at the bench top. Moreover, the unique properties of soft lithography allow nonplanar surfaces to be patterned, thus opening a completely new field of applications. Soft lithography limitations come basically from the use of a deformable stamp and hence counter-strategies involve other stamp materials with improved mechanical properties. For particular polymer applications, variations of soft lithography such may be suitable, such as, capillary force lithography, micromolding in capillaries, microtransfer molding, solvent-assisted molding, UV molding and forced soft lithography.

Acknowledgements

The financial support of the ISCIII through the FIS project PI071162 is gratefully acknowledged. E. Martinez also acknowledges the financial support of the Spanish Ministry of Science an Education through I3 program.

References

1 Whitesides, G.M., Ostuni, E., Takayama, S., Jiang, X.Y., and Ingber, D.E. (2001) Soft lithography in biology and biochemistry. *Annu. Rev. Biomed. Eng.*, **3**, 335–373.
2 Xia, Y. and Whitesides, G.M. (1998) Soft lithography. *Angew. Chem. Int. Ed.*, **37**, 550–575.
3 Kumar, A. and Whitesides, G.M. (1993) Features of gold having micrometer to centimeter dimensions can be formed through a combination of stamping with an elastomeric stamp and an alkanethiol "ink" followed by chemical etching. *Appl. Phys. Lett.*, **63** (14), 2002–2004.
4 Kumar, A., Abbott, N.L., Kim, E., Biebwck, H.A., and Whitesides, G.M. (1995) Patterned self-assembled monolayers and meso-scale phenomena. *Acc. Chem. Res.*, **28**, 219–226.
5 Mrksich, M. and Whitesides, G.M. (1996) Using self-assembled monolayers to understand the interactions of man-made surfaces with proteins and cells. *Ann. Rev. Biophys. Biomol. Struct.*, **25**, 55–78.
6 Mrksich, M. and Whitesides, G.M. (1995) Patterning self-assembled monolayers using microcontact printing - a new technology for biosensors. *Trends Biotechnol.*, **13**, 228–235.
7 Quake, S.R. and Scherer, A. (2000) From micro- to nanofabrication with soft materials. *Science*, **290**, 1536.
8 Park, T.H. and Shuler, M.L. (2003) Integration of cell culture and microfabrication technology. *Biotechnol. Progr.*, **19**, 243–253.

9 Michel, B., Bernard, A., Bietsch, A., Delamarche, E., Geissler, M., Juncker, D., Kind, H., Renault, J.P., Rothuizen, H., Schmid, H., Schmidt-Winkel, P., Stutz, R., and Wolf, H. (2001) Printing meets lithography: Soft approaches to high-resolution printing. *IBM J. Res. Dev.*, **45** (5), 697–719.

10 Kane, R.S., Takayama, S., Ostuni, E., Ingber, D.E., and Whitesides, G.M. (1999) Patterning proteins and cells using soft lithography. *Biomaterials*, **20**, 2363–2376.

11 Brehmer, M., Conrad, L., and Funk, L. (2003) New developments in soft lithography. *J. Dispers. Sci. Technol.*, **24**, 291–304.

12 Zhao, X.-M., Xia, Y., and Whitesides, G.M. (1997) Soft lithographic methods for nanofabrication. *J. Mater. Chem.*, **7** (7), 1069–1074.

13 Quist, A.P., Pavlovic, E., and Oscarsson, S. (2005) Recent advances in microcontact printing. *Anal. Bioanal. Chem.*, **381**, 591–600.

14 Chaudhury, M.K. and Whitesides, G.M. (1992) Correlation between surface free energy and surface constitution. *Science*, **255**, 1230–1232.

15 Ruiz, S.A. and Chen, C.S. (2007) Microcontact printing: A tool to pattern. *Soft Matter*, **3**, 168–177.

16 Odom, T.W., Love, J.C., Wolfe, D.B., Paul, K.E., and Whitesides, G.M. (2002) Improved pattern transfer in soft lithography using composite stamps. *Langmuir*, **18**, 5314–5320.

17 Tormen, M., Borzenko, T., Steffen, B., Schmidt, G., and Molenkamp, L.W. (2002) Using ultrathin elastomeric stamps to reduce pattern distortion in microcontact printing. *Appl. Phys. Lett.*, **81** (11), 2094–2096.

18 Xia, Y.N. and Whitesides, G.M. (1997) Extending microcontact printing as a microlithographic technique. *Langmuir*, **13**, 2059–2067.

19 Bar, G., Rubin, S., Parikh, A.N., Swanson, B.I., Zawodzinski, T.A., and Whangbo, M.H. (1997) Scanning force microscopy study of patterned monolayers of alkanethiols on gold. Importance of tip-sample contact area in interpreting force modulation and friction force microscopy images. *Langmuir*, **13**, 373–377.

20 Xia, Y., Mrksich, M., Kim, E., and Whitesides, G.M. (1995) Microcontact printing of octadecylsiloxane on the surface of silicon dioxide and its application in microfabrication. *J. Am. Chem. Soc.*, **117**, 9576–9577.

21 Guo, Q.J., Teng, X.W., Rahman, S., and Yang, H. (2003) Patterned Langmuir-Blodgett films of monodisperse nanoparticles of iron, oxide using soft lithography. *J. Am. Chem. Soc.*, **125**, 630–631.

22 Xu, C., Taylor, P., Ersoz, M., Fletcher, P.D.I., and Paunov, V.N. (2003) Microcontact printing of DNA-surfactant arrays on solid substrates. *J. Mater. Chem.*, **13**, 3044–3048.

23 Lange, S.A., Benes, V., Kern, D.P., Horber, J.K.H., and Bernard, A. (2004) Microcontact printing of DNA molecules. *Anal. Chem.*, **76**, 1641–1647.

24 Schmalenberg, K.E., Buettner, H.M., and Uhrich, K.E. (2004) Microcontact printing of proteins on oxygen plasma-activated poly(methyl methacrylate). *Biomaterials*, **25**, 1851–1857.

25 Roca-Cusachs, P., Rico, F., Martinez, E., Toset, J., Farre, R., and Navajas, D. (2005) Stability of microfabricated high aspect ratio structures in poly(dimethylsiloxane). *Langmuir*, **21** (12), 5542–5548.

26 Rogers, J.A., Paul, K.E., and Whitesides, G.M. (1998) Generating similar to 90 nanometer features using near-field contact-mode photolithography with an elastomeric phase mask. *J. Vac. Sci. Technol. B*, **16**, 88–97.

27 Sharp, K.G., Blackman, G.S., Glassmaker, N.J. *et al.* (2004) Effect of stamp deformation on the quality of microcontact printing: Theory and experiment. *Langmuir*, **20** (15), 6430–6438.

28 Biebuyck, H.A., Larsen, N.B., Delamarche, E., and Michel, B. (1997) Lithography beyond light: microcontact printing with monolayer resists. *IBM J. Res. Dev.*, **41**, 159.

29 Dusseiller, M.R., Schlaepfer, D., Koch, M., Kroschewski, R., and Textor, M. (2005) An inverted microcontact printing method on

topographically structured polystyrene chips for arrayed micro-3-D culturing of single cells. *Biomaterials*, **26**, 5917–5925.

30 Soolaman, D.M. and Yu, H.-Z. (2007) Monolayer-directed electrodeposition of oxide thin films: surface morphology versus chemical modification. *J. Phys. Chem. C*, **111**, 14157–14164.

31 Martinez, E., Rios-Mondragon, I., Pla-Roca, M., Rodriguez-Segui, S., Engel, E., Mills, C.A., Sisquella, X., Planell, J.A., and Samitier, J. (2007) Production of functionalised micro and nanostructured polymer surfaces to trigger mesenchymal stem cell differentiation. *J. Biotechnol.*, **131**, S67.

32 Embrechts, A., Feng, C.L., Mills, C.A., Lee, M., Bredebusch, I., Schnekenburger, J., Domschke, W., Vancso, G.J., and Schonherr, H. (2008) Inverted microcontact printing on polystyrene-block-poly(tert-butyl acrylate) films: a versatile approach to fabricate structured biointerfaces across the length scales. *Langmuir*, **24**, 8841–8849.

33 Spiering, V.L., Bouwstra, S., and Spiering, R. (1993) On chip decoupling zone for package-stress reduction. *Sens. Actuat. A*, **39**, 149–156.

34 Xia, Y., Mrksich, M., Kim, E., and Whitesides, G.M. (1995) Microcontact printing of octadecylsiloxane on the surface of silicon dioxide and its application in microfabrication. *J. Am. Chem. Soc.*, **117**, 9576–9577.

35 Bessueille, F., Pla-Roca, M., Mills, C.A., Martinez, E., Samitier, J., and Errachid, A. (2005) Submerged microcontact printing (SmCP): an unconventional printing technique of thiols using high aspect ratio, elastomeric stamps. *Langmuir*, **21**, 12060–12063.

36 Renault, J.P., Bernard, A., Bietsch, A., Michel, B., Bosshard, H.R., Delamarche, E., Kreiter, M., Hecht, B., and Wild, U.P. (2003) Fabricating arrays of single protein molecules on glass using microcontact printing. *J. Phys. Chem. B*, **107**, 703–711.

37 Schmid, H. and Michel, B. (2000) Siloxane polymers for high-resolution, high-accuracy soft lithography. *Macromolecules*, **33**, 3042–3049.

38 Pla-Roca, M., Fernandez, J.G., Mills, C.A., Martinez, E., and Samitier, J. (2007) Micro/nanopatterning of proteins via contact printing using high aspect ratio PMMA stamps and nanoimprint apparatus. *Langmuir*, **23**, 8614–8618.

39 Pisignano, D., Persano, L., Cingolani, R., Gigli, G., Babudri, F., Farinola, G.M., and Naso, F. (2004) Soft molding lithography of conjugated polymers. *Appl. Phys. Lett.*, **84** (8), 1365–1367.

40 Suh, K.Y. and Lee, H.H. (2002) Capillary soft lithography: large-area patterning, self organization and anisotropic dewetting. *Adv. Funct. Mater.*, **12** (6 + 7), 405–413.

41 Kim, Y.S., Suh, K.Y., and Lee, H.H. (2001) Fabrication of threedimensional microstructures by soft molding. *Appl. Phys. Lett.*, **79** (14), 2285–2287.

42 Innocenzi, P., Kidchob, T., Falcaro, P., and Takahashi, M. (2008) Patterning techniques for mesostructured films. *Chem. Mater.*, **20**, 607–614.

43 Yan, M. and Kapua, A. (2001) Fabrication of molecularly imprinted polymer microstructures. *Anal. Chim. Acta*, **435** (1), 163–167.

44 Choi, W.M. and Park, O.O. (2005) A soft-imprint technique for submicron structure fabrication via in situ polymerization. *Nanotechnology*, **15**, 135–138.

45 Ziaie, B., Baldi, A., Lei, M., Gu, Y., and Siegel, R.A. (2004) Hard and soft micromachining for BioMEMS: review of techniques and examples of applications in microfluidics and drug delivery. *Adv. Drug Deliv. Rev.*, **56**, 145–172.

46 Fernandez, J.G., Mills, C.A., Pla-Roca, M., and Samitier, J. (2008) Forced Soft Lithography (FSL): production of micro- and nanostructures in thin freestanding sheets of chitosan biopolymer. *Adv. Mater.*, **19**, 3696–3701.

47 Fernandez, J.G., Mills, C.A., and Samitier, J. (2009) Complex microstructured 3D surfaces using chitosan biopolymer. *Small*, **5** (5), 614–620.

Part Two
Writing and Printing

5
Transfer Printing Processes
Luciano F. Boesel

5.1
Introduction

Transfer printing (TP) methods combine molding and printing steps to generate and transfer polymer patterns onto a support material. The polymer is deposited on the mold, and, before demolding, transferred by printing to a different substrate. The critical step in TP is the detachment one, that is, the removal of the mold from the substrate surface, leaving the polymeric patterned features strongly adhered to the substrate surface while releasing cleanly from mold trenches and protrusions.

Transfer printing is very versatile and may be used to print both thermoplastics or thermoset resins on both rigid and flexible surfaces, and even on nonplanar or patterned ones. In some cases post-imprinting steps are required to remove the residual layer which may appear in some techniques.

Transfer printing starts by filling the mold with polymer precursors, polymer solutions or monomers. Mold filling occurs in the same way as explained for nanoimprint lithography (NIL) and soft lithography (see Chapters 1 and 4). Accordingly, molds used in TP are, with few exceptions, elastomeric poly(dimethyl siloxane) (PDMS) molds (as developed for soft lithography) or hard molds of silica/silicon (as developed for NIL). The mold can be completely or incompletely filled, leading to printing of positive or negative patterns on the surface.

For techniques where prepolymers or monomers are used as the precursor (e.g., microtransfer molding – µTM), soft molds are generally preferred because of low surface energy, easy demolding and because they allow processing of photo-curable precursors. On the other hand, techniques where a viscous polymer solution is used as the precursor make use of hard molds, since they will not deform during imprinting and maintain pattern resolution and fidelity. That is the case for reversal nanoimprinting (RNi) and variants.

Although applications of TP techniques are still scarce and mostly at a demonstration stage, the high resolution, low cost and capability of fabricating nonperiodic 3D structures and of patterning large areas in a single step, make it attractive for use in bio- and nanotechnological applications. Examples include fabrication of microfluidic devices for chemical and biological assays, waveguides and photonic band gap crystals to be used in integrated circuits, and micro- or nanometer masks for etching processes.

5.2
Techniques

5.2.1
Microtransfer Molding

In microtransfer molding (µTM) a PDMS mold is filled with liquid prepolymer and the excess liquid is removed by scraping or by blowing off with a stream of nitrogen (Figure 5.1a) [1–3]. The filled mold is then brought in contact with a substrate and subsequently heated or irradiated to cure the prepolymer. As PDMS is transparent, irradiation through the mold with a UV lamp may be used to crosslink the prepolymer [1]. After curing, the mold is peeled away, leaving a patterned microstructure on the surface of the substrate [2, 3].

Microtransfer molding is capable of generating both isolated and interconnected microstructures and, most interestingly, microstructures on nonplanar surfaces, being therefore advantageous over other microlithographic techniques for building 3D microstructures layer-by-layer. Recently, some variants of the technique have been developed that make use of different filling methods or that extend the capabilities of µTM.

A "wet-and-drag" filling method has been developed in which a metal blade is dragged at constant speed over the mold, resulting in filling of the channels without any residue on the raised features of the mold (Figure 5.1b) [4, 5]. The filling must be adjusted to the viscosity of the polymers and the shape of patterns on the mold. Too low or too high viscosity causes uneven filling due to rearrangement of the filled prepolymer within channels after dragging or formation of a thick layer over the channels, respectively [5].

Filling could also be performed by spin coating a polymer solution onto the mold in such a way that it resides only within the channels of the mold [6]. In another modification technique, filling of the mold cavities occurs by dipping the mold into the liquid prepolymer and then pulling up the mold at a constant drawing speed (Figure 5.1c). The drawing speed, which allows filling of the mold cavities while avoiding the presence of drops on the raised features, should be optimized taking into account the surface tension of the prepolymer and the pattern geometry [7].

5.2.2
Reversal Nanoimprinting

In reversal nanoimprinting (RNi) a polymer film is spin coated onto a mold, forming a replica of the mold patterns. This film is then transferred to a desired substrate by imprinting at a temperature close to the glass transition temperature (T_g) of the polymer (Figure 5.2a) [8, 9].

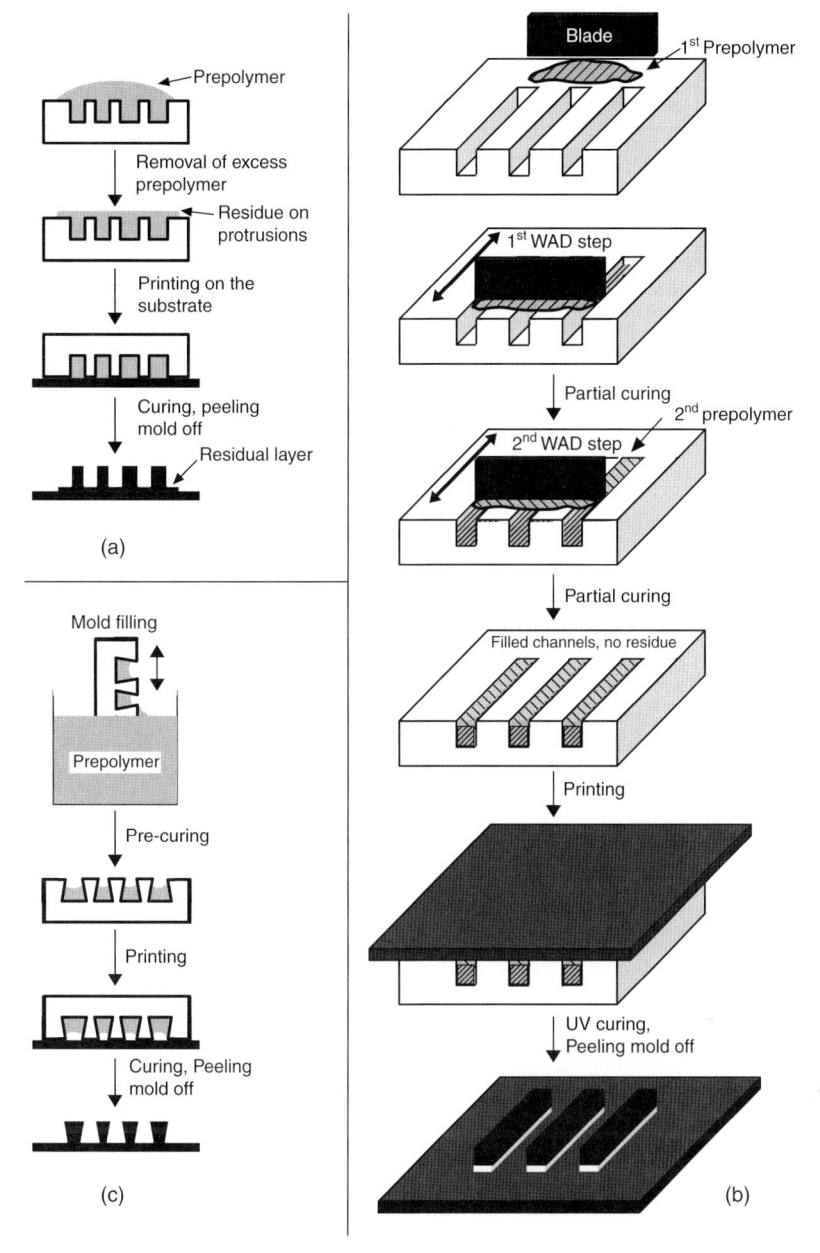

Figure 5.1 Schemes showing the steps in μTM [2] (a) and some variants; (b) filling by "wet-and-drag" method [42]; (c) filling by dipping method [7].

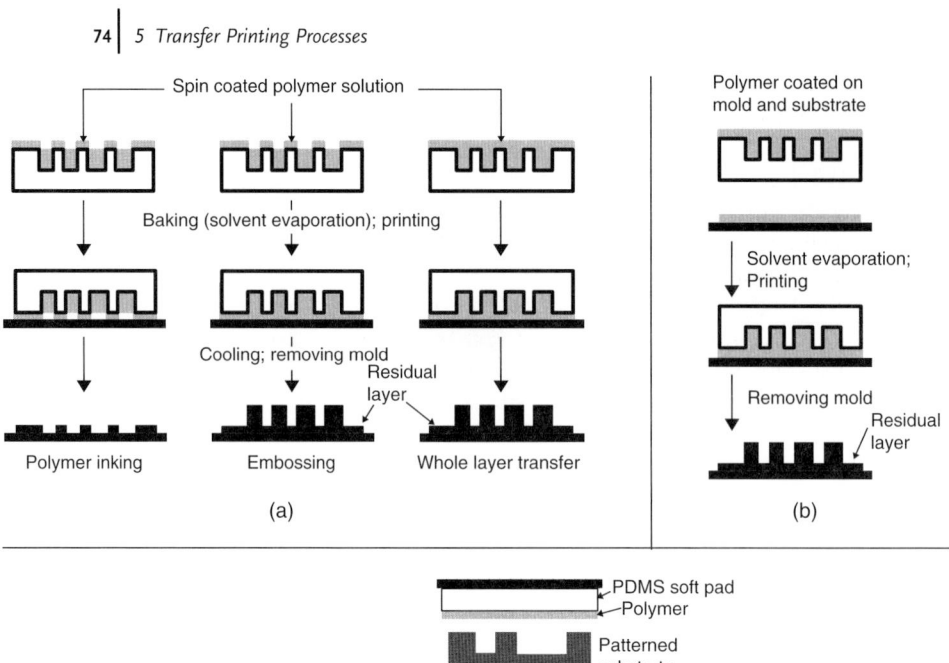

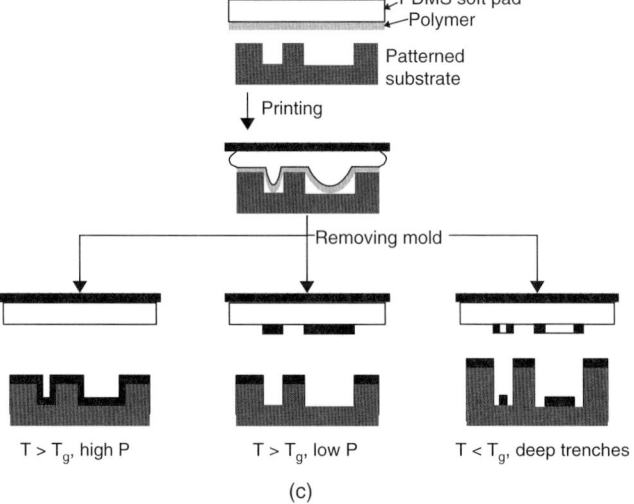

Figure 5.2 Schemes showing the steps in RNi [10] (a) and some variants; (b) coating of both mold and substrate [16]; (c) printing with a soft inkpad [20].

In practice, there are three variants of the method, depending on the geometry of the mold and pattern size, the thickness of the polymer film, polymer properties, and on the process conditions (temperature, pressure). If the depth of the recessed features is large compared to the thickness of the spin-coated film, the latter cannot planarize the relief features, leaving a large undulation on the coated polymer surface [9]. On the other hand, when the polymer thickness is comparable to the height of mold features, the polymer forms a planarized layer over the whole mold.

The "embossing" mode is similar to NIL in that a temperature well above T_g is required and, consequently, viscous flow of the polymer is observed. In this mode the spin-coated layer on the mold is not planarized, and high temperature and pressure are required to allow the polymer on the protruded areas to be squeezed into surrounding cavities during imprinting [10]. This mode, therefore, generates a negative replica of the mold on the substrate. In the "polymer inking" mode, by applying RNi at a temperature lower than or close to T_g of a nonplanarized polymer film, only the film coated on top of the raised features is transferred to the substrate [8–13]. When imprinting a planarized polymer film at low temperature (around T_g), the entire layer will be transferred without polymer flow, forming a negative replica of the mold ("whole-layer transfer") [8, 10, 14, 15].

A variant of RNi was developed where both the mold and the substrate are spin coated with a solution of a thermoplastic polymer. After evaporation of the solvent, both surfaces are brought into contact under heat and pressure; subsequently they are cooled down below T_g of the polymer and separated (Figure 5.2b) [16].

Reversal nanoimprinting was also adapted to the use of thermal- of photo-crosslinkable resins. In this variant, a resin containing the monomer and the crosslinker is drop cast [17] or spin coated [15, 18] on the mold, with the remaining steps being identical to RNi: the mold is brought into contact with the substrate, and imprinting is carried out under high temperature (to activate the polymerization process) and pressure [17]. Alternatively, if a photo-curable resin is used, the temperature may be lower, since crosslinking is activated by UV irradiation through the transparent mold [15, 18]. The transfer mode is equivalent to the "whole-layer transfer" (Figures 5.2a, right panel and 5.3a) [10, 17].

5.2.2.1 Reversal Nanoimprinting with Soft Inkpads

In reversal nanoimprinting with soft inkpads (RNsi), the polymer to be patterned is spin coated onto a hard substrate covered with a soft elastomer pad (e.g., PDMS). As usual a polymer film is coated on the stamp, which is then brought into contact with the patterned substrate under mild temperature ($T_g \pm 20$ °C) and pressure [9, 19, 20]. The main driving force for patterning and subsequent release is the difference in stiffness of the soft stamp (made of PDMS) and the rigid, glassy polymer film (e.g., poly(methyl methacrylate) – PMMA). Under pressure, elastic deformation of the PDMS pad around protrusions of the substrate gives rise to plastic deformation of the coated film. This deformation gives rise to localized rupture on the polymer layer at the edges of mold protrusions. After pressure release, the stamp recovers its original state while the film remains deformed and is selectively detached from the mold, leading to patterning of the substrate. The temperature, close to T_g of the polymer film, also plays a role, as it decreases the surface energy of PDMS [21] and thus facilitates the release without the requirement of releasing agents [19].

By tailoring temperature and pressure during the imprinting step, different pattern situations may be achieved [9, 19, 20]. With high pressure and temperature higher than T_g, the whole polymer film is transferred and covers both raised and

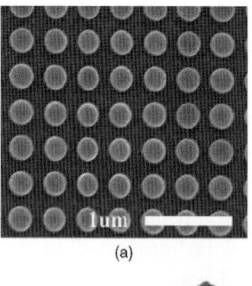

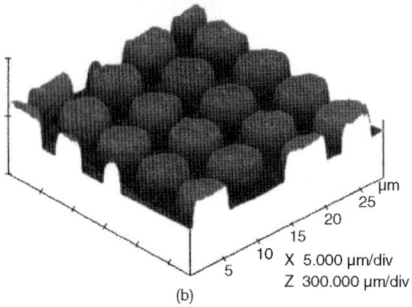

Figure 5.3 (a) PS pillars (250 nm diameter, aspect ratio 1) fabricated by RNi on a flexible substrate [17]; (b) PMMA patterns (160 nm thick, negative replica) prepared by PFP [19]. (Reproduced with permission from [17]. Copyright © (2006) American Chemical Society; Reproduced with permission from [19]. Copyright © (2003) American Institute of Physics.)

recessed features of the patterned substrate; if the pressure is low (e.g., 0.1 MPa), deformation of the stiff film is minimized and transfer occurs only onto protrusions of the substrate, leaving a negative replica on the stamp. Lastly, if the temperature is lower than T_g, the substrate has deep trenches and the stress along the contact edges of the film at protrusions or trenches exceeds the yield stress of the polymer, a discontinuous pattern transfer is achieved on both raised and recessed features. In this case, circular rings are formed as a negative image on the mold. In the last two variants, both replicas (on the stamp and on the mold) may subsequently be printed following the standard RNi technique (Figures 5.2c and 5.3b) [19]. With this technique, stacked polymer layers may be formed over a large area without the need to align the mold, what expands the capabilities of RNi.

5.2.3
Decal Transfer Microlithography

Decal transfer microlithography (DTM) is based on the transfer of elastomeric patterns onto a substrate by engineering the adhesion and release properties of a compliant polymer [22, 23]. This polymer, most commonly, is PDMS, which may be patterned on a variety of substrates, including on PDMS itself.

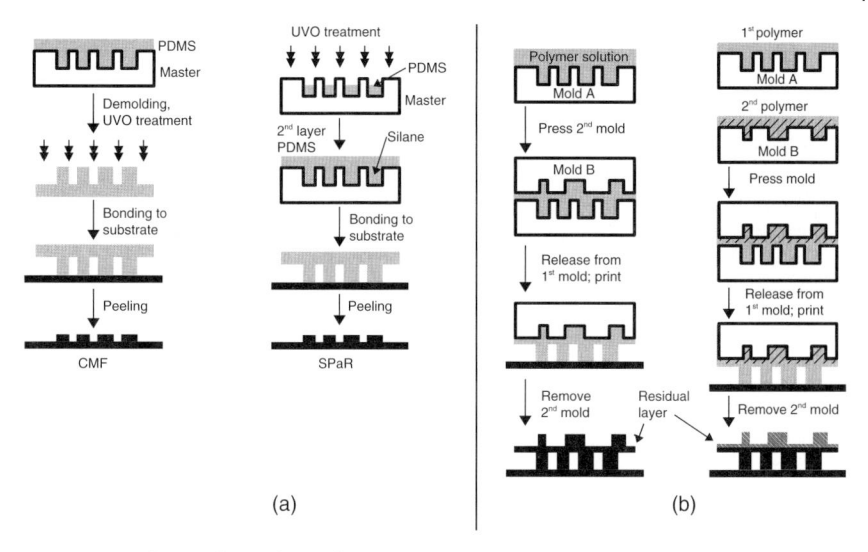

Figure 5.4 Schemes from other techniques: (a) DTM [22]; (b) DMI [27].

A key feature of the process is a UV/ozone pretreatment of the PDMS to allow the bonding to the substrate under mild conditions. In DTM, the adhesive bond created between a molded and pretreated PDMS stamp and an oxide-bearing surface is used to transfer PDMS in a patterned form from the mold to the substrate. The PDMS stamp is initially cast upon a master, then subjected to the UV/ozone treatment. After a variable "aging" period, the stamp is printed under pressure and temperature. Due to the previous treatment of PDMS stamp, the interfacial bond strength with the substrate gradually increases and ultimately becomes irreversible, completing the printing process (Figures 5.4a and 5.5a) [22].

The major disadvantage of DTM is the long processing time, specially imprinting time, since the interfacial bond strength between PDMS and the substrate increases only gradually, until it becomes irreversible [24, 25]. Nor is DTM well suited to patterning in the submicrometer scale since, due to the low modulus of PDMS, the patterns transferred by this mode would be very thin and their aspect ratio (AR) would be too low [26].

5.2.4
Duo-Mold Imprinting

Kong et al. [27] developed a TP technique where the substrate is also a patterned surface, forming therefore a polymeric film patterned on both surfaces. The technique is called duo-mold imprinting (DMI) and may be used to generate free-standing 3D structures as well as supported structures with sealed cavities and/or with two different polymers (Figure 5.5b) [28, 29].

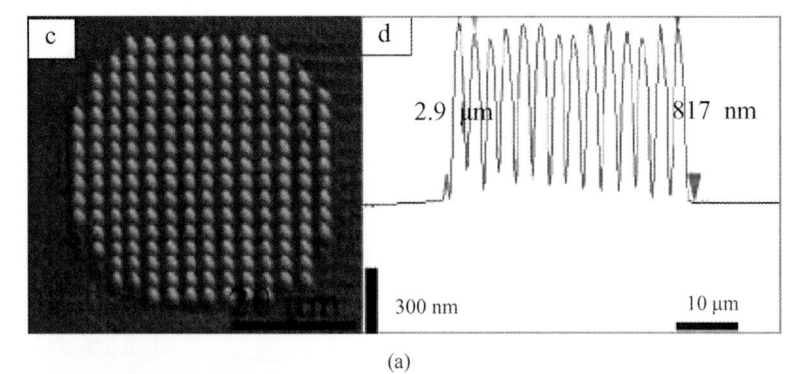

Figure 5.5 (a) AFM scan of a square array of PDMS posts (450 nm in height) prepared by DTM [24]; (b) complex 3D structures (inverted Z-shape) fabricated with PMMA using DMI [29]. (Reproduced with permission from [24]. Copyright © (2005) American Chemical Society; Reproduced with permission from [29]. Copyright © (2006) American Institute of Physics.)

In DMI, the polymer film may be spin coated on only one mold where it acquires the first 2D pattern, and the second mold is pressed against the first one (resulting in the transfer of another 2D pattern [29]), in a process resembling NIL. Alternatively, the polymer (or two polymers) may be spin coated on both molds, which are then pressed together at a temperature above the T_g of the polymer (or of both polymers). This method is similar to the RNi technique with coating of both mold and substrate (Section 5.2.2, and Figure 5.2b). In the first case, the process suffers from the same disadvantages as NIL, specifically in terms of high pressure and temperature (up to 75 °C above T_g) [27]. On the other hand, as it is carried out (for thermoplastics) well above the T_g, it ensures that the polymer will flow into the trenches of the embossing mold and fill them completely, avoiding the need for spin coating a planarized layer on the first mold [29]. The second method, on the other hand, requires lower imprint temperature and pressure, being therefore more advantageous for molds with deep features. After the pressing step, the system must be cooled down to below T_g in order to separate one of the molds, leaving the structure on the second one. The structure can subsequently be transferred to a suitable flat substrate at low pressure and below T_g of the polymer, forming a supported structure with both closed and open cavities (Figure 5.4b). This transfer step may result in deformation of the bottom patterns, depending on the feature size, thermo-mechanical properties of the polymer, temperature and pressure used [27].

The success of DMI depends on three factors: the accuracy of alignment of the two molds; the preferential demolding step, and the final pattern transfer to the substrate with minimal deformation of the polymer structures [29].

5.2.5
Printing on Topographies (Multilayer Printing) and on Flexible Substrates

One of the main features of TP methods is the ability to pattern on nonplanar or flexible surfaces (such as polymer membranes) and on topographies or previously patterned surfaces. The last feature may be repeated (with some processes, indefinitely), creating multilayered 3D structures and hierarchichal structures.

All techniques making use of elastomeric molds are suited to the patterning of nonplanar and/or flexible substrates, due to the flexibility of the used mold, to the low imprinting temperature of several processes, and to the fact the precursor is spin coated (or deposited) on the mold, not on the substrate. However, if the substrate has a large radius of curvature, the mold thickness cannot be too high in order to decrease the drawbacks of elastic deformation of the stamp resulting from conformal contact to a nonplanar surface [23]. The main drawback relates to a change in the pitch across the patterned area as the radius of curvature of the substrate increases.

Regarding imprinting on topographies, several approaches are possible. For most TP techniques, 3D structures may be built by sequentially applying the process to a previously patterned surface. Three-dimensional structures may therefore be created by simply repeating the process in a layer-by-layer fashion [8, 9]. DMI allows the creation of 3D structures in just one imprinting step, avoiding the need for precise alignment of the mold for the imprinting of each layer, a costly and difficult step required in layer-by-layer approaches. On the other hand, 3D patterns produced by DMI are usually not very high, and a multilayer approach may anyway be needed for high AR structures.

If a residual layer is present, the method is well suited to the formation of multilayers with closed channels or cavities, which do not require any additional processing step after each imprinting (Figure 5.6a). Methods which provide patterns without a residual layer, on the other hand, are more suitable to the fabrication of 3D structures with interconnected pores or hierarchical 3D structures. They are also the method of choice for polymers sensitive to plasma or chemical processing steps. When these techniques are applied onto a patterned substrate, the imprinting result strongly depends on the ratio between film thickness being imprinted and the distance between features on the substrate [30] (Figure 5.6b). If these two dimensions are comparable, the polymer film is imprinted over the whole area and a suspended layer is achieved (air-bridged structures) (Figure 5.7a). However, if the features on the substrate are separated by a distance much larger than the film thickness on the mold, selective transfer of structures will occur only over patterns on the substrate, as the suspended part of the film will break during mold separation (hierarchical structures) (Figure 5.7b) [30]. This behavior is also dependent on the mechanical strength of the

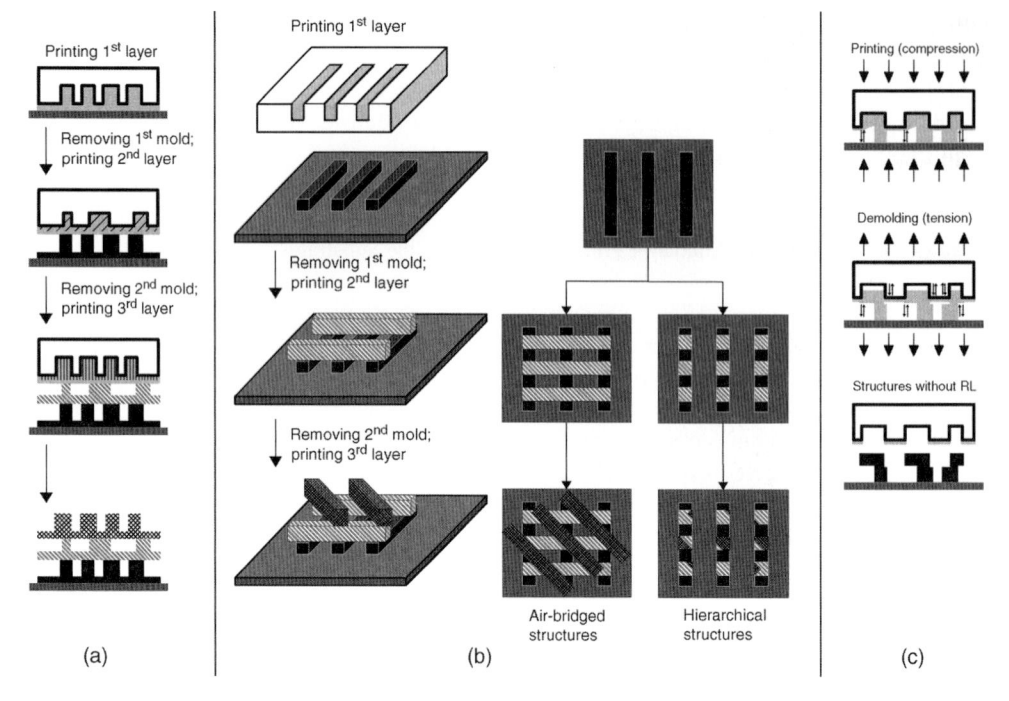

Figure 5.6 (a) Fabrication of multilayered structures with residual layer; (b) fabrication of multilayered structures without residual layer, showing the two stacking possibilities: air-bridged structures and hierarchical structures; (c) scheme showing the process of residual layer self-removal.

polymer: multilayer printing of brittle polymers is only possible for a relatively narrow gap, while tough and ductile polymers may form multilayers even across spacings that are much wider than the thickness of the film.

Theoretically, an indefinite number of layers may be superimposed onto one another in 3D structures. However, this capability is limited for processes in which the imprinting temperature is above T_g (RNi and variants). For such techniques, the polymer of each subsequent layer should have a T_g ∼30–40 °C lower than the previous one, in order to avoid distortion of the previously formed layers. That obviously limits the number of layers that can be built to three or four layers. These processes may be used for the preparation of multilayer patterns with a single polymer, as long as the imprinting temperature is kept in the range 10–20 °C below T_g [14, 30]. However, in this case the high pressure required and the proximity to the T_g do cause some deformation of the lower layer [30]. This method is therefore only suitable when dimension stability is not of critical concern.

A better alternative for multilayer printing of thermoplastics is the variant of RNi using a soft inkpad. Multilayers may be formed on the hard mold by simply repeating the process with freshly coated soft inkpads and finally transferring the stacked polymer structure from the mold to a substrate. One specific advantage of the method when compared with other patterning techniques is that it does not require alignment of mold and substrate when stacking several layers of polymer patterns,

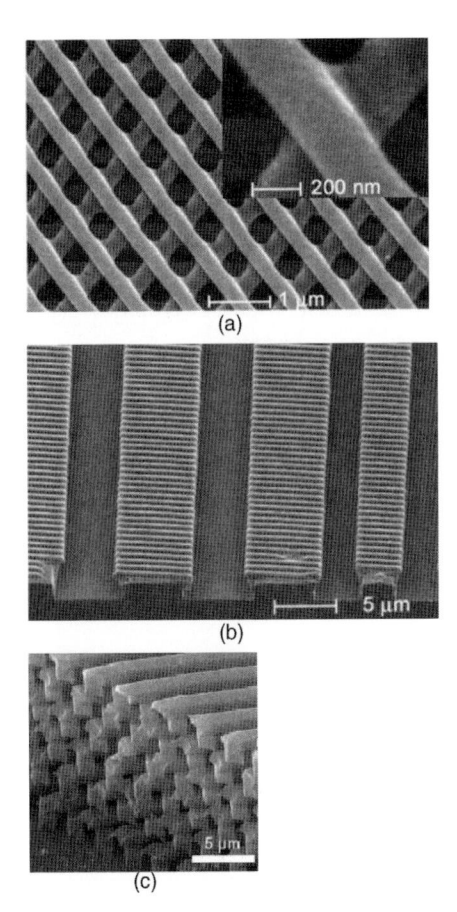

Figure 5.7 (a) Three-layer polymer structure fabricated by RNi: polymers are poly(butyl acrylate), PMMA, and PC from top to bottom [30]; (b) PMMA multilayered structures (700 nm lines on 3 µm SiO$_2$ patterns) fabricated by RNi [30]; (c) tilted view of a 12-layer structure fabricated by µTM with polyurethane/polymethacrylate prepolymers [5]. (Reproduced with permission from [30]. Copyright © (2002) American Institute of Physics;Reproduced with permission from [5]. Copyright © (2005) Wiley-VCH Verlag GmbH & Co. KGaA.)

overcoming the limitation of imprinting systems that do not have nanometer alignment capabilities [20]. However, such a method only allows stacking over protrusions, that is, one cannot fabricate closed channels or complex 3D structures with it.

Methods relying on curable precursors do not require pressure to transfer the polymer, therefore structures with a very large number of layers may be easily prepared without distortion of the previously fabricated structures using the same polymer for each layer (Figure 5.7c). In these methods, imprinting temperature should be below T_g and a pre-curing step should be applied before printing to minimize flow of prepolymer into the empty spaces of structures of previous layers and to allow successful bridging of such gratings. In fact, depending on the interfeature spacing of the underlying layer, the printing of the following one may occur only on the protrusions or on the whole surface (as explained above for TP, see Figure 5.6b) [15]. The high crosslinking degree and low imprinting temperature practically eliminates deformation of the underneath layer during printing of the next one. Moreover, due to the high crosslinking degree of the post-cured layer, there will be harder any diffusion of resin into it (a fact that could provide enough interfacial force to promote adhesion between layers). Therefore, the underneath layer must be surface treated (usually by oxygen plasma treatment [18]) in order to increase the surface energy of the crosslinked polymer. Additionally, if multilayered structures with open and/or interconnected cavities are needed, a residual layer removal step (oxygen RIE) may be required.

A particular situation occurs with DTM: the substrate with the previous layers must also be subjected to the UVO treatment in order to promote PDMS-to-PDMS adhesion. This step must be repeated for each new layer, which makes the process time-consuming due to the long treatment time and time needed for the development of the irreversible bonding [24].

5.3
Key Issues in Transfer Printing Methods

5.3.1
Surface Treatments of Mold and Substrate

The key for a successful transfer relies on the surface energy differences between the polymer, the mold and the substrate. The mold should have a lower surface energy than the substrate, so that the polymer film will adhere more strongly to the latter and could therefore be easily detached from the mold [8, 9]. This gets more critical as molds with small feature sizes are used. The larger contact area causes larger adhesive force with the polymer and difficult pattern transfer.

When working with hard molds, a selective treatment of the surface of mold and substrate to decrease and increase their surface energy, respectively, is required. The mold is usually coated with a perfluorinated layer, while the substrate is exposed to oxygen plasma and/or reactive coupling agents [17, 31]. The low energy perfluorinated surface has a drawback: it makes it very difficult to fill molds with a very low (lower than 5%) or very high (higher than 95%) protrusion area ratio [15]. Alternatively, electrostatic attraction between the polymer film and the substrate may be used to enhance selective adhesion [32]. The combination of a polycationic layer on the substrate and a negatively charged polymer layer on the mold ensures a highly effective adhesive junction between both surfaces.

The situation is more complex in DMI, because two molds are used, which should have different surface energies. The difference must be sufficient to allow one mold to separate cleanly from the 3D structures still attached to the second mold. Additionally, in the case of supported structures, the surface energy of the second mold must also allow an easy transfer to the final substrate [27]. For molds with similar pattern profiles, a strategy of employing different silane treatments for each mold is required in order to create the necessary difference in the work of adhesion [27]. As too large a contrast of surface energies between the two molds would reduce the contrast of surface energy between the second mold and the substrate, high yield transfer would be compromised, especially for low imprinting temperature. Another alternative is the use of two polymers. Thus, instead of surface energy differences between the molds, the difference in polymer–mold interactions are employed to guide the selective demolding from molds with the same surface energy. Therefore, the surface energy difference with respect to the substrate could be kept higher, facilitating the final transfer of supported structures [28].

The strategy of treating the mold with different silane coupling agents is also used in the "polymer inking" mode of RNi. This mode usually results in the formation of ragged pattern edges due to forced tearing of the polymer film upon mold separation [10]. In order to print polymer features with smoother edges, protruded surfaces and trenches may be treated to have different surface energies. The protrusions are first inked with a medium surface energy silane and, subsequently, the recessed areas and sidewalls are coated with a lower surface energy silane. After the polymer film is spin coated on such a modified mold and heated above T_g, it dewets easily from the sidewalls while still remaining attached to the protrusions, becoming discontinuous along the edges of the patterns. During the subsequent printing step, polymer patterns transferred to the substrate have much smoother edges [9, 12].

PDMS molds have low surface energy, allowing for easy detachment from the imprinted features and substrate. Therefore, in general no specific surface treatment is required. If that is the case (e.g., to facilitate filling of recessed features and/or of hydrophilic precursors), a simple oxygen plasma treatment may be applied to decrease the contact angle of the surface. Plasma process parameters and aging conditions may be adjusted to tailor the contact angle to specific values. Moreover, application of heat during the imprinting process may be used to reverse the state of oxygen plasma-treated PDMS mold back to a hydrophobic state, in order to make detaching of the film from the stamp easier [33].

A specific situation occurs in DTM, where substrates are required to have surface oxides, in order to form strong adhesive bonds to a UVO-treated PDMS. When that is not the case, surface modifications such as functionalization with self-assembled monolayers or deposition of thin (5–20 nm) adhesion layers of silica allows a wider range of materials to be employed as substrates for DTM [31].

5.3.2
Residual Layer

One drawback of several TP processes is the presence of a residual layer after imprinting, due to the polymer being present not only in the trenches, but also on the protrusions of the mold. Patterns transferred to the substrates will, in such cases, be connected by a thin film of the material, as it is not fully removed from the raised features of the mold. Typically, these layers are ~100 nm thick in µTM [2, 34], while in some variants of RNi much thinner layers (on the order of 4–5 nm) have been obtained [17]. Figures 5.1–5.4 show which methods and variants lead to the appearance of residual layers on the printed structure.

This residual layer may be removed in a oxygen reactive ion etching (RIE) post-processing step. This requires extra equipment and time, adds extra costs to the procedure, and may additionally degrade the imprinted polymer [20]. Lateral erosion of the imprinted patterns and, consequently, change in pattern size or formation of nonvertical sidewalls may also occur during RIE [13, 31]. If the thickness of the residual layer is not homogeneous, accelerated trimming and partial etching of the underneath layer may occur at locations where the layer is thinner.

With an adequate design of TP elements (mold, polymer, thickness of coated film, etc.), the residual layer may be self-removed. Residual layer self-removal works by inducing failure of the residual layer immediately adjacent to the patterned resist features. This is achieved through the application of both compressive and tensile stresses on the protrusions of the patterned resist, which in turn impose shear stresses on the supported residual layer, leading to failure along the feature boundaries (Figure 5.6c) [31].

In order to work efficiently, the mechanical properties of the layer should be tailored. A thin residual layer reduces the shear stress needed to induce failure, while the use of a brittle polymer allows for a cleaner failure of the layer and, consequently, smoother surface and sidewalls of the final transferred structures than if a ductile polymer is used. The quality of the process depends as well on mold design: molds with sharp edges induce failure of the residual layer in a specific location, while molds where protrusions edges are not sharp lead to nonspecific layer failure and, consequently, patterned features with rough edges [31].

Although self-removal has been demonstrated for both DMI and TP where only one mold is employed, it is better suited for the former case, since both tensile and compressive stresses are applied to the residual layer, increasing the effectiveness of the process. When only one mold is used, the failure of the layer is induced purely by tensile stresses.

While the process described above works well for thermoplastics spin coated onto a mold, a two-step dewetting method was developed to eliminate the residual layer in processes involving photocurable prepolymers or monomers [35]. After coating the mold with the precursor, a surface energy gradient is applied to the mold by scanning with an infrared lamp, resulting in residual film break-up and the formation of patches which are thinner than the original layer. The second step is a thermal dewetting, applied when mold and substrate are in contact but before UV curing of the prepolymer; during this step, the residual liquid in the thin patches is dewetted into the channels. After thermal dewetting, UV exposure causes curing of the resin, followed by a post-curing step after mold removal [35]. Parameters such as IR power, scan speed, and thermal dewetting time may influence the outcome of the process.

Lastly, some TP techniques are inherently residue-free, meaning that only the patterns are transferred to the substrate and no pre- or post-imprinting step is required to remove the residual layer. In this category are the "inking" mode of RNi, RNsi, and DTM.

5.4
Advantages and Disadvantages

Transfer printing techniques show a number of advantages for producing 3D microstructures such as low cost, capability of nonperiodic 3D structures, a wide range of materials compatibility, ability to pattern large areas (up to several cm^2) in one step, and flexibility in design [5, 34]. They also allow the fabrication of microstructures on contoured surfaces (an essential requirement for fabricating 3D structures layer-by-layer) [34] and are capable of reproducing re-entrant structures with overhangs due to the flexibility of PDMS [36].

Transfer printing techniques with thermoplastics are advantageous over NIL as they allow the use of reduced pressure and temperature and additionally reduce the impact of limited polymer transport on pattern formation [10, 16]. Those occur due to printing with solid polymers, therefore eliminating the dependence on polymer viscous flow to deform the polymer film and create the thickness contrast, a typical drawback of NIL. While NIL is typically carried out a temperature much higher than the T_g of the polymer being imprinted (70–80 °C higher) and under pressure as high as 10 MPa [10], some of the TP processes presented in this chapter can be performed at temperatures 30 °C below T_g and at pressures usually around 1 MPa, in some cases even 0.1 MPa. Another disadvantage of NIL is the need to spin coat the substrate, which is a difficult task when dealing with flexible ones, such as polymer membranes. As in almost all TP techniques the polymer resist or solution is not coated on the substrate, they may be employed to substrates that are not suitable for spin coating, that have surface topographies or prefabricated structures, or to flexible substrates.

Moreover, an interesting aspect of TP is its capability to generate patterns that are smaller than the original features of the mold. When the mold is heated above the T_g of the polymer, there is a retraction of the film on top of the protrusions [12]. This retraction tends to increase with increase in temperature, and values as high as 70% of lateral shrinkage have been observed for polycarbonate (PC) [12]. As the production of molds with sub-micrometer or nanometer feature size is costly and difficult, requiring complicated equipment and infrastructure, such a retraction effect offers a straightforward approach to obtain polymer patterns within that range by using molds with larger features. This feature has been demonstrated for RNi where the polymer, after heat treatment, formed a dome on the protrusions of the mold. The retraction behavior depends not only on temperature, but also on polymer properties: polymers with a more entangled network retract more strongly from the molecular orientation produced by the spin-coating process [12].

Besides the drawbacks related to the presence of a residual layer and the need for careful surface treatment of the mold, TP schemes have some other disadvantages that should be overcome to make the processes more interesting for general applications.

One major problem is related to the range of polymers used. Although it is claimed that the processes, altogether, may be applied to a wide variety of polymers, most of the published literature is restricted to three or four examples. The gold standard is, clearly, the acrylate family, both as thermoplastic (PMMA or blends thereof) or as thermoset prepolymer (acrylic or acrylate resins). Most of other polymers mentioned in Table 5.1 (with the exception of epoxy resin and polystyrene – PS) were basically used in only one work. The capabilities and limitations of TP processes may be fully exploited and generalized only if a wider range of polymers is demonstrated to work. That is an important step not only to understanding the fundamentals of printing techniques, but also to allow their advantages over other lithographic processes to be fully demonstrated. Both steps are essential to the development of concrete applications. As will be shown in the next section, applications of TP techniques are still very limited and mostly in a "potential" or "demonstration" stage, partly due to the problem discussed above.

Another critical issue, that may limit the application of TP, is the difficulty in generating high AR polymer structures, for the reasons that have already been discussed: difficulty of transferring high AR features due to the high work of adhesion with the mold, need for high temperature and pressure in some processes (therefore deforming the structures), difficulty of spin coating surface-treated molds with deep features, and so on. In deep molds, the coating thickness on the sidewalls is likely to be very small and may dewet spontaneously during spin coating, which results in discontinuous coverage of the mold [12]. Although the lateral resolution of printed features is clearly in the submicrometer or nanometer range (e.g., 700 nm for RNi [10, 30] and µTM [37], 450 nm for RNi with soft inkpad [20], 250 nm for DMI [28], and 50 nm for RNi where both mold and substrate were coated [16]), the reported AR for most of the techniques is close to or below one. Notable exceptions include some variants of RNi, with ARs between 4.4 and 5 [16, 20, 32]. In this sense, the use of soft inkpads (in RNi or RNsi) have the potential to generate high AR structures in a layer-by-layer fashion, once they do not require alignment between each imprinting step, therefore greatly facilitating the process.

In µTM, uncured filled prepolymer can smear out of the channels by capillary wicking [2, 5] when contacting a substrate. Such wicking deteriorates structural fidelity and requires an additional processing step to remove the uncontrolled polymer, such as blowing with nitrogen (if performed before curing) or reactive ion etching (after curing, see Section 5.3.2) [34]. The use of high pressure nitrogen to remove excess prepolymers from the mold may generate the appearance of tiny scattered droplets of prepolymer after blowing, which are very difficult to remove even using higher velocity of gas [5].

In most TP techniques, the inability to mold closed loops require that a layer-by-layer approach be adopted for the creation of 3D structures [38, 39]. This approach requires one molding step for each layer (and, sometimes, more than one mold) making registration a difficult task. Moreover, at each molding step a residue layer may be formed, requiring an additional step for its removal. Moving parts cannot be fabricated, because in each layer attachment points to the previous layer are required for release of the replicas from the mold. Lastly, the range of structures that may be fabricated is limited by the requirement that features in each layer be relatively dense to prevent sagging of the mold [38]. If PDMS is used as the mold, the inter-feature spacing should not be greater than five times the height of the feature [40]. The

Table 5.1 Main characteristics of transfer printing techniques.

Mold*	Polymer	Precursor[†]	Process Parameters and Steps[‡]			References
			T (°C)	P (MPa)	t (min)	
			Microtransfer Molding			
P	Epoxy	PP	65[p]	—	25	
			RT[i]	—	10	
			25[c]	—	24 h	[2, 34, 43]
P	PU	PP	UV	—	60	[1]
			With "wet-and-drag" filling method			
P	PU/PMA	PP	UV	—	5–30	[4, 42]
			With dipping filling method			
P	Acrylate	PP	UV[p]	—	3–15	
			80[i]	0–0,14	10–30	
			150/UV[c]	—	10–20	[35]
			Reversal Nanoimprinting			
G	PMMA	S	80–105[e], 75–180[i]	5[i]	5	[8–10, 12, 14, 30]
G	PC	S	150–160	5		[12, 30]
P	PEDOT	S	80	—	2	[33]
			Coating mold and substrate			
S	PMMA	S	200[e], 160[i]	10i	5	[16]
			With crosslinkable resins			
S	PS	M	110[i]	4	10	
			110[c]	—	60	[17]
G,S	Epoxy	PP	65–120[e]	—	2–5	
			40–85/UV[i]	1–6	2/10s	
			65–120[c]	—	2–5	[15, 18, 50]
			Printing with lithographic mask			
G	Epoxy	S	120[e]	—	5	
			80–90[i]	2–4	0,5	[51]
			Reversal nanoimprinting with soft inkpad			
G,S,P	PMMA	S	80[e]	—	3	
			110–120[in]	3–4	5–8	
			120[i]	4–5	5–8	[20]
P	PMMA	S	110–115	0,1–3	5	[19]
			Decal transfer microlithography			
S	PDMS	PP	—[u]	—	2,5	
			65–135[i]	—	20–40	[22, 23, 26, 46]
			Duo-mold imprinting			
G,S	PMMA	S	80–150[e]	—	2-3	
			120–180 m	4–6	4–10	
			65–95[i]	1–2	4–5	[27, 29]
S	PMMA/PS	S,M	80[e]	—	1	
			80[m]	4	120	
			180[i]	4	10	[28]
S	PS	M	110[m]	4	3	
			180[i]	4	5	[29]

*P = PDMS; G = glass; S = silicon; PUA = poly(urethane acrylate).
[†]PP = curable prepolymer; S = polymer solution; M = monomer.
[‡]p = pre-curing; [i] = imprinting; [c] = post-curing; [e] = solvent evaporation; [in] = inking; [m] = molding.

DTM variant combined with photolithography is also an effective solution to the problem, although it also requires long processing times, besides precise alignment of the mold.

5.5
Applications

One important application of μTM is in the fabrication of waveguides [1, 4, 41, 42], which are important components of sensors and switches (Figure 5.8). The main characteristics of μTM that make the technique suitable for such application are the possibility of producing multiple copies of complex microstructures, forming waveguides over large areas and on virtually any optically smooth surface (including nonplanar and flexible surfaces), and of fabricating structures that incorporate gratings and other optical elements in a single step (due to its 3D capability) [1]. However, the undesirable thin residual layer present between waveguides may limit their functionality. μTM can be also employed to prepare templates for 3D photonic band gap crystals operating in the infrared, optical and ultraviolet frequencies [43].

Reversal nanoimprinting has been applied to the preparation of organic thin film transistors with conductive poly(3,4-ethylenedioxythiophene)/poly(4-styrenesulfonate) (PEDOT/PSS) electrodes patterned on flexible substrates [33]. "Inking" is advantageous for the process, since it avoids the use of RIE or solvent-based postprinting steps that could damage the substrate film or its electrical properties. Polymer light emitting diodes have also been fabricated by RNi by printing a multilayer composed of PEDOT, light emitting polymer, and a metal cathode layer (Au/Al) onto an indium tin oxide coated glass substrate [44].

Due to the chemical and mechanical stability of patterns produced by RNi with crosslinkable resins, they are well suited for applications for chemical and biological assays [17]. If a functional monomer is included in the precursor mixture, reactions

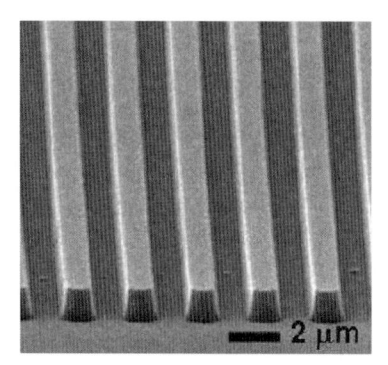

Figure 5.8 SEM micrograph of waveguides fabricated by μTM with polyurethane/polymethacrylate prepolymers. (Reproduced with permission from [42]. Copyright © (2006) SPIE.)

may be carried out only on the raised features of the patterned structure [17]. The use of a flexible polymer substrate allows easier handling of such nano-array systems, as it could be conveniently rolled up. Structures are also well suited to applications in biosensing and analysis, and photonic devices, due to the excellent mechanical stability of the crosslinked structures, transparency (allowing observation of particles or molecules flowing inside channels), and high fidelity of micro- and nanoscale features.

Decal transfer microlithography (DTM) is specially suited for applications in microfluidic devices for biological studies (such as controlling the micro-environment of neurons), or where a need exists to pattern very high ARs or 3D polymer structures (micro-electromechanical systems, photonics), for patterning materials on nonplanar substrate surfaces (sensors, optics), and for patterning micron scale or larger features inexpensively on large substrates (displays, bio-micro-electromechanical systems, etc.) [22, 23]. PDMS microfluidic devices prepared by DTM technique with open and closed-channel architecture have enabled the survival and differentiation of primary hippocampal neurons [45]. DTM is also suited to the creation of micrometer etch masks to guide RIE or wet etching processes [22, 23, 26], and for the fabrication of electronic devices such as silicon-based thin-film transistors [46].

A modified printing method was recently employed for the fabrication of biomimetic adhesives [47–49]. In this variant, the polymer is coated on a hard inkpad, while the mold is a soft, elastomeric one. After the "inking" step, the polymer on the protrusions is cured under different conditions to yield patterns on the mold with different tip geometries (spherical, spatular, etc.). The adhesion of such structures to sapphire surfaces is dependent on the geometrical parameters of the mold (AR and tip shape) [47–49]. Such structures are being developed to mimic gecko footpads; and are thoroughly discussed in Chapter 15 in this book.

References

1 Zhao, X.M., Smith, S.P., Waldman, S.J., Whitesides, G.M., and Prentiss, M. (1997) Demonstration of waveguide couplers fabricated using microtransfer molding. *Appl. Phys. Lett.*, **71** (8), 1017–1019.

2 Xia, Y. and Whitesides, G.M. (1998) Soft Lithography. *Angew. Chem. Int. Ed.*, **37**, 550–575.

3 Brehmer, M., Conrad, L., and Funk, L. (2003) New developments in soft lithography. *J. Disper. Sci. Technol.*, **24** (3–4), 291–304.

4 Lee, J.H., Kim, C.H., Constant, K., and Ho, K.M. (2006) Tailorable, 3D microfabrication for photonic applications: Two-polymer microtransfer molding. *Mater. Sci.*, **6128**, 612805.

5 Lee, J.H., Kim, C.H., Ho, K.M., and Constant, K. (2005) Two-polymer microtransfer molding for highly layered microstructures. *Adv. Mater.*, **17**, 2481–2485.

6 Moran, P.M. and Lange, F.F. (1999) Microscale lithography via channel stamping: relationships between capillarity, channel filling, and debonding. *Appl. Phys. Lett.*, **74** (9), 1332–1334.

7 Kim, M.J., Song, S., Kwon, S.J., and Lee, H.H. (2007) Trapezoidal structure for residue-free filling and patterning. *J. Phys. Chem. C*, **111** (3), 1140–1145.

8 Guo, L.J. (2007) Nanoimprint lithography: Methods and material requirements. *Adv. Mater.*, **19** (4), 495–513.

9 Guo, L. (2004) Recent progress in nanoimprint technology and its applications. *J. Phys. D Appl. Phys.*, **37** (11), R123–R141.

10 Huang, X.D., Bao, L.R., Cheng, X., Guo, L.J., Pang, S.W., and Yee, A.F. (2002) Reversal imprinting by transferring polymer from mold to substrate. *J. Vac. Sci. Technol. B*, **20**, 2872–2876.

11 Tan, L., Kong, Y.P., Pang, S.W., and Yee, A.F. (2004) Imprinting of polymer at low temperature and pressure. *J. Vac. Sci. Technol. B*, **22**, 2486–2492.

12 Bao, L.R., Tan, L., Huang, X.D., Kong, Y.P., Guo, L.J., Pang, S.W., and Yee, A.F. (2003) Polymer inking as a micro- and nanopatterning technique. *J. Vac. Sci. Technol. B*, **21** (6), 2749–2754.

13 Suh, D., Rhee, J., and Lee, H.H. (2004) Bilayer reversal imprint lithography: direct metal-polymer transfer. *Nanotechnology*, **15**, 1103–1107.

14 Ooe, H., Morimatsu, M., Yoshikawa, T., Kawata, H., and Hirai, Y. (2005) Three-dimensional multilayered microstructure fabricated by imprint lithography. *J. Vac. Sci. Technol. B*, **23**, 375–379.

15 Kehagias, N., Zelsmann, M., Sotomayor Torres, C.M., Pfeiffer, K., Ahrens, G., and Gruetzner, G. (2005) Three-dimensional polymer structures fabricated by reversal ultraviolet-curing imprint lithography. *J. Vac. Sci. Technol. B*, **23**, 2954–2957.

16 Borzenko, T., Tormen, M., Schmidt, G., Molenkamp, L.W., and Janssen, H. (2001) Polymer bonding process for nanolithography. *Appl. Phys. Lett.*, **79** (14), 2246–2248.

17 Zhao, W., Low, H.Y., and Suresh, P.S. (2006) Cross-linked and chemically functionalized polymer supports by reactive reversal nanoimprint lithography. *Langmuir*, **22**, 5520–5524.

18 Hu, W., Yang, B., Peng, C., and Pang, S.W. (2006) Three-dimensional SU-8 structures by reversal UV imprint. *J. Vac. Sci. Technol. B*, **24**, 2225–2229.

19 Tan, L., Kong, Y.P., Bao, L.R., Huang, X.D., Guo, L.J., Pang, S.W., and Yee, A.F. (2003) Imprinting polymer film on patterned substrates. *J. Vac. Sci. Technol. B*, **21**, 2742–2748.

20 Kong, Y.P., Tan, L., Pang, S.W., and Yee, A.F. (2004) Stacked polymer patterns imprinted using a soft inkpad. *J. Vac. Sci. Technol. A*, **22** (4), 1873–1878.

21 Fritz, J.L. and Owen, M.J. (1995) Hydrophobic recovery of plasma-treated polydimethylsiloxane. *J. Adhesion*, **54**, 33–45.

22 Childs, W.R. and Nuzzo, R.G. (2002) Decal transfer microlithography: a new soft-lithographic patterning method. *J. Am. Chem. Soc.*, **124**, 13583–13596.

23 Childs, W.R. and Nuzzo, R.G. (2004) Patterning of thin-film microstructures on non-planar substrate surfaces using decal transfer lithography. *Adv. Mater.*, **16**, 1323–1327.

24 Childs, W.R., Motala, M.J., Lee, K.J., and Nuzzo, R.G. (2005) Masterless soft lithography: patterning UV/ozone-induced adhesion on poly (dimethylsiloxane) surfaces. *Langmuir*, **21**, 10096–10105.

25 Childs, W.R. and Nuzzo, R.G. (2005) Large-area patterning of coinage-metal thin films using decal transfer lithography. *J. Vac. Sci. Technol. B*, **21**, 195–202.

26 Ahn, H., Lee, K.J., Shim, A., Rogers, J.A., and Nuzzo, R.G. (2005) Additive soft-lithographic patterning of submicrometer- and nanometer-scale large-area resists on electronic materials. *Nano Lett.*, **5** (12), 2533–2537.

27 Kong, Y.P., Low, H.Y., Pang, S.W., and Yee, A.F. (2004) Duo-mold imprinting of three-dimensional polymeric structures. *J. Vac. Sci. Technol. B*, **22** (6), 3251–3256.

28 Zhao, W. and Low, H.Y. (2006) Fabrication of hybrid bilayer nanostructure by duo-mold imprinting. *J. Vac. Sci. Technol. B*, **24**, 255–258.

29 Low, H.Y., Zhao, W., and Dumond, J. (2006) Combinatorial-mold imprint lithography: a versatile technique for fabrication of three-dimensional polymer structures. *Appl. Phys. Lett.*, **89**, 023109.

30 Bao, L.R., Cheng, X., Huang, X.D., Guo, L.J., Pang, S.W., and Yee, A.F. (2002) Nanoimprinting over topography and multilayer three-dimensional printing. *J. Vac. Sci. Technol. B*, **20**, 2881–2886.

31 Dumond, J. and Low, H.Y. (2008) Residual layer self-removal in imprint lithography. *Adv. Mater.*, **20**, 1291–1297.

32 Kim, Y.S., Baek, S.J., and Hammond, P.T. (2004) Physical and chemical nanostructure transfer in polymer spin-transfer printing. *Adv. Mater.*, **16** (7), 581–584.

33 Li, D. and Guo, L.J. (2006) Micron-scale organic thin film transistors with conducting polymer electrodes patterned by polymer inking and stamping. *Appl. Phys. Lett.*, **88**, 063513.

34 Zhao, X.M., Xia, Y., and Whitesides, G.M. (1996) Fabrication of three-dimensional micro-structures: microtransfer molding. *Adv. Mater.*, **8** (10), 837–840.

35 Kim, M.J., Song, S., and Lee, H.H. (2006) A two-step dewetting method for large-scale patterning. *J. Micromech. Microeng.*, **16** (8), 1700–1704.

36 LaFratta, C.N., Baldacchini, T., Farrer, R.A., Fourkas, J.T., Teich, M.C., Saleh, B.E.A., and Naughton, M.J. (2004) Replication of two-photon-polymerized structures with extremely high aspect ratios and large overhangs. *J. Phys. Chem. B*, **108**, 11256–11258.

37 Zhao, X.M., Xia, Y., and Whitesides, G.M. (1997) Soft lithographic methods for nano-fabrication. *J. Mater. Chem.*, **7** (7), 1069–1074.

38 LaFratta, C.N., Li, L., and Fourkas, J.T. (2006) Soft-lithographic replication of 3D microstructures with closed loops. *Proc. Natl. Acad. Sci. U.S.A.*, **103** (23), 8589–8594.

39 Gates, B.D., Xu, Q., Stewart, M., Ryan, D., Willson, C.G., and Whitesides, G.M. (2005) New approaches to nanofabrication: molding, printing, and other techniques. *Chem. Rev.*, **105**, 1171–1196.

40 Michel, B., Bernard, A., Bietsch, A., Delamarche, E., Geissler, M., Juncker, D., Kind, H., Renault, J.P., Rothuizen, H., Schmid, H., Schmidt-Winkel, P., Stutz, R., and Wolf, H. (2001) Printing meets lithography: soft approaches to high-resolution patterning. *IBM J. Res. Dev.*, **45** (5), 697–717.

41 Matsui, T., Komatsu, K., Sugihara, O., and Kaino, T. (2005) Simple process for fabricating a monolithic polymer optical waveguide. *Opt. Lett.*, **30** (9), 970–972.

42 Lee, J.H., Ye, Z., Constant, K., and Ho, K.M. (2006) Tailorable polymer waveguides for miniaturized bio-photonic devices via two-polymer microtransfer molding, in *Nanoengineering: Fabrication, Properties, Optics, and Devices III, Proc. of SPIE*, vol. **6327** (eds E.A. Dobisz and L.A. Eldada), p. 63270J, SPIE, Bellingham WA, USA.

43 Leung F W.Y., Kang, H., Constant, K., Cann, D., Kim, C.H., Biswas, R., Sigalas, M.M., and Ho, K.M. (2003) Fabrication of photonic band gap crystal using microtransfer molded templates. *J. Appl. Phys.*, **93** (10), 5866–5870.

44 Cardozo, B.L., and Pang, S.W. (2008) Patterning of polyfluorene based polymer light emitting diodes by reversal imprint lithography. *J. Vac. Sci. Technol. B*, **26** (6), 2385–2389.

45 Millet, L.J., Stewart, M.E., Sweedler, J.V., Nuzzo, R.G., and Gillette, M.U. (2007) Microfluidic devices for culturing primary mammalian neurons at low densities. *Lab on a Chip*, **7**, 987–994.

46 Ahn, H., Lee, K.J., Childs, W.R., Rogers, J.A., Nuzzo, R.G., and Shim, A. (2006) Micron and submicron patterning of polydimethylsiloxane resists on electronic materials by decal transfer lithography and reactive ion-beam etching: Application to the fabrication of high-mobility, thin-film transistors. *J. Appl. Phys.*, **100**, 084907.

47 del Campo, A., Greiner, C., Álvarez, I., and Arzt, E. (2007) Patterned surfaces with pillars with controlled 3D tip geometry mimicking bioattachment devices. *Adv. Mater.*, **19**, 1973–1977.

48 Greiner, C., del Campo, A., and Arzt, E. (2007) Adhesion of bioinspired micropatterned surfaces: Effects of pillar radius, aspect ratio, and preload. *Langmuir*, **23**, 3495–3502.

49 del Campo, A., Greiner, C., and Arzt, E. (2007) Contact shape controls adhesion of bioinspired fibrillar surfaces. *Langmuir*, **23**, 10235–10243.

50 Peng, C. and Pang, S.W. (2006) Hybrid mold reversal imprint for

three-dimensional and selective patterning. *J. Vac. Sci. Technol. B*, **6**, 2968–2972.

51 Kehagias, N., Reboud, V., Chansin, G., Zelsmann, M., Jeppesen, C., Schuster, C., Kubenz, M., Reuther, F., Gruetzner, G., and Torres, C.M.S. (2007) Reverse-contact UV nanoimprint lithography for multilayered structure fabrication. *Nanotechnology*, **18**, 175303–175306.

6
Direct-Write Assembly of 3D Polymeric Structures
Sara T. Parker and Jennifer A. Lewis

6.1
Introduction

The term "direct-write assembly" describes fabrication methods that employ a computer-controlled translation stage, which moves a pattern-generating device, for example, ink deposition nozzle, to create materials with controlled architecture and composition (Figure 6.1) [1–8]. Through careful control of ink composition, rheological behavior, and printing parameters, arbitrary 3D structures, including continuous solids, high aspect ratio (e.g., parallel walls) or self-supporting features that span gaps in an underlying layer(s), can be constructed. Of these, the latter structures offer the greatest challenge and interest, since they are not easily obtainable by other patterning techniques.

The development of inks that flow through fine deposition nozzles and then rapidly solidify is necessary for patterning filamentary features at the microscale. By carefully tuning both the ink viscosity and elastic properties, one can create suitable inks for direct-write assembly of complex 3D structures, such as microperiodic arrays. For the polymeric ink formulations highlighted here, solidification occurs either upon deposition into a coagulating reservoir or UV exposure (Figure 6.2). Typically, inks printed within a coagulating reservoir have viscosities of $\sim$1–10 Pa s over the shear rate range of interest ($\sim$ 40–400 s^{-1}). As formulated, these inks exhibit a liquid-like response with a shear elastic modulus on the order of 1 Pa. However, their elasticity rises dramatically ($\sim$ 4–5 orders of magnitude) upon printing within the coagulating reservoir. In this coagulated state, the ink is elastic enough to maintain its filamentary shape and span gaps across features patterned in the underlying layer(s). By contrast, photocurable inks must possess an inherent viscoelastic response due to the slower kinetics of the curing reaction. In this case, high molecular weight polymer chains are incorporated within the ink to induce a strongly shear thinning response. Typically, these inks possess a viscosity of $\sim$10^1–10^2 Pa s at the shear rates of $\sim$100–200 s^{-1} experienced during flow through the deposition nozzle. Once printed, the ink filaments have sufficient elasticity

---

Generating Micro- and Nanopatterns on Polymeric Materials. Edited by A. del Campo and E. Arzt
Copyright © 2011 WILEY-VCH Verlag GmbH & Co. KGaA, Weinheim
ISBN: 978-3-527-32508-5

Figure 6.1 Direct-write assembly stage and imaging camera.

($\sim 10^2$–10^4 Pa) to promote shape retention until the structure is strengthened by UV curing.

Representative examples of 3D lattices and radial arrays patterned by direct-write assembly are shown in Figure 6.3c and d. They include arbitrary shapes, as solid or porous walls, spanning (rod-like) filaments, and tight- or broad-angled features, revealing the flexibility of this fabrication approach. They possess features that are nearly two orders of magnitude smaller than those attained by other multilayer ink printing techniques [2, 9].

The following sections of this chapter describe the design of different inks for the direct-write assembly of 3D microperiodic architectures composed of filamentary arrays, their rheological properties, and potential applications. Future

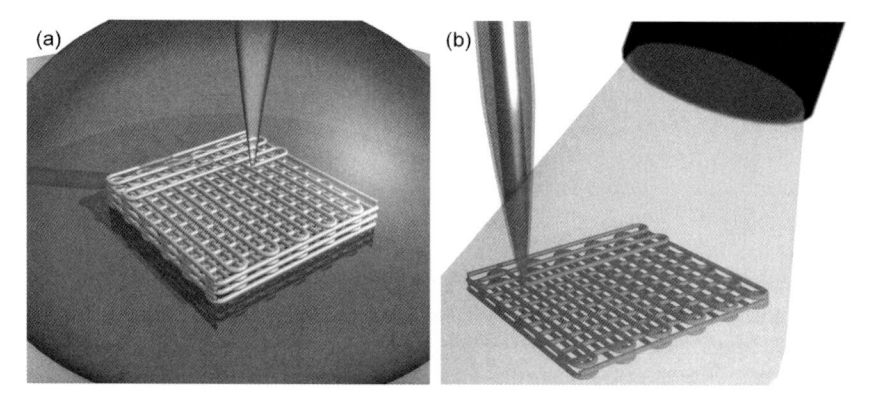

Figure 6.2 Schematic illustrations of direct-write assembly of 3D polymeric scaffolds deposited in (a) coagulating reservoir and (b) air and solidified *in situ* by UV curing. (Adapted with permission from [7] and [8]. Copyright (a) © 2008 and (b) © 2009 Wiley-VCH Verlag GmbH & Co. KGaA.)

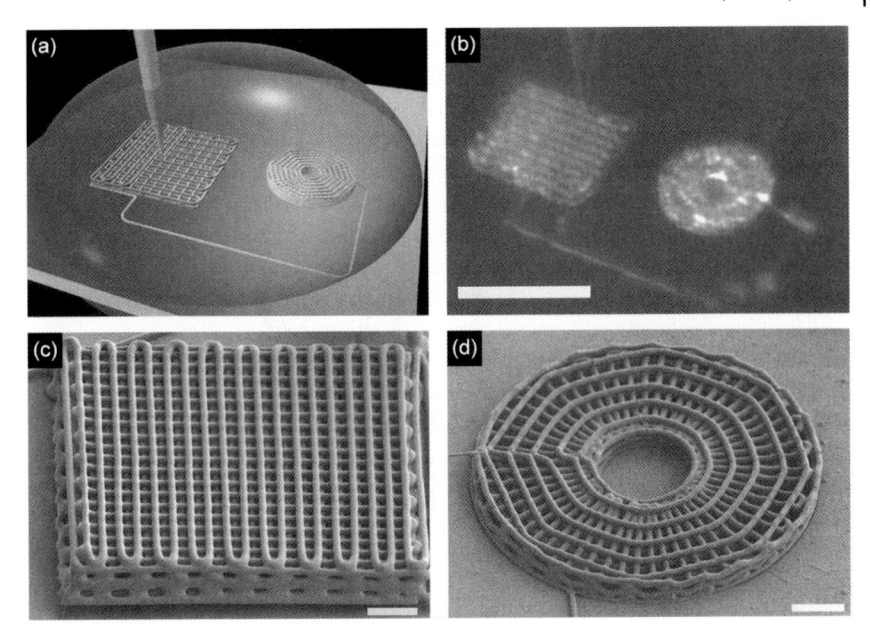

Figure 6.3 (a) Schematic illustration of the ink deposition process; (b) optical image acquired *in situ* during deposition reveals the actual features illustrated in (a) including the deposition nozzle that is currently patterning a 3D periodic lattice, and an image of a completed 3D radial array alongside this structure. This image is blurred, because the imaged features reside within the coagulation reservoir; (c) 3D periodic lattice with a simple tetragonal geometry (8 layers, 1 μm rod diameter); (d) 3D radial array (5 layers, 1 μm rod diameter). Scale bars for (b–d) are 100 μm, 10 μm, and 10 μm, respectively [2].

opportunities and current challenges of this novel patterning approach are also highlighted.

6.2
Polyelectrolyte Inks

6.2.1
Ink Design and Rheology

Nature provides the only example of "direct ink writing" of complex structures with micron-sized, self-supported features in the form of spider webs. Spiders create their webs by depositing a concentrated protein solution referred to as "spinning dope" [10]. This low-viscosity fluid undergoes solidification as it flows through a fine-scale spinneret to form silk filaments [10]. Depending on the type of spider, filament diameters ranging from the 10 nm to 10 μm range have been observed [10]. Even though this process is difficult to mimic, spinning dope has recently been harvested from spiders [11] or grown by bacteria [12] and formed into silk filaments via deposition into a coagulation reservoir.

Inspired by these natural and regenerative processes, concentrated solutions of polyelectrolyte complexes have been formulated. While less structurally complicated than spider silk, these systems are specifically tailored for direct-write assembly of 3D microperiodic scaffolds. They are composed of nonstoichiometric mixtures of polyanions and polycations [13]. A broad range of compositions have been demonstrated, including those based on a concentrated mixture of poly(acrylic acid), PAA ($M_w = 10\,000\,\text{g mol}^{-1}$), and poly(ethylenimine), PEI ($M_w = 600\,\text{g mol}^{-1}$), which is nominally 40 wt% polyelectrolyte in an aqueous solution. By regulating the ratio of anionic (COONa) to cationic (NH_x) ionizable groups and combining these species under solution conditions that promote polyelectrolyte exchange reactions [14], homogeneous fluids can be prepared over a broad compositional range. These solutions possess the requisite viscosities (see Figure 6.4a) needed to facilitate their flow through microcapillary nozzles of varying diameter ($D = 0.5$ to $5.0\,\mu m$).

The concentrated polyelectrolyte inks rapidly coagulate to yield self-supporting filaments (or rods) upon deposition into an alcohol/water reservoir (Figure 6.3a and b). The exact coagulation mechanism, driven by electrostatics in water-rich or solvent quality effects in alcohol-rich reservoirs, as well as the magnitude of ink elasticity depends strongly on reservoir composition [15]. As shown in Figure 6.4b, the PAA/PEI ink ([COONa]/[NH_x] = 5.7) exhibits a dramatic rise in ink elasticity from

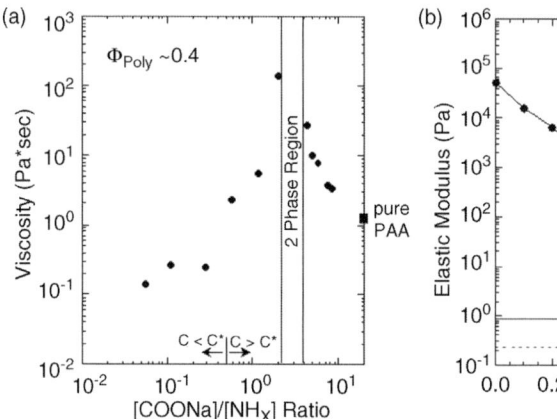

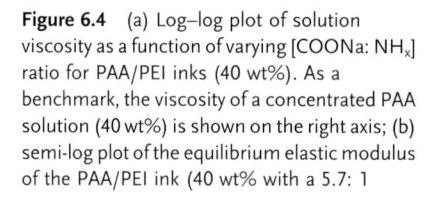

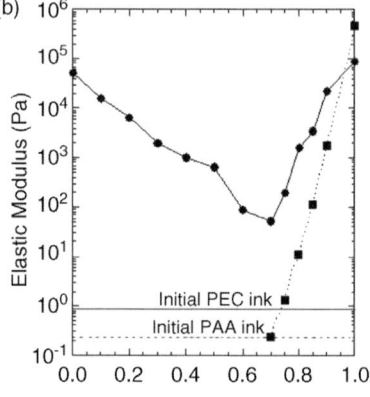

Figure 6.4 (a) Log–log plot of solution viscosity as a function of varying [COONa: NH_x] ratio for PAA/PEI inks (40 wt%). As a benchmark, the viscosity of a concentrated PAA solution (40 wt%) is shown on the right axis; (b) semi-log plot of the equilibrium elastic modulus of the PAA/PEI ink (40 wt% with a 5.7: 1 [COONa: NH_x] ratio) coagulated in reservoirs of varying composition for PAA/PEI ink (circles) and a pure PAA ink (40 wt%) (squares). (Reproduced with permission from [15]. Copyright © 2005 American Chemical Society.)

1 Pa (fluid phase) to nearly 10^5 Pa (coagulated phase) in a reservoir containing 83–88% isopropyl alcohol. Under these conditions, the deposited ink filament is elastic enough to promote shape retention, yet maintains sufficient flexibility for continuous flow and adherence to the substrate and underlying patterned layers. This ink design can be extended to other polyelectrolyte mixtures [2], including those based on biologically, electrically, or optically active species.

6.2.2
Polyelectrolyte Structures as Templates for Photonic and Biomimetic Applications

Since the concept of a photonic band gap (PBG) [16, 17] was first introduced, there has been intense interest in generating 3D microperiodic structures composed of alternating high and low refractive index materials [18]. Ideal materials for photonic crystals, such as silicon (Si, $n \sim 3.45$) and germanium (Ge, $n \sim 4.1$), have high refractive indexes and are optically transparent in the infrared region. The woodpile structure [19, 20], which consists of a 3D array of orthogonally stacked rods with center-to-center spacings on the order of the target wavelength, is particularly well suited for direct-write assembly [21] and has been obtained via ink writing with polyelectrolyte ink in a layer-by-layer build sequence (Figure 6.3c and d) [1]. However, the resulting polymeric structures lack the high refractive index contrast and mechanical integrity required for many applications. To overcome this limitation, the polymer woodpile structure is transformed to silicon hollow-woodpile structures via a sequential silica [22]/Si [23] chemical vapor deposition (CVD) process [13], and to germanium (Ge) inverse woodpile structures via a multistep atomic layer deposition (ALD) and CVD process [6]. Due to the high refractive index of these templated materials, the final structures exhibit increased reflectivity compared with the original polymer templates (Figure 6.5). In addition to photonic applications, these structures may serve as a starting point for other potential applications, including battery electrodes, microfluidic networks, low-cost MEMS, and other devices, where their hollow, thin-walled nature may be ideal for mass- and heat-transfer or reduced inertial forces.

In a different approach, 3D scaffolds obtained by direct-write assembly with polyamine-rich inks serve as templates for 3D hybrid silica–organic structures. Polymer scaffolds are fabricated, which closely mimic natural diatom shapes (Figure 6.6a and b), such as the *Arachnoidiscus ehrenbergii*. We explored the silicification of these scaffolds using either a single [24, 25] or sequential [26] immersion of the scaffold into a phosphate-buffered, silicic acid solution (Figure 6.6c). Silica condensation occurs uniformly, leading to scaffolds with enhanced thermal and mechanical stability [3]. Their original 3D shape is preserved with high precision, even when the organic phase is fully removed by heating them to 1000 °C (see Figure 6.6d–f). This approach can be further extended to incorporate natural or synthetic polypeptides into the polymer scaffold that allow a broader range of biomineralization strategies to be pursued, which, when coupled with the ability to pattern 3D structures of arbitrary shape and periodicity, may open up new avenues for soft-to-hard matter conversion under ambient conditions.

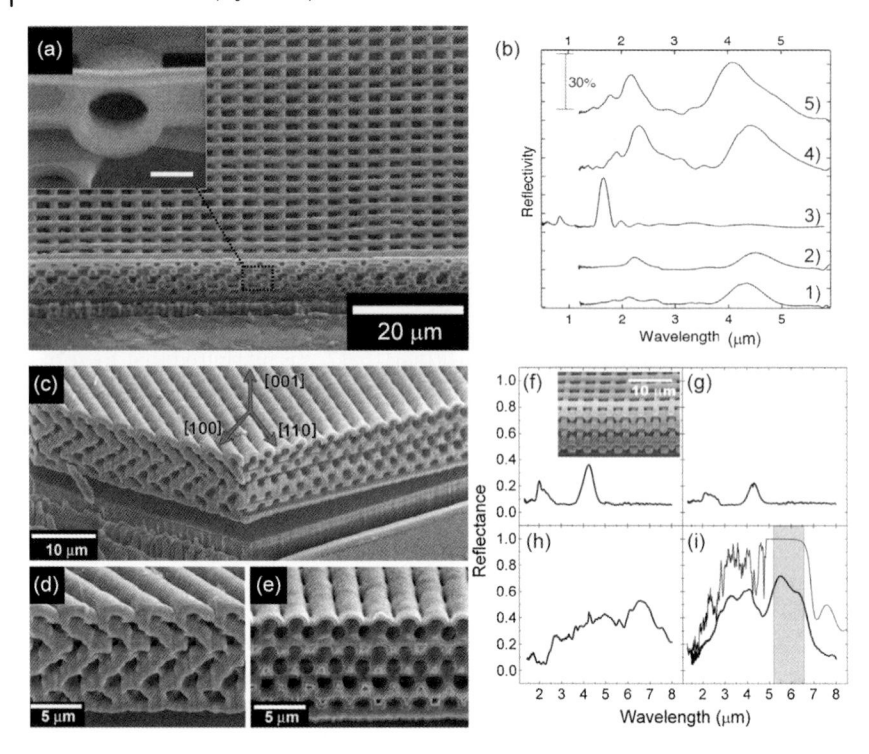

Figure 6.5 (a) SEM image of a Si/SiO$_2$/Si hollow-woodpile structure after focused ion beam milling (inset corresponds to higher magnification view of the interconnected hollow rods); (b) reflectance spectra for a 3D microperiodic structure (8 layers, 2.8 μm in-plane pitch, 1 μm rod diameter). (1) Polymer woodpile, (2) SiO$_2$/polymer woodpile, (3) SiO$_2$ hollow-woodpile, (4) Si/SiO$_2$/Si hollow-woodpile, (5) Si hollow-woodpile after HF etch; (c) SEM image of inverse Ge woodpile structure showing the surface of the samples with two edges exposed by FIB milling, with their orientations indicated; (d), (e) SEM higher resolution images of exposed (100) and (110) planes, respectively; (f–i) Reflectance spectra at different stages of fabrication: (f) Initial polymer woodpile structure (inset shows SEM cross section); (g) oxide-coated polymer woodpile structure; (h) Ge/oxide coated, polymer woodpile structure; (i) experimental (thick line) and simulated (thin line) Ge inverse woodpile structure. The lightly shaded region depicts the photonic band gap. (Adapted with permission from [4] and [6]. Copyright © 2006 and © 2007 Wiley-VCH Verlag GmbH & Co. KGaA.)

6.3
Silk Fibroin Inks

6.3.1
Ink Design and Rheology

Silkworms, such as the *Bombyx mori*, produce silk fibroin that can be readily transformed into a filamentary form [27]. Regenerated silk fibroin, obtained from *Bombyx mori* cocoons, is soluble in aqueous solutions. However, this ink possesses a limited shelf life and a narrow compositional range over which its rheological properties are suitable for filamentary printing. To achieve the desired rheology

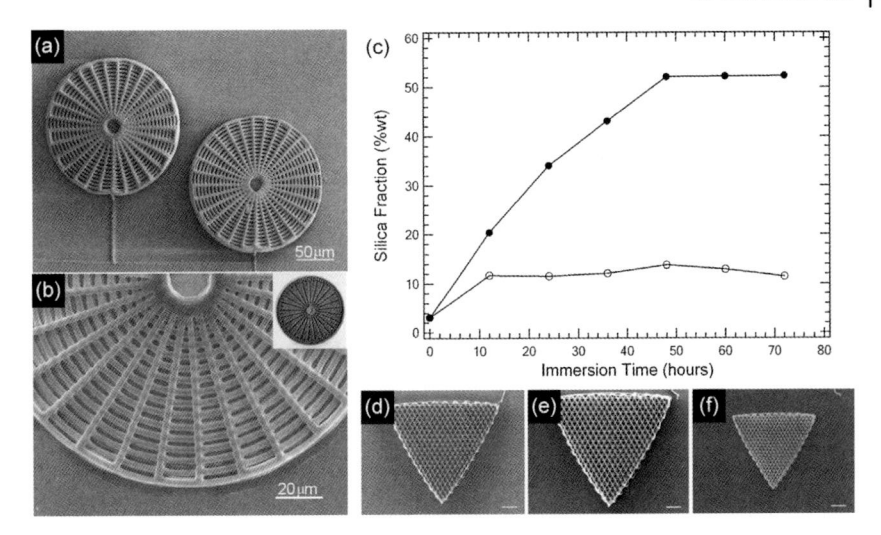

Figure 6.6 (a) SEM image of web-based polyamine rich scaffold fabricated by direct write assembly and (b) corresponding higher magnification view. Inset in (b) is natural diatom species *Arachnoidiscus ehrenbergii*; (c) the silica content (wt% of total mass) as a function of immersion time for both the single (open symbols) and sequential immersion methods (filled symbols). SEM images of triangular-shaped polyelectrolyte scaffolds showing their structural evolution; (d) as-patterned scaffold; (e) partially cross-linked scaffold after silicification by the sequential immersion method, and (f) silica scaffold after heat treatment at 1000 °C. Scale bars for (d–f) are 20 μm [3].

needed for filamentary printing, a regenerated silk fibroin solution is concentrated to ∼30 wt%. Above this value, the silk solution undergoes gelation and cannot be printed through micronozzles. The apparent viscosity of the optimal silk fibroin ink (∼29 wt%) is ∼2.9 Pa s (Figure 6.7a), which is similar to the viscosity of aforementioned polyelectrolyte inks [15]. Additionally, oscillatory measurements reveal a liquid-like response, where G'' exceeds G' over the experimental range probed (Figure 6.7b).

Due to the fluid nature of these silk inks, they must be coagulated in a reservoir to retain their filamentary shape during printing. We have shown that silk fibroin inks (∼29–30 wt%) flow readily through a 5 μm nozzle and retain their fiber-like morphology upon deposition into a methanol-rich reservoir. Methanol is a poor solvent for silk fibroin and thus likely induces aggregation (dehydration) that drives a structural transition from random coil to β-sheet [28, 29]. However, despite their rapid solidification, they also adhere well to underlying filaments [7].

6.3.2
Silk Scaffolds for Tissue Engineering Applications

The ability to pattern 3D biocompatible scaffolds may lead to new fundamental understanding of tissue development and remodeling as well as the formation of multi-cellular aggregates in complex micro-environments that better mimic the targeted tissues of interest. Using direct-write assembly, 3D silk scaffolds for tissue

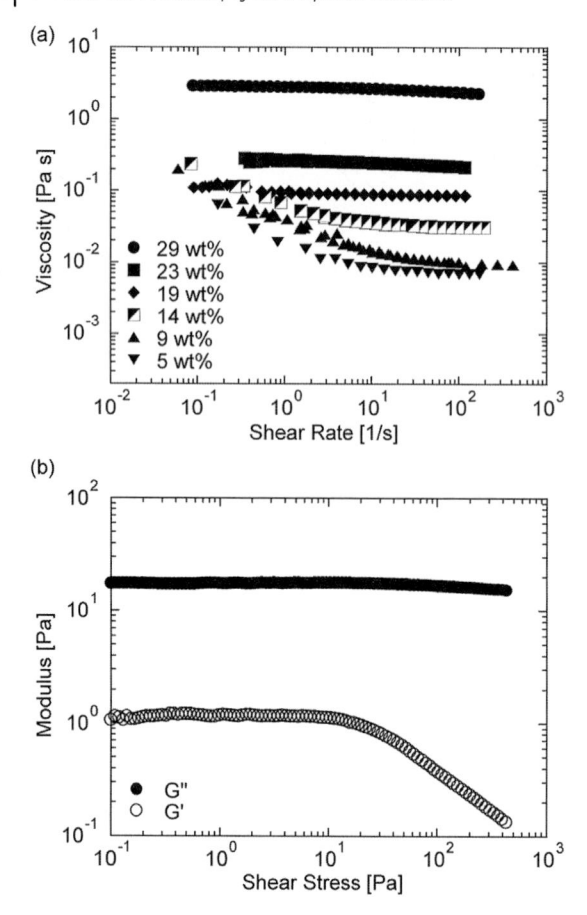

Figure 6.7 (a) Log–log plot of viscosity as a function of shear rate for silk fibroin solutions with varying concentrations from 5-29 wt%; (b) Log-log plot of shear elastic (G′) and viscous (G″) moduli as a function of shear rate for 29 wt% silk fibroin ink. (Reproduced with permission from [7]. Copyright © 2008 Wiley-VCH Verlag GmbH & Co. KGaA.)

engineering have been produced (Figure 6.8a–c). These scaffolds consist of a 3D microperiodic array of fibers approximately 5 μm in diameter with a center-to-center spacing in the x–y plane of 100 μm. They possess a large, interconnected porous network for cell growth and aggregation. Surface morphology at the intersection of printed fibers is shown in Figure 6.8c. The silk fibroin scaffolds have been cultured with human bone marrow-derived mesenchymal stem cells (hMSCs) in chondrogenic media and have proved to support the generation of 3D cartilaginous tissues (Figure 6.8d) [7].

By comparison with other methods for preparing 3D tissue engineering scaffolds from degradable polymeric systems, our approach confers important benefits such as finer fiber diameter, avoidance of harsh processing conditions of temperature or toxic organic solvents, and the ability to precisely control complex architectures. Direct-write assembly offers a unique path forward in support of biomaterial scaffolding for

Figure 6.8 Representative silk fibroin 3D structures of (a) square lattice and (b) circular web; (c) magnified image of direct write silk fiber. (d) 3D silk scaffold cultured with human bone marrow derived stem cells in chondrogenic media after 7 days *in vitro*. Cytoskeletal proteins actin and vinculin are stained red and green, respectively. (Adapted with permission from [7]. Copyright © 2008 Wiley-VCH Verlag GmbH & Co. KGaA.)

human disease models, complex tissue interfaces, and tissue gradients, all areas in need of new scaffold options where aqueous methods can be used to impart further functionalization.

6.4 Hydrogel Inks

6.4.1 Ink Design and Rheology

Direct-write assembly has also been extended to other polymer systems, including those based on photocurable chemistries. When printing photopolymerizable

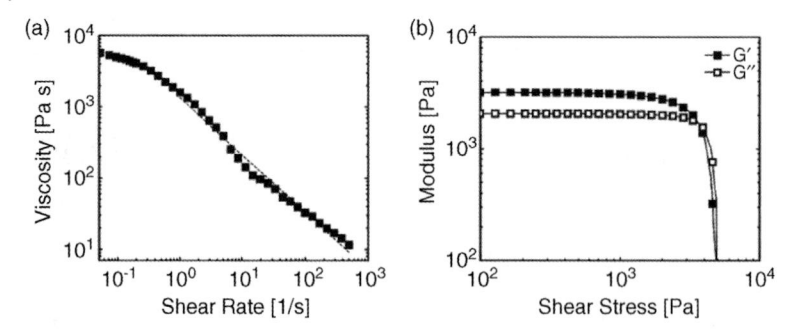

Figure 6.9 (a) Apparent viscosity as a function of shear rate for polyacrylamide ink; (b) shear elastic and viscous moduli as a function of shear stress for polyacrylamide ink. (Adapted with permission from [8]. Copyright © 2009 Wiley-VCH Verlag GmbH & Co. KGaA.)

hydrogels, the patterned ink filaments are cured *in situ* by UV exposure to obtain self-supporting features within the scaffolds. In this technique, the concentrated ink consists primarily of high molecular weight chains of polyacrylamide, as well as acrylamide (monomer), crosslinking agent, photo-initiator, glycerol, and water [8]. The high molecular weight chains impart the visco-elastic response needed to retain the desired filamentary form after ink deposition.

This polyacrylamide ink exhibits excellent printability due to its strong shear thinning behavior (Figure 6.9a). The ink viscosity decreases from 6500 Pa s at $0.1\,s^{-1}$ to 10 Pa s at $100\,s^{-1}$, which is comparable to the maximum shear rate experienced during deposition through a 5 μm nozzle at a printing speed of $0.5\,mm\,s^{-1}$. The high molecular weight chains also enhance ink elasticity, yielding a suitable shear elastic modulus of $\sim 3 \times 10^3$ Pa (Figure 6.9b). Even though physical entanglements between the polymer chains provide enough elasticity to facilitate assembly, the formation of a chemically crosslinked network via UV curing is necessary to prevent deformation of the patterned structures during the printing process.

6.4.2
Direct-Write Assembly of 3D Hydrogel Scaffolds

During printing, the hydrogel-based filaments are crosslinked *in situ* via exposure to UV light as they exit a metallic-coated, microcapillary nozzle (see Figure 6.2b). SEM micrographs of four-layer structures fabricated by this approach are shown in Figure 6.10a–c. The scaffold in Figure 6.10a and b is composed of 1 μm diameter filaments, whose center-to-center spacing is 5 μm. The image in Figure 6.10a illustrates that large-area patterns can be readily printed, while Figure 6.10b highlights the ability of individual filaments to span long distances that are unsupported. Figure 6.10c highlights a 3D hydrogel scaffold with filament diameters of 5 μm and spacing of 20 μm, feature sizes that are used for cell culture. These scaffolds are cultured with 3T3 murine fibroblast cells, which remain viable and integrate themselves into the porous "compartments" within the scaffolds (Figure 6.10d). Although the cells reside in distinct locations, adjacent fibroblast cells can easily

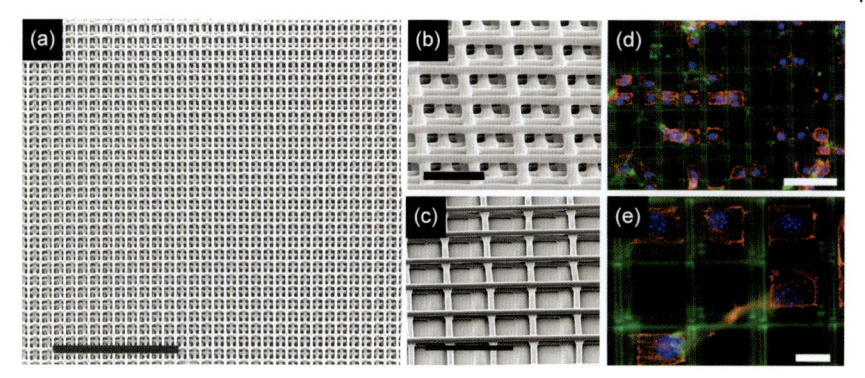

Figure 6.10 (a) SEM micrograph of 6 layer polyacrylamide hydrogel scaffold patterned using a 1 µm diameter nozzle with a filament pitch of 5 µm; (b) a higher magnification tilted view of scaffold in (a) demonstrates the filament spanning capabilities; (c) hydrogel scaffold printed with 5 µm diameter nozzle and a 20 µm pitch for cell culture; (d) optical fluorescence micrograph of 3T3 fibroblast incorporation into polyacrylamide hydrogel scaffold. In (e), a higher magnification image, cellular interaction between adjacent scaffold "compartments" is highlighted. In these images, rhodamine-phalloidin stains the actin red, DAPI stains the (DNA) nucleus blue, and fluorescein-O-acrylate stains the scaffolds green. Scale bars are 50 µm, 6 µm, 50 µm, 100 µm, 20 µm respectively. (Adapted with permission from [8]. Copyright © 2009 Wiley-VCH Verlag GmbH & Co. KGaA.)

communicate with one another, as revealed by filament interactions through the porous side walls shown in Figure 6.10e.

This photocurable ink design can be extended to many polymeric systems, including poly(2-hydroxyethyl methacrylate), a well-characterized, biocompatible hydrogel material [30]. We therefore envision many opportunities to further tailor the scaffold chemistry, mechanical properties, and structure to improve cell viability on these novel constructs.

6.5
Opportunities and Challenges

Direct-write assembly offers the ability to rapidly pattern soft materials in 3D functional architectures at the microscale. This low-cost, versatile approach enables one to create nearly arbitrary structures from a diverse array of materials that are inaccessible by traditional fabrication techniques. Several polymeric inks have recently been developed, whose solidification is induced via coagulation in a poor solvent reservoir [2, 3, 7, 15] or photopolymerization [8] during the printing process. Future advances will hinge on the development of new ink compositions, improved characterization and modeling of ink dynamics during deposition, and enhanced robotic control and ink delivery systems to allow three-dimensional writing with greater precision, speed, and local compositional specificity. For example, the development of efficient mixing strategies within the deposition nozzle that allow both composition and structure to be locally tailored during the assembly process would provide even greater control over the properties and functionality of the resulting patterned materials. Finally, to move from prototyping to large-scale production, more sophisticated printheads must be implemented, such as those based on microfluidic devices and/or multi-nozzle arrays, to facilitate the simultaneous creation of several components from a single printing platform. Given the rapid development to date, direct-write assembly of polymeric materials appears poised to deliver next generation designer materials for a wide range of technological applications.

Acknowledgments

We gratefully acknowledge the generous support for our work by the U.S. Department of Energy through the Frederick Seitz Materials Research Laboratory (Grant No. DEFG02-07ER46471) for polyelectrolyte inks/photonic templates, the Air Force Office of Scientific Research under Award No. FA9550-05-1-0092 (Subaward No. E-18-C45-G1) for biomimetic structures/mineralization, and the NSF Center for Nanoscale Chemical-Electrical-Mechanical Manufacturing Systems (DMI-0328162) for the hydrogel inks/scaffolds. We thank G. Gratson, R. Shepherd, M. Xu, F. Garcia-Santamaria, R. Barry, S. Ghosh, J. Hanson, R. Nuzzo, P. Braun, D. Kaplan, and P. Wiltzius for their valuable contributions to this research.

References

1 Lewis, J.A. and Gratson, G.M. (2004) *Mater. Today*, **7**, 32.
2 Gratson, G.M., Xu, M.J., and Lewis, J.A. (2004) *Nature*, **428**, 386.
3 Xu, M.J., Gratson, G.M., Duoss, E.B., Shepherd, R.F., and Lewis, J.A. (2006) *Soft Matter*, **2**, 205.
4 Gratson, G.M., Garcia-Santamaria, F., Lousse, V., Xu, M.J., Fan, S.H., Lewis, J.A., and Braun, P.V. (2006) *Adv. Mater.*, **18**, 461.
5 Lewis, J.A. (2006) *Adv. Funct. Mater.*, **16**, 2193.
6 Garcia-Santamaria, F., Xu, M.J., Lousse, V., Fan, S.H., Braun, P.V., and Lewis, J.A. (2007) *Adv. Mater.*, **19**, 1567.
7 Ghosh, S., Parker, S.T., Wang, X.Y., Kaplan, D.L., and Lewis, J.A. (2008) *Adv. Funct. Mater.*, **18**, 1883.
8 Barry, R.A., Shepherd, R.F., Hanson, J.N., Nuzzo, R.G., Wiltzius, P., and Lewis, J.A. (2009) *Adv. Mater.*, **21**, (23), 2407–2410.
9 Chrisey, D.B. (2000) *Science*, **289**, 879.
10 Vollrath, F. and Knight, D.P. (2001) *Nature*, **410**, 541.
11 Seidel, A., Liivak, O., and Jelinski, L.W. (1998) *Macromolecules*, **31**, 6733.
12 Lazaris, A., Arcidiacono, S., Huang, Y., Zhou, J.F., Duguay, F., Chretien, N., Welsh, E.A., Soares, J.W., and Karatzas, C.N. (2002) *Science*, **295**, 472.
13 Zezin, A.B. and Kabanov, V.A. (1982) *Russ. Chem. Rev.*, **51**, 833.
14 Zintchenko, A., Rother, G., and Dautzenberg, H. (2003) *Langmuir*, **19**, 2507.
15 Gratson, G.M. and Lewis, J.A. (2005) *Langmuir*, **21**, 457.
16 Yablonovitch, E. (1987) *Phys. Rev. Lett.*, **58**, 2059.
17 John, S. (1987) *Phys. Rev. Lett.*, **58**, 2486.
18 Joannopoulos, J.D., Villeneuve, P.R., and Fan, S.H. (1997) *Nature*, **386**, 143.

19 Ho, K.M., Chan, C.T., Soukoulis, C.M., Biswas, R., and Sigalas, M. (1994) *Solid State Commun.*, **89**, 413.
20 Sozuer, H.S. and Dowling, J.P. (1994) *J. Mod. Opt.*, **41**, 231.
21 Deubel, M., Von Freymann, G., Wegener, M., Pereira, S., Busch, K., and Soukoulis, C.M. (2004) *Nature Mater.*, **3**, 444.
22 Miguez, H., Tetreault, N., Hatton, B., Yang, S.M., Perovic, D., and Ozin, G.A. (2002) *Chem. Commun.*, 2736.
23 Blanco, A., Chomski, E., Grabtchak, S., Ibisate, M., John, S., Leonard, S.W., Lopez, C., Meseguer, F., Miguez, H., Mondia, J.P., Ozin, G.A., Toader, O., and van Driel, H.M. (2000) *Nature*, **405**, 437.
24 Menzel, H., Horstmann, S., Behrens, P., Barnreuther, B., Krueger, I., and Jahns, M. (2003) *Chem. Commun.*, 2994.
25 Pohnert, G. (2002) *Angew. Chem.-Int. Edit.*, **41**, 3167.
26 Harris, J.J., DeRose, P.M., and Bruening, M.L. (1999) *J. Am. Chem. Soc.*, **121**, 1978.
27 Jin, H.J. and Kaplan, D.L. (2003) *Nature*, **424**, 1057.
28 Nam, J. and Park, Y.H. (2001) *J. Appl. Polym. Sci.*, **81**, 3008.
29 Marolt, D., Augst, A., Freed, L.E., Vepari, C., Fajardo, R., Patel, N., Gray, M., Farley, M., Kaplan, D., and Vunjak-Novakovic, G. (2006) *Biomaterials*, **27**, 6138.
30 Wichterle, O. and Li'm, D. (1960) *Nature*, **185**, 117.

Part Three
Laser Scanning

7
Three-Dimensional Microfabrication by Two-Photon Polymerization

Tommaso Baldacchini

7.1
Introduction

Laser direct writing (LDW) exploits the interaction between a laser and a polymer substrate to create different effects: voids by means of ablation, material state modifications due to thermal cycles, and addition of diverse materials by means of chemical reactions or induced forward transfer procedures [1]. In all cases, LDW creates patterns in a "spot-by-spot" fashion where the target substrate is moved around a fixed laser beam and/or the laser beam is scanned following predetermined outlines. Although this serial approach to creating parts is slow compared with the more traditional one employed in photolithography where microfabrication is run in parallel [2], several characteristics of LDW render it attractive and competitive in many applications. Because of the laser high spatial coherence, patterns with resolutions that extend over four orders of magnitude from millimeters to nanometers can be fabricated. Moreover, by choosing the appropriate laser and experimental conditions, virtually any material can be processed, and finally, LDW processes make both surface and volume micromachining possible.

Both continuous wave and pulsed lasers have been successfully employed in LDW. Those most used thus far are infrared CO_2 gas lasers, ultraviolet excimer gas lasers, and visible solid state lasers. When interacting with these kinds of lasers, materials absorb light in a linear fashion and heating effects play an important role in the quality of the created patterns. If ultrashort (less than a picosecond) pulsed lasers are used for LDW, energy is deposited in the material in a time range shorter than the electron–phonon coupling time (a few picoseconds) rendering thermal effects negligible [3]. Moreover, the likelihood of inducing multiphoton absorption when using ultrashort pulsed lasers is such that the matter/light interaction can be spatially confined to sub-femtoliter volumes enabling high precision and high resolution patterning.

This chapter will focus on a particular LDW process that employs ultrashort pulsed lasers and photopolymerizable precursors: two-photon polymerization (TPP). Also known as two-photon induced photopolymerization and two-photon

Generating Micro- and Nanopatterns on Polymeric Materials. Edited by A. del Campo and E. Arzt
Copyright © 2011 WILEY-VCH Verlag GmbH & Co. KGaA, Weinheim
ISBN: 978-3-527-32508-5

stereolithography, TPP is a serial process for the fabrication of three-dimensional (3D) microstructures [4–17]. TPP is a unique tool in the arsenal of unconventional microfabrication methods now available to scientists because of its ability to produce geometries with no topological constraints with a resolution smaller that 100 nm. Although the first report in the scientific literature of photopolymerization induced by a multiphoton absorption process appeared more than 40 years ago [18], it was a pair of publications in 1990 that proposed and demonstrated the possibility of patterning by TPP [19, 20]. In 1997, a publication by Maruo and coworkers established TPP as a technique for the fabrication of 3D microstructures [21]. Since then, the number of publications based on TPP has grown rapidly, allowing researchers in fields diverse as photonics, microelectronics, and bioengineering to create and use devices that would previously have been impossible to manufacture with conventional microfabrication procedures.

7.2
Fundamentals

A molecule in a ground electronic state can be promoted to an excited vibronic state if it absorbs a photon with energy matching the energy gap between the two states. The same process can occur by quasi-simultaneous absorption of two lower energy photons via short-lived virtual states that have a lifetime of around 10^{-15} s, provided that the energy sum of the two photons is enough to reach the first excited electronic state [22]. Then, the excited molecule can relax by following different energetic pathways depending on its molecular structure and environment. In the case of a fluorophore for example, the excited molecule will first relax to the lower vibrational state and then return to the ground electronic state emitting a lower energy photon than the one absorbed.

Two-photon excitation can occur either by using two photons of different energies or two photons having the same energy. In the first case the process is called nondegenerate while in the second case it is called degenerate. Because of the simplicity of the experimental setup for degenerate two-photon excitation, most applications including TPP have employed this method.

Although Maria Göppert-Mayer predicted two-photon excitations in 1931 [23], the first experimental observation was not made until 1961 when Kaiser and Garret reported two-photon excitation fluorescence from CaF_2: Eu^{2+} crystals [24]. The reason for the long delay between the two events is that two-photon excitation is a rare event and in order to increase the probability of its occurrence intense, artificial sources of light must be used. For instance, if we take a molecule that is a good one- and two-photon absorber such as Rhodamine B and we measure the frequency of excitation events when exposed to sunlight, we will observe excitation by a one-photon process once per second, while excitation by a two-photon process will occur only once every 10 million years [25]. It is not surprising then that Kaiser and Garret's successful study occurred only after the invention of the laser, a source of light capable of providing sufficient photon densities for two-photon excitation [26].

Although two-photon excitation can be induced by using continuous wave lasers, pulsed lasers are preferable because they can achieve the same excitation efficiencies with much lower average powers due to their higher peak intensities. Specifically, the power of a continuous wave laser has to be $(\tau f)^{-1/2}$ higher than the average power of a pulsed laser, where τ and f are the pulse width and the repetition rate of the latter, respectively [27]. Currently, the most popular source of light for inducing two-photon excitation are Ti: sapphire based lasers [28]. They emit at wavelengths in the near-infrared (NIR) region of the spectrum and produce pulses shorter than 100 fs. Each pulse has a peak power on the order of a kilowatt, while at a typical repetition rate of 80 MHz the average power is on the order of milliwatts. Thus, a simple calculation demonstrates the major advantage of using pulsed lasers over continuous wave lasers for two-photon excitation: the fluorescence intensity emitted by a fluorophore will be equal, whether a 10 W of continuous wave laser power or 30 mW of average power from a typical, commercial femtosecond pulsed laser is employed.

Focusing is as important as the use of an intense source of light for inducing two-photon excitation. In order to produce high photon densities and hence increase the probability of two-photon excitation taking place, high numerical aperture objective lenses are used to focus the laser beam to a diffraction limited spot. Focusing the laser beam has a significant role in spatial confinement of two-photon excitation. Since the rate of absorption in two-photon excitation is proportional to the intensity squared of the excitation light, the probability of two-photon excitation is largest at the focus and diminishes rapidly in the regions in front of and behind it. Taking into consideration that the intensity of a focused laser beam scales down by the square of the distance from the focal plane, two-photon excitation falls off by the fourth power of the distance from the focal plane. This effect is displayed in Figure 7.1 where the spatial confinement of two-photon excitation is highlighted by the fluorescence emitted in a solution containing Rhodamine B when excited by a focused fs-pulsed laser.

The ability to confine light absorption to sub-femtoliter volumes by two-photon excitation is the key phenomenon that renders possible 3D microfabrication in TPP. If a photosensitive material (resin) can be solidified by two-photon excitation, then 3D microstructures with sub-micron resolution can be fabricated by means of accurate positioning of the laser focal point. As long as the solubility properties of the solidified and unsolidified resin are different, then unpolymerized material can be washed away to leave freestanding microstructures. A schematic diagram of the TPP process is shown in Figure 7.2a–c. First, the desired pattern is "written" by precise positioning of the laser focal point in the resin. Then, the unpolymerized remainder of the resin is washed away by means of an organic solvent, leaving only the newly fabricated microstructures on the substrate. Finally, after a drying step, the sample is tested with several imaging techniques, with scanning electron microscopy (SEM) being the most popular. As an example, the SEM image of a microstructure fabricated by TPP is shown in Figure 7.2d. The intricate topology of the microstructure highlights the fabrication capabilities of this laser direct-write technique.

Two-photon polymerization offers a unique combination of advantages. First, no topological constraints are present in the fabrication of a 3D microstructure. Second, sub-diffraction-limited resolution can be attained by employing laser intensities just

112 | *7 Three-Dimensional Microfabrication by Two-Photon Polymerization*

Figure 7.1 One-photon versus two-photon excitation fluorescence from a Rhodamine B solution in ethanol. The objective lens on the left focuses 405 nm from a continuous wave diode laser causing one-photon excitation in the dye solution. The objective lens on the right focuses 800 nm from a mode-locked Ti: sapphire laser causing two-photon excitation in the dye solution. Only when absorption occurs nonlinearly, is the excitation confined to the objective focal plane.

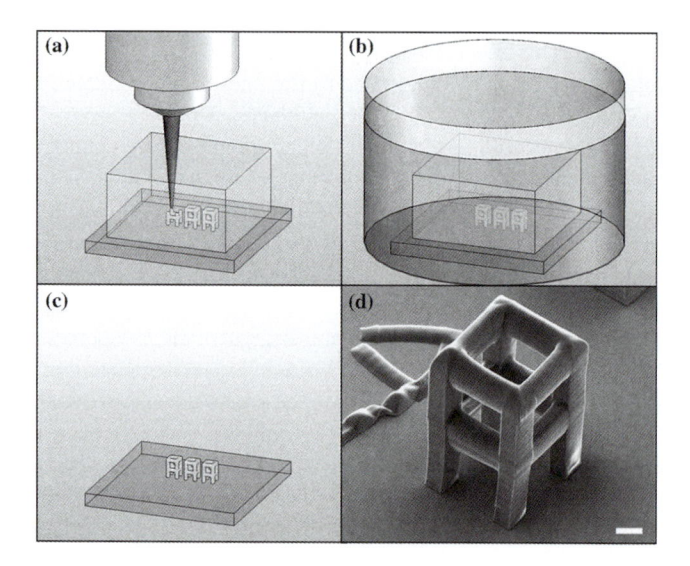

Figure 7.2 Free-standing 3D microstructures can be created by TPP. Schematic of the main steps involved in TPP: (a) patterning; (b) dissolving of unexposed resin; and (c) diagnosing of microstructures. The proportions of the objects depicted have been exaggerated for viewing clarity; (d) SEM image of a microstructure fabricated by TPP. The scale bar is 20 μm.

above the intensity threshold at which polymerization will occur. Third, moveable components can easily be fabricated without the use of sacrificial layers. And lastly, the carbon-based nature of the resins can be used as a chemical handle to fabricate microstructures with tunable physical and chemical properties such as hardness, shrinkage, refractive index, and chemical specificity.

7.3 Materials

Resins employed in photopolymerization processes consist of a mixture of photo-initiators, monomers and oligomers [29–31]. Photo-initiators absorb light and generate active species that start polymerization of monomers and oligomers. This results in a molecular network the solubility of which is different from the solubility of the starting materials. Other components are sometimes added to the resin mixture described above: solvents to aid in sample preparation, inhibitors to increase resolution, additives to introduce new functionalities to the final polymer, and fillers to increase the viscosity of the resin.

7.3.1 Photo-initiators

The initiation efficiency of TPP is governed by the characteristics of the photo-initiator. Photo-initiators with large two-photon cross-sections (δ) and large radical or cation quantum yields (Φ) are desirable to obtain resins with low intensity thresholds and large dynamic range of intensity. For a given resin, the intensity threshold is defined as the light intensity below which no polymerization is observed and the dynamic range is defined as the difference between the intensity threshold and the intensity of the excitation light above which uncontrollable polymerization occurs because of thermal effects. Low intensity thresholds enable rapid scanning speed during fabrication, an important benefit considering the serial nature of TPP processing. Large dynamic ranges allow dimensional control of the smallest volume element (voxel) that can be polymerized by TPP. Furthermore, the probability of disrupting microfabrication during TPP due to laser instability is reduced by using a resin with a large dynamic range.

Two-photon polymerization has been successfully applied to radical and cationic polymerization, although most published microstructures have been prepared through a radical-based polymerization mechanism. Examples of commercially available radical photo-initiators used in TPP are Irgacure 369, 184, and 819 (Ciba Specialty Chemicals Corp.), Lucirin TPO-L (BASF Inc.), BME (Sigma-Aldrich Corp.), H-Nu470 (Spectra Group Limited Inc.), Rose Bengal (Sigma-Aldrich Corp.), and ITX (Sigma-Aldrich Corp.) [7, 32–35].

Although commercially available photo-initiators have large Φ, they typically exhibit small δ since they have been optimized for single photon absorption only. For example, the photo-initiators described so far possess peak two-photon cross-sections on the order of a few GM (defined in SI units as $1\,\text{GM} = 10^{-58}\,\text{m}^4\,\text{s}$ photon^{-1} and where GM stands for the initials of the Nobel laureate physicist Göppert-Mayer) [36, 37]. For this reason, several research groups have synthesized and characterized rationally designed molecules with large δ which can be used as photo-initiators for TPP [38].

It has been shown that molecules having electron-donating (D) and electro-withdrawing (A) groups separated by π-conjugated systems have large δ values. An example of this class of molecules is 4,4′-bis(N,N-di-n-butylamino)-E-stilbene. Starting from this structural motif Marder and co-workers synthesized molecules with δ values as large as 4400 GM, three orders of magnitude greater than those of most commercially available photo-initiators [39–42]. When excited, these molecules go through a large electronic reorganization and become efficient reducing agents toward many substrates. The double bond of acrylic monomers is one such substrate, and upon electron transfer a radical is formed that initiates polymerization. Besides stilbene, other structural designs incorporating A and D groups have been synthesized possessing large δ such as biphenyl, bis(styryl)benzene, naphthalene, and fluorene [32, 43]. They all have been shown to be capable of initiating TPP at low laser intensities. An appealing feature of custom-made photo-initiators is the ability to shift the position of the two-photon cross-section maximum in the spectrum by small modifications in the molecular structure of the photo-initiator. Thus, photo-initiators can be tailored to be more efficient at specific excitation wavelengths.

Commercially available photo-initiators for cationic polymerization consist of triarylsulfonium and diaryliodionium salts such as CD1012 and CD1010 (Sartomer Inc.). Both kinds have been used successfully in TPP [32]. As for radical photo-initiators, custom-made cationic photo-initiators with large δ have been synthesized and used in TPP. By attaching groups containing sulfonium moieties to a bis(styryl) benzene molecule, Marder and coworkers have synthesized a photo-acid generator termed BSB-S$_2$ with δ value of approximately 700 GM [44]. Since oxygen cannot quench the cationic intermediates, cationic polymerization is insensitive to the presence of molecular oxygen. Therefore, cationic polymerization proceeds to a certain extent even after laser irradiation has stopped.

Molecular structures of representative commercial and custom-made radical and cationic photo-initiators are shown in Figure 7.3. Measured two-photon cross-section maxima and corresponding excitation wavelengths are included. Although custom-made photo-initiators possess several advantages relative to commercially available ones, the scarce accessibility of these molecules prohibits their widespread use. Moreover, the intensity thresholds required to induce TPP when using resins containing some of the commercial photo-initiators are low enough to make their use adequate.

7.3.2
Mixtures of Monomers and Oligomers

Mixtures of monomers and oligomers must possess two attributes in order to be practical for TPP. First, upon polymerization they should undergo as little shrinkage

(a) ~10 GM (640 nm)

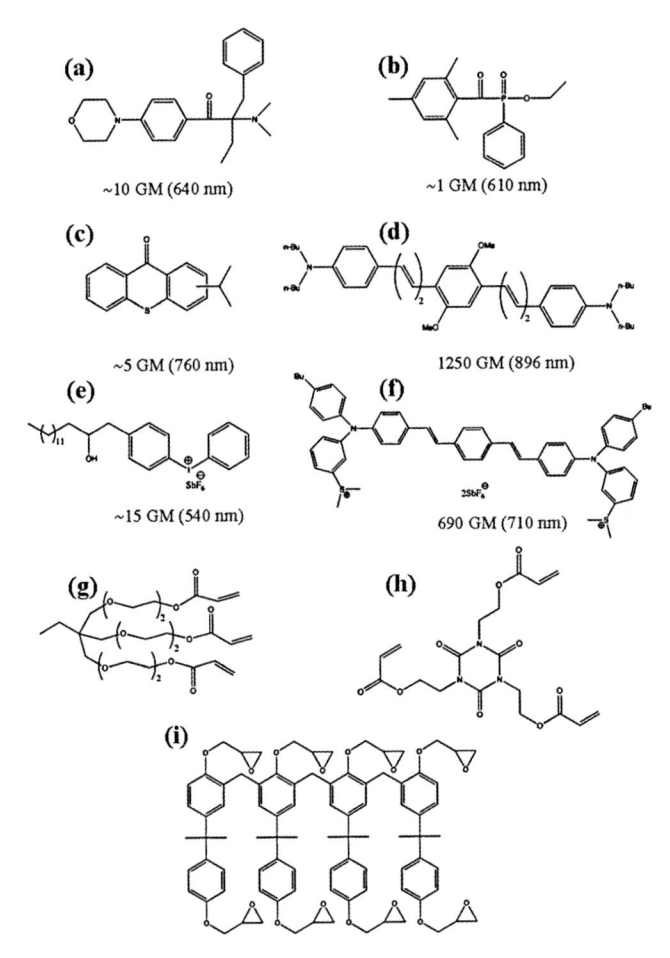

(b) ~1 GM (610 nm)

(c) ~5 GM (760 nm)

(d) 1250 GM (896 nm)

(e) ~15 GM (540 nm)

(f) 690 GM (710 nm)

(g)

(h)

(i)

Figure 7.3 Molecular structures of representative photo-initiators and monomers employed in TPP. Peak two-photon cross-sections and excitation wavelengths for the photo-initiators are also reported. While Irgacure 369 (a); Lucirin TPO-L (b); and ITX (c) are commercially available radical photo-initiators, the symmetric stilbene-like structure in (d) with a D-π-A-π-D motif is a custom-made radical photo-initiator. Two photo-initiators for cationic polymerization are also shown: commercially available CD1012 (e) and custom-made BSB-S$_2$ (f). Triacrylate monomers SR499 (g) and SR368 (h) are used in radical polymerization, and the epoxy-based monomer (i) is used for the cationic polymerization of SU-8.

as possible. Second, the resulting solid polymer must be hard enough to withstand the developing step without swelling. Only if these conditions are met can complex 3D microstructures be fabricated with the highest fidelity. Other characteristics that render mixtures of monomers and oligomers attractive for TPP are the ease with which they can be processed and their solubility in common solvents.

Most of the works reported in the scientific literature on TPP to date employ resins containing acrylic-based monomers and oligomers. By and large, this is because molecules with a wide variety of sizes and functionalities are readily available and relatively inexpensive in this class of materials. Moreover, since acrylates are extensively used in several industries for applications ranging from paints to car components, a large amount of information has been collected over time on their properties. By consuming a carbon–carbon double bond through a radical mechanism, acrylates form rapidly cross-linked polymers that are well-suited for TPP. Furthermore, among the monomers and oligomers that can be polymerized by either a radical or cationic mechanism, acrylates exhibit some of the highest reactivities.

Two-photon polymerization can be carried out by using commercially available pre-formulated acrylic-based resins such as SCR500 (Japan Synthetic Rubber Co.), a mixture of urethane acrylate monomers and oligomers [7]. Home-made mixtures, however, have the advantage of producing polymers with tailored physical and chemical properties [45]. For example, it was shown that microstructures with high mechanical strength, good optical quality, and low shrinkage were created by TPP using a combination of two triacrylate monomers (SR499 and SR368, Sartomer Inc.) and Lucirin TPO-L as photo-initiator [34]. While the cyclic structure of SR368 imparts rigidity to the polymer, the ethoxylated groups in SR499 offset the amount of shrinkage that occurs upon polymerization (molecular structures of the two monomers are shown in Figure 7.3). Other materials that have been successfully used for the fabrication of microstructures by TPP are hydrogels, silicon- and silicate-based resins. The last are particularly interesting since they combine the properties of both organic and inorganic materials [46–49].

Cationic polymerization has not reached the same extensive use as radical polymerization. The most common materials employed in cationic polymerization are epoxide monomers and oligomers. Two commercially available resins have been used for TPP, SU-8 (MicroChem Corp.) and SCR-701 (Japan Synthetic Rubber Co.) [50–53].

All the materials described up to this point are negative-tone resins in that the regions exposed to light harden, becoming insoluble in the post-exposure development step. In certain applications such as a microfluidics, it would be more appealing to use positive-tone resins where regions exposed to light can be washed away in the development step allowing for the formation of voids. Marder and co-workers have demonstrated the use of positive-tone resins in TPP [54]. The resin consisted of a random copolymer containing tetrahydropyranyl methacrylate units and of BSB-S_2 as photo-initiator. Upon excitation, the local increase of pH induces an ester-cleavage reaction in the tetrahydropyranyl moieties generating carboxylic acids. The newly formed polar groups make the volume in which they are present soluble in a basic aqueous solution, thus allowing for their removal.

7.4
Experimental Setup

Laser direct writing by TPP can be performed either by moving the laser beam inside the sample or by moving the sample around the fixed laser beam. In the first case, a set of Galvano-mirrors are used to scan the excitation beam in the x and y dimensions, while a piezo stage moves the sample or objective lens up and down. In the second case, the sample is moved in all dimensions with the aid of a three-axis stage. Although both techniques are effective, they have complementary advantages that have to be taken into consideration when choosing one system over the other. While processing times with scanning mirrors are shorter, the area they can pattern is limited by the microscope objective field of view. When using high magnification and numerical-aperture (NA) objectives (≥ 1.0) for example, the scanning range is restricted within an area only tens of microns wide before the effects of spherical aberration start to negatively affect the microfabrication resolution. On the other hand, when using a stationary laser beam, the total travel distance of the stage is the only boundary that determines the maximum area of the sample that can be patterned, resulting in processed surfaces many times larger than is possible with a set of scanning mirrors.

A schematic of a typical setup employed in TPP is shown in Figure 7.4a where microfabrication is carried out by translating the sample around a focused laser beam following predetermined patterns. A photograph of the TPP workstation used for the fabrication of some of the microstructures exhibited in this chapter is shown in Figure 7.4b. The source of light is a mode-locked Ti: sapphire laser producing pulses with duration of tens to hundreds of femtoseconds at a repetition rate on the order of 80 MHz. Even though average output power is largest at 800 nm, more than enough power to sustain TPP is emitted across the entire laser tuning range that stretches from less than 700 nm to greater than 1000 nm.

A half-wave plate (HWP) and a polarizer (GLP) are positioned after the oscillator and are used to variably attenuate the laser output power to the desired input power required by specific experiments. Using a beam sampler (M*), a small portion of the laser beam is directed into a beam diagnostic unit (AC). In it, the laser pulse is characterized both in the time and frequency domains by employing an autocorrelator and a spectrometer. The laser beam is then expanded to match or overfill the back aperture of the objective lens. This is accomplished using two positive lenses with the appropriate focal lengths. At the focal point of the first lens, a pinhole (SF) is carefully positioned to spatially filter the laser beam. An electro-mechanical shutter (S), used to control laser exposure times in the sample, is placed before this assembly. If exposure times shorter that a few milliseconds are required, faster response shutters such as acousto- or electro-optic modulators can be used.

A dichroic mirror (DM), which is reflective for NIR and transparent for visible radiation, deflects the laser beam into an objective lens (OL) that focuses the excitation light into the sample. The position of the sample in the objective field of view is adjusted by the aid of a three-axis computer-controlled stage (TS).

The coordinates of the desired microstructures, together with the velocities that have to be maintained, are fed to the stage assembly and shutter via a computer workstation that accurately synchronizes their activities through custom-made software. Three-dimensional microfabrication is executed by either one of the two following writing modes. In the first, a cubic volume is scanned layer-by-layer in the resin and the structure is formed inside the cube by opening and closing the shutter

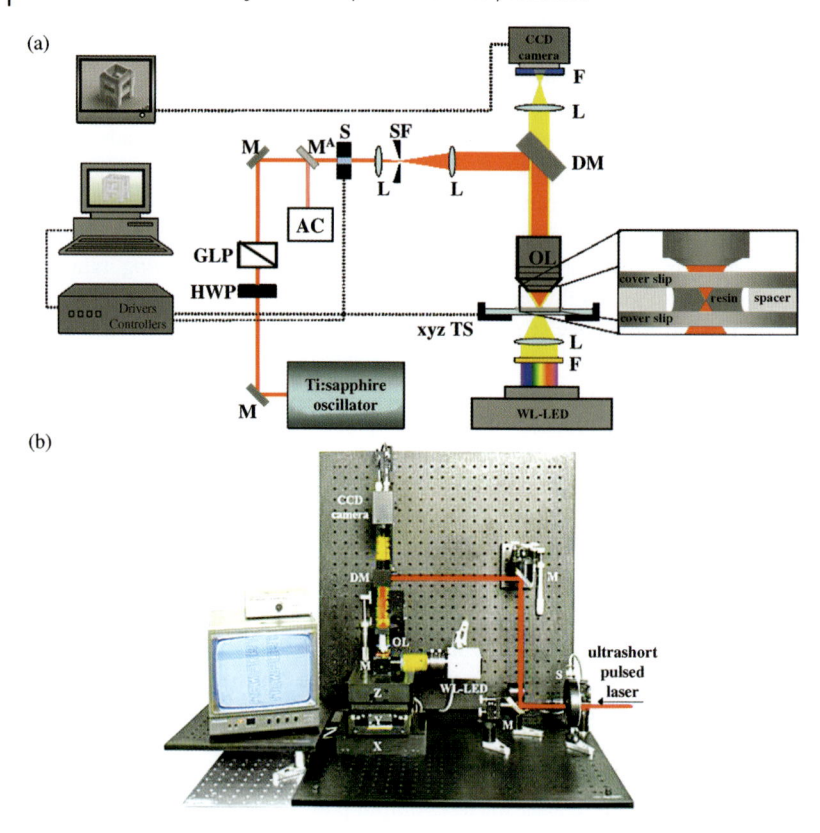

Figure 7.4 Schematic (a) and photograph (b) of an experimental setup used in TPP. HWP, half-wave plate; GLP, Glan-laser polarizer; M, mirror; M*, beam sampler; AC, autocorrelator; S, shutter; L, lens; SF, spatial filter; DM, dichroic mirror; F, filter; OL, objective lens; TS, translation stages; WL-LED, white light LED. The inset in (a) is a diagram of the sample assembly.

only where the volume of the microstructure overlaps with that of the cube. This mode is called raster-scan and in this way the entire volume of the microstructure is polymerized. In the second mode, the shutter remains open while the stage assembly moves the sample around the focused laser beam, tracing the profile of the microstructure. In this second mode, called vector-scan, only the surface of the microstructure is polymerized. When using negative-tone resins, the liquid resin trapped within the microstructure hard shell can be cured by UV irradiation after the development step. Although fabrication by vector-scan mode leads to shorter processing times, two drawbacks limit its use. First, vector-scan mode fabrication requires higher software sophistication than raster-scan mode fabrication. Second, upon UV irradiation, shrinkage of the unsolidified resin inside the microstructure can significantly alter its shape over time. This phenomenon is more likely to occur if the volume-to-surface ratio of the object is large.

A schematic of the sample assembly is shown in the inset in Figure 7.4a. The resin is sandwiched between two microscope cover slips. The thickness of the film is determined by the thickness of the spacer. If microfabrication starts at the resin/substrate interface furthest from the laser entrance, it is necessary to use a spacer thin enough to account for the limitation of the working distance of the objective lens.

When performing TPP, it is desirable to monitor patterning in real time. In this way, mistakes can be caught early on in the fabrication flow process and corrected accordingly. If the resin used in TPP undergoes a change of refractive index upon polymerization, bright-field transmission light microscopy is an effective way to monitor TPP. This is easily performed in the setup described in Figure 7.4 by assembling a basic microscope using a source of visible light, a CCD camera, and a small number of optical elements. Two measures must be taken when implementing bright-field microscopy into a TPP setup. One is to cover the imaging light source with a short-pass optical filter to prevent the blue part of the lamp emission spectrum from causing any undesired polymerization in the resin. Second, a long-pass optical filter must be placed in front of the CCD camera to prohibit laser back-reflections to reach and saturate the pixels of the camera.

A representative negative-tone resin employed in TPP consists of a mixture of SR499 and SR368 in a composition by weight of 49 and 48% respectively. The photoinitiator Lucirin TPO-L makes up the remaining 3% of the mixture [34]. After forming a homogenous solution by continuous mixing at 60 °C (SR368 is a wax at room temperature), the resin can be drop-cast easily to form the sample assembly represented in Figure 7.4a. Since the refractive index of the polymerized resin is larger than that of the unpolymerized resin ($\Delta n \sim 0.03$), real time imaging by bright-field microscopy is easily performed. After patterning is complete, the sample assembly is immersed in ethanol for few minutes where the unpolymerized resin is washed away and the microstructures revealed.

7.5
Resolution

Complex 3D microstructures are best created by TPP when the size and shape of the smallest polymerized volume element (voxel) is well characterized. It is indeed the accurate overlapping of these building blocks that permits predetermined geometries to form correctly while patterning. To appreciate the significance of resolving voxel dimensions and shape prior to laser writing, a montage of exquisite microstructures fabricated by TPP are shown in Figure 7.5 [34, 48, 55–58]. Several renowned sculptures such as the Aphrodite of Milos have been shrunk to microscopic proportions with high accuracy reproducing even the finest of details. These microstructures are without doubt remarkable when one considers the complexity of their surface topologies along with their sizes. For example, the reproduction of the Wall Street Bull in Figure 7.5a is similar in size to a red blood cell, and the letters in Figure 7.5g–i are only a tenth the thickness of a human hair. Although not functional,

Figure 7.5 SEM images of three-dimensional microstructures fabricated by TPP: (a) reproduction of the Wall Street Bull (reproduced with permission from [56], Copyright © (2001) Nature Publishing Group); (b) reproduction of the Aphrodite of Milos (reproduced with permission from [48], Copyright © (2003) Optical Society of America); (c) reproduction of the Sidney Opera House (reproduced with permission from [57], Copyright © (2004) Elsevier Limited); (d) a microdragon (reproduced with permission from reference [55], Copyright © (2006) Laurin Publishing Co); (e) a microbutterfly and (f) a microbeetle (reproduced with permission from [58], Copyright © (2004) SPIE); and (g-i) the word HAIR on top of a human hair at different magnifications (reproduced with permission from [34], Copyright © (2004) American Institute of Physics). The drawings in (g) and (h) outline the areas magnified in the successive images. The scale bars are 2 μm, 12 μm, 10 μm, 14 μm, 2 μm, 2 μm, 100 μm, 60 μm, and 10 μm for images from (a) to (i), respectively.

these microstructures are models that convey the unique capabilities of TPP in creating "true" 3D shapes.

Modification of the resin solubility while performing TPP occurs in the region with the highest concentration of active species. When considering a Gaussian beam focused by an objective lens, this region corresponds to the central part of the square of the light intensity distribution. As a consequence, voxels have the shape of spinning ellipsoids with the shorter axes oriented perpendicular to the optical axis [59]. Since the focusing of a Gaussian beam is described by diffraction optics, the volume of a voxel is proportional to the cube of the excitation wavelength (λ_{ex}) and inversely proportional to the fourth power of the objective's numerical aperture (NA). Thus, improving resolution in TPP can be achieved by using shorter λ_{ex} and larger NA. While microscope objectives with the highest NA (~1.45) are employed in TPP, the choice of λ_{ex} is limited by the tuning ranges of the light sources and most importantly by the photo-initiator efficiency [60].

Although TPP procedures for creating rods and fibers with diameters of about 30 nm have been devised, they represent more proof of principles than microfabrication techniques [61, 62]. They still lack in fact the control and reproducibility required for precise 3D microfabrication. Instead, voxels with a lateral resolution of 100 nm can be created and used consistently in TPP when using high NA objective lenses [63]. Considering that these results are obtained with an excitation wavelength centered at 800 nm, they indicate that microfabrication resolution in TPP is significantly higher than the diffraction limit (~$\lambda_{ex}/2$). Two factors contribute to this result. The first one is the nonlinear optical nature of the fabrication process. Since absorption of light by the resin is proportional to the intensity squared of the excitation light, the resolution is improved by a factor of $2^{-1/2}$ in respect to the resolution obtained with linear optical process, when comparing the same excitation wavelength. The second factor is material dependent and it is a consequence of the presence in the fabrication process of an intensity threshold below which no polymerization occurs [64]. By employing weaker and weaker light intensities, the fractions of the focal volume that exceeds the intensity threshold become smaller and smaller, hence producing smaller and smaller voxels. A diagram showing how the presence of an intensity threshold can be exploited to overcome the spatial limit imposed by diffraction is represented in Figure 7.6a.

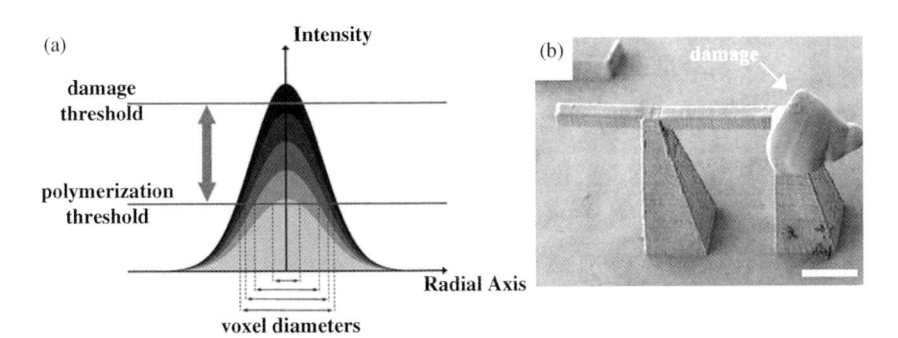

Figure 7.6 (a) Light intensity profiles at the focal plane of a focused beam used for TPP. Only the part of the beam with intensities higher than the intensity threshold causes polymerization. By using weaker and weaker laser intensities (from darker to lighter gray shades), voxels with smaller diameters can be created. The vertical double arrow represents the window of intensities that can be used for TPP. (b) SEM image of a 3D structure fabricated by TPP where in one position the intensity used was larger that the intensity damage threshold of the material. A "blob" of polymer caused by boiling of the resin is attached to the microstructure even after the developing step. The scale bar is 30 μm.

To understand the presence of an intensity threshold, one has to consider that in resins that polymerize through a chain-growth mechanism, the formation of large macromolecular units depends on delicate ratios between several processes such as initiation, propagation, recombination, and termination [30]. If active species are not generated above a certain concentration, polymerization is not sustainable. As a consequence, a light intensity threshold has to be surmounted in order to start TPP. Additionally, in the case of radical polymerization, the presence of molecular oxygen in the resin has a great impact on the intensity threshold. Oxygen indeed quenches the active state of the photo-initiator and interacts with the propagating radicals, slowing down the chain growth mechanism that ultimately would form a polymer.

Besides the presence of an intensity threshold below which no polymerization take place, all resins exhibit an upper intensity threshold above which damage will occur. There is only a certain amount of energy that any given material can accumulate before heating effects take over. Figure 7.6b is an SEM image of a cantilever created by TPP, exhibiting thermal damage. For a short amount of time during microfabrication, the light intensity used exceeded the intensity damage threshold of the material. This caused the resin to boil and consequently harden in an uncontrollable fashion. There is only a window of intensities (dynamic range) inside which microfabrication by TPP can be performed efficiently.

Since the polymerization intensity threshold depends on the concentration and nature of the photo-initiator, and the damage intensity threshold depends mostly on the nature of monomers and oligomers, one strategy for widening the window of laser intensities available for TPP is to employ highly efficient photo-initiators [34, 65]. The benefit arising from this approach is twofold. First, by increasing the gap between the polymerization and damage intensity thresholds a larger range of voxel sizes becomes accessible. Second, by increasing the resin sensitivity to light absorption, larger scanning speed can be used, enabling shorter processing times.

The shape of voxels created by TPP can vary from highly elongated ellipsoids to almost perfect spheres depending on experimental parameters such as the laser exposure time (t), and average power (P_{avg}). It has been observed that under near-threshold conditions the ratio between the length and the diameter of voxels (μ) deviates from linear exposure theory, that is the size of voxels depends only on the exposure dose ($t \times P_{avg}$). Specifically it was measured that μ approaches unity much faster by decreasing P_{avg} than by decreasing t. Furthermore, while μ keeps increasing with larger P_{avg}, it reaches a plateau with longer t [66–68]. To account for these trends, diffusion of active species during irradiation has been considered. Although diffusion plays an important role in determining voxel growth (as demonstrated by a recent work where TPP was performed at lower and lower temperatures and an increase of resolution with a rate of around $1\,\mathrm{nm\,°C^{-1}}$ was observed [69]), a more complete and general model is needed to fully explain the scaling laws governing voxels shapes and sizes, above all when considering the different properties of materials used in TPP.

Even if the progress achieved in TPP resolution is remarkable, it would be desirable for some applications to be able to write polymeric features with dimensions smaller than 100 nm. By employing sub-100 nm voxels for example, 3D photonics crystals and metamaterials operating in the visible region of the spectrum could be fabricated by TPP. A significant stride in this direction was recently achieved by Fourkas and coworkers [70]. They were able to achieve an axial resolution 20 times smaller than the excitation wavelength (40 nm) using a technique named resolution augmentation trough photo-induced deactivation (RAPID). In RAPID, two collinear laser beams are focused into the resin. One is used to initiate the polymerization by two-photon absorption, while the second is used to deactivate the polymerization process through linear absorption. By engineering the focal volume of the deactivation beam to a doughnut-like pattern, polymerization can be confined to volume much smaller than the excitation wavelength.

7.6
Microstructures: Properties and Characterization

With the aid of the proper hardware and software, TPP is a relatively straightforward method for the rapid fabrication of microstructures with complex shapes. If the same microstructures were to be produced successfully with conventional lithographic methods, it would require the use of various masks and sacrificial layers through a series of developing steps. This process is almost impractical to undertake for the majority of research laboratories from both an economical and instrumental point of view. Two-photon polymerization can be considered then as an enabling technology for rapid prototyping of microstructures with no geometrical constraints. To illustrate this point, a series of microstructures fabricated by TPP using the acrylic based resin described earlier [8, 34, 71] is shown in Figure 7.7. Microstructures with inclined planes such as the sides of a truncated pyramid with square sections (a), microstructures with large overhangs (b), and Russian doll type microstructures (an object within another object) (c) are all easily created by TPP. Microstructures with freely movable parts can also be patterned (d) and moreover, objects with large aspect ratios are attainable. For example, the SEM image in Figure 7.7e shows a free-standing tower that is 20 μm on each side and 1500 μm tall.

Even though the ability to fabricate complex 3D microstructures with high resolution is without doubt what distinguishes TPP from other LDW techniques, there are other aspects of TPP, perhaps less unique, which are nonetheless worthwhile pursuing. One of these is the ease with which two or more materials with different mechanical and/or optical properties can be deposited on the same substrate. This is a particularly attractive feature because it could lead to microstructures with "smart" features such as distinct responses to varied external stimuli, thus forming the basis for devices with sensing and actuating type of functions. To test this idea in TPP, the text NEWPORT shown in Figure 7.7f was fabricated by alternating letters made with two resins [71]. The resins were similar to those employed in creating the microstructures in Figure 7.7, with the exception that one was doped with Rhodamine B (resin 1), and the other one with Coumarin 334 (resin 2). Rhodamine B and Coumarin 334 are fluorophores that when excited emit a broad fluorescence with maximum intensities at 610 and 500 nm, respectively.

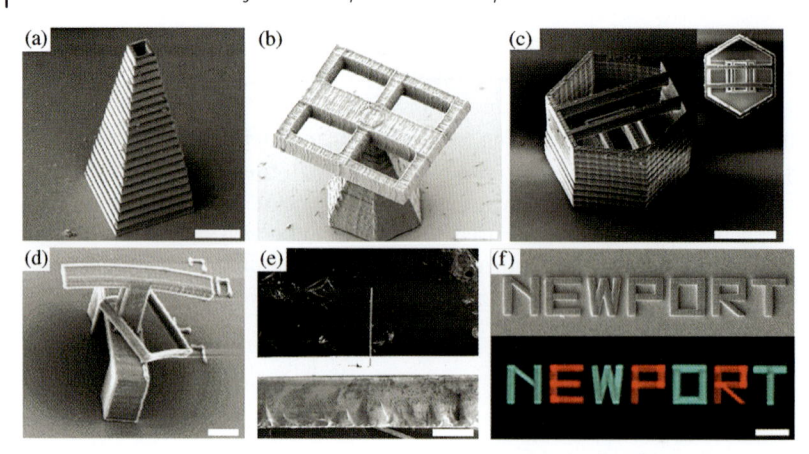

Figure 7.7 Assortment of SEM images of microstructures with diverse geometries: (a) inclined planes on a truncated pyramid; (b) large overhangs; (c) spatially separated structure within a structure (the inset is a top view of the same object); (d) freely movable objects; (e) high aspect-ratio tower (reproduced with permission from [34], Copyright © (2004) American Institute of Physics); and (f) two dimensional patterns written with different materials (see test for description). The top image in (f) is an SEM micrograph while the bottom one is a fluorescence image of the same microstructure. The scale bars are 20 μm, 20 μm, 20 μm, 10 μm, 500 μm, and 20 μm for images (a) to (f), respectively.

Upon polymerization, the fluorophores are entrapped in the polymer, leading to highly fluorescent microstructures. First, patterning was performed on the substrate using resin 1. A series of reference markers were also fabricated in key locations at this stage of the pattering process. Then, after washing away the unsolidified part of resin 1, resin 2 was poured onto the substrate and a second set of microstructures were fabricated by TPP in pre-defined locations relative to the microstructures fabricated with resin 1. This was accomplished by carefully aligning the substrate with the aid of the markers. The assembly was then immersed in the developing solvent to remove the unexposed resin. Finally, the sample was imaged using fluorescence and SEM. While detailed information on size and surface roughness can be gathered from the SEM image, no indication of the different nature of the materials is revealed. When using instead fluorescence microscopy with the appropriate excitation wavelengths and collection filters, the image reveals which elements of the pattern were made with resin 1 and which ones with resin 2. Each letter in the final microstructure is 20 μm wide and 25 μm long. Although the spaces left between the first set of letters were only 30 μm wide, the high accuracy of TPP allowed for precise "writing" of the remaining letters in the correct positions. By employing miniaturized sample assemblies such as those used in microfluidics, one can imagine writing patterns of different materials on the same substrate without the need to ever move it from the microfabrication setup.

Although the body of work collected so far on TPP has demonstrated how valuable this technique is in many applications, several fundamental aspects of TPP still remain ambiguous. For example, how do scanning patterns influence microstructure mechanical properties? How can microfabrication processing time be optimized while maintaining structural integrity? How does solvent permeation affect microstructure rigidity? At which dimensions do the properties of the polymeric microstructures differ from those of the bulk polymer?

Imaging techniques such as SEM and fluorescence microscopy have been traditionally used to characterize microstructures fabricated by TPP [72]. However, the information they provide is limited to confirming or not the success of TPP in creating complex geometries, and certainly they are inadequate to answer the aforementioned questions. An optimal diagnostic tool for TPP will preferably possess the following requisites. First, it needs to be sensitive to the degree of chemical conversion of the resin used in TPP. Second, the probing time should be comparable with the relevant time scale of TPP, which is in the range of microseconds to milliseconds per voxel. Third, the probing volume needs to overlap with the voxel addressed in TPP. A diagnostic technique that will answer all these specific demands could reveal a great deal of information about microstructures fabricated by TPP.

It has been shown recently that coherent anti-Stokes Raman scattering (CARS) microscopy is perhaps the diagnostic tool that can provide more information on TPP than any other technique previously used [71, 73]. CARS microscopy in fact provides 3D imaging with contrast based on molecular vibration, and given that signals generated by CARS microscopy are orders of magnitude stronger than spontaneous Raman signals, high imaging rates are accessible down to the microsecond time scale [74]. In addition, the 3D probing volume in CARS microscopy resembles the size of voxels in TPP. More importantly, CARS signals can be detected on the same platform used for TPP manufacturing, offering opportunities for *in-situ* and real-time probing during microfabrication.

CARS microscopy derives its analytical capability from the Raman active modes of molecules [75]. There are two specific vibrational modes in the Raman spectra of acrylic based polymers that can give insights on the polymerization process and the structural integrity of microstructures fabricated via TPP. One is around $3000\,\mathrm{cm}^{-1}$ and arises from the stretching of both aliphatic and aromatic carbon–hydrogen bonds. The other one is around $1640\,\mathrm{cm}^{-1}$ and is due to the carbon–carbon double bond resonant vibration with the carbonyl group present in the acrylic units of the starting materials [73, 76].

Representative images of microstructures fabricated by TPP using CARS microscopy in these two spectral regions are shown in Figures 7.8 and 7.9. In the first example, the test structure consists of a series of hanging cantilevers 5 μm wide and 50 μm long, suspended at a height of 40 μm through a rectangular shaped tower. Each cantilever was prepared by overlapping 5 μm long lines (one laser pass each) side-by-side for the entirety of its length. All the experimental conditions were kept the same during fabrication but the spacing between the polymeric lines that compose the cantilevers. In particular, from top to bottom they were 0.1 μm, 0.2 μm, 0.4 μm, 0.8 μm, and 1.0 μm.

From the SEM images of this structures it is noticeable that the shape of the last cantilever was distressed during the washing process of the unsolidified resin,

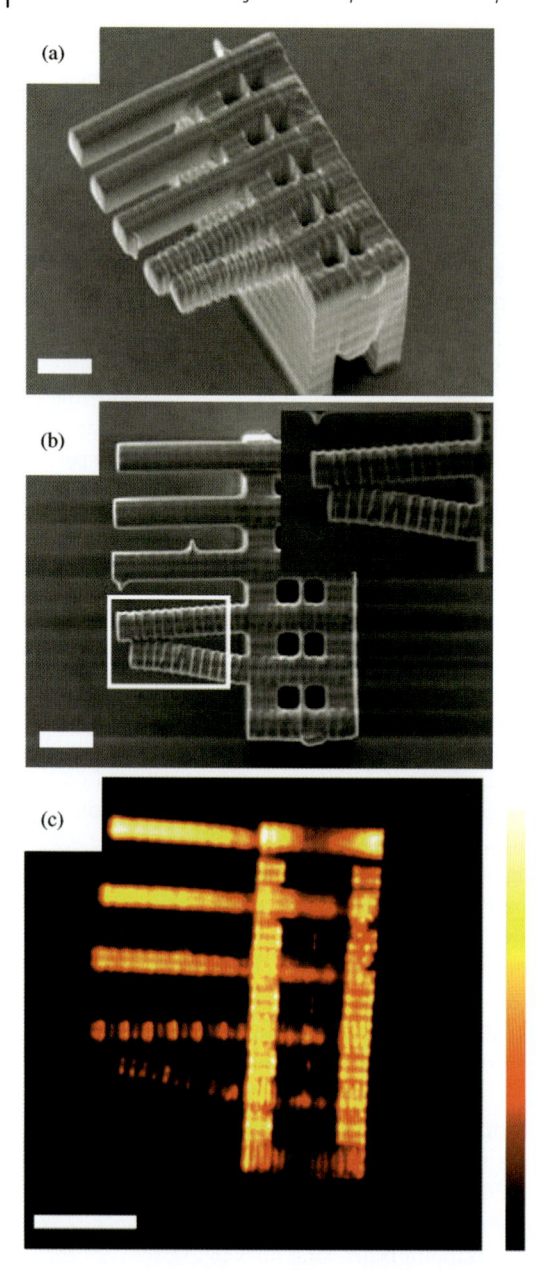

Figure 7.8 Three-dimensional microstructure fabricated by TPP and imaged by SEM and CARS microscopy. SEM images of the microstructure tilted at (a) 45° and (b) 0°. The inset in (b) is an enlarged viewed of the white outline. The scale bar is 10 μm in both images. CARS image (c) of the same microstructure recorded at around 3000 cm^{-1}. The measured signal originates from the stretching vibrations of carbon–hydrogen bonds present in the polymer chemical structure. Figure (c) is a cross-section image of the microstructure recorded at a height of 40 μm from the surface of the glass substrate to which it was attached. The scale bar in is 20 μm. LUT shown on far right.

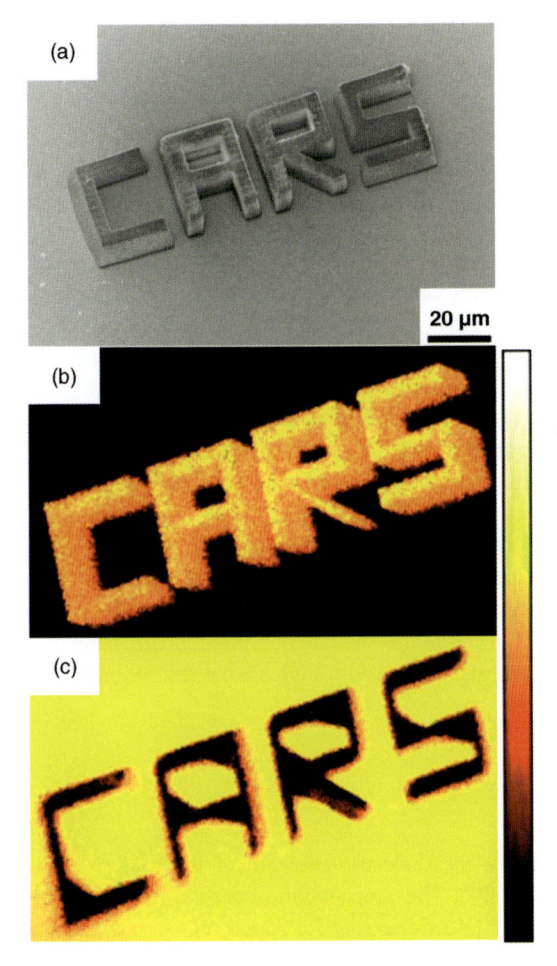

Figure 7.9 (a) SEM tilted view of a microstructure fabricated by TPP. Volume rendering of CARS microscopy signals of the same object still immersed in the unpolymerized resin carried out at (b) 1628 cm^{-1} and (c) 1643 cm^{-1}. The relative strengths between the signal of the unpolymerized resin and the signal of the microstructures are inverted by switching wavenumbers. The size of each letter is 20 µm by 25 µm. LUT shown on far right.

highlighting its weakened structural integrity (Figure 7.8a and b). Furthermore, while the first two cantilevers are straight and smooth, the last two show ridges on their surfaces. By imaging the same microstructure through CARS microscopy and collecting only the signal arising from the polymer vibrational modes around 3000 cm^{-1}, the inner structural flaws of the object can easily be identified (Figure 7.8c). This is because the denser the material, the higher the concentration of carbon–hydrogen bonds that consequentially give rise to stronger CARS signal. Although analysis of the microstructure by SEM depicts a solid object, imaging by

CARS microscopy unveils how inhomogeneous is the density of polymer in some parts the microstructure. Since density is related to the polymer cross-linking and polymer total conversion, CARS microscopy in the 3000 cm^{-1} range can be used to optimize the microfabrication experimental conditions.

In the second example, CARS imaging is performed in the spectral region around 1640 cm^{-1} (Figure 7.9). Since the carbon–carbon double bonds of the starting materials are consumed during polymerization, investigating microstructures fabricated by TPP using CARS microscopy at around 1640 cm^{-1} allows contrast imaging based on chemical differences between polymerized and unpolymerized portions of the sample. The dispersive spectral line shape of the CARS signal can in fact be used to generate images with enhanced chemical contrast based on a particular bond vibration. To prove this point, the word CARS was written by TPP onto a glass substrate. Each letter is 20 µm wide, 25 µm long and 10 µm tall. An SEM tilted view of this pattern is shown in Figure 7.9a. While still immersed in the bath of unpolymerized resin, the same microstructure was imaged by CARS microscopy. Images with completely reversed contrast where obtained when the Raman shift was tuned from 1643 to 1628 cm^{-1}. At 1643 cm^{-1}, the signal from the microstructure is stronger than the signal from the unpolymerized resin, while at 1628 cm^{-1}, the signal from the unpolymerized resin is stronger than the signal from the microstructure (Figure 7.9b and c). Thus, through this method, it is possible to distinguish between polymerized and unpolymerized resin with high spatial resolution.

7.7
Applications

Photonic crystals are objects where the dielectric constant of the material that constitute them varies periodically [77]. The propagation of certain modes of the electromagnetic spectrum though these devices is forbidden, creating what are known as photonic bandgaps. While the center frequency of the bandgap is dictated by the dimensions of the periodicity, its depth depends on the difference between the dielectric constants of the crystal and that of the medium surrounding it (generally air). When the periodicity is extended to all three spatial dimensions, the effect of the photonic bandgap becomes independent from the direction of light propagation. Since the action of photonic crystals to photons is analogous to the action of semiconductor crystals to electrons, these structures have the potential of being key elements in the realization of all-optical circuits. 3D photonic crystals can indeed be used among other applications, for low loss guiding, omni-directional reflection, and sharp bending of light. The fabrication of photonic crystals that operate in the NIR and visible part of the spectrum is quite challenging. To operate at these wavelengths, microstructures with sub-500 nm features must be created and, in the case of 3D photonic crystals, complicated patterns are required with high accuracy. TPP is a fitting choice for the fabrication of these microstructures. Not only does TPP fulfill the strict microfabrication requirements, but it holds a clear advantage over other techniques that can form periodic structures, which

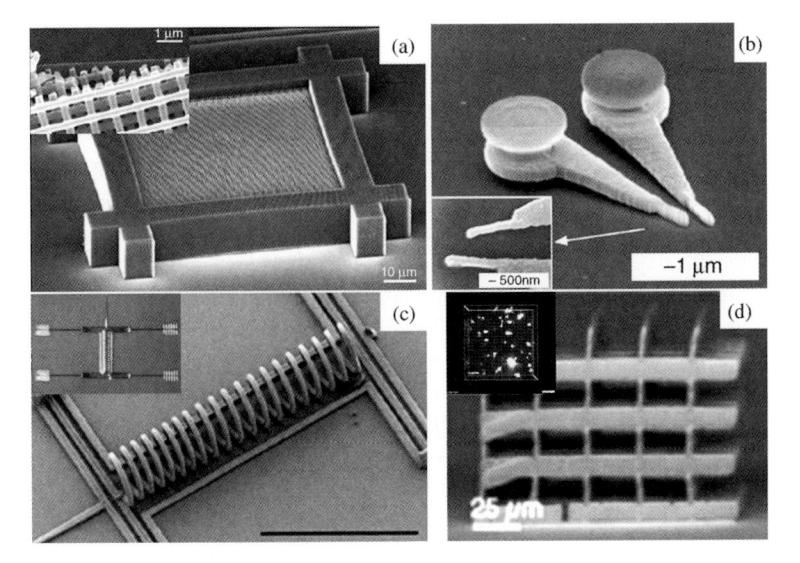

Figure 7.10 Microstructures fabricated by TPP for various applications. (a) 3D photonic crystal with woodpile-like configuration (reproduced with permission from [79], Copyright © (2004) Nature Publishing Group; (b) Optically attenuated nanotweezers (reproduced with permission from [104], Copyright © (2003) American Institute of Physics); (c) A microinductor (reproduced with permission from [15], Copyright © (2007) Elsevier Limited); and (d) a 3D scaffold for studying cells motility (reproduced with permission from [110], Copyright © (2008) Wiley). The insets in (a) and (b) are magnifications of the microstructure depicted in the figures. The inset in (c) is an optical micrograph recorded in reflective mode to enhance the metallic coating of the microinductor. The inset in (d) is a fluorescence micrograph of the microstructure with GFP-labeled cells. The scale bars in (a), (b), (c), and (d) are 10 μm, 1 μm, 100 μm, and 25 μm, respectively.

is the capability of inserting defects in arbitrary locations within 3D photonic crystals [78].

3D photonic crystals with several geometries have been created by TPP. Although the most common one is the so-called woodpile structure (Figure 7.10a), others consisting of diamond-like, spiral, quasi-crystal, and slanted-pore lattices have been fabricated [79–87]. The center wavelength of the photonic bandgaps produced with these structure ranges from 1.0 to 1.5 μm and their change in transmission is typically less than 50%. The latter value is a direct consequence of the small difference between the refractive index of the polymers used in TPP and air. In order to increase this difference and hence create a full photonic bandgap effect, several strategies have been undertaken. In one approach, monomers containing bromine atoms are employed, resulting in polymer with higher values of refractive index [88]. In another approach, the microstructure fabricated by TPP is used as a "skeleton" on top of which high refractive index materials are deposited [82, 89, 90]. The organic scaffold can then be removed by heating the assembly in high-temperature furnaces. Both conductors and dielectrics have been used in this method to create efficient

3D photonic crystals. Lastly, through a double inversion process, a pure silicon 3D photonic crystal was produced starting from a SU-8 master [91]. Because of the high refractive index of silicon, a full bandgap was created and it was centered at 2.35 μm.

Besides 3D photonic crystals, TPP has been applied to the fabrication of other photonic elements as well. For example, by doping the resin with fluorescent dyes, microstructures with lasing capabilities have been fabricated and their operation demonstrated [92, 93]. Moreover, planar and nonplanar waveguides, Mach-Zehnder interferometers, traditional and Fresnel type lenses, optical switches, and micro-ring resonators were also manufactured [69, 94–98]. The performance of these optical parts relies on smooth interfaces in order to minimize losses due to scattering. This task is easily achieved by TPP because of its high spatial resolution. By carefully overlapping voxels during microfabrication in fact, surface roughness between fifty and hundreds times smaller than visible and NIR wavelengths can be attained [99].

The ability to coat polymer microstructures with different materials is an attractive method for creating devices in other areas of research than photonics. When these materials are conductive, electrical components with intricate geometries can be fabricated by TPP such as the micro-inductor in Figure 7.10c [100]. By using wet-chemistry processes such as self-assembly and electroless plating, several metals (Ag, Au, and Cu) were grown to form conductive skins on the surfaces of 3D microstructures [101–103]. Particularly interesting is the process developed by Fourkas and coworkers because of its selectivity. They have shown for example that metallization can occur on microstructures composed of an acrylate backbone polymer while it is ineffective on microstructures composed of the methacrylate counterpart [100].

Two-photon polymerization has also been used for the fabrication of light-driven machines (Figure 7.10b). These microstructures, which contain freely movable parts, are created by TPP in a one step process without the need of sacrificial layers [50, 104]. The viscosity of the resin is usually high enough to keep in place the different components of the microstructure while fabrication occurs. Gears, rotors, needles, and manipulators were fabricated by TPP with high resolution and controlled in liquid environments using optical tweezers. Based on the same principles, two micropumps were fabricated by TPP for microfluidics applications [105, 106]. In the first one, two lobe-shaped rotors only 9 μm in diameter were fabricated within a microchannel and driven by a time-shared optical tweezer creating a flow as low as $1\,\mathrm{pL\,min^{-1}}$. In the second system, a single disk was rotated in a U-shaped microchannel creating a continuous flow without pulsation.

The elasticity of the material used in TPP is an important parameter to characterize, if microstructures fabricated with this technique are intended to be used as mechanical devices, such as for example valves, springs, cantilevers, and manipulators. The Young's modulus of polymeric microstructures larger than a few microns can be considered identical to that of the bulk polymer [107]. However, a significant difference between the elasticity of the bulk polymer and that of the microstructure was observed when at least one of its dimensions is smaller than 1 μm [108, 109].

The versatility of TPP in creating complex architectures with sub-micron resolution is advantageous for biological studies too. For example, fundamental information can be gathered by examining cell motility and adhesion in model structures that closely resemble the microscopic surroundings of cells *in vivo*. Recently, Tayalia and coworkers have accomplished this by studying cells' motility in 3D environments created by TPP (Figure 7.10d). By changing the pore size of the microstructures, several velocities and modalities of cells' movement were observed [110]. Future works where different materials, architectures, and cell lines will be used, could add a different way of studying important physiological and pathological processes such as tissue regeneration and cancer metastasis. Although some of the materials used in TPP are biocompatible with cell growth, in certain situations it would be desirable to use more biological friendly materials. To accomplish this, Campagnola and Shear have independently demonstrated the possibility of using TPP for cross-linking of biomolecules both *in vivo* and *in vitro* [35, 111–118]. A promising aspect of this research is the fact that proteins cross-linked by TPP retained enzymatic activities. In addition to fundamental research in biology, the use of TPP for the fabrication of medical devices is currently ongoing. Examples include microneedles for transdermal drug delivery and prostheses for ossicular replacement [119–122].

7.8
Limitations and Future Directions

Like all LDW techniques, TPP suffers from low throughput. This drawback is quite unfavorable for industrial production, and it constitutes an obstacle (probably the major one) for a wider use of TPP in research laboratories too. As a consequence, many efforts have been committed to finding solutions that can decrease TPP processing time. In one approach, polymerization is performed with a multipoint scheme by employing a microlens array to focus the excitation beam in the resin [123–125]. In this way identical microstructures are fabricated at different positions simultaneously. When the number of points is on the order of several hundreds, an amplified femtosecond laser is necessary to produce sufficient laser intensities at each fabrication point. This increased complexity could be avoided in the future if photo-initiators with larger initiation efficiencies were to be used.

In a second approach, periodic patterns that extend over large areas can be quickly formed using multibeam interference [126]. By focusing two or more laser beams in the same location within a resin, complex 3D patterns can be formed by changing the angle of incidence and polarization of each beam. Another technique based on light waves interference that can produce periodic patterns over large areas is proximity-field nanopatterning [127–129]. This time only one laser beam is required and interference patterns are generated with the aid of a conformable phase mask which is rested on top of the resin surface. Although interference lithography methods permit fast patterning of 3D polymeric microstructures over large areas, they are limited in the extent of depth within the resin they can pattern and by the nature of microstructure topologies.

Figure 7.11 SEM images of 3D microstructures fabricated by TPP (top row) and corresponding replicas created by microtransfer molding (bottom row). (a),(d) Master and replica of a microstructure with opposing 45° overhangs; (b),(e) of a microstructure with cavities and protrusions. (Reproduced with permission from [130]. Copyright © (2004) American Chemical Society); (c),(f) of a microstructure with a close-looped. (Reproduced with permission from [131]. Copyright © (2006) National Academy of Sciences). The scale bar in all pictures is 10 μm.

Fourkas and co-workers have approached this problem from a different perspective [130]. Instead of decreasing the time required for the fabrication of one set of microstructures, they have found a way to replicate the first set many times and quickly. The process is based on microtransfer molding and because of the nature of the material used as the mold, it permits it to reproduce fine features and complex geometries. Microstructures with high-aspect ratios, large overhangs, and re-entrant features were all replicated. Furthermore, by using a clever strategy, microstructures with closed loops were also replicated [131]. A collection of microstructures fabricated through this method is shown in Figure 7.11.

Although contributions from several research groups have allowed TPP to develop into a sophisticated prototyping technique, further improvements are desirable. For example, it is foreseeable that integrating different materials such as conductive polymers, photocurable glasses, photoresponsive polymers, polymers doped with metal nanoparticles, and polymers with chemical and environmental specificities could expand the range of applications where TPP can play an important role. Furthermore, it would be interesting to investigate the effect of excitation sources other than Ti: sapphire oscillators on TPP. A recent work has shown how large areas can be patterned quickly and precisely by employing the second harmonic of an Yb: KGW femtosecond laser working at a repetition rate of 312.5 kHz [132]. In the quest to optimize the TPP process, this result is quite promising. Finally, TPP microfabrication capabilities could multiply if it is used in conjunction with other microfabrication techniques as demonstrated by Braun and coworkers [133–135]. By combining colloidal self-assembly with the TPP, they were able to fabricate complex photonic devices.

The emergence of several technologies in the last two decades has stimulated the development of a set of unconventional microfabrication processes that go beyond the limits imposed by standard photolithography. TPP is one of these tools. Its distinctive capabilities to create complex 3D architectures with high resolution, positions TPP as a key enabling microfabrication method for future advancements.

References

1. Arnold, C.B. and Pique, A. (2007) Laser direct-write processing. *MRS Bull.*, **32** (1), 9–12.
2. Borodovsky, Y. (2006) Marching to the Beat of Moore's law. Advances in Resist Technology and Processing XXIII (ed. Q. Lin), Proc. SPIE, vol. 6153, pp. 6153011–61530119.
3. Cahill, D.G. and Yalisove, S.M. (2006) Ultrafast lasers in materials research. *MRS Bull.*, **31** (8), 594–600.
4. Wu, S.H., Serbin, J., and Gu, M. (2006) Two-photon polymerisation for three-dimensional micro-fabrication. *J. Photochem. Photobiol., A*, **181** (1), 1–11.
5. Park, S.H., Lim, T.W., Yang, D.Y., Yi, S.W., Kong, H.J., and Lee, K.S. (2005) Direct nano-patterning methods using nonlinear absorption in photopolymerization induced by a femtosecond laser. *J. Nonlin. Opt. Phys. Mater.*, **14** (3), 331–340.
6. Korte, F., Serbin, J., Koch, J., Egbert, A., Fallnich, C., Ostendorf, A., and Chichkov, B.N. (2003) Towards nanostructuring with femtosecond laser pulses. *Appl. Phys. A*, **77** (2), 229–235.
7. Sun, H.B. and Kawata, S. (2003) Two-photon laser precision microfabrication and its applications to micro-nano devices and systems. *J. Lightwave Technol.*, **21** (3), 624–633.
8. Baldacchini, T. and Fourkas, J.T. (2004) Three-dimensional nanofabrication using multiphoton absorption, in *Dekker Encyclopedia of Nanoscience and Nanotechnology* (eds J.A. Schwarz, I. Contenscu, and C.K. Putyera), Marcel Dekker, pp. 395–3915.
9. Sun, H.B., and Kawata, S. (2004) Two-photon photopolymerization and 3D lithographic microfabrication. *Adv. Polym. Sci.*, **170**, 169–273.
10. Yang, D., Jhaveri, S.J., and Ober, C.K. (2005) Three-dimensional microfabrication by two-photon lithography. *MRS Bull.*, **30** (12), 976–982.
11. Lee, K.S., Yang, D.Y., Park, S.H., and Kim, R.H. (2006) Recent developments in the use of two-photon polymerization in precise 2D and 3D microfabrications. *Polym. Advan. Technol.*, **17** (2), 72–82.
12. Jia, B.H., Li, J.F., and Gu, M. (2007) Two-photon polymerization for three-dimensional photonic devices in polymers and nanocomposites. *Aust. J. Chem.*, **60** (7), 484–495.
13. LaFratta, C.N., Fourkas, J.T., Baldacchini, T., and Farrer, R.A. (2007) Multiphoton fabrication. *Angew. Chem. Int. Ed*, **46** (33), 6238–6258.
14. Lee, K.S., Kim, R.H., and Prabhakaran, P. (2007) Two-photon stereolithography. *J. Nonlin. Opt. Phys. Mater.*, **16** (1), 59–73.
15. Li, L.J. and Fourkas, J.T. (2007) Multiphoton polymerization. *Mater. Today*, **10** (6), 30–37.
16. Maruo, S. and Fourkas, J.T. (2008) Recent progress in multiphoton microfabrication. *Laser & Photon. Rev.*, **2** (1–2), 100–111.
17. Ovsianikov, A. and Chichkov, B.N. (2008) Two-photon polymerization – high resolution 3D laser technology and its applications, in *Nanoelectronics and Photonics* (eds A. Korkin and F. Rosei), Springer Verlag, pp. 427–446.
18. Pao, Y.H., and Rentzepis, P.M. (1965) Laser-induced production of free radicals in organic compounds. *Appl. Phys. Lett.*, **6** (5), 93–95.
19. Strickler, J.H. and Webb, W.W. (1990) Two-photon excitation in laser scanning fluorescence microscopy. CAN-AM Easter '90 (eds R.L.J. Antos and

A. Krisiloff), Proc. of SPIE, vol. 1398, pp. 107–118.
20 Cabrera, M., Jezequel, J.Y., and Andre, J.C. (1990) Three-dimensional machining by laser photopolymerization, in *Laser in Polymer Science and Technology: Applications* vol. 3 (eds J.P. Fouassier and J.R. Rabek), CRC Press, pp. 73–95.
21 Maruo, S., Nakamura, O., and Kawata, S. (1997) Three-dimensional microfabrication with two-photon-absorbed photopolymerization. *Opt. Lett.*, **22** (2), 132–134.
22 He, G.S., Tan, L.S., Zheng, Q., and Prasad, P.N. (2008) Multiphoton absorbing materials: Molecular designs, characterizations, and applications. *Chem. Rev.*, **108** (4), 1245–1330.
23 Goppert-Mayer, M. (1931) Uber Elementarakte mit zwei Quantensprungen. *Ann. Phys.*, **9**, 273–295.
24 Kaiser, W. and Garrett, C.G.B. (1961) 2-Photon excitation in Caf2 − Eu2 + . *Phys. Rev. Lett.*, **7** (6), 229–231.
25 Denk, W. and Svoboda, K. (1997) Photon upmanship: Why multiphoton imaging is more than a gimmick. *Neuron*, **18** (3), 351–357.
26 Gratton, E. and van de Ven, M.J. (1995) Laser sources for confocal microscopy, in *Handbook of Confocal Microscopy* (ed. J.B. Pawley), Plenum Press, New York, pp. 69–97.
27 Diaspro, A., and Sheppard, C.J.R. (2002) Two-photon excitation fluorescence microscopy, in *Confocal and Two-Photon Microscopy: Foundations, Applications, and Advances* (ed. A. Diaspro), Wiley-Liss Inc., New York, pp. 39–73.
28 Spence, D.E., Kean, P.N., and Sibbett, W. (1991) 60-fsec pulse generation from a self-mode-locked Ti:sapphire laser. *Opt. Lett.*, **16** (1), 42–44.
29 Odian, G. (2004) *Principles of Polymerization*, John Wiley & Sons, Inc., Hoboken, USA.
30 Decker, C. (1994) Photoinitiated curing of multifunctional monomers. *Acta Polym.*, **45** (5), 333–347.
31 Decker, C. (1999) High-speed curing by laser irradiation. *Nucl. Instrum. Methods Phys. Res., Sect. B*, **151**, 22–28.
32 Belfield, K.D., Schafer, K.J., Liu, Y.U., Liu, J., Ren, X.B., and Van Stryland, E.W. (2000) Multiphoton-absorbing organic materials for microfabrication, emerging optical applications and non-destructive three-dimensional imaging. *J. Phys. Org. Chem.*, **13** (12), 837–849.
33 Belfield, K.D., Ren, X.B., Van Stryland, E.W., Hagan, D.J., Dubikovsky, V., and Miesak, E.J. (2000) Near-IR two-photon photoinitiated polymerization using a fluorone/amine initiating system. *J. Am. Chem. Soc.*, **122** (6), 1217–1218.
34 Baldacchini, T., LaFratta, C.N., Farrer, R.A., Teich, M.C., Saleh, B.E.A., Naughton, M.J., and Fourkas, J.T. (2004) Acrylic-based resin with favorable properties for three-dimensional two-photon polymerization. *J. Appl. Phys.*, **95** (11), 6072–6076.
35 Pitts, J.D., Campagnola, P.J., Epling, G.A., and Goodman, S.L. (2000) Submicron multiphoton free-form fabrication of proteins and polymers: Studies of reaction efficiencies and applications in sustained release. *Macromolecules*, **33** (5), 1514–1523.
36 Mendonca, C.R., Correa, D.S., Baldacchini, T., Tayalia, P., and Mazur, E. (2008) Two-photon absorption spectrum of the photo-initiator Lucirin TPO-L. *Appl. Phys. A*, **90** (4), 633–636.
37 Schafer, K.J., Hales, J.M., Balu, M., Belfield, K.D., Van Stryland, E.W., and Hagan, D.J. (2004) Two-photon absorption cross-sections of common photo-initiators. *J. Photochem. Photobiol., A*, **162** (2–3), 497–502.
38 Marder, S.R., Bredas, J.L., and Perry, J.W. (2007) Materials for multiphoton 3D microfabrication. *MRS Bull.*, **32** (7), 561–565.
39 Ehrlich, J.E., Wu, X.L., Lee, I.Y.S., Hu, Z.Y., Rockel, H., Marder, S.R., and Perry, J.W. (1997) Two-photon absorption and broadband optical limiting with bis-donor stilbenes. *Opt. Lett.*, **22** (24), 1843–1845.
40 Albota, M., Beljonne, D., Bredas, J.L., Ehrlich, J.E., Fu, J.Y., Heikal, A.A., Hess, S.E., Kogej, T., Levin, M.D., Marder, S.R., McCord-Maughon, D., Perry, J.W., Rockel, H., Rumi, M., Subramaniam, C.,

Webb, W.W., Wu, X.L., and Xu, C. (1998) Design of organic molecules with large two-photon absorption cross sections. *Science*, **281** (5383), 1653–1656.

41 Cumpston, B.H., Ananthavel, S.P., Barlow, S., Dyer, D.L., Ehrlich, J.E., Erskine, L.L., Heikal, A.A., Kuebler, S.M., Lee, I.Y.S., McCord-Maughon, D., Qin, J.Q., Rockel, H., Rumi, M., Wu, X.L., Marder, S.R., and Perry, J.W. (1999) Two-photon polymerization initiators for three-dimensional optical data storage and microfabrication. *Nature*, **398** (6722), 51–54.

42 Rumi, M., Ehrlich, J.E., Heikal, A.A., Perry, J.W., Barlow, S., Hu, Z.Y., McCord-Maughon, D., Parker, T.C., Rockel, H., Thayumanavan, S., Marder, S.R., Beljonne, D., and Bredas, J.L. (2000) Structure-property relationships for two-photon absorbing chromophores: Bis-donor diphenylpolyene and bis(styryl)benzene derivatives. *J. Am. Chem. Soc.*, **122** (39), 9500–9510.

43 Xing, J.F., Chen, W.Q., Gu, J., Dong, X.Z., Takeyasu, N., Tanaka, T., Duan, X.M., and Kawata, S. (2007) Design of high efficiency for two-photon polymerization initiator: combination of radical stabilization and large two-photon cross-section achieved by N-benzyl 3,6-bis(phenylethynyl)carbazole derivatives. *J. Mater. Chem.*, **17** (14), 1433–1438.

44 Zhou, W.H., Kuebler, S.M., Braun, K.L., Yu, T.Y., Cammack, J.K., Ober, C.K., Perry, J.W., and Marder, S.R. (2002) An efficient two-photon-generated photoacid applied to positive-tone 3D microfabrication. *Science*, **296** (5570), 1106–1109.

45 Nguyen, L.H., Straub, M., and Gu, M. (2005) Acrylate-based photopolymer for two-photon microfabrication and photonic applications. *Adv. Func. Mater.*, **15** (2), 209–216.

46 Watanabe, T., Akiyama, M., Totani, K., Kuebler, S.M., Stellacci, F., Wenseleers, W., Braun, K., Marder, S.R., and Perry, J.W. (2002) Photoresponsive hydrogel microstructure fabricated by two-photon initiated polymerization. *Adv. Func. Mater.*, **12** (9), 611–614.

47 Coenjarts, C.A. and Ober, C.K. (2004) Two-photon three-dimensional microfabrication of poly(dimethylsiloxane) elastomers. *Chem. Mater.*, **16** (26), 5556–5558.

48 Serbin, J., Egbert, A., Ostendorf, A., Chichkov, B.N., Houbertz, R., Domann, G., Schulz, J., Cronauer, C., Frohlich, L., and Popall, M. (2003) Femtosecond laser-induced two-photon polymerization of inorganic-organic hybrid materials for applications in photonics. *Opt. Lett.*, **28** (5), 301–303.

49 Houbertz, R., Frohlich, L., Popall, M., Streppel, U., Dannberg, P., Brauer, A., Serbin, J., and Chichkov, B.N. (2003) Inorganic-organic hybrid polymers for information technology: from planar technology to 3D nanostructures. *Adv. Eng. Mater.*, **5** (8), 551–555.

50 Maruo, S., Ikuta, K., and Korogi, H. (2003) Force-controllable, optically driven micromachines fabricated by single-step two-photon microstereolithography. *J. Microelectromech. Syst.*, **12** (5), 533–539.

51 Teh, W.H., Durig, U., Salis, G., Harbers, R., Drechsler, U., Mahrt, R.F., Smith, C.G., and Guntherodt, H.J. (2004) SU-8 for real three-dimensional subdiffraction-limit two-photon microfabrication. *Appl. Phys. Lett.*, **84** (20), 4095–4097.

52 Teh, W.H., Durig, U., Drechsler, U., Smith, C.G., and Guntherodt, H.J. (2005) Effect of low numerical-aperture femtosecond two-photon absorption on (SU-8) resist for ultrahigh-aspect-ratio microstereolithography. *J. Appl. Phys.*, **97** (5), 0549071.

53 Seet, K.K., Juodkazis, S., Jarutis, V., and Misawa, H. (2006) Feature-size reduction of photopolymerized structures by femtosecond optical curing of SU-8. *Appl. Phys. Lett.*, **89** (2), 024106.

54 Yu, T.Y., Ober, C.K., Kuebler, S.M., Zhou, W.H., Marder, S.R., and Perry, J.W. (2003) Chemically amplified positive resists for two-photon three-dimensional microfabrication. *Adv. Mater.*, **15** (6), 517–521.

55 Ostendorf, A. and Chichkov, B.N. (2006) Two-photon polymerization: A new

approach to micromachining. *Photon. Spectra*, **40** (10), 72–78.

56 Kawata, S., Sun, H.B., Tanaka, T., and Takada, K. (2001) Finer features for functional microdevices – Micromachines can be created with higher resolution using two-photon absorption. *Nature*, **412** (6848), 697–698.

57 Straub, M., Nguyen, L.H., Fazlic, A., and Gu, M. (2004) Complex-shaped three-dimensional microstructures and photonic crystals generated in a polysiloxane polymer by two-photon microstereolithography. *Opt. Mater.*, **27** (3), 359–364.

58 Ikuta, K., Maruo, S., Hasegawa, T., Itho, S., Korogi, H., and Takahashi, A. (2004) Light-drive biomedical micro tools and biochemical IC chips fabricated by 3D micro/nano stereolithography. Optomechatronic Micro/Nano Components, Devices, and Systems, (ed. Y. Katagiri), Proc. of SPIE, vol. 5604, pp. 52–66.

59 Sun, H.B., Tanaka, T., and Kawata, S. (2002) Three-dimensional focal spots related to two-photon excitation. *Appl. Phys. Lett.*, **80** (20), 3673–3675.

60 Haske, W., Chen, V.W., Hales, J.M., Dong, W.T., Barlow, S., Marder, S.R., and Perry, J.W. (2007) 65nm feature sizes using visible wavelength 3-D multiphoton lithography. *Opt. Express*, **15** (6), 3426–3436.

61 Juodkazis, S., Mizeikis, V., Seet, K.K., Miwa, M., and Misawa, H. (2005) Two-photon lithography of nanorods in SU-8 photoresist. *Nanotechnology*, **16** (6), 846–849.

62 Park, S.H., Lim, T.W., Yang, D.Y., Cho, N.C., and Lee, K.S. (2006) Fabrication of a bunch of sub-30-nm nanofibers inside microchannels using photopolymerization via a long exposure technique. *Appl. Phys. Lett.*, **89** (17), 173133.

63 Xing, J.F., Dong, X.Z., Chen, W.Q., Duan, X.M., Takeyasu, N., Tanaka, T., and Kawata, S. (2007) Improving spatial resolution of two-photon microfabrication by using photo-initiator with high initiating efficiency. *Appl. Phys. Lett.*, **90** (13), 131106.

64 Tanaka, T., Sun, H.B., and Kawata, S. (2002) Rapid sub-diffraction-limit laser micro/nanoprocessing in a threshold material system. *Appl. Phys. Lett.*, **80** (2), 312–314.

65 Kuebler, M.S., Rumi, M., Watanabe, T., Braun, K., Cumpston, H.B., Heikal, A.A., Erskine, L.L., Thayumanavan, S., Barlow, S., Marder, R.S., and Perry, W.J. (2001) Optimizing two-photon initiators and exposure conditions for three-dimensional lithographic microfabrication. *J. Photopolym. Sci. Technol.*, **14** (4), 657–668.

66 Sun, H.B., Takada, K., Kim, M.S., Lee, K.S., and Kawata, S. (2003) Scaling laws of voxels in two-photon photopolymerization nanofabrication. *Appl. Phys. Lett.*, **83** (6), 1104–1106.

67 Dong, X.Z., Zhao, Z.S., and Duan, X.M. (2008) Improving spatial resolution and reducing aspect ratio in multiphoton polymerization nanofabrication. *Appl. Phys. Lett.*, **92** (9), 091113.

68 Devoe, J.R., Kalweit, H., Leatherdale, A.C., and Williams, R.T. (2003) Voxel shapes in two-photon microfabrication. Multiphoton Absorption and Nonlinear Transmission Processes: Materials, Theory, and Applications (eds K.D. Belfield, S.J. Caracci, F. Kajzar, C.M. Lawson, and A.T. Yeates), Proc. of SPIE, vol. 4797, pp. 310–316.

69 Takada, K., Kaneko, K., Li, Y.D., Kawata, S., Chen, Q.D., and Sun, H.B. (2008) Temperature effects on pinpoint photopolymerization and polymerized micronanostructures. *Appl. Phys. Lett.*, **92** (4), 041902.

70 Li, L., Gattass, R.R., Gershgoren, E., Hwang, H., and Fourkas, J.T. (2009) Achieving $\lambda/20$ resolution by one-color initiation and deactivation of polymerization. *Science*, **324**, 910–913.

71 Baldacchini, T., Zimmerley, M., Potma, E.O., and Zadoyan, R. (2009) Chemical mapping of three-dimensional microstructures fabricated by two-photon polymerization using CARS microscopy. Laser Applications in Microelectronic and Optoelectronic Manufacturing VII (eds M. Meunier, A.S. Holmes, H. Niino, and B. Gu), Proc. of SPIE, vol. 7201.

72 Sun, H.B., Tanaka, T., Takada, K., and Kawata, S. (2001) Two-photon polymerization and diagnosis of three-dimensional microstructures containing fluorescent dyes. *Appl. Phys. Lett.*, **79** (10), 1411–1413.

73 Baldacchini, T., Zimmerley, M., Kuo, C.-H., Potma, E.O., and Zadoyan, R. (2009) Characterization of microstructures fabricated by two-photon polymerization using coherent anti-Stokes Raman scattering microscopy. *J. Phys. Chem. B*, **113**, 12663–12668.

74 Evans, C.L. and Xie, X.S. (2008) Coherent anti-Stokes Raman scattering microscopy: chemical imaging for biology and medicine. *Annu. Rev. Anal. Chem.*, **1**, 883–909.

75 Muller, M. and Zumbusch, A. (2007) Coherent anti-Stokes Raman scattering microscopy. *ChemPhysChem*, **8** (15), 2157–2170.

76 Esen, C., Kaiser, T., and Schweiger, G. (1996) Raman investigation of photopolymerization reactions of single optically levitated microparticles. *Appl. Spectrosc.*, **50** (7), 823–828.

77 Busch, K., von Freymann, G., Linden, S., Mingaleev, S.F., Tkeshelashvili, L., and Wegener, M. (2007) Periodic nanostructures for photonics. *Phys. Rep.*, **444** (3–6), 101–202.

78 Sun, H.B., Mizeikis, V., Xu, Y., Juodkazis, S., Ye, J.Y., Matsuo, S., and Misawa, H. (2001) Microcavities in polymeric photonic crystals. *Appl. Phys. Lett.*, **79** (1), 1–3.

79 Deubel, M., Von Freymann, G., Wegener, M., Pereira, S., Busch, K., and Soukoulis, C.M. (2004) Direct laser writing of three-dimensional photonic-crystal templates for telecommunications. *Nat. Mater.*, **3** (7), 444–447.

80 Mizeikis, V., Seet, K.K., Juodkazis, S., and Misawa, H. (2004) Three-dimensional woodpile photonic crystal templates for the infrared spectral range. *Opt. Lett.*, **29** (17), 2061–2063.

81 Sun, H.B., Matsuo, S., and Misawa, H. (1999) Three-dimensional photonic crystal structures achieved with two-photon-absorption photopolymerization of resin. *Appl. Phys. Lett.*, **74** (6), 786–788.

82 Serbin, J., Ovsianikov, A., and Chichkov, B. (2004) Fabrication of woodpile structures by two-photon polymerization and investigation of their optical properties. *Opt. Express*, **12** (21), 5221–5228.

83 Ledermann, A., Cademartiri, L., Hermatschweiler, M., Toninelli, C., Ozin, G.A., Wiersma, D.S., Wegener, M., and Von Freymann, G. (2006) Three-dimensional silicon inverse photonic quasicrystals for infrared wavelengths. *Nat. Mater.*, **5** (12), 942–945.

84 Seet, K.K., Mizeikis, V., Matsuo, S., Juodkazis, S., and Misawa, H. (2005) Three-dimensional spiral-architecture photonic crystals obtained by direct laser writing. *Adv. Mat.*, **17** (5), 541–545.

85 Seet, K.K., Mizeikis, V., Juodkazis, S., and Misawa, H. (2006) Three-dimensional horizontal circular spiral photonic crystals with stop gaps below 1 mu m. *Appl. Phys. Lett.*, **88** (22), 221101.

86 Kaneko, K., Sun, H.B., Duan, X.M., and Kawata, S. (2003) Submicron diamond-lattice photonic crystals produced by two-photon laser nanofabrication. *Appl. Phys. Lett.*, **83** (11), 2091–2093.

87 Deubel, M., Wegener, M., Kaso, A., and John, S. (2004) Direct laser writing and characterization of 'Slanted Pore' Photonic Crystals. *Appl. Phys. Lett.*, **85** (11), 1895–1897.

88 Murakami, Y., Coenjarts, C.A., and Ober, C.K. (2004) Preparation and two-photon lithography of a sulfur containing resin with high refractive index. *J. Photopol. Sc. Tech.*, **17** (1), 115–118.

89 Tal, A., Chen, Y.S., Williams, H.E., Rumpf, R.C., and Kuebler, S.M. (2007) Fabrication and characterization of three-dimensional copper metallodielectric photonic crystals. *Opt. Express*, **15** (26), 18283–18293.

90 Duan, X.M., Sun, H.B., Kaneko, K., and Kawata, S. (2004) Two-photon polymerization of metal ions doped acrylate monomers and oligomers for three-dimensional structure fabrication. *Thin Solid Films*, **453**, 518–521.

91 Tetreault, N., von Freymann, G., Deubel, M., Hermatschweiler, M., Perez-Willard, F., John, S., Wegener, M., and Ozin, G.A.

(2006) New route to three-dimensional photonic bandgap materials: silicon double inversion of polymer templates. *Adv. Mat.*, **18** (4), 457–460.

92 Yokoyama, S., Nakahama, T., Miki, H., and Mashiko, S. (2003) Fabrication of three-dimensional microstructure in optical-gain medium using two-photon-induced photopolymerization technique. *Thin Solid Films*, **438-439**, 452–456.

93 Li, C.F., Dong, X.Z., Jin, F., Jin, W., Chen, W.Q., Zhao, Z.S., and Duan, X.M. (2007) Polymeric distributed-feedback resonator with sub-micrometer fibers fabricated by two-photon induced photopolymerization. *Appl. Phys. A - Mat. Sci. & Proc.*, **89** (1), 145–148.

94 Klein, S., Barsella, A., Leblond, H., Bulou, H., Fort, A., Andraud, C., Lemercier, G., Mulatier, J.C., and Dorkenoo, K. (2005) One-step waveguide and optical circuit writing in photopolymerizable materials processed by two-photon absorption. *Appl. Phys. Lett.*, **86** (21), 211118.

95 Sherwood, T., Young, A.C., Takayesu, J., Jen, A.K.Y., Dalton, L.R., and Chen, A.T. (2005) Microring resonators on side-polished optical fiber. *IEEE Phot. Tech. Lett.*, **17** (10), 2107–2109.

96 Guo, R., Xiao, S.Z., Zhai, X.M., Li, J.W., Xia, A.D., and Huang, W.H. (2006) Micro lens fabrication by means of femtosecond two photon photopolymerization. *Opt. Express*, **14** (2), 810–816.

97 Chen, Q.D., Wu, D., Niu, L.G., Wang, J., Lin, X.F., Xia, H., and Sun, H.B. (2007) Phase lenses and mirrors created by laser micronanofabrication via two-photon photopolymerization. *Appl. Phys. Lett.*, **91** (17), 171105.

98 Mendonca, C.R., Baldacchini, T., Tayalia, P., and Mazur, E. (2007) Reversible birefringence in microstructures fabricated by two-photon absorption polymerization. *J. Appl. Phys.*, **102** (1), 013109.

99 Takada, K., Sun, H.B., and Kawata, S. (2005) Improved spatial resolution and surface roughness in photopolymerization-based laser nanowriting. *Appl. Phys. Lett.*, **86** (7), 071122.

100 Farrer, R.A., LaFratta, C.N., Li, L.J., Praino, J., Naughton, M.J., Saleh, B.E.A., Teich, M.C., and Fourkas, J.T. (2006) Selective functionalization of 3-D polymer microstructures. *J. Am. Chem. Soc.*, **128** (6), 1796–1797.

101 Chen, Y.S., Tal, A., Torrance, D.B., and Kuebler, S.M. (2006) Fabrication and characterization of three-dimensional silver-coated polymeric microstructures. *Adv. Func. Mater.*, **16** (13), 1739–1744.

102 Formanek, F., Takeyasu, N., Tanaka, T., Chiyoda, K., Ishikawa, A., and Kawata, S. (2006) Selective electroless plating to fabricate complex three-dimensional metallic micro/nanostructures. *Appl. Phys. Lett.*, **88** (8), 0831101.

103 Chen, Y.S., Tal, A., and Kuebler, S.M. (2007) Route to three-dimensional metallized microstructures using cross-linkable epoxide SU-8. *Chem. Mater.*, **19** (16), 3858–3860.

104 Maruo, S., Ikuta, K., and Korogi, H. (2003) Submicron manipulation tools driven by light in a liquid. *Appl. Phys. Lett.*, **82** (1), 133–135.

105 Maruo, S. and Inoue, H. (2006) Optically driven micropump produced by three-dimensional two-photon microfabrication. *Appl. Phys. Lett.*, **89** (14), 1441011.

106 Maruo, S. and Inoue, H. (2007) Optically driven viscous micropump using a rotating microdisk. *Appl. Phys. Lett.*, **91** (8), 0841011.

107 Bayindir, Z., Sun, Y., Naughton, M.J., LaFratta, C.N., Baldacchini, T., Fourkas, J.T., Stewart, J., Saleh, B.E.A., and Teich, M.C. (2005) Polymer microcantilevers fabricated via multiphoton absorption polymerization. *Appl. Phys. Lett.*, **86** (6), 0641051.

108 Sun, H.B., Takada, K., and Kawata, S. (2001) Elastic force analysis of functional polymer submicron oscillators. *Appl. Phys. Lett.*, **79** (19), 3173–3175.

109 Nakanishi, S., Shoji, S., Kawata, S., and Sun, H.B. (2007) Giant elasticity of photopolymer nanowires. *Appl. Phys. Lett.*, **91** (6), 0631121.

110 Tayalia, P., Mendonca, C.R., Baldacchini, T., Mooney, D.J., and Mazur, E. (2008) 3D cell-migration studies using two-photon

engineered polymer scaffolds. *Adv. Mater.*, **20** (23), 4494–4498.

111 Basu, S. and Campagnola, P.J. (2004) Enzymatic activity of alkaline phosphatase inside protein and polymer structures fabricated via multiphoton excitation. *Biomacromolecules*, **5** (2), 572–579.

112 Basu, S., Wolgemuth, C.W., and Campagnola, P.J. (2004) Measurement of normal and anomalous diffusion of dyes within protein structures fabricated via multiphoton excited cross-linking. *Biomacromolecules*, **5** (6), 2347–2357.

113 Basu, S., Cunningham, L.P., Pins, G.D., Bush, K.A., Taboada, R., Howell, A.R., Wang, J., and Campagnola, P.J. (2005) Multiphoton excited fabrication of collagen matrixes cross-linked by a modified benzophenone dimer: bioactivity and enzymatic degradation. *Biomacromolecules*, **6** (3), 1465–1474.

114 Basu, S., Rodionov, V., Terasaki, M., and Campagnola, P.J. (2005) Multiphoton-excited microfabrication in live cells via Rose Bengal cross-linking of cytoplasmic proteins. *Opt. Lett.*, **30** (2), 159–161.

115 Allen, R., Nielson, R., Wise, D.D., and Shear, J.B. (2005) Catalytic three-dimensional protein architectures. *Anal. Chem.*, **77** (16), 5089–5095.

116 Kaehr, B., Ertas, N., Nielson, R., Allen, R., Hill, R.T., Plenert, M., and Shear, J.B. (2006) Direct-write fabrication of functional protein matrices using a low-cost Q-switched laser. *Anal. Chem.*, **78** (9), 3198–3202.

117 Hill, R.T., Lyon, J.L., Allen, R., Stevenson, K.J., and Shear, J.B. (2005) Microfabrication of three-dimensional bioelectronic architectures. *J. Amer. Chem. Soc.*, **127** (30), 10707–10711.

118 Kaehr, B., Allen, R., Javier, D.J., Currie, J., and Shear, J.B. (2004) Guiding neuronal development with in situ microfabrication. *Proc. Nat. Acad. Sci. USA*, **101** (46), 16104–16108.

119 Ovsianikov, A., Schlie, S., Ngezahayo, A., Haverich, A., and Chichkov, B.N. (2007) Two-photon polymerization technique for microfabrication of CAD-designed 3D scaffolds from commercially available photosensitive materials. *J. Tissue Eng. Regen. Med.*, **1** (6), 443–449.

120 Schlie, S., Ngezahayo, A., Ovsianikov, A., Fabian, T., Kolb, H.A., Haferkamp, H., and Chichkov, B.N. (2007) Three-dimensional cell growth on structures fabricated from ORMOCER (R) by two-photon polymerization technique. *J. Biomater. Appl.*, **22** (3), 275–287.

121 Ovsianikov, A., Chichkov, B., Mente, P., Monteiro-Riviere, N.A., Doraiswamy, A., and Narayan, R.J. (2007) Two photon polymerization of polymer-ceramic hybrid materials for transdermal drug delivery. *Int. J. Appl. Ceram. Technol.*, **4** (1), 22–29.

122 Ovsianikov, A., Chichkov, B., Adunka, O., Pillsbury, H., Doraiswamy, A., and Narayan, R.J. (2007) Rapid prototyping of ossicular replacement prostheses. *Appl. Surf. Sci.*, **253** (15), 6603–6607.

123 Kato, J., Takeyasu, N., Adachi, Y., Sun, H.B., and Kawata, S. (2005) Multiple-spot parallel processing for laser micronanofabrication. *Appl. Phys. Lett.*, **86** (4), 0441021.

124 Matsuo, S., Juodkazis, S., and Misawa, H. (2005) Femtosecond laser microfabrication of periodic structures using a microlens array. *Appl. Phys. A*, **80** (4), 683–685.

125 Formanek, F., Takeyasu, N., Tanaka, T., Chiyoda, K., Ishikawa, A., and Kawata, S. (2006) Three-dimensional fabrication of metallic nanostructures over large areas by two-photon polymerization. *Opt. Express*, **14** (2), 800–809.

126 Sun, H.B., Nakamura, A., Shoji, S., Duan, X.M., and Kawata, S. (2003) Three-dimensional nanonetwork assembled in a photopolymerized rod array. *Adv. Mater.*, **15** (23), 2011–2014.

127 Jeon, S., Menard, E., Park, J.U., Maria, J., Meitl, M., Zaumseil, J., and Rogers, J.A. (2004) Three-dimensional nanofabrication with rubber stamps and conformable photomasks. *Adv. Mater.*, **16** (15), 1369–1373.

128 Jeon, S., Malyarchuk, V., White, J.O., and Rogers, J.A. (2005) Optically fabricated three dimensional nanofluidic mixers for microfluidic devices. *Nano Lett.*, **5** (7), 1351–1356.

129 Jeon, S., Malyarchuk, V., Rogers, J.A., and Wiederrecht, G.P. (2006) Fabricating three dimensional nanostructures using two photon lithography in a single exposure step. *Opt. Express*, **14** (6), 2300–2308.

130 LaFratta, C.N., Baldacchini, T., Farrer, R.A., Fourkas, J.T., Teich, M.C., Saleh, B.E.A., and Naughton, M.J. (2004) Replication of two-photon-polymerized structures with extremely high aspect ratios and large overhangs. *J. Phys. Chem. B*, **108** (31), 11256–11258.

131 LaFratta, C.N., Li, L.J., and Fourkas, J.T. (2006) Soft-lithographic replication of 3D microstructures with closed loops. *Proc. Nat. Acad. Sci. USA*, **103** (23), 8589–8594.

132 Malinauskas, M., Purlys, V., Rutkauskas, M., and Gadonas, R. (2009) Two-photon polymerization for fabrication of three-dimensional micro- and nanostructures over a large area. Micromachining and Microfabrication Process Technology XIV, (eds M. Maher, J. Chiao, and P. Resnick), Proc. of SPIE, vol. 7204, p. 72040C-1.

133 Rinne, S.A., Garcia-Santamaria, F., and Braun, P.V. (2008) Embedded cavities and waveguides in three-dimensional silicon photonic crystals. *Nat. Photon.*, **2** (1), 52–56.

134 Pruzinsky, S.A. and Braun, P.V. (2005) Fabrication and characterization of two-photon polymerized features in colloidal crystals. *Adv. Funct. Mat.*, **15** (12), 1995–2004.

135 Nelson, E.C., Garcia-Santamaria, F., and Braun, P.V. (2008) Lattice-registered two-photon polymerized features within colloidal photonic crystals and their optical properties. *Adv. Funct. Mat.*, **18** (13), 1983–1989.

8
Laser Micromachining of Polymers

Chantal G. Khan Malek, Wilhelm Pfleging, and Stephan Roth

8.1
Introduction

Lasers are unique energy sources characterized by high spectral purity, spatial and temporal coherence, and high peak intensity. They are used for macro- and micromachining applications in numerous industries such as automotive, home appliances, electronics, and medical and biotechnology [1]. Laser-based micromachining technologies are nontraditional techniques which have developed rapidly in recent years and their use in the manufacture of microsystems is increasing [2]. They are precise, efficient, and multitask tools for fast, flexible, and contamination-free micromanufacturing, from product development to finishing. Recent developments have also extended laser-based methods to the nanoscale [3].

Laser micromachining comprises ablative and additive (or generative) techniques. Ablation is the removal of material from the surface or bulk; it is suitable for machining a wide range of materials. It includes surface modification techniques, which generate chemical patterns, for example, for bio-interfaces, and sub-surface modification such as waveguide writing and micro-joining techniques. Additive microfabrication and rapid prototyping are applied in the design and development of products as well as small batch manufacturing. These laser techniques generate microstructures layer by layer using techniques such as microstereolithography and selective laser sintering.

This chapter is organized in five sections. The first reviews the various lasers used in laser micromachining and the principles of interaction of laser light with matter. The second section gives examples of laser ablation of polymers and their applications. Two other sections are devoted to surface modification and generative laser processes based on layered manufacturing. Finally, a number of other laser-based processes are presented, before concluding.

Generating Micro- and Nanopatterns on Polymeric Materials. Edited by A. del Campo and E. Arzt
Copyright © 2011 WILEY-VCH Verlag GmbH & Co. KGaA, Weinheim
ISBN: 978-3-527-32508-5

8.2
Principles of Beam-Matter Interaction in Ablation Processes

Laser radiation can be produced in different operating modes (continuous, pulsed) and in different media (solid state, gas, dye). Pulse durations range from milliseconds to femtoseconds, as in nanosecond pulsed excimer gas lasers which radiate in the deep ultraviolet (UV) (wavelength 157 to 351 nm) and femtosecond or millisecond pulsed solid state lasers (266 to 1070 nm). Continuous wave (cw) operation CO_2 gas lasers radiate in the IR (10.6 µm).

The properties of lasers are characterized by the following quantities:

Laser fluence

$$\varepsilon = \frac{E}{A} \ [\text{Jcm}^{-2}]$$

Pulse energy

$$E = \int_0^\tau P(t)dt \ [\text{J}]$$

Maximum peak intensity

$$I = \frac{P}{A} \ [\text{Wcm}^{-2}]$$

where $P(t)$ is the laser power as function of time t, P is the maximum peak power, A the focal spot area, and τ the pulse duration.

The laser ablation process involves laser–substrate interactions that are dependent on the properties of the laser beam and of the substrate. Typically, in laser direct writing a highly collimated, monochromatic and coherent light beam is generated and focused to a small spot. If the laser wavelength is chosen from a region where the substrate material absorbs, a high energy density is deposited in a small volume. A wavelength with a minimum absorption depth should be selected. This will help to ensure a high energy deposition in a small volume for rapid and complete ablation.

Further important laser parameters are the pulse duration and the laser repetition rate. A short pulse duration maximizes the peak power and can reduce the thermal conduction to the surrounding material. The repetition rate can also influence the thermal impact on the material. If it is too low, the energy loss by heat conduction dominates and all of the energy not used for ablation will leave the ablation zone. Higher repetition rates lead to an increase of the average surface temperature ΔT, which can be estimated by solving the one-dimensional heat equation for a rectangular laser pulse (Equation 8.1) [4].

$$\Delta T = 2(1-R)\varepsilon v_{\text{rep}} \sqrt{\frac{t}{\varrho c_p K \pi}}, \tag{8.1}$$

where R is the reflectance, t the laser processing time [s], ϱ the density [g cm^{-3}], c_p the specific heat capacity [J g^{-1} K^{-1}] and K the thermal conductivity [W cm^{-1} K^{-1}].

A higher surface temperature in turn can induce an increase of the material quantity removed per laser pulse. This can make the ablation process more efficient. Nevertheless thermal damage of the material or melt and debris formation must be avoided.

Finally, the beam quality is also important for laser ablation. It is described by a beam quality factor, M^2. The M^2 factor of a laser beam defines the possible focus diameter for a given beam divergence angle. Together with the optical laser power, the beam quality factor M^2 determines the radiance ("brightness") of a laser beam [5].

The beam energy is of no use if it cannot be efficiently delivered to the ablation region. High energy density is not always appropriate; especially for microfabrication technologies, when the best surface qualities are obtained for small laser fluences (approximately 2 to 3 times above the ablation threshold). Laser micromachining becomes an efficient method when high repetition rates or large area processing can be established.

Two modes of laser micromachining can be distinguished:

1) In the serial mode for direct writing, the surface is moved relative to a fixed laser beam. Serial machining is advantageous for prototyping because there is no requirement for a mask. For deep structures or 3D shapes, layer-by-layer ablation becomes necessary and an appropriate process strategy has to be applied in order to reduce the surface roughness in the ablated region.
2) In the parallel mode, the laser beam is imaged (projected) through a mask onto the workpiece, generally using a de-magnifying projection lens. Mask projection is generally preferred for large area patterning.

Polymer ablation depends on the absorptivity and thermal diffusivity of the materials. Laser ablation usually relies on the strong absorption of laser photons by the sample material, allowing a high energy deposition into a small volume to ensure complete and fast ablation, hence the wavelength of the laser has to be chosen carefully for maximum absorption. With ultrashort pulse lasers, however, ablation takes place as a result of multiphoton absorption at high peak intensities; therefore even materials which are normally transparent to the laser wavelength can be machined on the surface and even within the transparent material. The laser energy must also exceed a threshold value, which depends on material, wavelength, and pulse duration. In many polymers, the ablation begins only after multiple pulses (incubation effect). The incubation effect causes the ablation threshold to change as a function of pulse number and the absorption length to decrease with an increasing number of pulses. This phenomenon is due to a chemical or physical modification of the polymer material by the first few laser pulses. For example, in the case of poly(methyl methacrylate) (PMMA) and 248 nm laser wavelength, the incubation effect leads to an increased ablation rate (above the ablation threshold) or to an increased refractive index (below the ablation threshold). Incubation is mainly observed for polymers with low effective absorption coefficients α_{eff} ($<1\,\mu m^{-1}$) [6, 7].

Laser micromachining of polymers involves ablation, in which the matter is ejected in the form of species such as atoms, molecules, ions, and clusters because of the interaction with an intense laser pulse. The macroscopic effects of ablation include plasma formation, acoustic shocks, and cratering of the surface. It was first reported in 1982 by Srinivasan and Mayne-Banton [8] that when pulsed UV laser radiation falls on a polymer surface, material is spontaneously etched away from a depth of 0.1 μm to a few microns. Subsequently, the responses of numerous polymers to UV lasers have been investigated and reported; some polymers are easier to ablate than others.

The ablation mechanism is a complex combination of photochemical and photothermal processes and is dependent on the laser characteristics and materials properties. In general, the high photon energies of UV lasers are capable of direct bond breaking in an organic material which strongly absorbs at the wavelength of the laser emission. This leads to ablative photodecomposition, a process distinct from thermal ablation in which the laser primarily heats the material. Photochemical etching is a relatively "cold" process with minimum collateral thermal damage to the substrate; longer wavelengths mainly initiate temperature rises that will first melt and then decompose the material, leaving a void in the workpiece.

The potential of ultrashort laser pulses for material processing is based primarily on the localization of the energy deposition to dimensions smaller than the diffraction limit of the focusing optics, both on the surface and in the bulk through nonlinear absorption mechanisms. Another advantage is the reduction of residual damage by minimizing thermal effects. Femtosecond laser ablation is capable of machining any kind of material, in particular transparent materials. The physical aspects of lasers and details of laser micromachining processes are beyond the scope of this review and can be found in various books and articles dedicated to the subject [9–11]. Some general features concerning modes of operation and types of laser will however be recalled for the sake of clarity.

The choice of wavelength depends on the optical properties of the substrate material such as absorption and reflection characteristics and on the minimum structure size to be achieved. Since short wavelengths are diffracted less than longer wavelengths, optical beam delivery systems can have greater resolution that allows smaller lateral feature sizes to be machined. Theoretically, minimal feature sizes down to the minimum achievable focal spot (about twice the laser wavelength λ) or the optical diffraction limit $\lambda/(2 \cdot NA)$ are possible (NA: numerical aperture of the optical system).

In general, optimal laser micromachining is obtained when photons are absorbed at submicron depth on the surface of a material. Furthermore, if these photons are delivered in a short duration burst (<100 ns), a miniexplosion is created, ejecting solid and gaseous particulates from the irradiated site. Thermal conduction to the surrounding substrate material is then minimized and the heat-affected damage of the material is reduced.

In the case of ultrashort pulses ($\tau < 10$ ps), the high intensity delivered by the focused laser beam can induce strong nonlinear optical absorption of photons in materials that might otherwise be highly transparent to photons at much lower intensities. Srinivasan *et al.* demonstrated for the first time that the high power of ultrashort pulses of UV light can produce photochemical etching due to multiphoton excitation to dissociative states in PMMA; PMMA has negligible absorption at 308 nm wavelength for one-photon excitation [12].

Thermal laser ablation involves typically a series of steps encompassing initial absorption of the laser energy by the solid material, followed by local heating of the irradiated volume. The molecular weight or polymer chain length is a key parameter for the ablation. Short polymer chains are volatile whereas longer ones melt. Material is therefore removed by vaporization and/or transport of melted material from ablated spots. Evaporation from the substrate results in an ablation plume (material vapor plasma) containing molecular fragments, ions, free electrons, neutral particles and chemical products from reactions in the plume between the components of the plasma and the atmosphere. The laser-induced material vapor plasma can attenuate the laser beam. For small laser fluences the plasma could result in better thermal coupling between the workpiece and the laser beam in comparison to the absorption at the laser wavelength without plasma formation. Nevertheless, for high laser fluences the initiated plasma can result in a complete shielding of the workpiece from the laser irradiation preventing thermal coupling to the material. The ablation depth is determined by the ability of the substrate material to absorb the laser energy and is a direct function of the beam fluence (energy per unit area), wavelength, repetition rate and pulse duration.

Material damage achieved on the target is determined to a major extent by the heat affected zone (HAZ). From an application standpoint, thermal damage to the surrounding material needs to be minimal as otherwise spatial resolution and quality of ablation deteriorate. Therefore photochemical processes leading to "cold" ablation are most desirable. With new laser sources (UV laser, ultrashort pulsed laser), ablation of polymers with no thermal damage of surrounding material and reduced debris formation is possible, leading to high accuracy and improved lateral resolution.

In addition to machining by melting or vaporization, most polymers can also be finely micromachined by photolytic decomposition at wavelengths below 300 nm. In fact, most of the works on polymer ablation have concentrated on the application of UV lasers, which enable the removal of material by vaporization, without involving a melting phase.

Ultraviolet lasers, excimer and frequency tripled or quadrupled Nd: YAG lasers, typically operated in pulsed mode, ionize and decompose polymer materials. High energy photons in the UV range with wavelengths between 157 and 355 nm are highly absorbed by many polymers inducing efficient interaction with matter, without the necessity to add chromophores in the polymer matrix to produce efficient photo-absorption. Excimer lasers and frequency tripled or quadrupled Nd: YAG solid-state lasers are used depending on the absorption characteristics of the polymers. The capability of a UV laser beam to ablate a polymer depends on the absorption characteristics of the polymer at that wavelength (single-photon absorption where the photon energy is near the material band gap energy). The short wavelength is linearly absorbed in the top layer of the polymer to a depth determined by the effective absorption coefficient α_{eff} of the material, as obtained from Beer–Lambert's law [13]. It is important to mention here that this law is normally used without ablation. Therefore, for the ablation regime, an "effective" absorption coefficient needs to be defined which is determined by the relation between the ablation rate d (ablation depth per pulse) and the laser fluence ε (Equation 8.2) [14]. For small laser fluences (typically $<1\,\mathrm{J\,cm^{-2}}$), the ablation rate d increases with the logarithm of the laser fluence ε.

$$d = \frac{1}{\alpha_{eff}} \ln\left(\frac{\varepsilon}{\varepsilon_t}\right) \tag{8.2}$$

where ε_t denotes the ablation threshold and α_{eff} is called "effective absorption coefficient".

For laser fluences of about $1\,\mathrm{J\,cm^{-2}}$ up to $2\,\mathrm{J\,cm^{-2}}$ saturated excimer laser absorption is observed (Equation 8.3) with

$$d \propto \frac{1}{\alpha_{eff}}(\varepsilon-\varepsilon_0), \tag{8.3}$$

where ε_0 denotes a characteristic laser fluence usually not equal to the threshold laser fluence. For $\varepsilon \gg 2\,\mathrm{J\,cm^{-2}}$ the ablation rate in general tends to saturate because of laser beam attenuation by the generated plasma plume.

The electronic transitions caused by the absorption break chemical bonds in the large molecules typical of polymer and lead to smaller (often gaseous) molecular fragments. Since the smaller molecules have lower density, the irradiated volume expands rapidly; this produces a shock wave which causes photo-ablated decomposition products to be ejected, leaving a photo-ablated cavity. A characteristic of the process is that much of the excess energy is carried away in the form of kinetic energy by the ejected material. The thermal impact to the sample surface is drastically reduced, an important point for thermally delicate materials. A rough estimate of the start of the ablation process is given by the lifetime of the excited states (e.g., 35 ps for polyimide (PI) at 355 nm, [15]) whereas a rough estimate of the "ablation time" is given by the laser pulse width. Shielding of laser energy can play a major role in ablation. The laser-induced material vapor plasma can attenuate the laser beam and shield the laser energy, reducing the interaction time. Therefore an appropriate laser fluence should be selected in order to reduce the shielding effect and to increase ablation efficiency. This was investigated in detail for polymers in Schmidt et al. [14].

Photochemical ablation leads to the possibility of machining microstructures with high dimensional accuracy and reduced defects in the surface layer in comparison to a thermally driven ablation process.

An interesting technical approach in laser ablation of polymers consists in creating new polymers with properties especially designed for a desired technique. Ideally polymers designed to decompose mainly to gaseous products would reduce or eliminate the problem of nongaseous products that redeposit on the substrate (debris) [16]. Incomplete removal of polymer debris requires additional cleaning steps and mitigates one of the big advantages of laser ablation. These specially designed polymers are now used for laser-induced forward transfer (LIFT) and for the development of organic light emitting diode (OLED) [17] or for the microtransfer of biological material [18].

8.3
Laser Ablation of Polymers

Laser ablation of polymers has been known since 1982 [8, 19]. Many aspects of polymer ablation and laser processing, in general, have been described by Bäuerle [20]. More recently Lippert and Dickinson [21] reviewed in detail the chemical and spectroscopic aspects of polymer ablation and new directions. Many types of polymers can be laser machined, the most common ones being PI, PMMA, polyethylene (PE), polycarbonate (PC), poly(ethylene terephthalate) (PET) and polyetheretherketone (PEEK). Other polymers include polytretrafluoroethylene (PTFE), S-U8 resist, other photoresists and acrylics.

8.3.1
Ultraviolet Laser

Excimer laser ablation is an established technique for drilling microvias as small as 10 μm in PI dielectrics for electrical interconnection in micro-electronic devices. Ultraviolet laser ablation also proved a key technology in optical board manufacturing to reach the stringent coupling tolerances imposed by the integration of optical interconnections on a printed circuit board. For example, fabrication of multimode polymer waveguides, micro-mirrors, alignment features, and microlenses was carried out by laser ablation using a frequency-tripled Nd: YAG and KrF excimer laser [22].

Dopant-assisted UV laser modification of polymer substrates has been reported repeatedly in the literature in the near-UV range (308 nm); where PMMA shows little absorption, doping is essential for producing well-defined structures. Formation of microlenses with diameters of 15 μm diameter of PMMA (Figure 8.1) that is suitably doped with small amounts of diphenyltriazene has been reported using single-shot projection with excimer laser radiation at 308 nm [23]. Naessens *et al.* fabricated microlenses in PC material using a direct-write technique based on scanning excimer laser ablation [24]. Submicron texturing in the form of grating (330 nm period) on large surface areas was produced in PI and polyethersulfone (PES) using a grating interferometer and KrF laser ablation at 248 nm [25].

The SU-8 resist, optimized for photolithography at 365 nm wavelength (I-line), was found to be suitable for excimer ablation using a 248 nm KrF excimer laser. Structures with high aspect ratio, that is the ratio of the height to the width of a microstructure were demonstrated and used as a template for electroplating (alternative laser-LIGA) [26].

Many groups employed UV lasers to manufacture microfluidic systems and to modify the resulting surface. For example, Roberts *et al.* [27] demonstrated the fabrication of channels with micrometer-sized features, straight side walls and high aspect ratio in a variety of polymer substrates including polystyrene (PS), PC, cellulose acetate, and PET.

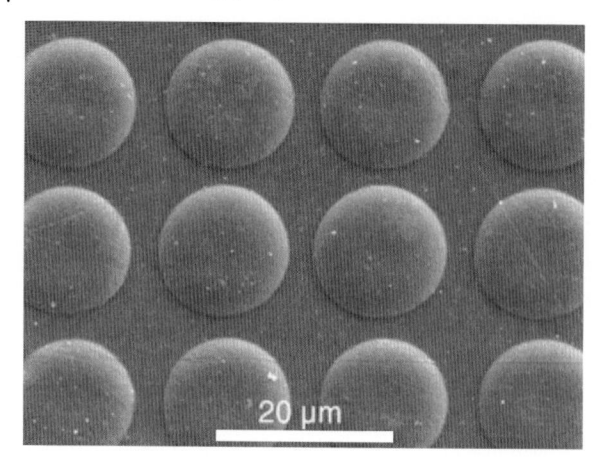

Figure 8.1 SEM micrograph of a relief lens array produced by single-shot irradiation of doped PMMA. Diameter of each lens: 15 μm. Irradiation parameters: fluence F: 3 J cm^{-2}, wavelength: 308 nm, pulse duration: 30 ns. (Reproduced with permission from [23]. Copyright © (1999) Springer.)

Numerous studies have highlighted the increase of the surface roughness induced by the photo-ablation process, in particular its dependence on the laser fluence. It is also known that the photo-ablation process is not only able to create structures within polymer substrates but also can substantially change the surface properties of the resulting surfaces, such as wetting properties, surface charge, formation of new functional groups, compared to the surfaces of the original material. Roberts et al. [27] reported that, relative to the original polymer samples, the photo-ablated surfaces showed an increase in their rugosity and hydrophilicity, as well as a negative charge capable of generating capillary flow or even electro-osmotic (EO) flow when filled with a buffer and placed in high electrical fields. Rossier et al. [28] pointed out that a homogeneous PET ablated surface with a high degree of crystallinity and poor wettability could be obtained using a static ablation mode, whereas dynamic ablation provided an inhomogeneous and hydrophilic surface. These differences were attributed in part to the effects of redeposited fragments. Bianchi et al. [29] extended this work and demonstrated control of the surface modification in a microchannel by alternating dynamic and static ablation procedures. They also demonstrated that well defined static patterning in a microchannel can significantly reduce the EO flow and that this patterning could be used as an alternative to photochemical treatments to locally modify the surface potential. Laser-induced changes in the chemical properties of polymer surfaces, if sufficiently controlled, can be engineered on the microscale to affect protein adsorption or to introduce functional groups allowing easy covalent attachment of particular affinity reagent [30]. Waddel et al. [31] also used a KrF laser (248 nm) to create channels in a number of polymer substrates: PMMA, poly(vinyl chloride) (PVC), poly(ethylene terephthalate glycol) (PETG), and PC. It was observed that the physical morphology of the ablated region was dependent on the temporal profile, the spatial profile, and the wavelength of the laser pulse. The longer the pulse length, the higher the fluence required to ablate the substrate. Similarly, for a given substrate material, different ablation wavelengths afford different limiting aspect ratios. The physical morphology of the channels was also found to be a function of the local atmosphere at the ablation site. In addition to these physical changes, they also observed changes in the surface chemistry and surface charge of ablated polymer substrates. For example, ablation of PMMA under nitrogen or methanol resulted in a rectangular channel profile, whereas ablation under water resulted in a wedge-shaped profile. Ablation of PC under different atmospheres yielded different EO flow characteristics of the resulting microchannels. Johnson et al. [32] compared the surface modifications in hot embossed and UV laser photo-ablated microchannels within a PMMA substrate. Using a 248 nm excimer laser system to modify the PMMA surface, they confirmed that changes in the surface properties were not limited to ablation level laser pulses; instead, a measurable change in the surface chemistry of the polymer occurred when the surface was exposed to UV pulses below the ablation threshold, which induced an increased surface charge. Surface charges were also dependent on the irradiation strategy and atmosphere of ablation (nitrogen and oxygen) and post-treatment. Finally, Pugmire et al. [33] investigated changing the atmosphere during the excimer laser ablation of polymer microchannels to fabricate the channels and to control their surface properties and EO mobility in one step. The substrates that exhibited non-negligible absorption at this energy, namely, PETG, PVC, and PC, showed significant changes in surface chemical composition and EO mobility with varying ablation atmospheres. Ablation under nitrogen or argon resulted in low EO mobilities with a loss of the well-defined chemical structures of the native surfaces; ablation under oxygen, on the contrary, yielded surfaces that retained native chemical structures and supported higher EO mobilities. In contrast, the ablated surfaces of PMMA were very similar to those of the native material, regardless of ablation atmospheres due to the negligible absorption of 248 nm light by that polymer. Atkin et al. [34] demonstrated the suitability of a direct-write method using a frequency-tripled Nd: YAG laser at 355 nm to machine a PET film. This method allowed for rapid prototyping of a biochip and for complex geometric shapes to be realised with channel dimensions as small as 10 µm. While untreated PET is hydrophobic, the authors showed that the cutting process caused an increase in the hydrocarbon content at the surface; this resulted in a decrease in surface charge and hence reduced EO flow. However, with chemical treatment (saponification) it was possible to render the ablated surface more hydrophilic, improving EO flow and providing an increase in carboxylate ions that allow a tethering platform for DNA oligonucleotides. This methodology provided a cost effective process for constructing PET biochips with tailored surface chemistry.

Excimer laser micromachining at 248 and 193 nm was used as a flexible, stand-alone technique or in conjunction with photolithography and thermal molding to manufacture a bio-factory-on-a-chip device. Channels were generated in PC or PMMA, poly(dimethylsiloxane) (PDMS) membranes were thinned, and insulation layers between electrodes as well as encapsulation layers were machined [35]. Lin et al. [36] used UV excimer laser micromachining in bulk PI or PC substrates and lamination process to fabricate a complete microfluidic analytical device. Gillner et al. [37] used F_2 lasers at 157 nm to pattern PTFE microstructures for microfluidic reactor and medical components. A range of microfluidic structures such as a particle micropore PI filter, fluid manifold, channels, and nozzles were manufactured and incorporated in a droplet dispenser for integrated thermal inkjet printheads [38].

Excimer lasers have also been used to manufacture novel composite membranes to be used as an effective transducer for the selective transfer and sensing of molecular ions [39]. Matson et al. [40] also employed excimer laser direct patterning at 248 nm to produce membranes for solvent separators by a step and drill method but they also developed a mask patterning process to create multiple pores of small size. McNeely et al. [41] developed a rapid prototyping technique to fabricate passive hydrophobic microfluidic systems integrated with macroscopic external devices aimed at highly parallel sample analysis. Sabbert et al. [42] machined cycloolefin copolymer (COC) with no redeposition effects, smooth surface and ablation rates smaller than for PMMA using an ArF excimer laser (193 nm).

Numerous studies have reported that PI has better photo-ablation properties than PMMA under UV radiation. High quality laser ablation of PI using frequency-tripled Nd: YAG laser at 355 nm is used industrially for via drilling in microelectronics, e. g. in high density interconnects and flexible circuits [43]. It is also used for drilling ink-jet nozzles [44, 45]. Several groups used a KrF laser (248 nm) to machine microfluidic systems in Kapton® [46–48]. Moss et al. [48] used the KrF laser to ablate fine features as small as 4 µm and a CO_2 laser to ablate quickly and coarsely big features. A commercial high performance liquid chromatography chip made of PI layers was manufactured by direct-write UV laser ablation using a solid state laser at 355 nm followed by a lamination process [49]. Pfleging et al. [4] used various UV lasers, KrF (248 nm) and ArF (193 nm) excimer and frequency-tripled and quadrupled Nd: YAG laser sources emitting at 355 nm and 266 nm respectively, to directly ablate and locally modify polymers, in particular PMMA and PS. The best surface quality was obtained for 193 nm which induced a pure photolytical ablation, superior to that of the frequency-quadrupled Nd: YAG at 266 nm, though in the latter case the higher repetition rate made the process faster. The authors also reported direct ablation of PI channels or micro-gears (see Figure 8.2) with a frequency-tripled Nd: YAG laser. Thermal damage of the polymer material could be avoided even for a pulse repetition rate of 2 kHz by using pulse durations of 500 ps. Yao et al. [50] used a KrF laser at 248 nm to manufacture a flow-through PCR chip in PMMA. Finally, Yu et al. [51] manufactured embedded channels by multistep inclined exposure in a single layer SU-8 film using the third harmonic of Nd: YAG laser at 355 nm. Jensen et al. [52] used excimer KrF laser (248 nm) to produced moulds in PEEK.

Laser micromachining is also used to fabricate biodegradable microdevices for biomedical applications [53]. Chen et al. [54] reviewed the advancements of laser micromachining for etching biodegradable polymers and concluded that deep UV lasers operating at 193 nm were a better choice to minimize the photothermal effect.

Figure 8.2 SEM image of micro-gears after laser cutting of polyimide with short-pulsed frequency-tripled Nd: YAG laser (pulse width 500 ps).

8.3.2
CO_2 Laser

CO_2 laser micromachining provides a flexible and low-cost means for the rapid prototyping and manufacturing of miniaturized polymer systems such as PMMA microfluidic chip devices. Klank et al. [55] investigated the fabrication of such devices, achieving very short fabrication times of about two hours. The best results were obtained by far with PMMA as it combines: (i) high absorbance in the infrared with low heat capacity and low heat conductance, which means that any absorbed heat results in a rapidly rising temperature (see Equation 8.1); (ii) thermal decomposition into volatile products (MMA monomer, carbon dioxide and other gases), leading to the formation of clean structures which are not contaminated by degradation components; by contrast, most other polymers burn and produce large amount of soot, which can be difficult to remove. Structures could be cut accurately with high speed and microprecision even through thick materials (30 mm or more). Typical channel depths were between 100 and 300 μm while their widths were typically 250 μm, the narrowest being 85 μm wide. The smoothness of the wall surfaces of the channels was estimated to be 1 to 2 μm. However, resolidified protrusions with diameter of 15 μm were created by small droplets of recondensed polymers on the edges of the structures. Snakenborg et al. [56] demonstrated that even for a single polymer type such as PMMA, the material properties can differ considerably depending on the specific grade of PMMA and manufacturer, from a slight change in surface roughness to a completely altered surface with irregularities, rendering it no longer transparent. They developed a simple model to relate the laser-cut channel depth to three variables: velocity, power and number of passes of the laser system.

Additionally they explored the effects of processing sequences and number of passes, in particular the effects of cooling time, channel width change and profile. Liu et al. [57] also used laser machining in PC using a CO_2 laser in combination with thermal bonding and adhesive bonding to fabricate low-cost, disposable, and monolithic genetic assay devices integrating PCR, hybridization, and hybridization wash functions. Cheng et al. [58] fabricated channels in PMMA with varying aspect ratios on the same substrate, with width from 100 to 900 μm, and typical trench width of 140 μm wide with an aspect ratio of 7. They developed a process to smooth the channel surface using a one-step thermal annealing treatment after machining (Figure 8.3). Cheng et al. also developed a versatile laser micromachining platform for efficient mass production of PMMA devices. The authors claimed that thousands of copies could be produced in one workday using one system. Wang et al. [59] also used a simple industrial CO_2 laser marker to craft microchannels into Vivak co-polyester chips. They developed a low power process in a nearly anaerobic environment to avoid overheating the substrates, using dry ice in the marking chamber to significantly suppress the partial pressure of oxygen. The typical laser-ablated channel, though not perfectly uniform, was 300–350 μm wide with an aspect ratio from 0.5 to 2. Wang et al. fabricated two-dimensional microchannels at moderate throughput of approximately tens of chips per day. Yuan et al. [60] investigated experimentally and theoretically laser micromachining of channels in PMMA using a continuous-wave CO_2 laser with low power. They focused on the fabrication of narrower channels (from 44 to 240 μm width and depths ranging from 22 to 130 μm). Their models incorporated the threshold fluence of PMMA and predicted the channel depth and profile for a specific choice of laser power and scanning speed. Pfleging et al. [61] manufactured functional capillary electrophoresis (CE) chips made of PMMA by CO_2 laser ablating channel widths down to 30 μm.

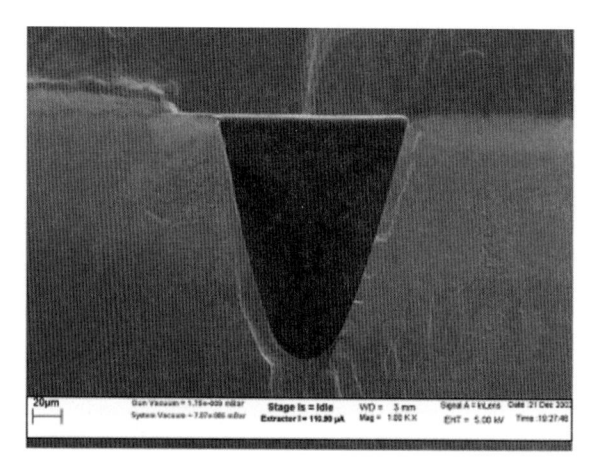

Figure 8.3 Close-up of a thermally annealed sealed channel carved with a CO_2 laser at 1.5 W laser power and at a speed of 36 mm s^{-1}. (Reproduced with permission from [58]. Copyright © (2006) Elsevier.)

8.3.3
Femtosecond Laser

Femtosecond pulsed lasers have been used to manufacture photonic devices (splitters, interferometers, etc.) and gratings. An ultrafast-laser driven micro-explosion method using a tightly focused femtosecond laser was exploited to fabricate three-dimensional (3D) void-based diamond lattice photonic crystals in a low refractive index polymer material (solid resin) [62].

Relatively smooth channels ranging from 2 to 20 µm diameter and a maximum length of 10 mm were formed in PMMA. To prevent clogging of channels by debris and rough, fractured channel walls, Farson et al. used a gas-assisted material removal concept while laser-micromachining the internal channels [63]. Wolfe et al. [64] directly patterned PDMS using a Ti: sapphire laser generating femtosecond pulses to produce stamps for microcontact printing, as well as to customize generic microfluidic channels. Kim et al. [65] employed femtosecond laser to add unmoldable features to microfluidic devices by combining it with soft lithography, in particular replica casting of PDMS. In addition, they integrated a microcapillary with a diameter as small as 0.5 µm and aspect ratio as high as 800 : 1 with a molded PMDS microfluidic network. They also created spiral-shaped 3D capillaries. Gomez et al. [66] used a frequency-doubled Ti: sapphire laser to directly machine passive microfluidic components in polymers, PMMA, PI (Kapton®). Day and Gu [67] used femtosecond laser pulses at 750 nm to etch straight microchannels in PMMA.

Microstructures with a high aspect ratio are difficult to produce using laser ablation. The walls of holes generally start to taper with increasing depth owing to the large numerical aperture (NA) of the demagnification optics. Using femtosecond lasers, high quality holes of aspect ratio 10 were ablated using 100 fs laser pulses at a wavelength of 800 nm in PMMA [68]. Figure 8.4 shows a typical hole with a diameter of 40 µm and depth about 400 µm (aspect ratio of 10) with a well-defined and sharp edge at the entrance, almost straight walls, good surface quality and a relatively flat bottom. Holes with diameters as small as 2 µm over a depth greater than 10 mm (aspect ratio 5000) could also be generated using projection patterning. Beam homogeneity, NA and laser wavelength did not appear to be important criteria for achieving those high aspect ratio (Gaussian beam and large NA optics were used). One of the key issues in drilling high aspect ratio capillaries, in particular in the bulk of materials, is finding an efficient way to remove debris. Water-assisted drilling of small and high aspect ratio holes seems to help prevent clogging.

Lasers with ultrashort pulses demonstrate their ability to effectively ablate even difficult-to-machine polymers. Femtosecond laser micromachining has succeeded in machining fluoro-compounds with good quality. For example, Niino and Yabe [69] reported the successful ablation of several UV-transparent fluoropolymers such as PTFE, tetrafluoroethylene-hexafluoropropylene (FEP). More recently, polyvinylidene fluoride (PVDF) film, coated with NiCu on both sides, was machined using a femtosecond laser to fabricate a vibration microsensor prototype [70]. Such high precision and damage-free results are almost impossible to obtain when using longer laser pulses with the possible exception of 157 nm F_2 excimer lasers [71].

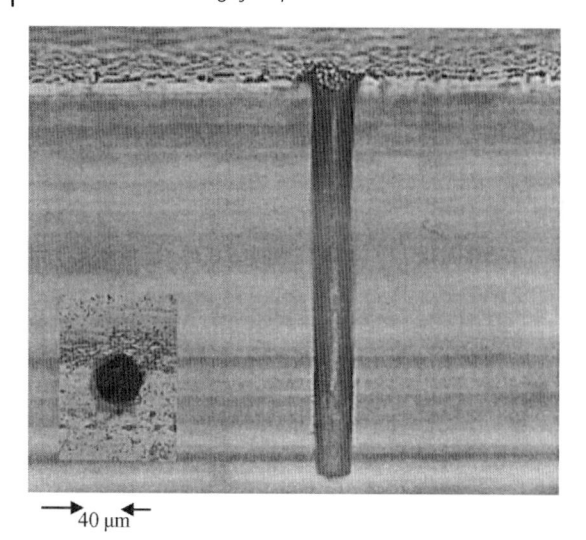

Figure 8.4 Microscope image of the side and top view (inset) of a typical hole (diameter of 40 μm) machined in PMMA using 100 fs laser pulses at a wavelength of 800 nm. The hole has a diameter of 40 μm and depth about 400 μm (aspect ratio of 10) and was machined by 0.4 mJ pulses at a repetition rate of 50 Hz for 5 s using a beam focused by an $f = 40$ mm plano-convex lens. (Reproduced with permission from [68]. Copyright © (2002) Elsevier.)

Concerning the machining of biodegrable polymers for biomedical engineering, Aguilar *et al.* used both a Ti: sapphire femtosecond pulsed laser and an ArF excimer laser for direct patterning of micron-sized channels and holes in poly(ε-caprolactone) and poly(glycolic acid) [72]. They reported that both lasers could successfully pattern biodegradable polymers leaving the bulk properties intact. Figure 8.5 shows an example of a laser-cut biodegradable polymer stent [73].

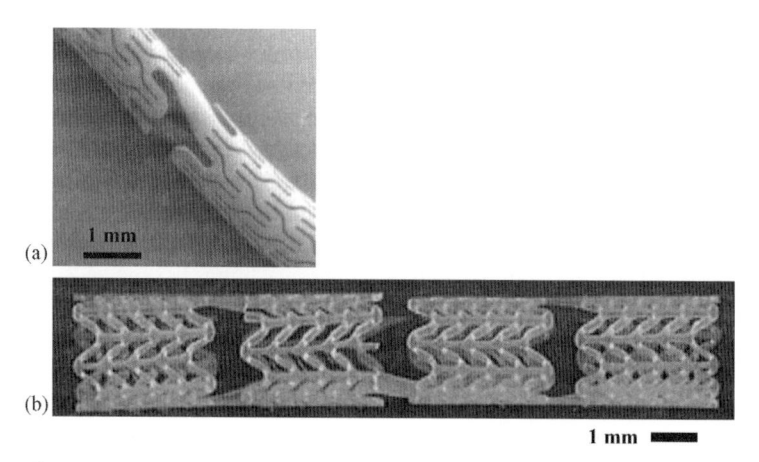

Figure 8.5 (a) Detail of a femtosecond laser cut biodegradable polymer stent; (b) stent after expansion by a balloon catheter. (Reproduced with permission from [73]. Copyright © (2002) SPIE.)

Finally, femtosecond pulse laser ablation is used to machine sensitive biological materials such as collagen polymer [74]. It has been used to section a number of biological tissues like human corneas for a long time [75]. There are current approaches in medical technology to apply ultrashort pulse laser for high-precision treatment of tissue. It is the aim of these endeavours to keep the relevant treatment as locally confined and with as few side effects as possible by means of nonthermal and tissue-conserving ablation. In ophthalmology the precise treatment of the transparent tissue of the ocular lens or the cornea has become possible. The use of femtosecond lasers avoids shockwaves that might damage the tissue. For example, the process is being examined for the treatment of glaucoma, where an increased intraocular pressure can lead to injuries of the retina or the optic nerve [75].

The extensive industrial application of ultrashort pulse lasers is, however, still inhibited by the currently high costs of the beam sources and the low average powers and repetition rates. Pico and especially femtosecond lasers are applied only if their cost is justified by economic or scientific added value of the relevant process step.

8.4
Laser-Induced Roughening

Laser-induced roughening of the polymer surfaces can be observed below, near and above the ablation threshold. Below the threshold, it can be initiated by photolytically driven reactions, for example, by photo-oxidation or photobleaching. Near the ablation threshold, the change of surface topography plays an important role, for example, with respect to wettability and surface energy. Above the threshold, laser-induced roughening (pores, cracks) can be observed depending on material and laser parameters. A number of applications combine both laser roughening and laser patterning via ablation.

Laser roughening has been used extensively to produce random microstructures on the surface of polymer films, which help to improve adhesive bonding, control surface friction and align liquid crystals [76–79]. Superhydrophobic PDMS surfaces with lotus-type micro- and nanoscale roughening have been generated by various groups using different laser types [80–83]. Bremus-Köbberling and Gillner [82] studied the influence of different topographies (regular ripplings, scales, wells and pillars) with dimensions from 50 to 250 μm to modify the surface properties of polymers. They reported that the hydrophobic polymers like PDMS and perfluoro-aloxy-copolymer (PFA) became superhydrophobic (lotus effect), while polymers containing polar functional groups like PEEK, PC, PMMA, and PET, or polymers with aromatic groups like PS strongly increased in hydrophilicity. They used laser wavelength of 193 nm and laser fluences of $0.8–1.5\,\mathrm{J\,cm^{-2}}$, which is clearly above the ablation threshold of the treated polymers. Pfleging et al. obtained microstructured PS surfaces nearby the ablation threshold using an excimer laser at 193 nm (ArF) at laser fluences of $90\,\mathrm{mJ\,cm^{-2}}$ (Figure 8.6) [84, 85]. The wettability of the surface was switched between superhydrophobic and superhydrophilic only by changing the processing gas. With oxygen as processing gas and for moderate laser pulse

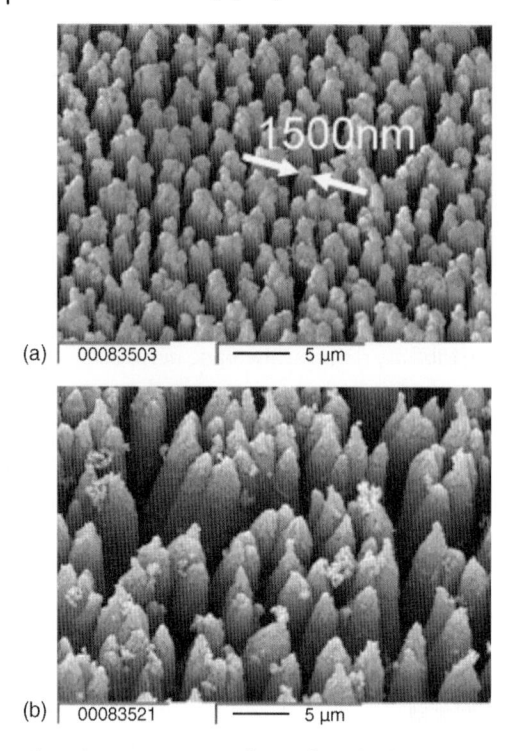

Figure 8.6 SEM images of PS surface after laser treatment with 400 laser pulses in He atmosphere (a) and O_2 atmosphere (b). (Reproduced with permission from [84]. Copyright © (2007) Elsevier.)

numbers, superhydrophilic surfaces were produced, while with He as processing gas the topographical surface change led to a superhydrophobic, lotus-like behavior with contact angles of nearly 160°. The changes in the surface hydrophobicity and hydrophilicity have also been exploited for controlling cell attachment and proliferation to polymer surfaces [85–88].

The physico-chemical changes of polymer surfaces upon laser treatment can be tailored for sensing applications for fast response, sensitivity, repeatability and chemical selectivity. As an example, the absorbance of PMMA thin films irradiated at 157 nm was shown to increase as a result of chemical modification of the material surface. Polymer hydrophobicity, which is responsible for the polymer volume expansion during sorption, was enhanced after irradiation. The methodology was used to enhance sensor detection sensitivity of ethanol and methanol analytes [89].

If the photon energy is high enough, material modification can be induced below the ablation threshold. The high energy of the photons dissociates the bonds in the molecules which lead to new molecule structures after reorganization processes at the surface and possible molecule desorption leading to a local change of refractive index. In particular, the refractive index of some polymers can be locally increased in a controllable way by irradiation using UV lasers [e.g., excimer or UV ns laser (like fourth harmonic of Nd: YAG at 266 nm)] or femtosecond lasers. For example, Wochonowski et al. [90] generated integrated-optical waveguiding and dispersive structures in the surface of a planar PMMA chip using an excimer laser with a mask, leading to an integrated Bragg sensor. These are well-suited for lab-on-chip devices or opto-fluidic chips with integrated optical sensors. Pfleging et al. [91] have directly microstructured PMMA using excimer laser at 193 nm in order to generate microchannels of 100 µm width and more than 100 µm depth in a first step, and in a second step generated an optical waveguide of 7 µm width in the transverse direction to the channel by direct exposure at 248 nm. The functionality of this device was successfully applied for optical detection measurements with 670 nm. Optical waveguides have also been fabricated on the surface of silicone $[(SiO(CH_3)_2)_n]$ rubber by photochemical modification of silicone rubber into silica with F_2 excimer laser radiation (157 nm) [92].

Femtosecond lasers have emerged in recent years as a powerful tool for materials processing, with waveguide and grating technologies being two of the foremost applications. Due to the high peak powers of the femtosecond laser pulses, refractive index changes in the host material are induced via nonlinear effects and therefore photosensitivity is not a prerequisite for inscription of waveguides. Waveguide writing offers striking advantages: it is a direct, single step, maskless fabrication technique; it is a 3D technique, enabling the generation of waveguides at arbitrary depths inside the material. Current techniques used for waveguide fabrication, such as photolithography, reactive ion etching, and high-energy ion implantation, are inherently planar technologies that require numerous processing steps and most often the prior design and fabrication of a mask. In particular, PMMA is an inexpensive and widely used polymer for the cores of communications-grade polymer optical fibers. Femtosecond laser-induced refractive index change in polymers has been explained by a photochemical modification, resulting in a negative or positive index change depending on the parameters used to write the waveguides [93, 94]. High-repetition-rate low-energy femtosecond laser pulses were used to fabricate high quality optical waveguides inside biocompatible hydrogel polymers [95]. Refractive-index gratings have been photo-induced in bulk PMMA or azodye-doped PMMA using femtosecond lasers [96, 97]. Using femtosecond laser, waveguides can be realized not only on the surface but also deep inside the volume of the polymer substrate [98].

Femtosecond laser was also used to ablate polypropylene film with micropatterns specifically tailored in size, location and number for breathable microperforated packaging films [99].

With the rapid development of high-pulse-energy and high-peak-power lasers, particularly at UV wavelengths and in the femtosecond range, interest in laser direct writing of holographic gratings for holographic data storage has increased. Relief gratings on the surface and inside PMMA have been written by ablation and change of refractive index respectively by holography using two beam interference of individual femtosecond pulse at 800 nm [100].

8.5
Generative Laser Processes

Laser-based generative processes such as stereolithography and selective laser sintering are part of a group of techniques commonly known as "layered manufacturing". These are rapid prototyping techniques that build up a 3D object (or physical model) layer by layer. They are used in fields like mechanical engineering, and more recently in medicine and health care, as they are fast and cost effective techniques for the manufacture of 3D parts.

8.5.1
Microstereolithography

Stereolithography, also known as "laser direct write polymerization", is a well-established direct manufacturing technique for 3D rapid prototyping in the macro-world. This technique relies on light-induced, space-resolved polymerization or curing of a liquid curable resin which solidifies when selectively exposed to visible or UV light. Each layer fuses to the one below, allowing the creation of complex structures. Software is available to generate sacrificial support structures for regions such as overhangs, which need support during the building phase. Microstereolithograhy (μSL), also called micro-photoforming, spatial forming, 3D optical modeling, or 3D-photopolymerization, is a general designation of various microfabrication technologies with high resolution based on the principle used in stereolithography [101]. An active domain of application is the photopolymerization of hydrogel for biological and biomedical applications (e.g. guiding cells and tissue formation) [102, 103].

In microstereolithography, structures can be constructed using direct 3D writing process in a point-by-point or line-by-line mode (vector-by-vector microstereolithography) by scanning an accurately focused laser beam on the surface of a photosensitive resin [104] or in a layer-by-layer mode through single exposure steps using the projection method (integral microstereolithography) [105]. Integral microstereolithography is faster than vector-by-vector microstereolithography. The height of the structure is determined by the exposure length and dose. Zhang et al. [106] fabricated a microtube with an aspect ratio of 16 (inner diameter of 50 μm and height of 800 μm) on a silicon substrate (Figure 8.7) by solidifying multiple layers of 20 μm thickness from a 1.6–hexanediol diacrylate based resin mixed with a photoinitiator using a UV laser (argon laser at 364 nm).

The stereolithographic method can be extended to sub-micrometer resolution (below the diffraction limit of the light used) using a two-photon polymerization process [107, 108]. The structure is then created by exposing a transparent liquid resin locally to two high intensity light beams, usually from one (with a split beam) or two laser sources. Using two beams allows the light of each single beam to penetrate the liquid without generating photo-induced crosslinking as the photopolymer is transparent at the laser wavelength. Only at the location where the two beams meet is the light intensity high enough to cross-polymerize the photoresist. The quadratic

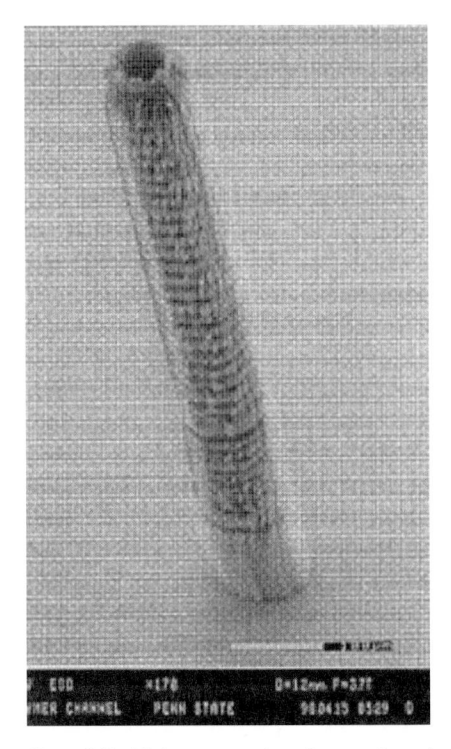

Figure 8.7 High aspect ratio polymer microtube with inner diameter of 50 μm and height of 800 μm (aspect ratio 16) fabricated by multiple layer μSL. (Reproduced with permission from [106]. Copyright © (1999) Elsevier.)

dependence of the two-photon absorption rate on the light intensity confines the absorption to the vicinity of the focal point of the laser beams, thus generating a 3D volume element, called a voxel, of solidified resist. As the two-photon transition rate is extremely small, the power of the light source has to be extremely high. Such high power light sources can be achieved through ultrashort pulse lasers. The advantage of this method is the simplicity of the process and the short time to produce the device [109, 110]. In recent years, microstereolithography based on two-photon photopolymerization has been studied to increase the 3D spatial resolution. Nanofabrication has been demonstrated by a number of groups. In particular, layer-by-layer or woodpile photonic-crystal structures with a resolution down to 100 nm have been produced by two-photon polymerization in a number of groups [111, 112]. Multiphoton 3D photopolymerization processes are well adapted to producing organic–inorganic composite polymers or organic polymers loaded with high inorganic content (green ceramic parts) from resins filled with ceramic particles. These have been used in photonic applications [113, 114] as well as in biomedicine [115].

There are, of course, many advantages to creating a 3D, complex micro-object directly in the resin: no time is spent spreading the liquid on the part being manufactured, which potentially can speed-up the process significantly. Freely movable structures can be fabricated without the need of sacrificial layers. The method enables more freedom for fabrication of 3D lithographic structures than holographic or gray tone lithography. There are numerous resins with special additives developed for special applications. The approach can easily incorporate different polymer solutions for each layer (or even for partial layers). This allows μSL the unique ability to generate heterogeneous polymer structures by stacking different polymerized layers. It can also eliminate the need for microassembly and bonding [116, 117]. Also, structures can be integrated directly with wafers and with other components [118, 119].

8.5.2
Selective Laser Sintering

Selective laser sintering (SLS) involves the selective use of a laser to build up a physical model layer by layer from a fine polymer powder. The powder particles soften or melt, adhere and solidify (or sinter) under laser illumination. Functional prototypes produced via laser sintering already address the sub-millimeter region with resolutions in the low micrometer range. The manufacturing process is very suitable for small scale, micro-featured functional polymer parts because of its ability to generate complex and fine shaped geometries including undercuts and pores.

The parts are built up layer-by-layer from 3D data generated by CAD programs or by micro-computer tomography as a copy of already existing parts. The process of laser sintering can be divided into four steps. First a thin and uniform polymer powder layer with a high density is applied over the whole building platform or over polymer layers that have been previously applied. Afterwards the newly applied powder (which has commonly been pre-heated before applying it on the platform) has to be heated as fast as possible above the crystallization temperature of the material. The pre-heating avoids high temperature gradients in the already sintered material, which can cause shrinkage and deformation of the parts laying in the powder bed which have been manufactured so far. After preparing the preconditions the laser treatment starts. The power of the scanner guided laser beam which sinters the polymer material has to be adapted in reference to the absorption of the used material, so that only the amount of heat energy which is needed to overcome the melting enthalpy of the material is absorbed. During fabrication, the object is supported and embedded by the surrounding unprocessed powders and has to be extracted from the powder bed after fabrication.

Selective laser sintering can perfectly well produce structures with complex external and internal geometries such as containing channels and overhanging features. Other advantages of SLS for polymer processing include the fact that it is solvent free and does not require any secondary binder system, hence minimizing any risk of material contamination. It can also incorporate multiple materials. Process parameters such as laser power, scan speed and part bed temperature used during SLS fabrication determine the amount of heat energy delivered onto the powders and hence the quality of sintering attained during SLS fabrication. Since the powders are subjected to low compaction forces during their deposition to form new layers, SLS-fabricated objects are usually porous. The pores created in the parts are dependent on the particle size of the powder stock used and the compaction pressure exerted onto the powder bed while depositing the powder layers, and have sizes that are distributed over a wide range of values. Kruth *et al.* provides a survey of laser-induced consolidation mechanisms for polymer and composites powders [120].

Since the SLS technique involves high processing temperatures, it is limited to the processing of thermally stable polymers. Up to now the production of prototypes and functional parts has been dominated by polyamide, a semi-crystalline polymer. This thermoplastic material covers a wide range of applications but requirements concerning a high temperature resistance above 200 °C and tensile strengths above $50\,\text{N}\,\text{mm}^{-2}$ are not fulfilled despite glass fillings. Further materials which are investigated include other thermoplastic materials, PC, PS [121], PEEK [122], PE [123]. In particular, SLS of biocompatible polymers such as PEEK, polyvinyl alcohol (PVA), polycaprolactone, poly(L-lactic acid) [124–127] is an area of active research for manufacturing scaffolds for tissue engineering.

8.6
Conclusion

Laser machining and laser-based processing are well established techniques for polymer processing. They provide flexibility and obvious advantages in terms of rapid manufacturing and customization of devices. Laser processes will very likely become more prevalent as lasers offer more cost-effective solutions.

Laser ablation is suitable for direct etching of materials. It presents several advantages compared to other technologies: the process is data-driven, that is, patterns can be generated without the need for masks, resist materials or chemicals. It offers a single step method of direct writing of features of various geometries, depths and aspect ratios. Laser processing parameters, such as fluence and wavelength, can be adjusted to suit a specific application. Lasers allow rapid prototyping and small-batch manufacturing. They can also be used to pattern moving substrates, permitting fly-processing of large areas at reasonable speed. Different types of laser processes such as ablation, modification and welding can be successfully combined in order to enable a high grade of functionality. Ultraviolet lasers are favored for precise and debris-free machining of polymers; for faster operation, thermally driven laser processes using NIR and IR laser radiation could be increasingly attractive for a real rapid manufacturing.

Femtosecond processing is likely to remain a niche technology because nanosecond lasers are more mature, simpler, cheaper for most applications. Because of cost and lack of robustness, femtosecond sources will remain the preserve of the research laboratory. Some of the advantages of femtosecond sources, such as reduced thermally induced damage, can often be minimized by appropriate materials processing techniques. But their advantages in terms of ultimate precision and 3D capability will create new functionalities. The development of high average power femtosecond lasers with much higher repetition rate will lead to faster machining rates, making it more competitive for a number of processes.

Generative laser-based techniques for 3D microfabrication are widely used and their importance is likely to grow as the need for rapid prototyping and manufacturing as well as customization of a variety of products is increasing.

Acknowledgements

This work was carried out within the framework of the European Union (EU) Network of Excellence "Multi-Material Micro Manufacture: Technology and Applications (4M)" (EC funding FP6-500274-1; http://www.4m-net.org).

References

1 Meijer, J. (2004) Laser beam machining (LBM), state of the art and new opportunities. *J. Mater. Process. Tech.*, **149** (1–3), 2–17.
2 Booth, H.J. (2004) Recent applications of pulsed lasers in advanced materials processing. *Thin Solid Films*, **453–454**, 450–457.
3 Ali, M., Wagner, T., Shakoor, M., and Molian, P.A. (2008) Review of laser nanomachining. *J. Laser Appl.*, **20** (3), 135–191.
4 Pfleging, W., Przybylski, M., and Brückner, H.J. (2006) Excimer laser material processing – State of the art and new approaches in microsystem technology. *Proc. SPIE*, **6107**, 61070G1–61070G15.
5 Siegman, A.E. (1993) Defining, measuring, and optimizing laser beam quality. in Laser Resonators and Coherent Optics: Modeling, Technology, and Applications, (ed. A. Bhowmik) Proc. SPIE, 1868, 2–12.
6 Graubner, V.-M., Jordan, R., Nuyken, O., Lippert, T., Hauer, M., Schnyder, B., and Wokaun, A. (2002) Incubation and ablation behaviour of poly (dimethylsiloxane) for 266 nm irradiation. *Appl. Surf. Sci.*, **197-198**, 786–790.
7 Gomez, D. and Goenaga, I. (2006) On the incubation effect on two thermoplastics when irradiated with ultrashort laser pulses: Broadening effects when machining microchannels. *Appl. Surf. Sci.*, **253**, 2230–2236.
8 Srinivasan, R. and Mayne-Banton, V. (1982) Self-developing photoetching of poly(ethylene terephthalate) films by far-ultraviolet excimer laser radiation. *Appl. Phys. Lett.*, **41** (6), 576–578.
9 Gattass, R.R. and Mazur, E. (2008) Femtosecond laser micromachining in transparent materials. *Nature Photonics*, **2**, 219–225.
10 Sugioka, K. (2010) *Ultrafast Laser Processing of Glass Down to the Nano-Scale*, vol. 130 Springer Series in Materials Science, Springer, Berlin Heidelberg, pp. 279–293. doi: 10.1007/978-3-642-03307-0.
11 Sugioka, K. (2009) Three-dimensional femtosecond laser micromachining of photosensitive glass for biomicrochips. *Laser & Photon. Rev.*, 1–15. doi: 10.1002/lpor.200810074.
12 Srinivasan, R., Sutcliffe, E., and Braren, B. (1987) Ablation and etching of polymethylmethacrylate by very short (160fs) ultraviolet (308nm) laser pulses. *Appl. Phys. Lett.*, **51**, 1285–1287.
13 Pettit, G.H. and Sauerbrey, R. (1993) Pulsed ultraviolet laser ablation. *Appl. Phys. A*, **56** (1), 51–63.
14 Schmidt, H., Ihlemann, J., Wolff-Rottke, B., Luther, K., and Troe, J. (1998) Ultraviolet laser ablation of polymers: spot size, pulse duration, and plume

attenuation effects explained. *J. Appl. Phys.*, **83**, 5458–5468.

15 Frisoli, J.K., Hefertz, Y., and Deutsch, T.F. (1991) Time-resolved UV absorption of polyimide. *Appl. Phys. B*, **52**, 168–172.

16 Lippert, T., Dickinson, J.T., Langford, S.C., Furutani, H., Fukumura, H., Masuhara, H., Kunz, T., and Wokaun, A. (1998) Photopolymers designed for laser ablation – photochemical ablation mechanism. *Appl. Surf. Sci.*, **127–129**, 117–121.

17 Fardel, R., Nagel, M., Nüesch, F., Lippert, T., and Wokaun, A. (2007) Fabrication of organic light-emitting diode pixels by laser-assisted forward transfer. *Appl. Phys. Lett.*, **91**, 061103. doi:

18 Doraiswamy, A., Narayan, RJ., Lippert, T., Urech, L., Wokaun, A., Nagel, M., Hopp, B., Dinescu, M., Modi, R., Auyeung, R.C.Y., and Chrisey, D.B. (2007) Excimer laser forward transfer of mammalian cells using a novel triazene absorbing layer. *Appl. Phys. Lett.*, **91**, 061103. doi: 10.1063/1.2759475.

19 Kawamura, Y., Toyoda, K., and Namba, S. (1982) Effective deep ultraviolet photoetching of polymethyl methacrylate. *Appl. Phys. Lett.*, **40**, 374–375.

20 Bäuerle, D. (2000) *Laser Processing and Chemistry*, Springer.

21 Lippert, T. and Dickinson, J.T. (2003) Chemical and spectroscopic aspects of polymer ablation: Special features and novel directions. *Chem. Rev.*, **103**, 453–485.

22 Van Steenberge, G., Geerinck, P., Van Put, S., Van Koetsem, J., Ottevaere, H., Morlion, D., Thienpont, H., and Van Daele, P. (2004) MT-compatible laser-ablated interconnections for optical printed circuit boards. *J. Lightwave Technol.*, **22** (9), 2083–2090.

23 Beinhorn, F., Ihlemann, J., Luther, K., and Troe, J. (1999) Micro-lens arrays generated by UV laser irradiation of doped PMMA. *Appl. Phys. A*, **68** (6), 709–713.

24 Naessens, K., Ottevaere, H., Baets, R., Van Daele, P., and Thienpont, H. (2003) Direct writing of microlenses in polycarbonate with excimer laser ablation. *Appl. Opt.*, **42**, 6349–6359.

25 Bekesi, J., Meinertz, J., Ihlemann, J., and Simon, P. (2008) Fabrication of large-area grating structures through laser ablation. *Appl. Phys. A*, **93**, 27–31.

26 Ghantasala, M.K., Hayes, J.P., Harvey, E.C., and Sood, D.K. (2001) Patterning, electroplating and removal of SU-8 moulds by excimer laser micromachining. *J. Micromech. Microeng.*, **11**, 133–139.

27 Roberts, M.A., Rossier, J.S., Bercier, P., and Girault, H. (1997) UV laser machined polymer substrates for the development of microdiagnostic systems. *Anal. Chem.*, **69**, 2035–2042.

28 Rossier, J.S., Bercier, P., Schwarz, A., Loridant, S., and Girault, H.H. (1999) Topography, crystallinity and wettability of photoablated PET surfaces. *Langmuir*, **15**, 5173–5178.

29 Bianchi, F., Chevelot, Y., Mathieu, H.J., and Girault, H.H. (2001) Photomodification of polymer microchannels induced by static and dynamic excimer ablation: effect on the electroosmotic flow. *Anal. Chem.*, **73**, 3845–3853.

30 Schwarz, A., Rossier, J.S., Roulet, E., Mermod, N., Roberts, M.A., and Girault, H.H. (1998) Micropatterning of biomolecules on polymer substrates. *Langmuir*, **14**, 5526–5531.

31 Waddell, E.A., Locascio, L.E., and Kramer, G.W. (2002) UV laser micromachining of polymers for microfluidic applications. *JALA*, **7** (1), 78–82.

32 Johnson, T.J., Waddell, E.A., Kramer, G.W., and Locascio, L.E. (2001) Chemical mapping of hot-embossed and UV-laser-ablated microchannels in poly(methyl methacrylate) using carboxylate specific fluorescent probes. *J. Appl. Surf. Sci.*, **181**, 149–159.

33 Pugmire, D.L., Waddell, E.A., Haasch, R., Tarlov, M.J., and Locascio, L.E. (2002) Surface characterization of laser-ablated polymers used for microfluidics. *Anal. Chem.*, **74** (4), 871–878.

34 Atkin, M., Hayes, J.P., Brack, N., Poetter, K., Cattrall, R., and Harvey, E.C. (2002) Disposable biochip fabrication for DNA diagnostics. *Proc. SPIE*, **4937**, 125–135.

35 Burt, P., Goater, A.G., Hayden, C.J., and Tame, J.A. (2002) Laser micromachining of biofactory-on-a-chip devices. *Proc. SPIE*, **4637**, 305–317.

36 Lin, Y., Wen, J., Fan, X., Matson, D.W., and Smith, R.S. (1999) Laser micromachined isoelectric focusing devices on polymer substrate for electrospray mass spectrometry. *Proc. SPIE*, **3877**, 28–35.

37 Gillner, A., Bremus-Koebberling, E.A., Wehner, M., Russek, U.A., and Berden, T. (2001) Laser processing of components for polymer mircofluidic and optoelectronic products. *Proc. SPIE*, **4274**, 411–419.

38 Andrews, J.R. and Gerner, B. (2003) Laser processes for prototyping and production of novel microfluidic structures. *Proc. SPIE*, **5345**, 147–158.

39 Lee, H.J., Beattie, P.D., Seddon, B.J., Osborne, M.D., and Girault, H. (1997) Amperometric ion sensors based on laser-patterned composite polymer membranes. *J. Electroanal. Chem.*, **440**, 73–82.

40 Matson, D.W., Martin, P.M., Bennett, W.D., Stewart, D.C., and Johnston, J.W. (1997) Laser-micromachined microchannel solvent separator. *Proc. SPIE*, **3223**, 253–259.

41 McNeely, M.R., Spute, M.K., Tusneem, N.A., and Oliphant, A.R. (1999) Hydrophobic microfluidics. *Proc. SPIE*, **3877**, 210–220.

42 Sabbert, D., Landsiedel, J., Bauer, H.-D., and Ehrfeld, W. (1999) ArF-excimer laser ablation experiments on cycloolefin copolymer (COC). *Appl. Surf. Sci.*, **150**, 185–189.

43 Zheng, H., Gan, E., and Lim, G.C. (2001) Investigation of laser via formation technology for the manufacturing of high density substrates. *Opt. Laser Eng.*, **36** (4), 355–371.

44 Riccardi, G., Cantello, M., Mariotti, F., and Giacosa, P. (1998) Micromachining xith excimer laser. *Ann. CIRP*, **47** (1), 145–148.

45 Gower, M. and Rizvi, N. (2000) Applications of laser ablation to microengineering. *Proc. SPIE*, **4065**, 452–460.

46 Kim, J. and Xu, X. (2003) Laser-based fabrication of micro-fluidic components and systems. *Proc. SPIE*, **4982**, 73–82.

47 Barrett, R., Fauchon, M., Lopez, J., cristobal, G., Destremaut, F., DDDodge, A., Guillot, P., Laval, P., Masson, C., and Salmon, J.-B. (2006) X-ray microfocussing combined with microfluidivs for on-chip X-ray scattering measurements. *Lab-on-Chip*, **6**, 494–499.

48 Moss, E.D., Han, A., and Frazier, A.B. (2007) A fabrication technology for multi-layer polymer-based Microsystems with integrated fluidic and electrical functionality. *Sensor. Actuat. B*, **121**, 689–697.

49 Yin, H., Killeen, K., Brennen, R., Sobek, D., Werlich, M., and van de Goor, T. (2005) Microfluidic chip for peptide analysis with an integrated HPLC column, sample enrichment column, and nanoelectrospray tip. *Anal. Chem.*, **77** (2), 527–533.

50 Yao, L., Liu, B., Chen, T., Liu, S., and Zuo, T. (2005) Micro flow-through PCR in a PMMA chip fabricated by KrF excimer laser. *Biomed. Microdevices*, **7** (3), 253–257.

51 Yu, H., Balogun, O., Li, B., Murray, T.W., and Zhang, X. (2005) Rapid manufacturing of embedded microchannels from a single layered SU-8 and determining the dependence of SU-8 Young's modulus on exposure dose with a laser acoustic technique. IEEE Proc. MEMS'05 pp. 654–657.

52 Jensen, M.F., McCormack, J.E., Helbo, B., Christensen, L.H., Christensen, T.R., and Geschke, O. (2004) Rapid prototyping of polymer microsystems via excimer laser ablation of polymeric moulds. *Lab-on-Chip*, **4**, 391–395.

53 Kancharla, V.V. and Chen, S. (2002) Fabrication of biodegradable polymeric devices using laser micromachining. *Biomed. Microdevices*, **4** (2), 105–109.

54 Chen, S., Kancharla, V.V., and Lu, Y. (2003) Laser-based microscale patterning of biodegradable polymers for biomedical applications. *Int. J. Mater. Prod. Tech.*, **18** (4–5), 457–468.

55 Klank, H., Kutter, J.P., and Geschke, O. (2002) CO_2 laser micromachining and back-end processing for rapid production

56 Snakenborg, D., Klank, H., and Kutter, J.P. (2004) Microstructure fabrication with a CO_2 laser system. *J. Micromech. Microeng.*, **14**, 182–189.

57 Liu, Y., Rauch, C.B., Stevens, R.L., Lenigk, R., Yang, J., Rhine, D.B., and Grozinski, P. (2002) DNA amplification and hybridization assays in integrated plastic monolithic devices. *Anal. Chem.*, **74**, 3063–3070.

58 Cheng, J.-Y., Wei, C.W., Hsu, K.-H., and Young, T.-H. (2004) Direct-write laser micromachining and universal surface modification of PMMA for device development. *Sensor. Actuat. B*, **99**, 186–196.

59 Wang, S.-C., Lee, C.-Y., and Chen, H.-P. (2006) Thermoplastic microchannel fabrication using carbon dioxide laser ablation. *J. Chromatogr. A*, **1111** (2), 252–257.

60 Yuan, D. and Das, S. (2007) Experimental and theoretical analysis of direct-write laser micromachining laser ablation of polymethyl methacrylate by CO2. *J. Appl. Phys.*, **101** (2), 024901.1–024901.6.

61 Pfleging, W., Kohler, R., Schierjott, P., and Hoffmann, W., (2009) Laser patterning and packaging of CCD-CE-Chips made of PMMA. *Sensor. Actuat. B*, **138** (1), 336–343, doi: 10.1016/j.snb.2009.01.036.

62 Zhou, G., Ventura, M.J., Vanner, M.R., and Gu, M. (2005) Fabrication and characterization of face-centred-cubic void dots photonic crystals in a solid polymer material. *Appl. Phys. Lett.*, **86** (1), 011108 (3 pages).

63 Farson, D.F., Choi, H.W., Lu, C., and Lee, L.J. (2006) Femtosecond laser bulk micromachining of microfluidic channels in poly(methylmethacrylate). *J. Laser Appl.*, **18** (3), 210–215.

64 Wolfe, D.B., Ashcom, J.B., Hwang, J.C., Schaffer, C.B., Mazur, E., and Whitesides, G.M. (2003) Customization of poly(dimethylsiloxane) stamps by micromachining using a femtosecond-pulsed laser. *Adv. Mater.*, **15** (1), 62–65.

65 Kim, T.N., Campbell, K., Groisman, A., Kleinfeld, D., and Schaffer, C.B. (2005) Femtosecond laser-drilled capillary integrated into a microfluidic device. *Appl. Phys. Lett.*, **86**, 201106-1–201106-3.

66 Gomez, D., Goenaga, I., Lizuain, I., and Ozaita, M. (2005) Femtosecond laser ablation for microfluidics. *Opt. Eng.*, **44** (5), 051105-1–051105-8.

67 Day, D. and Gu, M. (2005) Microchannel fabrication in PMMA based on localized heating by nanojoule high-repetition rate femtosecond pulses. *Opt. Express*, **13** (16), 5939–5946.

68 Zhang, Y., Lowe, R.M., Harvey, H., Hannaford, P., and Endo, A. (2002) High aspect-ratio micromachining of polymers with an ultrafast laser. *Appl. Surf. Sci.*, **186**, 345–351.

69 Niino, H. and Yabe, A. (2001) Laser ablation of transparent materials by UV fs-laser irradiation. *J. Photopolym. Sci. Tech.*, **14** (2), 197–202.

70 Lee, S., Bordatchev, E.V., and Zeman, M.J.F. (2008) Femtosecond laser micromachining of polyvinylidene fluoride (PVDF) based piezo films. *J. Micromech. Microeng.*, **18** (4), 045011. (8p).

71 Obata, K., Sugioka, K., and Midorikawa, K. (2005) F_2 laser ablation of UV transparent polymer material. *JLMN*, **1** (1), 28–32.

72 Aguilar, C.A., Lu, y., Mao, S., and Chen, S. (2005) Direct-patterning of biodegradable polymers using ultraviolet and femtosecond lasers. *Biomaterials*, **26**, 7642–7649.

73 Ostendorf, A., Bauer, T., Korte, F., Howorth, J., Momma, C., Rizvi, N., Saviot, F., and Salin, F. (2002) Development of an industrial femtosecondlaser micro-machining system. in Commercial and Biomedical Applications of Ultrafast and Free-Electron Lasers, (eds G.S. Edwards, J. Neev, A. Ostendorf, J.C. Sutherland), Proc. SPIE, 4633, 128–135.

74 Liu, Y., Sun, S., Singha, S., Cho, M.R., and Gordon, R.J. (2005) 3D femtosecond laser patterning of collagen for directed cell attachment. *Biomaterials*, **26**, 4597–4605.

75 Kautek, W., Mitterer, S., Kruger, J., Husinsky, W., and Grabner, G. (1994)

Femtosecond-pulse laser ablation of human corneas. *Appl. Phys. A*, **58** (5), 513–518.

76 Niino, H., Kawabata, Y., and Yabe, A. (1989) Application of excimer laser polymer ablation to alignment of liquid crystals: periodic microstructure on polyethersulfone. *Jpn. J. Appl. Phys.*, **28** (12), L2225–L2227.

77 Petit, S., Laurens, P., Amourox, J., and Arefi-Khonsari, F. (2000) Excimer laser treatment of PET before plasma metallization. *Appl. Surf. Sci.*, **168**, 300–303.

78 Knittel, D. and Schollmeyer, E. (1998) Surface structuring of synthetic polymers by UV-laser irradiation. Part IV. Applications of excimer laser induced surface modification of textile materials. *Polym. Intern.*, **45** (1), 110–117.

79 Horn, H., Beil, S., Wesner, D.A., Weichenhain, R., and Kreutz, E.W. (1999) Excimer laser pretreatment and metallization of polymers. *Nucl. Instrum. Methods Phys. Res. B*, **151** (1), 279–284.

80 Khorasani, M.T., Mirzadeh, H., and Sammes, P.G. (1996) Laser induced surface modification of polydimethylsiloxane as a super-hydrophobic material. *Radiat. Phys. Chem.*, **47** (6), 881–888.

81 Khorasani, M.T. and Mirzadeh, H. (2004) Laser surface modification of silicone rubber to reduce platelet adhesion *in vitro*. *J. Biomat. Sci.-Polym. E.*, **15** (1), 59–72.

82 Bremus-Köbberling, E. and Gillner, A. (2003) Laser structuring and modification of surfaces for chemical and medical micro components. *Proc. SPIE*, **5063**, 217–222.

83 Yoon, T.O., Shin, H.J., Jeoungand, S.C., and Park, Y.-I. (2008) Formation of *superhydrophobic* poly(dimethysiloxane) by ultrafast laser-induced surface modification. *Opt. Express*, **16** (17), 12715–12725.

84 Pfleging, W., Bruns, M., Welle, A., and Wilson, S. (2007) Laser-assisted modification of polystyrene surfaces for cell culture applications. *Appl. Surf. Sci.*, **253** (23), 9177–9184.

85 Pfleging, W., Torge, M., Bruns, M., Trouillet, V., Welle, A., and Wilson, S. (2009) Laser- and UV-assisted modification of polystyrene surfaces for control of protein adsorption and cell adhesion. *Appl. Surf. Sci.*, **255**, 5453–5457.

86 Heitz, J., Olbrich, M., Mototz, S., Romanin, C., Svorcik, V., and Bäuerle, D. (2005) Surface modification of polymers by UV-irradiation: applications in micro- and biotechnology. *Proc. SPIE*, **5958**, 466–471.

87 Hopp, B., Smausz, T., Papdi, B., Bor, Z., Szabó, A., Kolozsvári, L., Fotakis, C., and Nógrádi, A. (2008) Laser-based techniques for living cell pattern formation. *Appl. Phys. A*, **93** (1), 45–49.

88 Brayfiel, C.A., Marra, K.G., Leonard, J.P., Cui, X.T., and Gerlach, J.C. (2008) Excimer laser channel creation in polyethersulfone hollow fibers for compartmentalized in vitro neuronal cell culture scaffolds. *Acta Biomatriala*, **4**, 244–255.

89 Sarantopoulou, E., Kollia, Z., Cefalas, A.C., Manoli, K., Sanopoulou, M., Goustouridis, D., Chatzandroulis, S., and Raptis, I. (2008) Surface nano/micro functionalization of *PMMA* thin films by 157nm irradiation for sensing applications. *Appl. Surf. Sci.*, **254** (6), 1710–1719.

90 Wochnowski, C., Abu-El-Qomsan, M., Pieper, W., Meteva, K., Metev, S., Wenke, G., and Vollertsen, F. (2005) UV-laser assisted fabrication of Bragg sensor components in a planar polymer chip. *Sensor. Actuat. A*, **120** (1), 44–52.

91 Pfleging, W., Adamietz, R., Brückner, H.J., Bruns, M., and Welle, A. (2007) Laser-assisted modification of polymers for microfluidic, microoptics and cell culture applications. *Proc. SPIE*, **6459**, 645911-1–645911-9.

92 Okoshi, M., Li, J., and Herman, P.R. (2005) 157nm F2 laser writing of silica optical waveguides in silicone rubber. *Opt. Lett.*, **30** (20), 2730–2732.

93 Zoubir, A., Lopez, C., Richardson, M., and Richardson, K. (2004) Femtosecond laser fabrication of tubular waveguides in

poly(methyl methacrylate). *Opt. Lett.*, **29**, 1840–1844.
94 Sowa, S., Watanabe, W., Tamaki, T., Nishii, J., and Itoh, K. (2006) Symmetric waveguides in poly(methyl methacrylate) fabricated by femtosecond laser pulses. *Opt. Express*, **14**, 291–297.
95 Ding, L., Blackwell, R.I., Künzler, J.F., and Knox, W.H. (2008) Femtosecond laser micromachining of waveguides in silicone-based hydrogel polymers. *Appl. Optics*, **47** (17/10), 3100–3108.
96 Scully, P.J., Jones, D., and Jaroszynski, D.A. (2003) Femtosecond laser irradiation of polymethylmethacrylate for refractive index gratings. *J. Opt. A*, **5**, S92–S96.
97 Si, J., Qiu, J., Zhai, J., Shen, Y., and Hirao, K. (2002) Photoinduced permanent gratings inside bulk azodye-doped polymers by the coherent field of a femtosecond laser. *Appl. Phys. Lett.*, **80** (3), 359.
98 Wochnowski, C., Meteva, K., Metev, S., Sepold, G., Vollertsen, F., Cheng, Y., Hanada, Y., Sugioka, K., and Midorikawa, K. (2006) Fs-laser-induced Fabrication of Polymeric Optical and Fluidic Microstructures. *JLMN*, **1** (3), 195–200.
99 Sohn, I.-B., Noh, Y.-C., Choi, S.-C., Ko, D.-K., Lee, J., and Choi, Y.-J. (2008) Femtosecond laser ablation of polypropylene for breathable film. *Appl. Surf. Sci.*, **254**, 4919–4924.
100 Li, Y., Yamada, K., Ishizuka, T., Watanabe, W., Itoh, K., and Zhou, Z. (2002) Single femtosecond pulse holography using polymethyl methacrylate. *Opt. Express*, **10** (21), 1173–1178.
101 Ikuta, K. and Hirowatari, K. (1993) Real three dimensional microfabrication using stereolithography and metal molding. Micro Electro Mechanical Systems, Proc. IEEE MEMS, Feb 1993, pp 432–447.
102 Akselrod, G.M., Timp, W., Mirsaidov, U., Zhao, Q., Li, C., Timp, R., Timp, K., Matsudaira, P., and Timp, G. (2006) Laser-guided assembly of heterotypic three-dimensional living cell microarrays. *Biophys. J.*, **91** (9), 3465–3473.
103 Yuan, D., Lasagni, A., Shao, P., and Das, S. (2008) Rapid prototyping of microstructured hydrogels via laser direct-write and laser interference photopolymerisation. *Virtual and Physical Prototyping*, **3** (4), 221–229.
104 Nakamoto, T., Yamaguchi, K., Abraha, P.A., and Mishima, K. (1996) Manufacturing of three-dimensional micro-parts by UV laser induced polymerization. *J. Micromech. Microeng.*, **6** (2), 240–253.
105 Ha, Y.M., Choi, J.W., and Lee, S.H. (2008) Mass production of 3-D microstructures using projection microstereolithography. *J. Mech. Sci. Tech.*, **22**, 514–521.
106 Zhang, X., Jiang, X.N., and Sun, C. (1999) Micro-stereolithography of polymeric and ceramic microstructures. *Sensor. Actuat. A*, **72** (2), 149–156.
107 Maruo, S., Nakamura, O., and Kawata, S. (1997) Three dimensional microfabrication with two-photon-absorbed photopolymerization. *Opt. Lett.*, **22**, 132–134.
108 Kawata, S., Sun, H., Tanaka, T., and Takada, K. (2001) Finer features for functional microdevices - Micromachines can be created with higher resolution using two-photon absorption. *Nature*, **412**, 697–698.
109 Sun, H.B. and Kawata, S. (2004) Two-photon polymerization and 3D lithographic microfabrication. *Adv. Polym. Sci.*, **170**, 169–173.
110 Sodian, R., Loebe, M., Hein, A., Martin, D.P., Hoerstrup, S.P., Potapov, EV., Hausmann, H., Lueth, T., and Hetzer, R. (2002) Application of stereolithography for scaffold fabrication for tissue engineering. *ASAIO J.*, **48** (1), 12–16.
111 Sun, H.B., Matsuo, S., and Misawa, H. (1999) Three-dimensional photonic crystal structures achieved with two-photon polymerization of resin. *Appl. Phys. Lett.*, **74**, 786–788.
112 Nguyen, L.H., Straub, M., and Gu, M. (2005) Acrylate-based photopolymer for two-photon microfabrication and photonic applications. *Adv. Funct. Mater.*, **15** (2), 209–216.
113 Serbin, J., Ovsianikov, O., and Chichkov, B. (2006) Fabrication of woodpile

structure by two-photon polymerization and investigation of their optical properties. *Opt. Lett.*, **31**, 1307–1309.

114 Chen, W., Kirihara, S., and Miyamoto, Y. (2007) Fabrication of three-dimensional micro photonic crystals of resin-incorporating TiO_2 particles and their terahertz wave properties. *J. Am. Ceram. Soc.*, **90** (1), 92–96.

115 Narayan, R.J., Jin, C., Doraiswamy, A., Mihailescu, I.N., Jelinek, M., Ovsianikov, A., Chichkov, B., and Chrisey, D.B. (2005) Laser processing of advanced bioceramics. *Adv. Eng. Mater.*, **7**, 1083–1098.

116 Hasegawa, T., Nakashima, K., Omatsu, F., and Ikuta, K. (2008) Multi-directional micro-switching valve chip with rotary mechanism. *Sensor. Actuat. A*, **143**, 390–398.

117 Kang, H.W., Lee, I.H., and Cho, D.W. (2004) Development of an assembly-free process based on virtual environment for fabricating 3D microfluidic systems using microstereolithography technology. *J. Manuf. Sci. Eng.*, **126**, 766–771.

118 Han, A., Graff, M., Wang, O., and Frazier, A.B. (2005) An approach to multilayer microfluidic systems with integrated electrical, optical and mechanical functionality. *IEEE Sensors J.*, **5** (1), 82–89.

119 Länge, K., Blaess, G., Voigt, A., Götzen, R., and Rapp, M. (2006) Integration of a surface acoustic wave biosensor in a microfluidic polymer chip. *Biosens. Bioelectron.*, **22**, 227–232.

120 Kruth, J.P., Levy, G., Klocke, F., and Childs, T.H.C. (2007) Consolidation phenomena in laser and powder-bed-based layered manufacturing. *Ann. CIRP*, **56**, 730–759.

121 Shi, Y.S. and Li, Z.C. (2004) Effect of the properties of polymer materials on the quality of selective laser sintering parts. *J. Mater. Des. Appl.* **218/L3**, 247–252.

122 Rechtenwald, T., Krauss, H.J., Pohle, D., and Schmidt, M. (2007) Small scale and micro featured functional prototypes generated by laser sintering of polyetheretherketone. in Micromachining Technology for Micro-Optics and Nano-Optics V and Microfabrication Process Technology XII., Bellingham (eds M.-A. Maher, H.D., Stewart, J.-C., Chiao, T.J., Suleski, E.G., Johnson and G.P. Nordin) Proc. SPIE, 6462, 646203.

123 Rimell, J.T. and Marquis, P.M. (2000) Selective laser sintering of ultra high molecular weight polyethylene for clinical applications. *J. Biomed. Mater. Res.*, **53** (4), 414–420.

124 Das, S., Hollister, S.J., Flanagan, C., Adeunmi, A., Bark, K., Chen, C., Ramaswamy, K., Rose, D., and Widjaja, E. (2003) Freeform fabrication of nymon-6 tissue engineering scaffolds. *Rapid Prototyping J.*, **9** (1), 43–49.

125 Tan, K.H., Chua, C.K., Leong, K.F., Cheah, C.M., Gui, W.S., Tan, W.S., and Wiria, F.E. (2005) SLS of biocompatible polymers for applications in tissue enginerring. *Bio-Med. Mater. Eng.*, **15**, 113–124.

126 Williams, J.M., Adewunmi, A., Schek, R.M., Flanagan, C.L., Krebsbach, P.H., Feinberg, S.E., Hollister, S.J., and Das, S. (2005) Bone tissue engineering using polycaprolactone scaffolds fabricated via selective laser sintering. *Biomaterials*, **26** (23), 4817–4827.

127 Antonov, E.N., Bagratashvili, V.N., Howdle, S.M., Konovalov, N., Popov, V.K., and Panchenkov, V.Ya. (2006) Fabrication of polymer scaffolds for tissue engineering using surface selective laser sintering. *Laser Phys.*, **16** (5), 774–787.

Part Four
Self-Organization

9
Colloidal Polymer Patterning
Eoin Murray, Philip Born, and Tobias Kraus

9.1
Introduction

Monodispersed lattices contain a sizable number of particles with uniform diameter. Their well-defined geometry forms the basis for larger patterns. Thus, instead of defining a number of identical substructures during patterning, only their arrangement is controlled; the substructure is efficiently defined by the particle geometry. A serial step is replaced by a parallel, low-cost method, namely, the emulsion synthesis of polymer particles.

Polymer lattices are widespread in polymer formulation [1], and the idea of using monodispersed lattices for patterning purposes is an old one. In 1968, the director of the Paint Research Institute, Raymond R. Mayers, suggested that *"opalescent coatings are almost a reality"* [2]. Today, optical applications are the main driving force behind particle-based polymer patterning, because particles with diameters in the range of optical wavelengths are easily accessible. Researchers from many other fields also increasingly use polymer particle structures, because the commercial availability of monodispersed latex suspensions (originally sold for calibration purposes in electron microscopy and particle sizing) has allowed groups without expertise in emulsion polymerization to use such particles for their structures.

Colloidal patterns are excellent templates that can mask parts of a surface during coating processes, provide a porous structure that can be filled with a second compound, or occupy parts of a volume so that pores can later be opened. Depending on the polymer and the other material involved, the templates can subsequently be removed by thermal decomposition or dissolution in an appropriate solvent. The particle surfaces can be tuned using a wide range of chemical functional groups to improve compatibility with different matrices.

While the production of monodispersed particles takes place in a well-understood process, their deposition in regular arrangements is less well-researched and remains challenging to perform on a large scale. In this chapter, we will describe the state-of-the-art of this field. The preparation of polymer particles via emulsion polymerization and the forces and mechanisms that can contribute to the arrangement of the particles will be described first. Different implementations of particle-

Generating Micro- and Nanopatterns on Polymeric Materials. Edited by A. del Campo and E. Arzt
Copyright © 2011 WILEY-VCH Verlag GmbH & Co. KGaA, Weinheim
ISBN: 978-3-527-32508-5

based polymer patterning are discussed, covering not only templated assembly, artificial opals, and the important application of colloidal lithography, but also the use of nonpolymer particles for polymer structuring. Some of the setups that have been used to aid the particle assembly are presented together with a number of applications that routinely use particle-based polymer structures.

9.2
Emulsion Polymerization

In emulsion polymerization, the polymerization process (typically radical initiated) takes place in micellar reactors composed of monomer droplets stabilized by surfactant molecules and dispersed in water (Figure 9.1). A colloidally stable polymer dispersion or latex is formed in this reaction by a complex mechanism consisting of three distinct intervals termed Smith–Ewert intervals [3–5]. On addition of a dispersed phase soluble monomer to the surfactant/solvent system, the system contains monomer-swollen small surfactant micelles ($\approx$ 10 nm in diameter) and large emulsion droplets of monomer. On the subsequent addition of a continuous phase soluble initiator, free radical species form which diffuse into the micelles. The monomer quickly polymerizes in the micelle and, as diffusion of monomer from the emulsion droplet to the micelle is rapid on the timescale of polymerization, the micelles contain both monomer and polymer. As the concentration of free monomer reduces to zero, the polymerization of the remaining monomer in the latex particles takes place, ending the reaction. Monodispersity is retained throughout the reaction to the final product as all polymerization takes place within the surfactant micelles [6, 7].

Emulsion and microemulsion polymerization are the most common ways to produce polymer dispersions and generate latex particles with diameters between 0.05 and 0.5 μm. In contrast to bulk radical polymerization it is possible to obtain high molecular weights at high polymerization rates. A large number of monodisperse polymer lattices with diameters between 20 and 100 nm have been synthesized in oil-in-water emulsions, most importantly for research and industrial processes, from

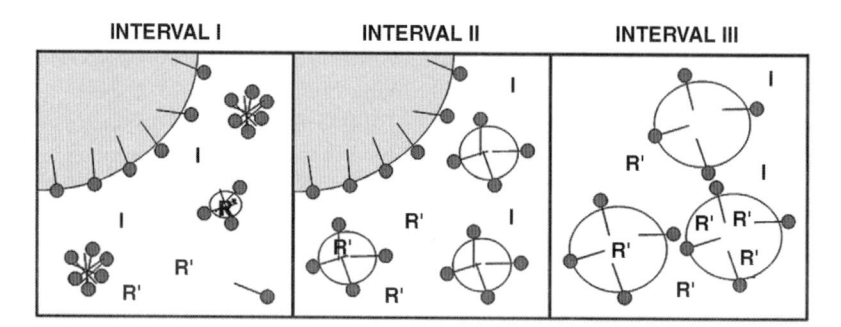

Figure 9.1 The three stages of emulsion polymerization. Surfactant molecules stabilize large monomer droplets and form small micelles, in which the polymerization takes place. (Reproduced with permission from [3]. Copyright © (2007) Elsevier.)

polystyrene (PS) [8–10], but also from a variety of polyalkylacrylates [11–13]. Recently nanolattices of PS were produced with a particle diameter of 2–4 nm in a microemulsion [14].

To produce monodisperse nanoparticles with very small sizes, high surfactant concentrations (up to 10–15 wt%) are usually required. However, colloidal crystal assemblies require a surfactant-free polymer surface. Surfactant-free emulsion polymerization processes have been developed for PS and a number of acrylate-based polymers [15–21]. In this process a poorly water-soluble monomer and water-soluble initiator, usually peroxodisulfate, are dispersed in water. Polymerization is initiated by the sulfate radical and charged oligomers are formed, which, when above the critical micelle concentration, aggregate into micelles. Polymerization further takes place in these micelles, resulting in monodisperse polymer nanoparticles on the scale of the original micelles.

9.3
Forces and Mechanisms in Polymer Dispersions

Particle behavior in colloidal suspensions is dominated by forces and mechanisms different from that relevant for macroscopic objects. Their behavior will be an interplay of sedimentation, random Brownian motion, viscous drag, and interparticle interactions. Thus, in order to use colloidal particles as building blocks for patterning in the micrometer and sub-micrometer range, the balance between these forces must be controlled to provide a method for the robust and reproducible placement of the particles. In the following a short description of the forces will be given.

9.3.1
Brownian Motion

A common characteristic of small scale colloidal systems is the constant thermal movement of the components. Kinetic energy is transferred to suspended particles by collisions of solvent molecules, causing a random motion usually called Brownian motion. This motion of particles causes diffusion, which is a net transport of randomly moving particles along density gradients. The mean displacement of a particle by Brownian motion over time leads to the diffusion coefficient, D (Equation 9.1) [22]:

$$\langle x^2 \rangle = \frac{k_B T}{3\pi \eta r} \cdot t = 2 \cdot D \cdot t, \tag{9.1}$$

where $\langle x^2 \rangle$ is the mean displacement of an isolated sphere after a time t. For example, the mean displacement of a 1 μm PS bead in water will be several particle radii in few seconds.

Brownian motion is a crucial factor in particle assembly. The mass transport from a source, the bulk colloid, to a drain, the pattern, can be accomplished by diffusion. After assembly, the particles have to be trapped to prevent further random motion.

9.3.2
Sedimentation and Viscous Drag

The Peclet number for sedimentation can be used to estimate the influence of gravity on a colloidal system. It specifies the ratio between gravitational settling and Brownian motion (Equation 9.2) [23]:

$$Pe \equiv \frac{8\pi r^3 \Delta \varrho g}{9 k_B T}, \qquad (9.2)$$

where r is the particle radius, $\Delta \varrho$ the difference in densities, g the gravitational acceleration, k_B the Boltzmann's constant, and T the absolute temperature. $Pe \gg 1$ indicates that the particle behavior is dominated by gravitation, while for $Pe \ll 1$ Brownian motion prevails. For 1 μm PS beads in water the Peclet number approaches unity.

Similarly, the Reynolds number (Re) measures the ratio between inertia and viscous drag, and has values below unity for PS beads in water (Equation 9.3) [24]:

$$\text{Re} \equiv \frac{\varrho v r}{\eta}, \qquad (9.3)$$

where v is the dynamic velocity and η the viscosity of the carrier fluid. A $\text{Re} \ll 1$ indicate that inertia is negligible for the system under examination and thermal motion results in random motion of the particles.

9.3.3
Particle–Particle and Particle–Surface Interactions

The following list is an overview of important particle–particle and particle–surface interactions. The various interactions superpose to an effective interaction potential E. To use colloids for patterning, this interaction potential must be tailored appropriately. Unbalanced strong attractions will destabilize the colloid, while long-ranged repulsion might prevent trapping of particles or destabilize close-packed structures. The interactions are summarized in Table 9.1.

1) The van der Waals interactions produce an omnipresent attraction among matter. The magnitude of the attraction depends on the difference in dielectric permittivities of the interacting bodies and the surrounding media, which is approximated by the Hamaker coefficient, A_H. For equal, perfect spheres, the van der Waals interactions decay with a d^{-6}-dependence (Equation 9.4) [25]:

$$E_{vdW} = -\frac{16}{9} \frac{A_H \cdot r^6}{d^6}, \qquad (9.4)$$

where d is the particle surfaces' distance.

2) If surface groups of the particles bear charges, repulsive electrostatic interactions are present. The charged surfaces collect layers of counter ions from the solvent

9.3 Forces and Mechanisms in Polymer Dispersions

Table 9.1 Characteristic potentials and typical ranges.

Interaction	Distance dependence	Typical range
van der Waals	$\sim d^{-6}$	1 nm
Electrostatic	$\sim e^{-\kappa d}$	1 nm … 100 nm
Steric	$\sim e^{-d^2}$	1 nm
Depletion	$\sim (\text{const.} - d^2)$	1 nm
Capillary	$\sim d^{-1}$	1 mm

which shield the Coulomb interaction such that the electrostatic potential decays exponentially (Equation 9.5) [26]:

$$E_c = E_{c,0} \cdot e^{-\kappa d}, \tag{9.5}$$

where $E_{c,0}$ is the potential at the particle surface and κ^{-1} is Debye thickness of the ion layers. Thus, the range of the electrostatic interaction can be tuned by the ion concentration in the solvent.

3) Steric interactions arise from long molecules such as polymers adsorbed to the particle surface. The surface groups must be lyophilic, that is soluble in or strongly interacting with the continuous phase so that the molecules are stretched out. If the particles then approach each other or a surface, the molecules have to change their conformation, resulting in a repulsive potential. The potential decays quickly (Equation 9.6), and has little effect above the mean chain length [27]:

$$E_s = \frac{2NkT}{A} \cdot e^{-\frac{3d^2}{2lL}}, \tag{9.6}$$

where N is the density of polymer chains, A the contact area, l the monomer length, and L the chain length.

4) Entropic forces are relevant for systems that contain particles or molecules with disparate sizes. Counterintuitively, these forces are attractive for the larger particles. The volume excluded for the smaller species is minimized for densely packed larger species, giving the smaller particles more free space to move. The potential decays linearly with the spacing d and disappears for spacings in the range of the particle radius (Equation 9.7) [28]:

$$E_e = -\frac{3}{2} kT \phi \beta x^2, \tag{9.7}$$

where ϕ is particle density of the smaller species, β the size ratio between the bigger and the smaller species, and $x = r_1 + r_2 - d$ states the dependency on the particle spacing.

5) Capillary forces are longer ranging interactions. They appear when the contact of the particles with the interface of the carrier medium causes perturbation in the interfacial shape. The description of the force between two particles resembles that of Coulomb interaction, where the angle of the meniscus slope plays the role

of the charge (Equation 9.8) [29]:

$$F_{cap} = -2\pi\sigma \frac{Q_1 Q_2}{d}, \tag{9.8}$$

where σ is the surface tension and $Q = r \sin\alpha$ the dependency on the slope angle α.

Colloidal systems are often governed by van der Waals interactions. This class of dipole, induced dipole, and dispersion interactions causes a ubiquitous attractive potential between the particles and the container walls and between the particles themselves, which must be balanced to produce a stable colloid. Stabilization of the suspension can be achieved using functional surface groups on the particles. They can induce repulsive electrostatic and steric interactions, which counterbalance the attractive potential. A self-contained description of electrostatically stabilized colloids in polar solvents was first given in the classical DLVO theory by Derjaguin, Landau, Verwey and Overbeek [26, 30].

9.3.4
Mechanisms and Techniques of Structure Formation

Hard spheres undergo a disorder-to-order transition when they occupy approximately 50% of the available volume (see Figure 9.2a). This structural transition of spheres from an unordered state to a long-range ordered state was first predicted by Kirkwood and proven by Alder [31]. The transition involves no exchange of heat or change in the internal energy and is solely driven by entropy. Counterintuitively, in a close-packed arrangement each individual sphere has increased its degrees of freedom compared to a random packing, where the available volume per sphere is much smaller. Although the packing densities in both structures are identical, it is generally

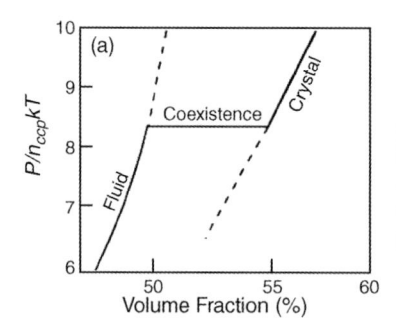

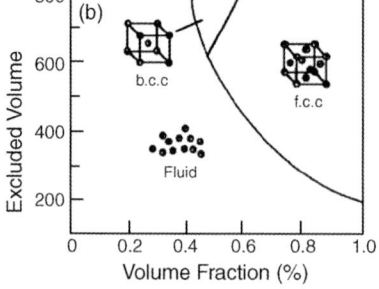

Figure 9.2 Phase diagrams for colloidal spheres whose constituent units are treated as (a) "hard" and (b) "soft" spheres. The vertical axis of (a) represents the pressure P normalized by the thermal energy kT and the number density n_{ccp} of the ccp structure. As shown in (b), the colloidal spheres can order into either bcc or fcc structures. In either case, the disorder–order (Kirkwood–Alder) transition occurs at lower volume fractions for soft spheres than in the hard-sphere system. (Reproduced with permission from [35]. Copyright © (1998) American Institute of Physics.)

accepted that the particles preferentially crystallize into a face-centered cubic (fcc) (also called cubic close-packed, ccp) structure rather than into a hexagonal close-packed (hcp) arrangement due to the slightly higher entropy of the fcc structure [32]. When an attractive potential overlays the repulsion among the spheres ("soft spheres"), a body-centered cubic structure is also observed (see Figure 9.2b). Size uniformity is essential for ordering. With increasing polydispersity of the particles, the volume of the crystalline structure increases linearly with the size dispersion, while an unordered phase can compensate the deviations, making it increasingly energetically favorable. Theoretical considerations indicate that crystallization is suppressed if the polydispersity of the particles exceeds 12% [33]. However, mixing different monodisperse samples to obtain bimodal or ternary distributions can eventually provide more complicated regular structures [34].

To induce a Kirkwood–Alder transition for the production of ordered particle arrays, it is necessary to increase the particle concentration locally or globally above the transition threshold in the colloid. Examples of techniques to achieve such a concentration increase are sedimentation of the particles in the gravitational field, centrifugation, evaporation of the solvent, attracting charged particles to electrodes in DC fields, confining uncharged particles with AC fields, confining the particles at a liquid surface in a Langmuir–Blodgett trough, or confining the particles in a fluid flow at a semipermeable membrane. A detailed description of the techniques for particle assembly is given in the following sections. In most cases, shearing the sample during the assembly process by shaking or sonicating enhances the crystallinity of the array.

Another approach to fabricating ordered arrays of particles where the Kirkwood–Alder transition is not relevant is to use capillary forces in thin liquid films or on liquid–gas interfaces. In an evaporating thin film of a suspension, the thickness of the liquid film will eventually decrease below the particle diameter. Strong capillary interactions emerge from the conformation of the liquid–gas interface, which guide the particles into a close-packed structure. When the film locally evaporates completely, a meniscus forms. The particle deposits can pin this meniscus, preventing it from receding. As evaporation continues, liquid from the bulk suspension replenishes the loss of material. The convecting fluid strongly influences the particle ordering. The voids between the already deposited particles funnel the fluid flow, directing approaching particles towards the voids, where they get trapped by the neighboring particles. The combination of steady convective flow and capillary attraction between the particles enables a steady deposition of close-packed layers. Appropriate thin liquid films can be deposited by spin-coating, dip-coating or by drop-casting onto wetting substrates.

Combining the above-mentioned deposition methods with pre-patterned substrates, which provide pinning sites for particles, enables a wider variety of possible arrangements beyond close-packed structures. The pinning sites must be designed such that they trap only a specified amount of particles at the desired position, but hold these particles strongly enough to prevent Brownian motion. The pinning sites should be chosen according to the colloidal particles and the deposition techniques used. For example, electrostatically stabilized particles can be trapped from bulk suspensions by local charges provided by functionalized surface molecules, while surface asperities distort the surface of thin liquid films and can trap any particle with capillary interactions in spin- or dip-coating processes. The assembly process can be divided into a deposition step, where a sufficient number of particle must be transported close enough to the pinning sites to be trapped, and a removal step, where the untrapped particles are removed. For an example, in dip-coating a non-wetting patterned substrate, both steps occur simultaneously. With the withdrawal of the substrate, a meniscus forms. Evaporation at the contact line causes a convective flow enhancing the particle concentration, which increases the trapping probability at the pinning sides. At the same time, the receding meniscus transports the untrapped particles away.

An approach which combines pre-patterning and confining particles in AC fields is to assemble using interfering laser beams. The dielectric response of the particles to the alternating electric field of the laser light creates a force parallel to the gradient of the electric field. The light pushes the particles into the minima. Their arrangement is thus determined by the interference pattern.

9.3.5
Setups

The variety of strategies available to assemble colloidal polymer particles and the manifold applications of particle assemblies has lead to the design of a large range of setups to fabricate colloidal polymer patterns. The fabrication setups can be divided into two main groups: setups to produce bulk colloidal assemblies and setups to deposit thin two-dimensional (2D) arrays of particles.

9.3.5.1 Fabrication of Three-Dimensional Arrays

Sedimentation is a convenient way to produce bulk assemblies for some particle classes. In the simplest case, the colloid is left standing in a cuvette. After a period of time the concentration at the bottom will have increased beyond the disorder–order transition threshold. Quenching the system and removing the supernatant liquid results in stable three-dimensional (3D) arrays. Although this method is straightforward, it has drawbacks: Equation 9.2 implies that settling effectively occurs only for large polymer particles. For 1 µm PS beads in water, the sedimentation speed is about 10^{-3} µm s^{-1}, leading to sedimentation times of the order of weeks to months. Additionally, in pure sedimentation processes, due to multiple nucleation events the domain size of the assembled colloidal crystal measures only a few micrometers.

Figure 9.3 shows a setup to improve the structure formation in such a sedimentation process [36]. A container with the colloidal suspension was mounted onto a rigid stage. To increase the settling rate, solvent was removed from the system by evaporation through a filter membrane in the bottom of the container. In order to improve the crystallinity, the authors used the principle of vibratory compacting, a technique well-known from the packing of granular matter, on the sediment by applying sinusoidal lateral or vertical motion to the stage. With this setup, the sedimentation speed of micrometer sized particles could be increased to 0.07 µm s^{-1},

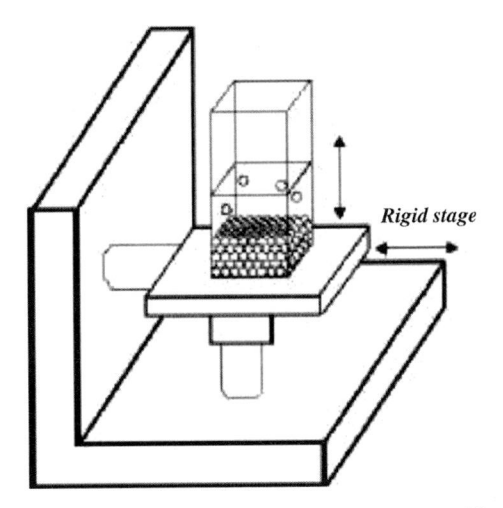

Figure 9.3 The experimental setup used to assemble core-shell latex microbeads with sedimentation under oscillatory shear. (Reproduced with permission from [36]. Copyright © (2000) Wiley-VCH Verlag GmbH & Co. KGaA.)

leading to 3 mm thick sediments after 20 h of sedimentation. To obtain stable colloidal arrays, sediments were dried and annealed to particulate films. Polycrystalline structures were gained with domain sizes of 200 µm². As the total size of the sample is not limited, the method is potentially applicable for the large-area production of templates of porous samples and ordered polymer-based nanocomposites. However, the method is still slow and applicable only to sedimenting colloids, which, together with the filter membrane, limits the usable particle size.

A more versatile setup is shown in Figure 9.4. It consists of a flow cell out of a square frame of photoresist and two glass substrates [37]. Very shallow channels were embossed into the frame via photolithography. The chamber was filled with the particle suspension, and the liquid was pushed out through the channels with nitrogen. Additional sonication aided the assembly process. After the liquid had been pushed out, the sample was dried and the glass slides were removed. In this setup, PS spheres with diameters from 60 nm up to 3 µm were successfully assembled. The height of the resulting structures was controlled by the thickness of the photoresist film, and crystalline structures with domains covering the whole setup ($\sim$1 cm²).

9.3.5.2 Fabrication of Two-Dimensional Arrays

The production of 2D thin layers of particles requires much better control over the thickness of the deposit. In contrast to settling and draining setups, thin liquid film setups deposit large-area films suitable for deposition of particle monolayers.

Figure 9.5 shows a setup where a Teflon® ring with an inner diameter of 14 mm is pressed onto the substrate [38]. A drop of the colloidal suspension is cast onto the substrate, spreads and forms a concave layer as the liquid wets the Teflon ring. Evaporation thins the liquid film in the center of the ring further, until eventually the thickness reaches the diameter of the particles. The ordering is then driven by

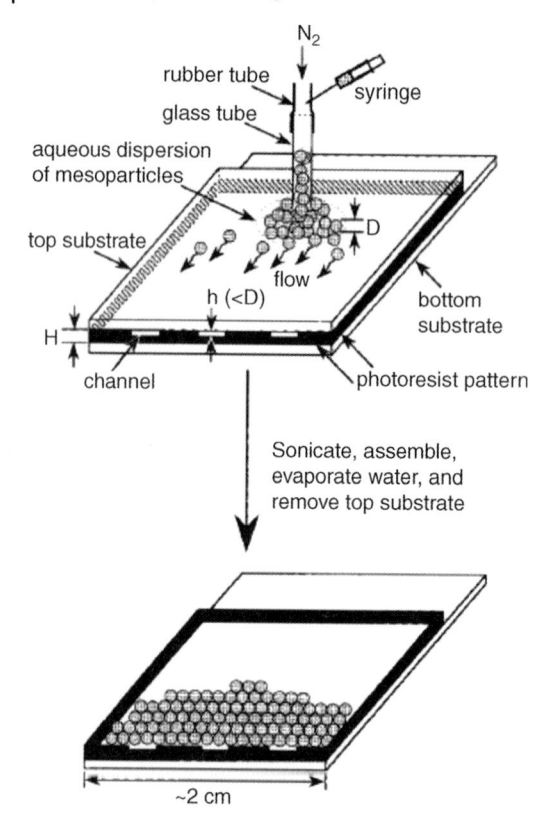

Figure 9.4 Schematic outline of the experimental procedure that was used to assemble polystyrene spheres in confined geometries. (Reproduced with permission from [37]. Copyright © (1999) American Chemical Society.)

convective flow and capillary attraction of the particles. This experimental approach is conceptually similar to dip-coating: material from a bulk reservoir – here, the liquid gathered at the Teflon ring – is deposited on a substrate via a thin liquid film during evaporation. Further experiments have shown that with horizontal or vertical dip-coating approaches continuous deposition of ordered layers of micrometer and sub-micrometer particles over large areas can be facilitated [39, 40].

A different approach is depicted in Figure 9.6. Using electron beam lithography, a substrate was patterned with gold electrodes spaced a few micrometers up to several millimeters apart. A small container containing particle suspension was then mounted on top of the patterned substrate. AC square wave signals with megahertz frequencies were applied to the electrodes. The dielectric response of the particles suspended in the liquid squeezed the particles out of the bulk onto the substrate, where they assembled into ordered structures, as could be observed in the diffraction pattern of a laser beam. In the experiments, particles with diameters in the range of micrometers were used. A key feature of this approach is the switchability of the assembly process, as the particles disassemble when the AC field is turned off.

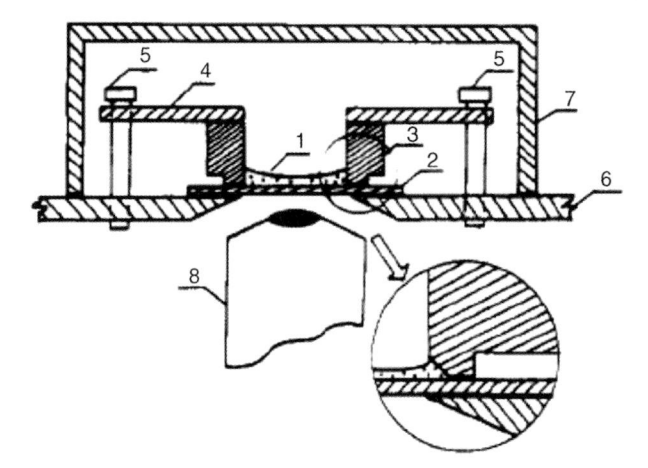

Figure 9.5 Scheme of the experimental cell to produce 2D arrays using capillary interaction, (1) latex suspension, (2) glass plate, (3) Teflon ring, (4) brass plate, (5) screws, (6) microscopy table, (7) glass cap, (8) microscope objective. (Reproduced with permission from [38]. Copyright © (1992) American Chemical Society.)

The use of dielectrophoretic assembly for switchable 2D photonic crystals [41] and biosensors [42] has been proposed.

9.4
Polymer Patterns from Colloidal Suspensions

Ideally, a simple coating step using a well-formulated particle suspension would result in a well-defined polymer structure that has the desired geometry. In reality,

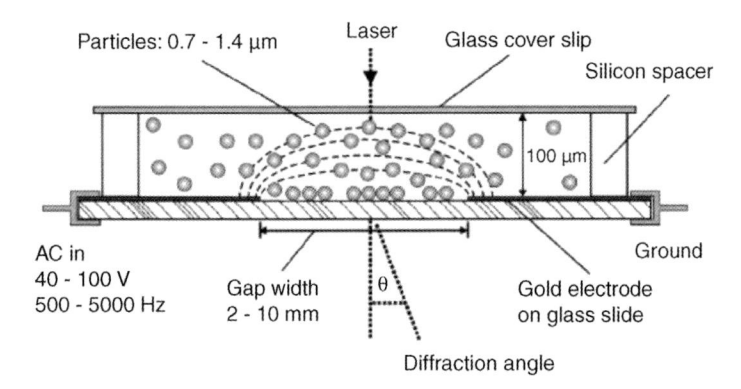

Figure 9.6 Scheme of the experimental setup to assemble particles between gold electrodes with dielectrophoretic mechanisms. The diffraction pattern from the laser beam is collected on a screen below the chamber (Reproduced with permission from [43]. Copyright © (2003) American Institute of Physics.)

only simple geometries can be formed at present, most prominently densely packed monolayers and artificial opals. More complex structures might be achievable via multimodal particle mixtures or by using particles with anisotropic interactions – some experimental results [44] and simulations [45] suggest a broad variety of attainable structures – but today, complex arrangements are usually formed with the help of templates. The most widespread applications of regular particle monolayers are colloidal lithography, while 3D artificial opals are usually used for optical applications. Polymers can also be patterned using colloids that are not themselves polymeric, but interact with the polymer to create predictable geometries.

9.4.1
Polymer Particle Assembly

Monodispersed sub-micron polymer particles in evaporating wetting films tend to form regular, dense mono- and multilayers composed of "grains" of single crystals with domain sizes up to approximately 100 μm (Figure 9.7). Their formation takes place far from equilibrium and is strongly dependent on the fluid mechanics of the convective assembly process. Both fluid flow and capillary forces direct the particles into lattice positions of densely packed layers [46]. Those layers have been used extensively for patterning as discussed in the next section, but it is often desirable to be more flexible in arranging the polymer particles.

Since the assembly process is kinetically driven, a promising strategy to control the particle deposition process is the use of templates that influence the hydrodynamics of the process. As first shown by Xia [48], simple topographical structures on a surface cause local pinning of the receding meniscus of a particle suspension. Particles can be trapped in the topography and assemble regularly inside the confinement in a process called "capillary assembly". Thus, the arrangement is not controlled directly by the substrate, but rather indirectly through fluid motion, and relatively complex structures form even in simple template structures (Figure 9.8). Particularly interesting are template shapes that are incommensurate with the particle diameters and

Figure 9.7 A crystalline lattice of 215 nm polystyrene beads (imaged by SEM): (a) top view; (b) cross-sectional view (Reproduced with permission from [47]. Copyright © (1999) Wiley-VCH Verlag GmbH & Co. KGaA.)

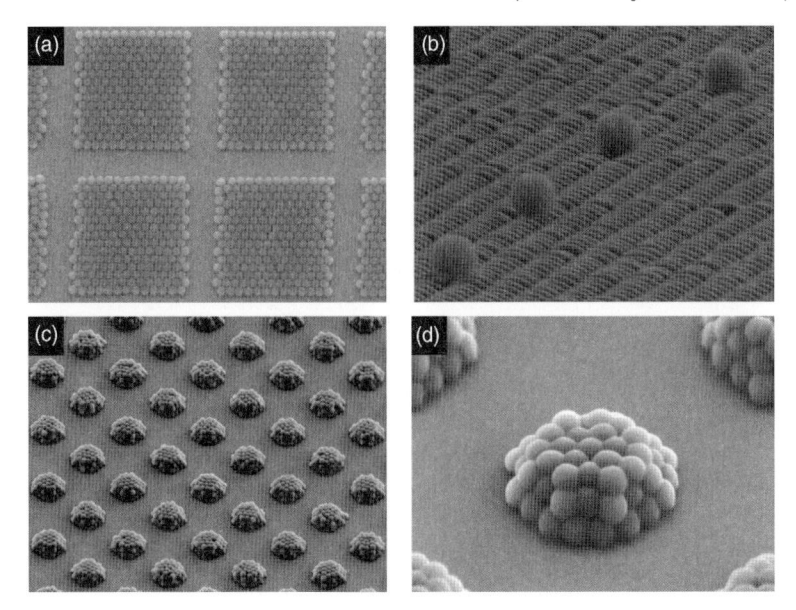

Figure 9.8 Superstructures assembled from polystyrene particles (500 nm diameter) by templated assembly. Possible arrangements include (a) parts of crystalline monolayers, (b) aligned single particles on top of full monolayers and (c-d) three-dimensional pieces of densely packed particles. (Electron micrograph reproduced with permission from [50]. Copyright (C) (2005) Wiley-VCH Verlag GmbH & Co. KGaA.)

lead to "defect structures" [49] through which particles relax the unaccommodating constraints.

Fluid motion can also be controlled by chemically patterned templates which modulate the wetting of the surface [51]. Liquid is pinned in a manner similar to that occurring in topographical templates, although the effect is less pronounced and can be easily disturbed by self-pinning of the particle suspension due to unspecific deposition [52]. A general limitation of this kinetic approach is the systematic influence of the direction that the particle suspension is moved across the surface, an effect more pronounced on chemical than on topographical templates.

Instead of exploiting wetting phenomena for the assembly of particles on chemically structured surfaces, specific adsorption offers a viable alternative. Surfaces are patterned with areas that specifically interact with particles. The particles are adsorbed in these areas [53]. Depending on the strength of the particle–surface interaction, random sequential adsorption (RSA) can occur inside the adsorbing areas so that the particles inside are entirely disordered (with a density of approximately 55% [54]). If the particles can still diffuse laterally or desorb, the particle density inside the areas can increase until it reaches the limit of a densely packed layer. Specific particle–substrate interactions are achieved using electrostatically charged monolayers [53], molecular coatings that cause supramolecular interactions, such as cyclodextrins [55] and biological molecules such as DNA [56], which can be tuned to cause a defined binding strength and specifically bind only an appropriately modified particle type. The functional molecules can be patterned on the template using methods such as micro contact printing [57], dip-pen lithography [58], photolithography, and many others. Even for highly accurate surface patterns, however, the particle assembly step remains critical, and it is not easy to fully cover the functionalized parts with a dense particle layer and retain the arrangement throughout the drying steps, in particular if groups of small numbers or single particles are part of the desired pattern.

The assembly of single particles with high precision is challenging. Defining a very sharp energy minimum with a template is costly if it requires high-resolution patterning. The energy well that is formed has to be deep enough to limit Brownian motion of the particles. Even then, the low particle concentrations in typical colloids and the small diffusion constant of the particles limits the yield and the rate of assembly. Different strategies have been pursued to amend such problems. The particle concentration can be increased locally to increase the assembly probability, an effect that is likely to occur in many processes where particles assembly at a meniscus [59]. If multiple forces act at the same time or at different stages of the particle assembly, additional focusing can occur and lead to assembly with an accuracy that exceeds that of template features. For example, the combination of electrostatic and hydrodynamic forces during the evaporation of particle suspensions on chemically patterned substrates can lead to single particles being assembled at the center of certain surface patterns [60]. This effect can also ameliorate the statistical problem of assembling a small, but precise number of particles: in the absence of a geometrical constraint, it is hard to exactly define the number of particles that are to be assembled at a given position [59]. If the particles interact, however, a single particle can be enough to shield the binding site from any other particle.

Templates also apply to 3D particle assemblies. Planar templates can impress a structure onto the 3D colloidal crystal that is growing on top of it [61], 3D templates can confine the growth of crystals to small spaces [49]. In three dimensions, the growth of crystals by simple settling of particles is applicable (particle monolayers are hard to produce using this route), and the most straightforward templating method is to have particles settle onto a template, possibly in the presence of an external field ("colloidal epitaxy") [62]. If liquid motion is exploited for the "convective" assembly of thick particle layers, for example by evaporating a solvent film, planar templates can influence the flow and thus strongly influence the deposited multilayers. This templating effect is particularly pronounced if the liquid film is moved over the substrate.

An alternative route to particle arrangements beyond dense packings are binary supercrystals. Such structures contain two particle populations, each of which is monodispersed with a different mean particle size. Complex crystals can be formed from such mixtures by concurrent sedimentation [63, 64], but this process is slow and unreliable. Convective particle assembly that exploits convective forces and capillary interactions has been shown to yield large, ordered domains [65] (Figure 9.9). The small particles fill the interstices between the larger ones, with the detailed structure depending both on the particle size ratio $\gamma = a_s/a_t$ and the ratio of the particle concentrations. Local fluctuations in the concentration ratio lead to domains of different structure [66]. Mixtures with $\gamma > 0.3$ do not assemble into binary crystals in

Figure 9.9 Binary supercrystals assembled from polystyrene and silica particles having different diameters. Panels (a-c) are micrographs obtained with different diameter ratios. (Reproduced with permission from [65]. Copyright (C) (2003) Wiley-VCH Verlag GmbH & Co. KGaA.)

convective assembly, but form mixed random arrays [65]. Such crystals can be formed, however, using stepwise deposition of layers, for example using multiple drying steps [67] or spin-coating [68].

It is sometimes impossible or undesirable to use the particle structures on the substrate where they were originally assembled. If templates were used, the topographical or chemical structure of the template might be incompatible with the desired function. In other cases, the wetting properties required for the assembly might be inappropriate later on. In such cases, a transfer step can be useful to bring the polymer structure from the assembly substrate to the target material. Such transfer has been performed by exploiting differences in particle–substrate adhesion, caused either by differences in the contact area, the interaction strength, or both [50]. For thermoplastic polymer particles that have a moderate glass transition temperature, it is straightforward to slightly melt the particles and thus increase their contact area with a target substrate. An alternative is to use kinetic effects that render adhesion strength velocity-dependent, so that transfer can be performed by tuning the removal rate of a stamp [69].

9.4.2
Colloidal Lithography

Colloidal lithography uses a 2D array of colloidal nanoparticles as a lithographic template for etching and sputtering. Monodisperse nanospheres of materials such as PS will self-assemble to form ordered monolayers, generally with hexagonal close packed (hcp) alignment [70–72], resulting in a convenient route to large scale nanoparticle arrays. The shape, size and interparticle spacing of these nanoparticles can be easily controlled leading to large variety of nanoparticle layers [73, 74]. This route has a number of advantages over the more commonly used lithographic routes as it is a simple process using small amounts of low-cost, easy-to-produce colloids to fabricate relatively large ordered arrays without the requirement for specialized machinery [75].

A number of methods can produce ordered colloidal monolayers, the simplest of which is drop-casting on a flat substrate where the degree of order is governed by the attractive capillary forces between particles during slow solvent evaporation [46, 70, 71, 76–78]. This process occurs by nucleation around an ordered region followed by growth by convective transport. The nucleation-growth mechanism requires a very clean, chemically homogeneous surface and tight control over particle concentration and evaporation and, while it can produce uniform domains of over 100 μm, is susceptible to dislocations and stacking faults [73]. The deposition rate can be increased at the expense of defect-free assembly by the use of spin coating, although to achieve an ordered monolayer it is sometimes necessary to use a surfactant to improve wetting which can cause other problems [79, 80].

A number of assembly methods employ self-assembly of nanoparticles at the air–water interface. Monolayers (and subsequent multilayers) can be formed at the interface due particle interaction and transferred to a solid substrate by controlled dip-coating and vertical deposition methods similar to Langmuir–Blodgett film deposition [81–86]. Regular monolayers of polymer colloids can also be assembled via an electrohydrodynamic route, whereby electrophoretically deposited particles between two electrodes can be manipulated to cluster in the presence of an electric field. On application of an AC or DC field, contrary to electrostatic norms, the like-charged particles are observed to coalesce producing large close-packed 2D crystalline domains [87].

Two-dimensional colloidal crystals have been used for many years as lithographic patterning masks. In a close-packed hcp crystal there exists a connected void space that has been used for both etching of the substrate and deposition of both inorganic oxides and metals [80, 88, 89]. Recently this lithographic technique has been applied to form large areas of regularly patterned polymers. One approach to create polymer nanoarrays is by transferring colloidal patterns to polymeric substrates via reactive ion etching. Directed ion bombardment etches away the area surrounding the colloids creating both pits and columns in the polymer substrate which are accessible when the colloid is removed [73, 90–93]. Oxygen plasma has been used to create functional nanodomes from a monolayer of regularly-spaced PS nanoparticles on a polyacrylamide (PAA) thin layer. The plasma etches the uncovered PAA at a different rate than the PS. If the etch is interrupted before the PS layer is removed, a carboxylic functionalized plateau remains that can be used as a biosensor [94].

A relatively easy route to a variety of array patterns such as nanorings, nanodots, "bowls" and porous structures, is the injection of a monomeric precursor into a colloidal crystal with subsequent polymerization and colloid removal [89, 95]. The morphology of the produced array can be controlled by the precursor concentration. To avoid the problems associated with liquid phase deposition, including long polymerization times and surface tension effects [96], a solvent-free, initiated chemical vapor deposition (iCVD) route has been established which can deposit a regular thin layer of polymer on a colloidal crystal substrate at low energy [97]. Similar arrays can also be achieved by simple sintering of a colloidal crystal matrix consisting of a crosslinked copolymer when the two constituents have different solubility. Sintering creates necks that can be isolated by selective dissolution. This approach is limited mainly by the colloidal material restraints [98].

As monodispersed colloidal polymers are generally spherical, a number of methods have been developed to diversify the colloidal crystal shape to produce a larger range of nanoarrays. Colloidal crystal sensor devices and polymer materials use the tendency of polymer colloids to deform and change periodicity in response to a physical change, such as temperature or chemical environment change [99]. These can be then used as masks to form different array morphologies [100]. Recently, angle-resolved colloidal lithography (ARCL) has been used to fabricate a range of nanostructures from a monolayer colloidal crystal mask. Nanooverlaps, nanochains, nanodots, and crescents among others have been created by vapor deposition at various incident angles to the colloidal mask [101–103]. More complex shapes have been achieved by using a processed close-packed double-layer mask [104, 105].

9.4.3
Polymer Opals

Artificial polymer opals are long-range ordered, 3D colloidal crystals of monodisperse polymer colloid particles with diameters between 200 and 900 nm, formed by self-assembly into close-packed, usually fcc, structures. These are analogous to natural opals, which consist of domains of fcc-packed monodispersed SiO_2 particles, and exhibit Bragg reflection of light matching the periodicity of the assemblies [17, 106]. However, polymer opals have a number of advantages over silica colloidal crystals. For example, crystalline polymer domains over 100 μm can be easily obtained in polymer opals and physical properties such as refractive index and particle size and shape can be simply modified [21, 107]. The polymer particles can also be removed easily when used as a template or loaded with fluorescent dyes. The light propagation properties of the crystals can thus be tuned by bandgap engineering, which has attracted great interest for photonics (Section 9.4.4).

The simplest method to assemble opals from polymer colloids is by gravitational settling on a flat substrate. In this route the colloidal dispersion is spread over the substrate and dried slowly, where it undergoes a disorder to order phase transition forming large areas of a densely packed cubic lattice [21, 83, 107–109]. However, this route has a number of inherent drawbacks such as the lack of morphological control leading to polycrystalline regions or sub-fcc packing resulting in areas of differing densities due to parallel nucleation at a number of different sites and inhibited rearrangement [70, 72, 110, 111]. In addition, very long settling times are required to produce ordered crystals, which can be of the order of weeks for materials with density similar to that of the solvent or with particle sizes of less than 0.5, as these will preferably form stable dispersed solutions [70]. The settling process can be significantly accelerated by using centrifugation, resulting in reduced fabrication times albeit with inferior crystal structure [72].

Another uncomplicated method for the production of colloidal crystals from dispersed polymer emulsion systems is based on a vertical deposition method for 2D films [46, 76, 112]. In this method the substrate is vertically immersed into the dispersion and the solvent allowed to evaporate, forming a film on the substrate. In this sample setup, the particle concentration changes as the solvent evaporates affecting the layer thickness. Improvements such as withdrawing the substrate with a constant velocity lead to more control over deposition, resulting in consistent film thicknesses up to tens of layers over a number of centimeters [83, 110].

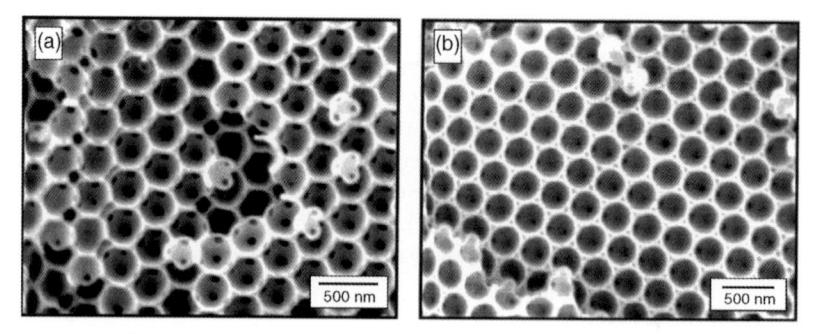

Figure 9.10 Inverse opal structures in silica, templated by PMMA (a) or polystyrene (b) spheres (electron micrograph reproduced with permission from [122]. Copyright © (2002) American Chemical Society.)

Colloidal crystals can also form when the colloid is physically confined by two solid boundaries [113–116]. This approach has been used to produce centimeter-scale close-packed crystals without defects with highly controllable thicknesses from polymer nanoparticles of varying sizes. This is achieved by filtering the colloidal suspension into a thin slit between two solid glass substrates under constant sonication. Similarly, close-packed, ordered structures can be formed by compression of the dry particles between two solid surfaces [72].

Polymer inverse opals (Figure 9.10) can be formed by partially or fully filling the void spaces of a colloidal crystal with a liquid precursor by capillary action. These precursors are generally UV or thermally-curable polymers, such as polyurethenes or polyacrylates, epoxies with initiators, sol–gel materials or more recently conducting polymers like polypyrrole [113, 117–122]. The liquid precursors are then solidified and the original template is removed. This results in long-range ordered porous material with a precise periodic structure which has found use in many applications such as photonics, membranes and biosensors.

9.4.4
Applications

Colloidally patterned polymers have been applied in a wide variety of fields such as biomaterials and cell adhesion, optical devices and, most notably, photonic band gap materials.

Photonic crystals are periodic structures, where spatially modulated regions of differing dielectric constants lead to coherent scattering of light on the lengthscale of these periodicities [99, 123]. Similar to the modulation of electron motion by the atomic structure of semiconductors, the periodic refractive index difference in photonic crystals, if great enough, can influence photon movement and induce a bandgap. Materials constructed from colloidally patterned polymers are useful for photonic band gap materials due to their easily controlled periodicity and facile chemical modification. However, the relatively low refractive index of polymers implies that polymer based materials do not possess a full band gap. Colloidally patterned polymers do exhibit Bragg diffraction of light in the optical region, useful for some photonic devices [124].

The wavelength of coherently scattered light in a photonic material is affected both by the dielectric constant difference and the diameter of the constituent particles. A wide variety of photonic crystal materials can be fabricated from colloidal crystal polymer arrays by the addition of a material with a different dielectric constant or by changing the size or shape of the colloid. A linear dependence of the wavelength of coherently scattered light on the size of the colloid particles was observed in a close-packed polymer opal of PS beads [113, 115, 125]. By sintering of other colloidal materials the wavelength and intensity of scattered light could be fine-tuned and controlled [126].

A range of sensors has been developed based upon colloidal crystal polymer structures [99]. They take advantage of the ability of hydrogel polymers to change dimensions in the presence of certain stimuli, which induces shifts in Bragg reflection that can be observed in the visible range by a color change. These functionalized hydrogels are generally coated on self-organizing PS or poly(methyl methacrylate) (PMMA) colloidal crystals forming polymer opals, or deposited in the interstitial sites of these structures forming inverse opals. The hydrogel-coated photonic crystals have been tailored to show color changes when their periodicity is modulated by temperature [127], pH [128, 129], bulk material deformation [130] or the presence of chemical species such as ethanol [129], metal ions [131] and biomolecules [132].

Recently polymer photonic crystal materials have been applied as optical data storage media. An optically active, core-shell latex colloidal crystal with dye-labeled polymer cores was annealed into a material with a periodic array of fluorescent particles in an optically inert matrix. A two-photon laser scanning microscope was used to write information to, and read from, the material leading to increased optical storage density over conventional materials [133].

Diffraction of light in photonic crystals has also been used to enhance the photoluminescent properties of fluorophores [17, 99], and by loading the interstitial voids of polymer opals with highly fluorescent dyes, stimulated emission can be enhanced [134, 135]. For more complex opal structures the dye must be loaded into the colloid matrix by using water-soluble dyes [136] or by modified emulsion polymerization reactions for non-water soluble dyes [137, 138].

Surface enhanced Raman spectroscopy (SERS) is a surface sensitive technique that exploits the enhancement of Raman scattering by rough or patterned metal surfaces. Polymer nanosphere lithography and surface patterning has produced patterned metal surfaces for SERS analysis by trapping inorganic particles in the voids followed by the removal of the colloidal monolayer template [139]. In a similar process silica microspheres coated with silver were deposited in a close-packed monolayer on a polymer substrate and, following the etching of the silica, a silver nanoparticle doped polymer array remained [140].

Colloidally patterned polymers have also found use in a number of biomaterial applications, such as materials for biodevices and cell regulation. Surface topology

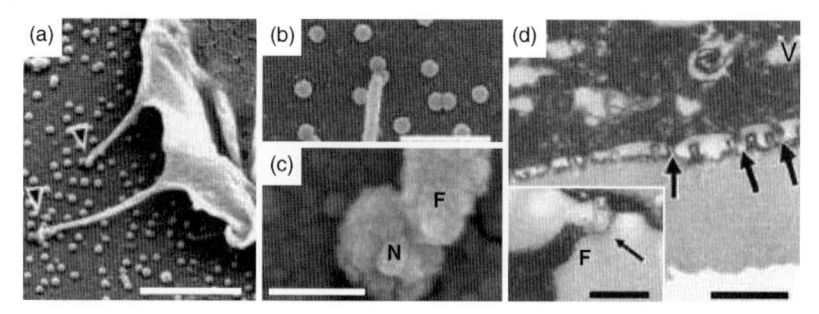

Figure 9.11 Human fibroblast filopodia interacting with nanocolumns produced by nanolithography. (a)–(c) Scanning electron micrograph (scale bars: 1 μm (a), 500 nm (b), 100 nm (c)) of filopodia interacting with nanocolumns (arrowheads). (d) Transmission electron micrograph of filopodia underneath the cells interacting with nanocolumns (arrows) (scale bar: 500 nm), inset at higher magnification (F: filopodia) (scale bar: 200 nm) (Reproduced with permission from [146]. Copyright © (2004) Elsevier.)

has an important role in cell interactions and hence, in the design of new substrates for biomaterials. These surfaces need to have regular features of the order of cell elements (<100 μm) across a large surface area, formed by reproducible low cost synthetic routes. Polymer patterns formed by colloidal lithography and similar methods fit these criteria [141] and have been used in a number of biological systems. They have been used to investigate protein adsorption on nanopatterned surfaces and cell–substrate adhesion was examined on polymer pillars formed from PMMA colloidal assembly and subsequent ion etching [142, 143]. A similar system was used to investigate fibroblast sensing and morphology and cell–cell contacts with a view to guiding cell response on biodevices (Figure 9.11) [142, 144, 145].

The highly regular 3D structure of inverted opals has also found use in biodevices. For example, a well ordered porous substrate, formed from the infiltration of the voids of a PS colloidal crystal and subsequent removal of the colloidal particles, has been used as a 3D scaffold for cell-growth and regulation [147, 148]. In addition a potentiometric biosensor for enzymes has been developed from inverted polymer opals [120]. The sensor was created by filling the interstices of a PS colloidal crystal with the conducting polymer, doped polypyrrole, by electrochemical growth. The PS colloid was subsequently removed and a highly regular 3D porous structure remained into which an enzyme which catalyses a pH modulating biochemical reaction was distributed. The material could be subsequently used for analyte detection using a potentiometric sensing response to changing pH.

Colloidal crystal polymers also have been used as templates for bioinspired materials. Geckos use closely packed arrays of hairs to create adhesion. The hairs can be scalably mimicked using colloidal crystal monolayers on silica substrates [149, 150]. The voids created by the PS colloid were etched to create nanotrenches in the substrate which, following dissolution of the polymer colloid, were filled with parylene. Subsequently the silica was removed, leaving a free-standing flexible parylene pillar array with enhanced adhesive properties.

9.5
Summary and Outlook

Polymer patterning via colloids is a promising method and a very active field of research. It is currently limited to a small number of geometries, for some of which it is the most efficient fabrication approach. Today, particle-based structures are arguably the most efficient source of optical bandgap materials and (via shadowed evaporation) a very promising route to large-area metal nanostructures.

Although the potential of particle-based patterning has been recognized and exploited in research, its application in large-scale production is very limited. The processes leading from dispersed particles to a stable, well-ordered polymer structure with predictable geometry and high quality in large quantities have not yet been developed beyond the laboratory stage. Clearly, if such methods are to be used for the production of novel, patterned materials, research into the scale-up is required.

The prospects are alluring. Colloidal polymer patterning is similar to very well-established painting and coating processes, so that existing equipment could be adapted for the production of coatings and materials with new properties. The polymer latex formulations that form the basis of the process are well-known in industry, and the step to monodispersed particles and controlled deposition seems viable. For applications where molding methods are inapplicable, particle-based polymer patterns could form the basis of industrial processes.

References

1. Keddie, J.L. (1997) Film formation of latex. *Mat. Sci. Eng. R*, **21** (3), 101–170.
2. Woods, M.E., Dodge, J.S., Krieger, I.M., and Pierce, P.E. (1968) Monodisperse lattices I. Emulsion polymerization with mixtures of anionic and nonionic surfactants. *J. Paint Technol.*, **40** (527), 541.
3. Thickett, S.C. and Gilbert, R.G. (2007) Emulsion polymerization: State of the art in kinetics and mechanisms. *Polymer*, **48** (24), 6965–6991.
4. Smith, W.V. and Ewart, R.H. (1948) Kinetics of emulsion polymerization. *J. Chem. Phys.*, **16** (6), 592–599.
5. Harkins, W.D. (1947) A general theory of the mechanism of emulsion polymerization. *J. Am. Chem. Soc.*, **69** (6), 1428.
6. Gan, L.M., Chew, C.H., Lee, K.C., and Ng, S.C. (1993) Polymerization of methyl-methacrylate in ternary oil-in-water microemulsions. *Polymer*, **34** (14), 3064–3069.
7. Pavel, FM. (2004) Microemulsion polymerization. *J. Disper. Sci. Technol.*, **25** (1), 1–16.
8. Guo, J.S., Elaasser, MS., and Vanderhoff, JW. (1989) Microemulsion Polymerization of Styrene. *J. Polym. Sci. Polym. Chem.*, **27** (2), 691–710.
9. Puig, J.E., Perezluna, V.H., Perezgonzalez, M., Macias, E.R., Rodriguez, BE., and Kaler, EW. (1993) Comparison of oil-soluble and water-soluble initiation of styrene polymerization in a 3-component microemulsion. *Colloid Polym. Sci.*, **271** (2), 114–123.
10. Capek, I. and Fouassier, J.P. (1997) Kinetics of photopolymerization of butyl acrylate in direct micelles. *Eur. Polym. J.*, **33** (2), 173–181.
11. Gan, L.M., Chew, C.H., Ng, S.C., and Loh, S.E. (1993) Polymerization of methyl-methacrylate in ternary-systems - emulsion and microemulsion. *Langmuir*, **9** (11), 2799–2803.

12 Capek, I. and Potisk, P. (1995) Microemulsion and emulsion polymerization of butyl acrylate 1. Effect of the initiator type and temperature. *Eur. Polym. J.*, **31** (12), 1269–1277.

13 Capek, I. (1999) Microemulsion polymerization of styrene in the presence of anionic emulsifier. *Adv. Colloid Interfac.*, **82** (1–3), 253–273.

14 Steytler, D.C., Gurgel, A., Ohly, R., Jung, M., and Heenan., R.K. (2004) Retention of structure in microemulsion polymerization: formation of nanolattices. *Langmuir*, **20** (9), 3509–3512.

15 Goodwin, J.W., Hearn, J., Ho, C.C., and Ottewill, R.H. (1974) Studies on preparation and characterization of monodisperse polystyrene lattices 3. Preparation without added surface-active agents. *Colloid Polym. Sci.*, **252** (6), 464–471.

16 Goodwin, J.W., Ottewill, R.H., Pelton, R., Vianello, G., and Yates, D.E. (1978) Control of particle-size in formation of polymer lattices. *British Polym. J.*, **10** (3), 173–180.

17 Lange, B., Fleischhaker, F., and Zentel, R (2007) Chemical approach to functional artificial opals. *Macromol. Rapid. Comm.*, **28** (12), 1291–1311.

18 Egen, M. and Zentel, R. (2004) Surfactant-free emulsion polymerization of various methacrylates: towards monodisperse colloids for polymer opals. *Macromol. Chem. Phys.*, **205** (11), 1479–1488.

19 Egen, M., Voss, R., Griesebock, B., Zentel, R., Romanov, S., and Torres, CS. (2003) Heterostructures of polymer photonic crystal films. *Chem. Mater.*, **15** (20), 3786–3792.

20 Egen, M., Braun, L., Zentel, R., Tannert, K., Frese, P., Reis, O., and Wulf, A. (2004) Artificial opals as effect pigments in clear-coatings. *Macromol. Mater. Engineer.*, **289** (2), 158–163.

21 Egen, M. and Zentel, R. (2002) Tuning the properties of photonic films from polymer beads by chemistry. *Chem. Mater.*, **14** (5), 2176–2183.

22 Einstein, A. (1906) Zur theorie der brownschen bewegung. *Ann. Phys.-Berlin*, **324**, 371–381.

23 Russel, W.B., Saville, D.A., and Schowalter, W.R. (1989) *Colloidal Dispersions*, Cambridge University Press.

24 Reynolds, O. (1883) An experimental investigation of the circumstances which determine whether the motion of water shall be direct or sinuous, and of the law of resistance in parallel channels. *Phil. Trans. R. Soc. Lond. B*, **174**, 935–982.

25 Parsegian, A.V. (2006) *van der Waals Forces*, Cambridge University Press.

26 Overbeek, J.Th.G. and Verwey, E.J.W. (1999) *Theory of the Stability of Lyophobic Colloids*, Courier Dover Publications.

27 Dolan, A.K. and Edwards, S.F. (1974) Theory of the stabilization of colloids by adsorbed polymer. *Proc. Roy. Irish Acad. A*, **337**, 509–516.

28 Asakura, S. and Oosawa, F. (1958) Interaction between particles suspended in solutions of macromolecules. *J. Polym. Sci.*, **33**, 183–192.

29 Kralchevsky, P.A. and Nagayama, K. (2000) Capaillary interactions between particles bound to interfaces, liquid films and biomembranes. *Adv. Colloid Interfac.*, **85**, 145–192.

30 Derjaguin, B. and Landau, L. (1941) Theory on the stability of strongly charged lyophobic sols. *Acta Physicochimica U.R.S.S.*, **14**, 633–622.

31 Alder, B.J., Hoover, W.G., and Young, D.A. (1968) Studies in molecular dynamics. v. high-density equation of state and entropy of hard disks and spheres. *J. Chem. Phys.*, **49**, 3688–3696.

32 Woodcock, L.V. (1997) Entropy difference between the face-centered cubic and hexagonal close-packed crystal structures. *Nature*, **385**, 141–143.

33 Phan, S.-E., Russel, W.B., Zhu, J., and Chaikin, P.M. (1998) Effect of polydispersity on hard sphere crystals. *J. Chem. Phys.*, **108**, 9789–9795.

34 Chen, Z. and O'Brien, S. (2008) Structure direction of ii-vi semiconductor quantum dot binary nanoparticle superlattices by

tuning radius ratio. *ACS Nano*, **2**, 1219–1229.
35 Gast, A.P. and Russel, W.B. (1998) Simple ordering in complex fluids. *Physics Today*, **51**, 24–30.
36 Vickreva, O., Kalinina, O., and Kumacheva, E. (2000) Colloidal crystal growth under oscillatory shear. *Adv. Mater.*, **12**, 110–112.
37 Park, S.H. and Xia, Y. (1999) Assembly of mesoscale particles over large areas and its application in fabricating tunable optical filters. *Langmuir*, **15**, 266–273.
38 Denkov, N.D., Velev, O.D., Kralchevsky, P.A., Ivanov, I.B., Yoshimura, H., and Nagayama, K. (1992) Mechanism of formation of two-dimensional crystals from latex particles on substrates. *Langmuir*, **8**, 3183–3190.
39 Dimitrov, A.S. and Nagayama, K. (1996) Continuous convective assembling of fine particles into two-dimensional arrays on solid surfaces. *Langmuir*, **12**, 1303–1311.
40 Malaquin, L., Kraus, T., Schmid, H., Delamarche, E., and Wolf, H. (2007) Controlled particle placement through convective and capillary assembly. *Langmuir*, **23**, 11513–11521.
41 Lumsdon, S.O., Kaler, E.W., and Velev, O.D. (2004) Two-dimensional crystallization of microspheres by a coplanar ac electric field. *Langmuir*, **20**, 2108–2116.
42 Kaler, E.W. and Velev, O.D. (1999) In situ assembly of colloidal particles into miniaturzed biosensors. *Langmuir*, **15**, 3693–3698.
43 Lumsdon, S.O., Kaler, E.W., Williams, J.P., and Velev, O.D. (2003) Dielectrophoretic assembly of oriented and switchable two-dimensional photonic crystals. *Appl. Phys. Lett.*, **82**, 949–951.
44 Shevchenko, E.V., Talapin, D.V., Kotov, N.A., O'Brien, S., and Murray, C.B. (2006) Structural diversity in binary nanoparticle superlattices. *Nature*, **439** (7072), 55–59.
45 Zhang, Z.L. and Glotzer, S.C. (2004) Self-assembly of patchy particles. *Nano Lett.*, **4** (8), 1407–1413.
46 Denkov, N.D., Velev, O.D., Kralchevsky, P.A., Ivanov, I.B., Yoshimura, H., and Nagayama, K. (1992) Mechanism of formation of 2-dimensional crystals from latex-particles on substrates. *Langmuir*, **8** (12), 3183–3190.
47 Park, S.H., Gates, B., and Xia, Y. (1999) A three-dimensional photonic crystal operating in the visible region. *Adv. Mater.*, **11**, 462–466.
48 Xia, Y.N., Yin, Y.D., Lu, Y., and McLellan, J. (2003) Template-assisted self-assembly of spherical colloids into complex and controllable structures. *Adv. Func. Mater.*, **13** (12), 907–918.
49 Vanapalli, S.A., Iacovella, C.R., Sung, K.E., Mukhija, D., Millunchick, J.M., Burns, M.A., Glotzer, S.C., and Solomon, M.J. (2008) Fluidic assembly and packing of microspheres in confined channels. *Langmuir*, **24** (7), 3661–3670.
50 Kraus, T., Malaquin, L., Delamarche, E., Schmid, H., Spencer, N.D., and Wolf, H. (2005) Closing the gap between self-assembly and microsystems using self-assembly, transfer, and integration of particles. *Adv. Mater.*, **17** (20), 2438.
51 Fan, F.Q. and Stebe, K.J. (2004) Assembly of colloidal particles by evaporation on surfaces with patterned hydrophobicity. *Langmuir*, **20** (8), 3062–3067.
52 Fustin, C.A., Glasser, G., Spiess, H.W., and Jonas, U. (2004) Parameters influencing the templated growth of colloidal crystals on chemically patterned surfaces. *Langmuir*, **20** (21), 9114–9123.
53 Chen, K.M., Jiang, X.P., Kimerling, L.C., and Hammond, P.T. (2000) Selective self-organization of colloids on patterned polyelectrolyte templates. *Langmuir*, **16** (20), 7825–7834.
54 Feder, J. (1980) Random sequential adsorption. *J. Theor. Biol.*, **87** (2), 237–254.
55 Maury, P., Escalante, M., Reinhoudt, D.N., and Huskens, J. (2005) Directed assembly of nanoparticles onto polymer-imprinted or chemically patterned templates fabricated by nanoimprint lithography. *Adv. Mater.*, **17** (22), 2718.
56 Le, J.D., Pinto, Y., Seeman, N.C., Musier-Forsyth, K., Taton, T.A., and Kiehl, R.A. (2004) Dna-templated self-assembly of

metallic nanocomponent arrays on a surface. *Nano Lett.*, **4** (12), 2343–2347.

57 Michel, B., Bernard, A., Bietsch, A., Delamarche, E., Geissler, M., Juncker, D., Kind, H., Renault, J.P., Rothuizen, H., Schmid, H., Schmidt-Winkel, P., Stutz, R., and Wolf, H. (2001) Printing meets lithography: Soft approaches to high-resolution printing. *IBM J. Res. Dev.*, **45** (5), 697–719.

58 Piner, R.D., Zhu, J., Xu, F., Hong, SH., and Mirkin, CA. (1999) "Dip-pen" nanolithography. *Science*, **283** (5402), 661–663.

59 Kraus, T., Malaquin, L., Schmid, H., Riess, W., Spencer, N.D., and Wolf, H. (2007) Nanoparticle printing with single-particle resolution. *Nature Nanotechnology*, **2** (9), 570–576.

60 Aizenberg, J., Braun, P.V., and Wiltzius, P. (2000) Patterned colloidal deposition controlled by electrostatic and capillary forces. *Phys. Rev. Lett.*, **84** (13), 2997–3000.

61 Dziomkina, N.V. and Vancso., G.J. (2005) Colloidal crystal assembly on topologically patterned templates. *Soft Matter*, **1** (4), 265–279.

62 Dziomkina, N.V., Hempenius, M.A., and Vancso, G.J. (2005) Symmetry control of polymer colloidal monolayers and crystals by electrophoretic deposition onto patterned surfaces. *Adv. Mater.*, **17** (2), 237.

63 Bartlett, P., Ottewill, R.H., and Pusey, P.N. (1992) Superlattice formation in binary-mixtures of hard-sphere colloids. *Phys. Rev. Lett.*, **68** (25), 3801–3804.

64 Eldridge, M.D., Madden, P.A., and Frenkel, D. (1993) Entropy-driven formation of a superlattice in a hard-sphere binary mixture. *Nature*, **365** (6441), 35–37.

65 Kitaev, V. and Ozin, G.A. (2003) Self-assembled surface patterns of binary colloidal crystals. *Adv. Mater.*, **15** (1), 75.

66 Cong, H.L. and Cao, W.X. (2005) Array patterns of binary colloidal crystals. *J. Phys. Chem. B*, **109** (5), 1695–1698.

67 Velikov, K.P., Christova, C.G., Dullens, R.P.A., and van Blaaderen, A. (2002) Layer-by-layer growth of binary colloidal crystals. *Science*, **296** (5565), 106–109.

68 Wang, D.Y. and Mohwald, H. (2004) Rapid fabrication of binary colloidal crystals by stepwise spin-coating. *Adv. Mater.*, **16** (3), 244.

69 Meitl, M.A., Zhu, Z.T., Kumar, V., Lee, K.J., Feng, X., Huang, Y.Y., Adesida, I., Nuzzo, R.G., and Rogers, J.A. (2006) Transfer printing by kinetic control of adhesion to an elastomeric stamp. *Nature Mater.*, **5** (1), 33–38.

70 Xia, Y.N., Gates, B., Yin, Y.D., and Lu, Y. (2000) Monodispersed colloidal spheres: Old materials with new applications. *Adv. Mater.*, **12** (10), 693–713.

71 Lopez, C. (2003) Materials aspects of photonic crystals. *Adv. Mater.*, **15** (20), 1679–1704.

72 Velev, OD. and Lenhoff, AM. (2000) Colloidal crystals as templates for porous materials. *Curr. Opin. Colloid In.*, **5** (1–2), 56–63.

73 Hanarp, P., Sutherland, D.S., Gold, J., and Kasemo, B. (2003) Control of nanoparticle film structure for colloidal lithography. *Colloids and Surfaces A-Physicochemical and Engineering Aspects*, **214** (1–3), 23–36.

74 Hua, F., Shi, J., Lvov, Y., and Cui, T. (2002) Patterning of layer-by-layer self-assembled multiple types of nanoparticle thin films by lithographic technique. *Nano Lett*, **2** (11), 1219–1222.

75 Cox, J.K., Eisenberg, A., and Lennox, RB. (1999) Patterned surfaces via self-assembly. *Curr. Opin. Coll. Int. Sci.*, **4** (1), 52–59.

76 Denkov, N.D., Velev, O.D., Kralchevsky, P.A., Ivanov, I.B., Yoshimura, H., and Nagayama, K. (1993) 2-dimensional crystallization. *Nature*, **361** (6407), 26.

77 Rakers, S., Chi, LF., and Fuchs, H. (1997) Influence of the evaporation rate on the packing order of polydisperse latex monofilms. *Langmuir*, **13** (26), 7121–7124.

78 Dimitrov, AS. and Nagayama, K. (1996) Continuous convective assembling of fine particles into two-dimensional arrays on solid surfaces. *Langmuir*, **12** (5), 1303–1311.

79 Jiang, P. and McFarland, MJ. (2004) Large-scale fabrication of wafer-size

colloidal crystals, macroporous polymers and nanocomposites by spin-coating. *J. Am. Chem. Soc.*, **126** (42), 13778–13786.
80 Hulteen, J.C., Treichel, D.A., Smith, M.T., Duval, M.L., Jensen, T.R., and Van Duyne, RP. (1999) Nanosphere lithography: Size-tunable silver nanoparticle and surface cluster arrays. *J. Phys. Chem. B*, **103** (19), 3854–3863.
81 Fulda, K.U. and Tieke, B. (1994) Langmuir films of monodisperse 0.5-Mu-M spherical polymer particles with a hydrophobic core and a hydrophilic shell. *Adv. Mater.*, **6** (4), 288–290.
82 Gu, Z.-Z., Wang, D., and Moehwald, H. (2007) Self-assembly of microspheres at the air/water/air interface into free-standing colloidal crystal films. *Soft Matter*, **3** (1), 68–70.
83 Jiang, P., Bertone, J.F., Hwang, K.S., and Colvin, V.L. (1999) Single-crystal colloidal multilayers of controlled thickness. *Chem. Mater.*, **11** (8), 2132–2140.
84 Shimmin, R.G., DiMauro, A.J., and Braun, P.V. (2006) Slow vertical deposition of colloidal crystals: A Langmuir-Blodgett process? *Langmuir*, **22** (15), 6507–6513.
85 Lenzmann, F., Li, K., Kitai, AH., and Stover, HDH. (1994) Thin-film micropatterning using polymer microspheres. *Chem. Mater.*, **6** (2), 156–159.
86 Burmeister, F., Schafle, C., Matthes, T., Bohmisch, M., Boneberg, J., and Leiderer, P. (1997) Colloid monolayers as versatile lithographic masks. *Langmuir*, **13** (11), 2983–2987.
87 Trau, M., Saville, DA., and Aksay, IA. (1996) Field-induced layering of colloidal crystals. *Science*, **272** (5262), 706–709.
88 Deckman, HW. and Dunsmuir, JH. (1982) Natural lithography. *Appl. Phys. Lett.*, **41** (4), 377–379.
89 Li, Y., Cai, W., and Duan, G. (2008) Ordered micro/nanostructured arrays based on the monolayer colloidal crystals. *Chem. Mater.*, **20** (3), 615–624.
90 Agheli, H. and Sutherland, D.S. (2006) Nanofabrication of polymer surfaces utilizing colloidal lithography and ion etching. *IEEE Trans. on Nanobioscience*, **5** (1), 9–14.
91 Denis, F.A., Hanarp, P., Sutherland, D.S., and Dufrene, Y.F. (2004) Nanoscale chemical patterns fabricated by using colloidal lithography and self-assembled monolayers. *Langmuir*, **20** (21), 9335–9339.
92 Hanarp, P., Sutherland, D., Gold, J., and Kasemo, B. (1999) Nanostructured model biomaterial surfaces prepared by colloidal lithography. *Nanostruct. Mater.*, **12** (1–4), 429–432.
93 Jiang, P. (2006) Large-scale fabrication of periodic nanostructured materials by using hexagonal non-close-packed colloidal crystals as templates. *Langmuir*, **22** (9), 3955–3958.
94 Valsesia, A., Colpo, P., Silvan, M.M., Meziani, T., Ceccone, G., and Rossi, F. (2004) Fabrication of nanostructured polymeric surfaces for biosensing devices. *Nano Lett.*, **4** (6), 1047–1050.
95 Yan, F. and Goedel, WA. (2004) A simple and effective method for the preparation of porous membranes with three-dimensionally arranged pores. *Adv. Mater.*, **16** (11), 911–915.
96 Johnson, S.A., Ollivier, P.J., and Mallouk, T.E. (1999) Ordered mesoporous polymers of tunable pore size from colloidal silica templates. *Science*, **283** (5404), 963–965.
97 Trujillo, N.J., Baxamusa, S.H., and Gleason, K.K. (2009) Grafted functional polymer nanostructures patterned bottom-up by colloidal lithography and initiated chemical vapor deposition (iCVD). *Chem. Mater.*, **21** (4), 742–750.
98 Yi, D.K. and Kim, D.Y. (2003) Polymer nanosphere lithography: fabrication of an ordered trigonal polymeric nanostructure. *Chem. Commun* (8), 982–983.
99 Paquet, C. and Kumacheva, E. (2008) Nanostructured polymers for photonics. *Mater. Today*, **11** (4), 48–56.
100 Kosiorek, A., Kandulski, W., Glaczynska, H., and Giersig, M. (2005) Fabrication of nanoscale rings, dots, and rods by combining shadow nanosphere lithography and annealed polystyrene nanosphere masks. *Small*, **1** (4), 439–444.

101 Haynes, C.L. and Van Duyne, R.P. (2001) Nanosphere lithography: A versatile nanofabrication tool for studies of size-dependent nanoparticle optics. *J. Phys. Chem. B*, **105** (24), 5599–5611.

102 Lu, Y., Liu, G.L., Kim, J., Mejia, Y.X., and Lee, L.P. (2005) Nanophotonic crescent moon structures with sharp edge for ultrasensitive biomolecular detection by local electromagnetic field enhancement effect. *Nano Lett.*, **5** (1), 119–124.

103 Kosiorek, A., Kandulski, W., Chudzinski, P., Kempa, K., and Giersig, M. (2004) Shadow nanosphere lithography: simulation and experiment. *Nano Lett.*, **4** (7), 1359–1363.

104 Zhang, G., Wang, D., and Moehwald, H. (2007) Ordered binary arrays of Au nanoparticles derived from colloidal lithography. *Nano Lett.*, **7** (1), 127–132.

105 Zhang, G. and Wang, D. (2009) Colloidal lithography-the art of nanochemical patterning. *Chemistry-An Asian J.*, **4** (2), 236–245.

106 Jones, J.B., Sanders, J.V., and Segnit, E.R. (1964) Structure of opal. *Nature*, **204**, 1151.

107 Muller, M., Zentel, R., Maka, T., Romanov, SG., and Torres, CMS. (2000) Dye-containing polymer beads as photonic crystals. *Chem. Mater.*, **12** (8), 2508–2512.

108 van Blaaderen, A., Ruel, R., and Wiltzius, P. (1997) Template-directed colloidal crystallization. *Nature*, **385** (6614), 321–324.

109 Philipse, AP. (1989) Solid opaline packings of colloidal silica spheres. *J. Mater. Sci. Lett.*, **8** (12), 1371–1373.

110 Gu, Z.Z., Fujishima, A., and Sato, O. (2002) Fabrication of high-quality opal films with controllable thickness. *Chem. Mater.*, **14** (2), 760–765.

111 Salvarezza, R.C., Vazquez, L., Miguez, H., Mayoral, R., Lopez, C., and Meseguer, F. (1996) Edward-wilkinson behavior of crystal surfaces grown by sedimentation of SiO2 nanospheres. *Phys. Rev. Lett.*, **77** (22), 4572–4575.

112 Yamaki, M., Higo, J., and Nagayama, K. (1995) Size-dependent separation of colloidal particles in 2-dimensional convective self-assembly. *Langmuir*, **11** (8), 2975–2978.

113 Park, S.H., Qin, D., and Xia, Y. (1998) Crystallization of mesoscale particles over large areas. *Adv. Mater.*, **10** (13), 1028.

114 Park, S.H. and Xia, Y.N. (1999) Assembly of mesoscale particles over large areas and its application in fabricating tunable optical filters. *Langmuir*, **15** (1), 266–273.

115 Gates, B., Qin, D., and Xia, YN. (1999) Assembly of nanoparticles into opaline structures over large areas. *Adv. Mater.*, **11** (6), 466.

116 Pieranski, P. (1983) Colloidal crystals. *Contemp. Phys.*, **24** (1), 25–73.

117 Park, SH. and Xia, YN. (1998) Fabrication of three-dimensional macroporous membranes with assemblies of microspheres as templates. *Chem. Mater.*, **10** (7), 1745.

118 Xia, Y.N., Kim, E., and Whitesides, G.M. (1996) Micromolding of polymers in capillaries: applications in microfabrication. *Chem. Mater.*, **8** (7), 1558–1567.

119 Miguez, H., Meseguer, F., Lopez, C., Lopez-Tejeira, F., and Sanchez-Dehesa, J. (2001) Synthesis and photonic bandgap characterization of polymer inverse opals. *Adv. Mater.*, **13** (6), 393–396.

120 Cassagneau, T. and Caruso, F. (2002) Conjugated polymer inverse opals for potentiometric biosensing. *Adv. Mater.*, **14** (24), 1837–1841.

121 Xu, L., Wang, J., Song, Y., and Jiang, L. (2008) Electrically tunable polypyrrole inverse opals with switchable stopband, conductivity, and wettability. *Chem. Mater.*, **20** (11), 3554–3556.

122 Schroden, R.C., Al-Daous, M., Blanford, C.F., and Stein, A. (2002) Optical properties of inverse opal photonic crystals. *Chem. Mater.*, **14** (8), 3305–3315.

123 Joannopoulos, JD. (2001) Self-assembly lights up. *Nature*, **414** (6861), 257–258.

124 Hiltner, P.A. and Krieger, I.M. (1969) Diffraction of light by ordered suspensions. *J. Phys. Chem.*, **73** (7), 2386.

125 Park, S.H., Gates, B., and Xia, YN. (1999) A three-dimensional photonic crystal operating in the visible region. *Adv. Mater.*, **11** (6), 462.

126 Miguez, H., Meseguer, F., Lopez, C., Blanco, A., Moya, J.S., Requena, J., Mifsud, A., and Fornes, V. (1998) Control of the photonic crystal properties of fcc-packed submicrometer SiO2 spheres by sintering. *Adv. Mater.*, **10** (6), 480.

127 Ueno, K., Matsubara, K., Watanabe, M., and Takeoka, Y. (2007) An electro- and thermochromic hydrogel as a full-color indicator. *Adv. Mater.*, **19** (19), 2807.

128 Lee, Y.J. and Braun, P.V. (2003) Tunable inverse opal hydrogel pH sensors. *Adv. Mater.*, **15** (7–8), 563–566.

129 Xu, M., Goponenko, A.V., and Asher, S.A. (2008) Polymerized polyHEMA photonic crystals: pH and ethanol sensor materials. *J. Am. Chem. Soc.*, **130** (10), 3113–3119.

130 Viel, B., Ruhl, T., and Hellmann, G.P. (2007) Reversible deformation of opal elastomers. *Chem. Mater.*, **19** (23), 5673–5679.

131 Holtz, J.H. and Asher, S.A. (1997) Polymerized colloidal crystal hydrogel films as intelligent chemical sensing materials. *Nature*, **389** (6653), 829–832.

132 Cassagneau, T. and Caruso, F. (2002) Inverse opals for optical affinity biosensing. *Adv. Mater.*, **14** (22), 1629–1633.

133 Siwick, B.J., Kalinina, O., Kumacheva, E., Miller, R.J.D., and Noolandi, J. (2001) Polymeric nanostructured material for high-density three-dimensional optical memory storage. *J. Appl. Phys.*, **90** (10), 5328–5334.

134 Gaponenko, S.V., Bogomolov, V.N., Petrov, E.P., Kapitonov, A.M., Yarotsky, D.A., Kalosha, I.I., Eychmueller, A.A., Rogach, A.L., McGilp, J., Woggon, U., and Gindele, F. (1999) Spontaneous emission of dye molecules, semiconductor nanocrystals, and rare-earth ions in opal-based photonic crystals. *J. Lightwave Technol.*, **17** (11), 2128–2137.

135 Yoshino, K., Lee, S.B., Tatsuhara, S., Kawagishi, Y., Ozaki, M., and Zakhidov, A.A. (1998) Observation of inhibited spontaneous emission and stimulated emission of rhodamine 6G in polymer replica of synthetic opal. *Appl. Phys. Lett.*, **73** (24), 3506–3508.

136 Romanov, S.G., Maka, T., Torres, C.M.S., Muller, M., and Zentel, R. (1999) Photonic band-gap effects upon the light emission from a dye-polymer-opal composite. *Appl. Phys. Lett.*, **75** (8), 1057–1059.

137 Lange, B., Zentel, R., Ober, C., and Marder, S. (2004) Photoprocessable polymer opals. *Chem. Mater.*, **16** (25), 5286–5292.

138 Fleischhaker, F. and Zentel, R. (2005) Photonic crystals from core-shell colloids with incorporated highly fluorescent quantum dots. *Chem. Mater.*, **17** (6), 1346–1351.

139 Jang, S.G., Choi, D.-G., Heo, C.-J., Lee, S.Y., and Yang, S.M. (2008) Nanoscopic ordered voids and metal caps by controlled trapping of colloidal particles at polymeric film surfaces. *Adv. Mater.*, **20** (24), 4862.

140 Chen, Z.M., Gang, T., Yan, X., Li, X., Zhang, J.H., Wang, Y.F., Chen, X., Sun, Z.Q., Zhang, K., Zhao, B., and Yang, B. (2006) Ordered silica microspheres unsymmetrically coated with Ag nanoparticles, and Ag-nanoparticle-doped polymer voids fabricated by microcontact printing and chemical reduction. *Adv. Mater.*, **18** (7), 924.

141 Wood, M.A. (2007) Colloidal lithography and current fabrication techniques producing in-plane nanotopography for biological applications. *J. R. Soc. Interface*, **4** (12), 1–17.

142 Dalby, M.J., Riehle, M.O., Sutherland, D.S., Agheli, H., and Curtis, A.S.G. (2004) Use of nanotopography to study mechanotransduction in fibroblasts - methods and perspectives. *Eur. J. Cell Biol.*, **83** (4), 159–169.

143 Norman, J. and Desai, T. (2006) Methods for fabrication of nanoscale topography for tissue engineering scaffolds. *Ann. Biomed. Eng.*, **34** (1), 89–101.

144 Andersson, A.S., Backhed, F., von Euler, A., Richter-Dahlfors, A., Sutherland, D., and Kasemo, B. (2003) Nanoscale features influence epithelial cell morphology and cytokine production. *Biomaterials*, **24** (20), 3427–3436.

145 Wood, M.A., Wilkinson, C.D.W., and Curtis, A.S.G. (2006) The effects of

colloidal nanotopography on initial fibroblast adhesion and morphology. *IEEE Trans. on Nanobioscience*, **5** (1), 20–31.

146 Dalby, M.J., Berry, C.C., Riele, M.O., Sutherland, D.S., Agheli, H., and Curtis, A.S.G. (2004) Attempted endocytosis of nano-environment produced by colloidal lithography by human fibroblasts. *Exp. Cell Res.*, **295**, 287–394.

147 Kotov, N.A., Liu, Y.F., Wang, S.P., Cumming, C., Eghtedari, M., Vargas, G., Motamedi, M., Nichols, J., and Cortiella, J. (2004) Inverted colloidal crystals as three-dimensional cell scaffolds. *Langmuir*, **20** (19), 7887–7892.

148 Liu, Y.F., Wang, S.P., Lee, J.W., and Kotov, N.A. (2005) A floating self-assembly route to colloidal crystal templates for 3D cell scaffolds. *Chem. Mater.*, **17** (20), 4918–4924.

149 Kim, S., Kustandi, T.S., and Yi, D.K. (2008) Synthesis of artificial polymeric nanopillars for clean and reusable adhesives. *J. Nanosci. Nanotech.*, **8** (9), 4779–4782.

150 Kustandi, T.S., Samper, V.D., Yi, D.K., Ng, W.S., Neuzil, P., and Sun, W. (2007) Self-assembled nanoparticles based fabrication of gecko foot-hair-inspired polymer nanofibers. *Adv. Func. Mater.*, **17** (13), 2211–2218.

10
Directed Self-Assembly of Block Copolymer Films
Gordon S.W. Craig and Paul F. Nealey

10.1
Introduction

Block copolymers are composed of two linked polymer chains able to microphase separate into equilibrated morphologies containing discrete, nanoscale domains [1–5]. While block copolymers can generate domains on a length scale of 10s of nm, for lithographic purposes it is necessary that these domains are well organized and regularly oriented on an underlying substrate, with minimal defect density [6]. To achieve these requirements, it is necessary to direct the self-assembly of the block copolymer morphology by introducing surface constraints, either topographical or chemical, which then drives the orientation and location of the domains as the block copolymer equilibrates. In the case of topographically directed assembly, physical features, such as trenches [7] or posts [8], serve to confine block segregation and orient and align the domains. Directing the self-assembly with a chemical pattern involves creating a chemical pattern with alternating regions that are preferential to the different blocks of the block copolymer, such that one block will wet one set of regions of the chemical pattern, and the other block will wet the remaining area of the chemical pattern [9–11]. In this chapter we will focus on directed self-assembly by chemical patterns, and refer to this type of directed self-assembly as "directed assembly".

Directed self-assembly shows promise in advanced lithography and a variety of other applications that have less complex requirements. For example, directed self-assembly could be used for enhancing etch selectivity, placing dopants in ordered arrays, or generating high-density, close-packed electrodes in capacitor arrays [6]. Additionally, the assembled nanostructures could be used for fabricating densely packed porous templates [12–14] or membranes [15, 16] at the nanoscale. Other potential applications of assembled block copolymer thin films include the fabrication of MOSFETs (metal-oxide-semiconductor field-effect transistors) [17], quantum dots [18], high surface area devices [19, 20], photovoltaic devices [21], and bit patterned media [22–24].

Generating Micro- and Nanopatterns on Polymeric Materials. Edited by A. del Campo and E. Arzt
Copyright © 2011 WILEY-VCH Verlag GmbH & Co. KGaA, Weinheim
ISBN: 978-3-527-32508-5

Apart from its potential technological significance, directed self-assembly of block copolymer films also provides researchers with a tool to advance our understanding of the formation of block copolymer domains in the vicinity of a surface. Even without a chemical pattern on a surface, such as on a chemically homogeneous substrate, the reconstruction of the morphology of a block copolymer in the presence of the surface can lead to a variety of morphologies not readily obtainable by other means. Lamellae-forming block copolymers in thin films can form lamellar bends, disclinations, and missing rows [25, 26]. Cylinder-forming block copolymers have been shown to form an even greater array of morphologies such as lamella, spheres, perforated lamella, perpendicular and parallel cylinders, as well as combinations of these structures [27–30]. Phase diagrams of morphologies achieved during surface reconstruction point to the importance of film thickness on the morphology [31, 32].

Directed assembly has enabled researchers to examine surface reconstruction on heterogeneous surfaces, in addition to the results on homogeneous surfaces listed above. Research on directed assembly has shown that the same molecular weight (M_n) block copolymer will form lamellae of different dimensions as it equilibrates in the presence of chemically nanopatterned surfaces of varying pattern spacings [33]. Other research has found that potentially tortuous morphologies, never before observed in the bulk copolymer, were created when a block copolymer was forced to equilibrate on a surface pattern with a spacing period that was completely mismatched with that of the copolymer morphology [34]. All of the surface reconstruction studies, whether on homogeneous or chemically patterned surfaces, point to the important role the surface plays in the assembly of the block copolymer as it equilibrates in the presence of the surface.

10.2
Energetics of the Basic Directed Assembly System

In the bulk, the thermodynamics of microphase separation of an A-B block copolymer in the bulk are driven by two competing factors: an enthalpic free energy term, F_{A-B}, that favors a decreased number of A-B domain interfaces, balanced by an entropic free energy term, F_{ent}, caused by the limited extensibility of each chain of the block copolymer [4]. The balance of these two terms leads to different morphologies [5] and domain spacings that are determined primarily by the total overall degree of polymerization of the block copolymer, N, the volume fraction of component A, f, and the Flory–Huggins interaction parameter between the A and B repeat units, χ. For a lamellar morphology of a symmetric block copolymer ($f = 0.5$), the free energy per copolymer chain of the microphase separated block copolymer in the bulk in the strong segregation limit, F_{bulk}, can be expressed as Equation 10.1 [35, 36]:

$$F_{bulk} = F_{ent} + F_{A-B} = kT\left(\frac{3L^2}{8l^2 N} + \frac{2lN}{L}\sqrt{\frac{\chi}{6}}\right), \tag{10.1}$$

where k is Boltzmann's constant, T is the absolute temperature, l is the Kuhn segment length, and L is the period of the lamellar morphology. If L_o is defined as the equilibrium bulk lamellar period, then F_{bulk} is at its equilibrium value, and therefore minimized, when $L = L_o$. The equation for F_{bulk} assumes that the role of the surfaces of the block copolymer is negligible.

In the case of a thin film of a block copolymer on a chemically patterned substrate, the role of the surfaces of the film cannot be neglected. To understand their role that these surfaces play in the equilibration of the block copolymer, it is first necessary to review the process of directed assembly on a chemical pattern [33, 37], outlined in Figure 10.1. In general, the directed assembly process starts with the grafting of a polymer brush, such as the polystyrene (PS) brush shown in Figure 10.1, to a silicon wafer substrate. Photolithography, followed by an oxygen plasma treatment, is used to pattern the polymer brush. The oxygen plasma treatment generates regions with higher surface energy. In the case of the process shown in Figure 10.1, the pattern consists of alternating lines of untreated and oxygen-plasma-treated polymer brush with a pattern period of L_s. A thin ($\leq$ 200 nm) film of block copolymer is coated onto the chemical pattern, and subsequently annealed. In the case of a block copolymer that has a microphase separated morphology in the bulk [4], its domains can align with and register to the pattern if the different blocks of the copolymer selectively wet alternating regions of the pattern.

For a thin film of block copolymer on a chemical pattern, the effect of the surfaces can be described as follows. In the absence of stronger driving forces, the energy of the free surface, F_{surf}, will lead the block with the lowest surface energy to populate the surface layer. F_{surf} per chain in the film can be expressed as Equation 10.2 [33, 38]:

$$F_{surf} = (f_{AS}\gamma_A + (1-f_{AS})\gamma_B)\left(\frac{M_n}{\varrho Nt}\right), \qquad (10.2)$$

where γ_A and γ_B are the surface energies of blocks A and B, respectively, f_{AS} is the fraction of the free surface covered by the A block, ϱ is the density of the block

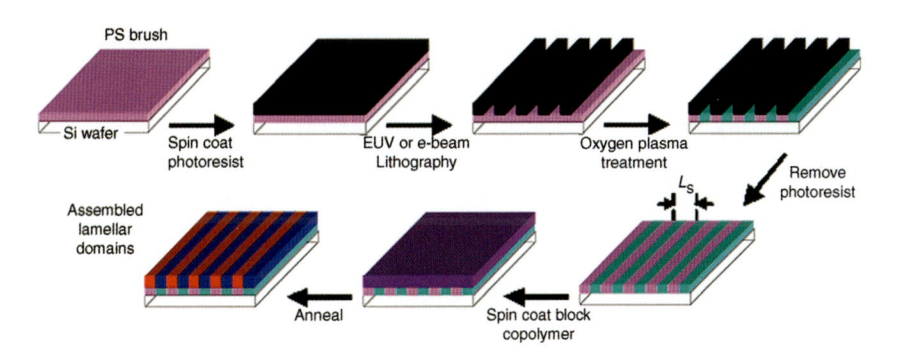

Figure 10.1 Schematic of the directed assembly of a block copolymer on a chemical pattern with pattern period L_s. In this case, a PS brush layer (pink) is lithographically patterned and treated with an oxygen plasma, resulting in alternating bands of PS-preferential (pink) and PMMA-preferential (light blue) surface chemistry. Lamellae-forming PS-b-PMMA is then spin coated onto the chemical pattern, and annealed to yield ordered, registered lamellar domains of PS (red) and PMMA (blue).

copolymer, and t is the thickness of the film. Similarly, for a chemical pattern having equal areas of the different regions of the chemical pattern, such as that shown in Figure 10.1, the interfacial energy per chain in the film, F_{int}, between the block copolymer and the chemically patterned interface can be expressed as Equation 10.3:

$$F_{int} = (f_{A1}\gamma_{A1} + (1-f_{A1})\gamma_{B1} + f_{A2}\gamma_{A2} + (1-f_{A2})\gamma_{B2})\left(\frac{M_n}{2\varrho Nt}\right), \quad (10.3)$$

where f_{A1} and f_{A2} are the fractions of the regions 1 and 2 of the chemical pattern, respectively, that are covered by the A block, and γ_{A1}, γ_{A2}, γ_{B1}, and γ_{B2} are the interfacial energies of the A and B blocks with each region of the chemical pattern.

For a chemical pattern to successfully direct the assembly of a block copolymer film, the desired assembly must be at thermodynamic equilibrium, and therefore, the total free energy of the film, F_{film}, expressed as Equation 10.4 [33]:

$$F_{film} = F_{A-B} + F_{ent} + F_{surf} + F_{int}, \quad (10.4)$$

must be at a minimum. Although the minimization of F_{film} incorporates all of the factors shown in Equation 10.4, we can gain some understanding of the effect of each group of factors by examining F_{bulk}, F_{surf}, and F_{int} individually. For any block copolymer directed to assemble on a chemical pattern with period L_s, F_{bulk} will be minimized when $L = L_s = L_o$, a result observed consistently throughout the experimental development of directed assembly on a chemical pattern [9, 11, 33, 37, 39–41]. Thus, the best examples of directed assembly in terms of the perfection of the assembly, and the film thickness through which the assembled domains persist, occur when the pattern spacing matches the bulk period of the block copolymer morphology.

The effect of the free surface can be viewed through the minimization of F_{surf}, which will lead to $f_{AS} = 1$ if $\gamma_A \ll \gamma_B$, and to $f_{AS} = 0$ if $\gamma_A \gg \gamma_B$. In these cases, there is a strong driving force for one particular block to reside at the surface, making it difficult for the chemical pattern to direct the assembly of the morphology at the free surface of the film, especially when the energetic driving force supplied by the chemical pattern is relatively weak. One would expect these assemblies to have cross-sectional structures similar to those observed in films of polystyrene-*block*-poly (methyl methacrylate) block copolymers (PS-*b*-PMMA) on unpatterned, neutral substrates [42, 43]. However, when $\gamma_{AS} \sim \gamma_{BS}$, as is the case for PS-*b*-PMMA [44], the effect of F_{surf} on the total free energy of the film can be minimal. As a result, the energetics of the underlying chemical pattern can overwhelm the preference of the free surface for one block and, for example, force some of block A to reside on the surface, even though γ_{AS} may be slightly more than γ_{BS}.

The properties of the chemical pattern that are necessary to achieve a pattern with a high degree of perfection can be determined from the factors in Equation 10.3, coupled with factors that would cause F_{surf} and F_{bulk} to have a minimal negative impact on the assembly. In general, there are three primary factors to consider. First, as mentioned above, perfect assemblies can be achieved more readily when $L_s = L_o$, although it has been shown that block copolymers can be directed to assemble on patterns in which L_s differs from L_o by as much as 10% [33]. Second, to limit the effect of F_{surf}, it is beneficial to have $\gamma_{AS} \sim \gamma_{BS}$. Finally, in terms of F_{int}, it is advantageous to maximize the interfacial energy contrast between each of the blocks of the copolymer with the alternating regions of the chemical pattern, such that the A block has a strong preference to reside on one region of the pattern, and the B block has a strong preference to reside on the neighboring region of the chemical pattern. In terms of the separate interfacial energies, directed assembly is especially effective when $\gamma_{A1} \ll \gamma_{B1}$, and $\gamma_{B2} \ll \gamma_{A2}$ (or conversely, $\gamma_{B1} \ll \gamma_{A1}$, and $\gamma_{A2} \ll \gamma_{B2}$).

When there is a strong interfacial energy contrast between neighboring regions of the chemical pattern, the chemical pattern can direct the assembly of the block copolymer even when $L_s \neq L_o$, as has been demonstrated with an array of patterns and surface treatments, shown in Figure 10.2 [33]. In the work presented in Figure 10.2, PS-b-PMMA was assembled on a striped chemical pattern consisting of alternating regions of oxidized polymer brush, which is preferential to the poly(methyl methacrylate) (PMMA) block, and a polystyrene-*random*-poly(methyl methacrylate) copolymer brush (PS-r-PMMA). The fraction of PS in the PS-r-PMMA brush was systematically varied from 50 to 100% (going from left to right in Figure 10.2), thereby providing a way to vary the interfacial energy contrast of the chemical pattern with the overlying PS-b-PMMA film. With each brush, chemical patterns were made having L_s values ranging from 40 nm to 50 nm in increments of 2.5 nm. A thin film of PS-b-PMMA with $L_o = 48.0$ nm was spin coated onto each chemical pattern and then annealed. The scanning electron micrographs (SEMs) of the assembled block copolymer films, shown in Figure 10.2, reveal the interaction of interfacial energy contrast with the commensurability of L_s and L_o in terms of determining the pattern quality of the assembled block copolymer film.

The role of interfacial energy contrast can be observed when examining the SEMs corresponding to a single L_s value. For example, for the films made on the chemical patterns with $L_s = 45.0$ nm, the most perfect assemblies were achieved on the chemical pattern made from PS-r-PMMA brush containing 100% PS (a PS homopolymer brush). The chemical pattern made from the PS homopolymer brush had the greatest interfacial energy contrast. For that brush, the oxidized PS regions of the chemical pattern preferred to be wetted by PMMA, whereas the untreated PS brush regions preferred to be wetted by PS. In contrast, when the PS-r-PMMA brush contained 58% PS, it was non-preferential for PS and PMMA [44]. While the oxidized brush still preferred to be wetted by PMMA, the untreated brush had no wetting preference. As a result, there was less of a driving force to create a perfect assembly when the PS content was 58% than when it was 100%, and the most perfect assemblies were formed on substrates from brushes containing 100% PS.

The effect of the commensurability of L_s and L_o is also evident in Figure 10.2, and matches previous reports that determined the best assemblies were achieved when $L_s = L_o$ [9, 11]. However, when the chemical interfacial energy contrast was increased by using a PS brush instead of a PS-r-PMMA brush, good assemblies could also be achieved when $L_s \neq L_o$, such as when $L_s = 50$ nm or 45 nm on the chemical pattern made with 100% PS. In that case, the chemical pattern was able to supply enough energy to the block copolymer film to force it to assemble with a period different from that of its naturally occurring bulk value.

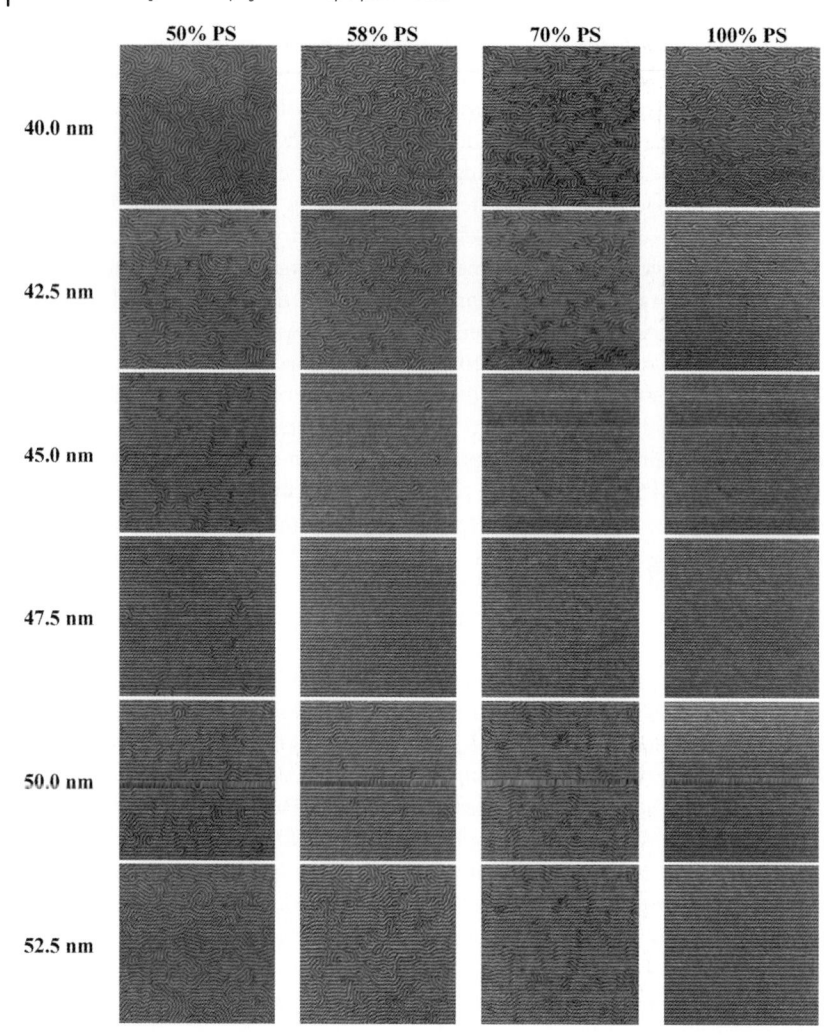

Figure 10.2 SEM images of equilibrated morphologies of symmetric PS-*b*-PMMA ($L_o = 48$ nm) on chemically patterned substrates as a function of pattern period (40 nm $< L_s <$ 52.5 nm) and S: MMA composition of a PS-*r*-PMMA brush used to create the chemical pattern. Each image depicts a 2 μm × 2 μm area.

Although the phenomenological explanation that accompanied the work presented in Figure 10.2 assumed F_{surf} to be negligible because $\gamma_A \sim \gamma_B$ for PS-*b*-PMMA, we can expand the methodology developed in that work for analyzing the effect of L_s and interfacial energy contrast [33] and apply it to a broader class of A-B block copolymer films on chemical patterns with regions labeled 1 and 2. We arbitrarily assume that $\gamma_{AS} < \gamma_{BS}$, such that block A normally resides on the free surface. Also, we select the chemistry of pattern regions 1 and 2 so that $\gamma_{A1} < \gamma_{B1}$ and $\gamma_{B2} < \gamma_{A2}$, where region 1 is preferentially wet by block A, and region 2 is preferentially wet by block B. We select as a reference state a thin film of microphase-separated, symmetric A-B block copolymer coated on a chemically patterned substrate, which has block A residing on the free surface and blocks A and B randomly distributed across the chemical pattern. The free energy of the reference state F^* is then Equation 10.5:

$$F^* = kT\left(\frac{3L_o^2}{8l^2 N} + \frac{2lN}{L_o}\sqrt{\frac{\chi}{6}}\right) + \frac{\gamma_A M_n}{\varrho Nt} + (\gamma_{A1} + \gamma_{A2} + \gamma_{B1} + \gamma_{A2})\frac{M_n}{4\varrho Nt}. \quad (10.5)$$

To optimize the ability of the chemical pattern to direct the assembly of the block copolymer, we will assume that $L_s = L_o$. This implies that $\Delta F_{bulk} = 0$, and the ΔF corresponding to perfect epitaxial assembly of the block copolymer is Equation 10.6:

$$\Delta F = (\gamma_B - \gamma_A)\frac{M_n}{2\varrho Nt} + \{(\gamma_{A1} - \gamma_{A2}) + (\gamma_{B2} - \gamma_{B1})\}\frac{M_n}{4\varrho Nt}. \quad (10.6)$$

Perfect epitaxial assembly throughout the thickness requires $\Delta F < 0$, or, rearranging Equation 10.6 to give Equation 10.7 or (10.8):

$$(\gamma_{A2} - \gamma_{A1}) + (\gamma_{B1} - \gamma_{B2}) > 2(\gamma_B - \gamma_A), \quad (10.7)$$

or

$$\Delta\gamma_{int} > 2\Delta\gamma_{surf}, \quad (10.8)$$

where $\Delta\gamma_{int}$ is the combined difference of the interfacial energies of both blocks on both regions of the chemical pattern (the left side of Equation 10.7), and $\Delta\gamma_{surf}$ is the difference of the surface energy of the two blocks. The condition specified by Equation 10.8 is not a sufficient condition to guarantee perfect directed assembly. For example, low energy defects could occur, especially in thicker films, due to ΔF_{int} scaling inversely with t. However, Equation 10.8 does provide a necessary criterion for a chemical pattern to direct the assembly of a block copolymer across the entire thickness of the film. Simply put, the difference in interfacial energies of the two regions of the chemical pattern for the different blocks of the block copolymer together must provide more energy to the block copolymer than is lost by forcing some of the B block to reside at the surface.

One challenge in analyzing the energetics of a block copolymer film on a chemical pattern is that the dependence of γ on M_n and T limits the applicability of γ values obtained from the literature [44]. Literature values [45, 46] often do not cover the range of M_n values (typically ~100 kg/mol) or annealing temperatures (typically ~190 °C, in the case of PS-b-PMMA) used in directed assembly. The limited γ data is even more of an issue for interfacial energy values, for which four different values are required, as shown in Equation 10.7. However, we can compare the γ and χ values that are available, and draw qualitative conclusions on pattern formation. Table 10.1 presents values in the literature for the surface energy, interfacial energy, and χ for PS-b-PMMA, polystyrene-block-poly(2-vinylpyridene) (PS-b-P2VP), and polystyrene-block-polydimethylsiloxane (PS-b-PDMS).

Most of the published research on directed assembly of block copolymers has focused on PS-b-PMMA, so we can use that as a baseline to which we can compare

Table 10.1 Surface and interfacial energies (in erg cm^{-2}) and Flory–Huggins interaction parameter values, χ, for three block copolymer systems. The temperature (in °C) for each value is given in parentheses. References are given in square brackets.

Block A	Block B	χ	γ_A	γ_B	$\gamma_{A/B}$
PS	PMMA	0.0368 (170) [47]	29.9 (170) [46]	30.02 (170) [44]	1.4 (170) [48]
			29.1 (183) [45]		1.2 (190) [50]
PS	P2VP	0.106 (200) [52]	27.7 (200) [46]	34.0 (200) [54]	2.5 (200) [54]
		0.099 (200) [53]	27.9 (200) [45]	34.1 (200) [45]	
PS	PDMS	0.26 (140) [55]	32 (140) [45, 46]	14 (140) [55]	4.85 (140) [55]

other block copolymer systems. Assuming PS-*b*-PMMA is to be assembled on a chemically patterned PS brush, we would need to know the following four interfacial energies to apply Equation 10.7: that of the PS block on the PS brush ($\gamma_{S/S}$); the PS block on the oxidized PS brush ($\gamma_{S/O}$); the PMMA block on the PS brush ($\gamma_{S/MMA}$, which we assume to be equal to the corresponding $\gamma_{A/B}$ in Table 10.1); and the PMMA block on the oxidized PS brush ($\gamma_{MMA/O}$). Previous work on random copolymer brushes determined a value of 0.01 erg cm^{-2} for $\gamma_{S/S}$ [44]. The interfacial energy values for PS or PMMA on the oxidized PS brush are not readily available. For the sake of a qualitative analysis, we will assume the oxidized surface to be similar to SiO$_x$. It has been shown in previous research on the interfacial energy of PMMA with SiO$_x$ that the interaction of the two materials is somewhat attractive, with a value of $-0.8 kT$ per segment, or -3.5 erg cm^{-2} at 190 °C. Given the hydrophobicity of PS relative to PMMA, it is likely that $\gamma_{S/MMA} \leq \gamma_{S/O} \leq -\gamma_{MMA/O}$. Assuming $\gamma_{S/O} = \gamma_{S/MMA}$, $\Delta\gamma_{int} \sim 6.3$ erg cm^{-2}, whereas the penalty associated with forcing PMMA to the free surface, represented by $\Delta\gamma_{surf}$, is only ~ 0.1 erg cm^{-2}. Even if $\gamma_{MMA/O}$ is set to 0, the energy of the chemical pattern (~ 2.8 erg cm^{-2}) still is much greater than 2 $\Delta\gamma_{surf}$ (~ 0.1 erg cm^{-2}), for the case of PS-*b*-PMMA.

While the similar surface energy values for PS and PMMA make it an ideal choice for investigating the directed assembly of block copolymers by a chemical pattern, other block copolymers are of interest for improving the etch selectivity of the domains, or increasing the functionality of the assembled block copolymer. PS-*b*-PDMS, and other block copolymers containing heteroatoms, offer a potential route to increase etch selectivity [54]. PS-*b*-P2VP is another possible alternative to PS-*b*-PMMA for directed assembly [55]. Comparing the parameters of PS-*b*-PDMS and PS-*b*-P2VP shown in Table 10.1 to those of PS-*b*-PMMA, there are two significant differences. First, χ of both PS-*b*-PDMS and PS-*b*-P2VP is larger than χ of PS-*b*-PMMA, implying that shorter chains (smaller N), compared to PS-*b*-PMMA, are necessary to achieve microphase separation, resulting in smaller domains than are achievable with PS-*b*-PMMA. Second, and more importantly from the point of view of directing the assembly of the block copolymers, $\Delta\gamma_{surf}$ values for PS-*b*-PDMS and PS-*b*-P2VP are likely to be larger (much larger, in the case of PS-*b*-PDMS) than the contribution of the interfacial energies provided by the chemical pattern, based on the relatively low value of $\gamma_{A/B}$. As a result, achieving an assembled block copolymer film with vertical domains that traverse the thickness of the film, as shown in Figure 10.1, will be difficult with these two block copolymers, especially in thicker films. For example, topographically directed self-assembly of PS-b-PDMS required a CF_4 plasma etch of the PDMS layer at the surface of the film [54], and achieving vertically oriented domains of PS-b-P2VP across the entire thickness of a film required either an oxygen plasma etch back of the free surface of the film, or sandwiching the film between two chemically neutral substrates [55].

10.3 Examples of Directed Assembly

10.3.1 Non-Bulk Morphologies

Although much interest has focused on directing the assembly of block copolymers to yield structures that naturally occur in the bulk [9, 56, 57], the majority of assemblies that form are non-bulk morphologies. These non-bulk structures can be beneficial in advanced lithographic applications, where angled structures [37] and the range of semiconductor design elements, both individually and in arrays [58], are desired. However, potential applications of non-bulk structures go beyond advanced lithography to include a range of applications [59], such as separation membranes, filtration, catalysis, and photonics [60–62]. Additionally, such structures could be used in the fabrication of quantum dots [18], metallic nanodots [62, 63], and magnetic nanowires [65].

These non-bulk structures also provide a window for the study of surface reconstruction of block copolymers [31]. It has been shown that surface reconstruction of the domains in thin films of cylinder-forming block copolymers on a chemically homogeneous surface can lead to a variety of structures, such as perforated lamellae, undulating cylinders, and spherical nodules [31, 66]. Additionally, separate research demonstrated that cylindrical domains can assume very tortuous, three-dimensional (3D) structures in films on chemically homogeneous substrates as the film thickness is increased beyond a first layer of cylinders adjacent to the substrate [29].

A simple example of non-bulk structures that are a result of a block copolymer assembling on a chemical pattern is an angled or curved one-dimensional array of spheres. Previous research demonstrated the use of directed assembly to form cylindrical domains of polystyrene-*block*-poly(*t*-butylacrylate) (PS-b-PtBA) on a non-regular chemical pattern to form angled or curved cylinders parallel to the substrate, which could then be converted thermally to polystyrene-*block*-poly(acrylic anhydride) (PS-b-PAA), resulting in ordered, non-regular, one-dimensional arrays of spherical domains, as shown in Figure 10.3 [67]. The spherical domains could then be used as host sites for nanoparticle formation [68]. The transformation from cylinders to spheres showed that it is possible to use directed assembly to lead to a geometric transformation from the substrate surface to the film surface. In their

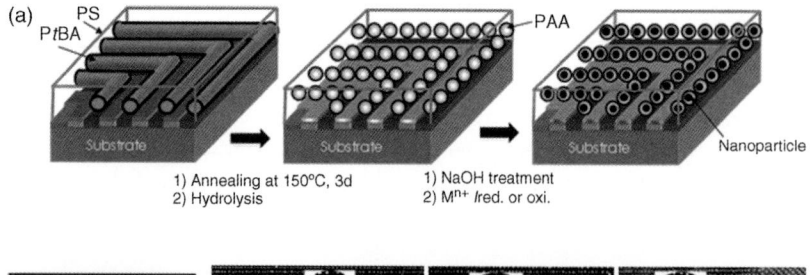

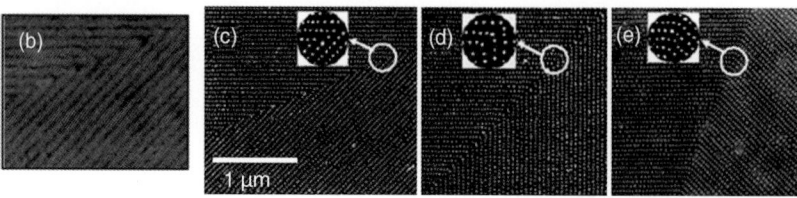

Figure 10.3 Schematic (a) and SEM images (b–e) of the use of directed assembly to form one-dimensional arrays of nanoparticles. Cylindrical domains are ordered parallel to the chemical pattern on the substrate (a, b). Thermal processing and the resulting domain volume reduction converts the domains to spheres, which then serve as host sites for nanoparticle formation (c–e).

case, an array of lines on the patterned substrate was transformed into a spotted pattern in the film.

A similar geometric transformation was demonstrated in a separate study on the directed assembly of spherical domains on a linear pattern [69]. In that work, the equilibration of the spherical domains of PS-b-PMMA on the linear pattern led to the formation of a knurled, linear domain on the chemical pattern, as shown in Figure 10.4. The knurls seem to play an important role in the reconstruction of the spherical bulk morphology from the chemically patterned lines on the substrate. They appeared to be sections of spheres connected to the PMMA wetting layer on the surface, and simultaneously ordered with overlying spherical domains. The knurls had a spacing along the lines that was closely matched to the sphere spacing in the bulk. Additionally, the lines of the chemical pattern directed the overlying spheres to arrange in linear arrays running in the same direction as the pattern lines, with a hexagonal arrangement of spheres across the assembled area. For contrast, the same block copolymer was spin coated onto a homogeneous substrate and annealed to yield a self-assembled structure, shown in Figure 10.4c.

Surface reconstruction is also evident in the directed assembly of a cylinder-forming PS-b-PMMA on a chemical pattern comprised of a square array of spots [70]. The cylinders aligned with and registered to the spots in the thinnest films, as seen in Figure 10.5a and b. However, even in the thinnest (22 nm) film, some defects were evident, such as domains located in the center of a square unit cell, which were dubbed "semi-cylinders". These pattern defects resulted from the block copolymer beginning to reconstruct its bulk hexagonal ordering of cylindrical domains. In the 34-nm-thick film (Figure 10.5b), more semi-cylinder defects were apparent, as well as

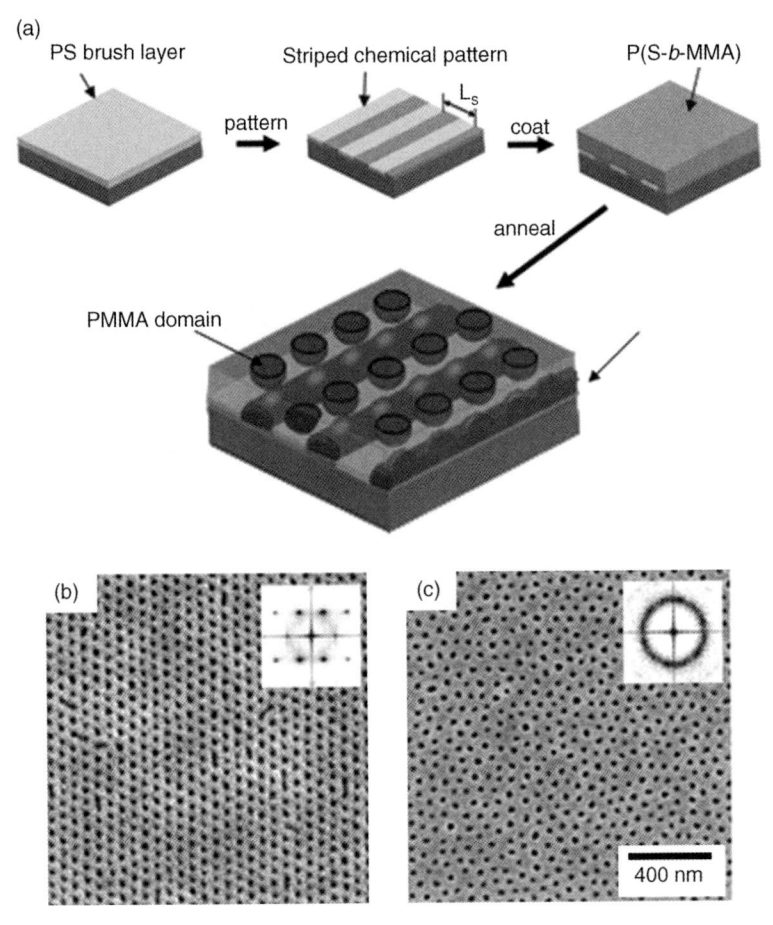

Figure 10.4 Schematic (a) and SEM image (b) of the directed assembly of spherical block copolymer domains on a linear chemical pattern, with pattern lines in (b) running from top to bottom. The SEM image in (c) shows the self-assembly of the same block copolymer on a homogeneous pattern. Fourier transforms of the SEM images are shown in the upper right hand corner of each image. (Reproduced with permission from [69] Copyright © (2008) by the American Chemical Society.)

"loop cylinders" which connect two of the cylinders. As the film thickness is increased further (Figure 10.5c), many more defects are apparent, and finally, in the case of 73-nm-thick film (Figure 10.5d), no order remains in the morphology, and the block copolymer has reconstructed a disordered bulk morphology.

The effect of film thickness can be understood in terms of the simple energetics model outlined above (Equations 1–3). Although the model focused on lamellae, and did not include the energy associated with the arrangement of the domains (hexagonal versus square, in this case), the inverse dependence of F_{int} on film thickness t shown in Equation 10.3 applies to the directed assembly of the cylindrical domains, as seen in Figure 10.5. The chemical pattern was able to direct the assembly

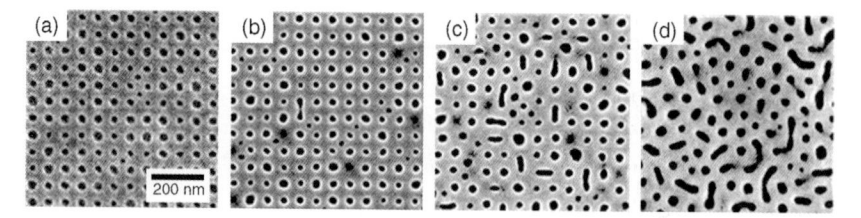

Figure 10.5 The effect of film thickness on the ability of a chemical pattern to direct the assembly of block copolymer cylinders is seen in these SEM images of cylinder-forming PS-*b*-PMMA films with varying thickness directed to assemble on a chemically patterned square array of spots. Film thicknesses were (a) 22 nm, (b) 34 nm, (c) 53 nm, and (d) 73 nm. (Reproduced with permission from [63] Copyright © (2007) by the American Chemical Society.)

of the block copolymer only over a very limited thickness range compared to lamellar domains directed to assemble on a linear chemical pattern. From this one can conclude that the hexagonal organization of the cylinders must contribute a significant reduction to F_{film} of cylinder-forming block copolymers.

While the preceding example resulted from a cylinder-forming block copolymer being directed to assemble on a square array of spots, one of the most striking examples of the ability of a chemical pattern to form non-bulk structures appears when a lamellae-forming system is directed to assemble on a square array of spots. In such a case, both experimental results and single chain in mean field (SCMF) simulations have shown the formation of a tortuous, bicontinuous quadratically perforated lamella (QPL) when a lamellae-forming symmetric ternary blend of PS-*b*-PMMA/PS/PMMA is assembled on a square array of spots [34, 71]. The results of SCMF simulations, shown in Figure 10.6 [34], present an image of how the morphology of the assembled ternary blend begins to reconstruct its bulk morphology as the film thickness is increased. In the cut-away view in Figure 10.6 the blend takes the form of interlocking PS and PMMA pillars, with the PS pillars rising up from the substrate, and the transparent PMMA pillars pointing down and connecting with the chemically patterned spots on the substrate. As the film gets thicker, the PS pillars become a more connected, tortuous morphology.

10.3.2
Lithographic Applications

The use of chemical and topographical patterns to direct the assembly of block copolymers provides a method to register pattern features to the underlying substrate, and to increase the order of the assembled structure [72]. While order and registration are necessary features for lithographic and template-forming applications, to achieve the broadest application of block copolymers in directed assembly, it is necessary to possess each of the following attributes: (i) the ability to create the essential fabric architectures used in semiconductor design; (ii) critical dimension control equal to or better than the lithographic patterning tool; and (iii) a route to

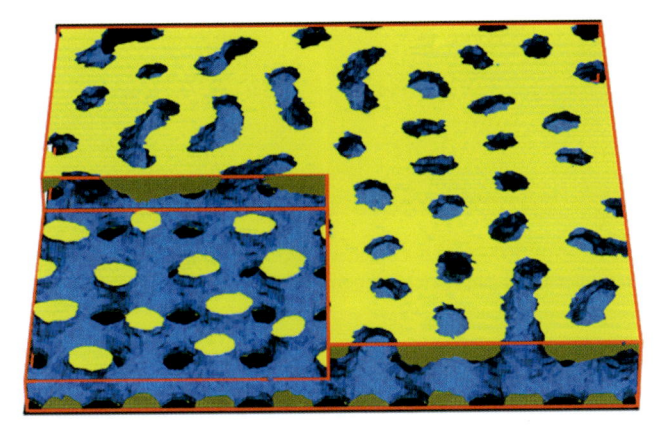

Figure 10.6 SCMF simulation of the equilibrated, bicontinuous, quadratically perforated morphology formed in a thin film (thickness $= 1.27 L_o$) of a lamellae-forming symmetric ternary blend. In the image, only the top view of the PS-rich domains (white/yellow online) and the interface between PS and PMMA (dark gray/blue online) are shown. Regions on top of the spots are PMMA-rich (transparent) and one looks through to the substrate (black). In the lower left corner the upper half of the film is removed for better viewing of the film closer to the chemical pattern. (Reproduced with permission from [34]. Copyright © (2006) by the American Physical Society.)

improve the resolution provided by the lithographic tool. Work over the past decade has demonstrated the capability of directed assembly of block copolymers to achieve each of these three objectives, as shown below.

The ability to create the essential semiconductor fabric architectures with directed assembly of block copolymers is challenging because it is easiest to achieve highly ordered, assembled structures when the chemical pattern matches the size and shape of the bulk morphology. Thus, the earliest examples of directed assembly on chemical patterns used linear patterns with periods that matched the lamellar spacing of symmetric block copolymers, and focused on improving the quality of the assembled structure [9, 11, 73]. Although these assembled lamellar patterns were scientifically interesting, their range of technological applications was limited. More recent work has focused on broadening the range of possible applications of assembly by demonstrating the formation of semiconductor fabric architecture elements, such as jogs, bends, T-junctions, periodic and isolated lines, and periodic and isolated spots [58].

The first demonstration of the use of directed assembly to generate non-regular, device-oriented structures focused on the formation of assembled PS-*b*-PMMA domains in an angled pattern [37]. In this initial work it was found that blending PS and PMMA homopolymers with the PS-*b*-PMMA facilitated the formation of the angled structures. The primary benefit of the presence of the homopolymer occurred along the line of vertices, which naturally has a longer period than the period of the lamellae not at the vertex (by a factor of $\sqrt{2}$ in the case of the right-angled structure in Figure 10.7a). Subsequent research with PS-*b*-PMMA broadened the range of

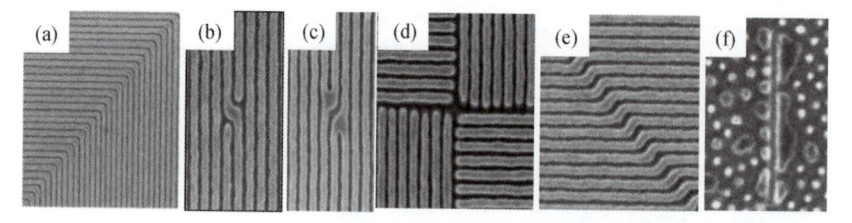

Figure 10.7 SEM structures of semiconductor fabric architectures formed by the directed assembly of PS-b-PMMA: (a) bends, (b) line terminations, (c) junctions, (d) T-junctions, (e) jogs, and (f) an isolated line.

features that could be generated with directed assembly to include all of the structures listed in the preceding paragraph, as shown in Figure 10.7 [58].

The 2007 International Technology Roadmap for Semiconductors focusing on lithography cites critical dimension (CD) control as one of the most challenging issues in the advancement of lithography [74]. The insertion of block copolymers into advanced lithographic processes offers a route for improved CD control because the process for the formation of the desired structures is thermodynamically driven, as shown in the free energy equations above, and not kinetically driven, as is the case with photoresists. The kinetically driven process of exposing and developing photoresists can increase the line edge roughness (LER) and affect the CD control in a pattern [75–79].

In the case of the directed assembly of block copolymers, issues such as LER and poor CD control in the underlying photoresist can be decreased by annealing the assembled block copolymer, so that it achieves its thermodynamically preferred dimensions. This self-correcting effect was demonstrated by the directed assembly of PS-b-PMMA on a chemical pattern with constant L_s but varying the line width W [80]. As shown in Figure 10.8, the assembled domains correct for the change in W by forming domains with a trapezoidal cross-section, such that the average width in all 3 cases is constant at $L_s/2$.

Two recent publications provide additional examples of the CD control abilities of directed assembly of block copolymers, but also demonstrate for the first time the opportunity for resolution enhancement with block copolymer films directed to assemble on chemical patterns [81, 82]. In one case, cylinder-forming PS-b-PMMA was directed to assemble on a chemical pattern comprised of hexagonally arranged

Figure 10.8 Results of SCMF simulations of thin films of PS-b-PMMA on chemically nanopatterned substrates with varying normalized (W/L_s) linewidths equal to (a) 0.3, (b) 0.45, and (c) 0.65, with $L_s = L_o = 80$ nm. Red domains are styrene and blue domains are methyl methacrylate.

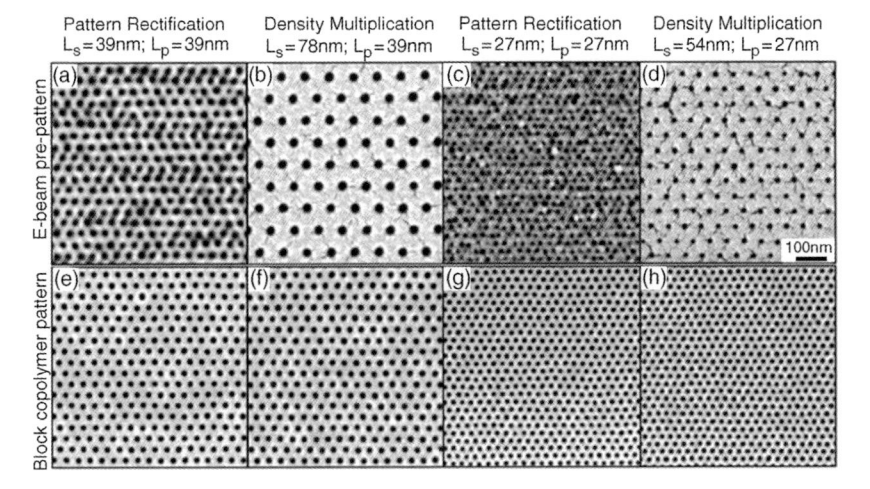

Figure 10.9 Examples of pattern rectification and density multiplication with cylinder-forming PS-b-PMMA. (a–d) SEM images of developed e-beam resist with L_s values shown in the top row. (e–h) SEM images of the block copolymer film on top of the pre-pattern defined by the corresponding e-beam pattern above. (Reproduced with permission from [81]. Copyright © (2008) AAAS.)

spots with a pattern pitch L_s either equal to the bulk cylinder pitch L_c or equal to $2L_c$ [81]. Directed assembly of the block copolymer resulted in structures (bottom row of Figure 10.9) that demonstrated rectification (when $L_s = L_c$) or density multiplication (when $L_s = 2L_c$) of the chemical pattern, which was made via e-beam lithography patterning (top row of Figure 10.9). In similar work, lamellae-forming PS-b-PMMA was directed to assemble on chemical patterns with $L_s = nL_o$ ($n = 2$, 3, or 4) to form ordered arrays of lamellae (Figure 10.10) [82]. In both cases, when the spacing of the chemical pattern was double the bulk block copolymer domain spacing, there was still a sufficient thermodynamic driving force supplied by F_{int} to force PMMA

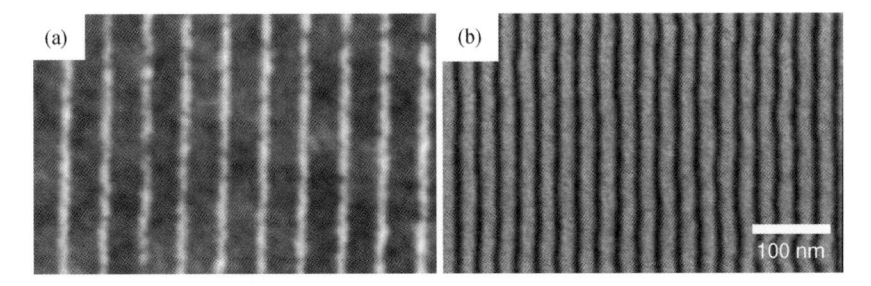

Figure 10.10 Density multiplication of PS-b-PMMA lamellae directed to assemble on a chemical pattern with $L_s = 2L_o$. (a) Atomic force micrograph height image of stripes of e-beam resist on a non-preferential PS-r-PMMA brush. (b) SEM of PS-b-PMMA directed to assemble on the chemical pattern formed from the material shown in (a). (Reproduced with permission from [82]. Copyright © (2008) Wiley-VCH Verlag GmbH & Co. KGaA.)

domains to align with and register to the spots or lines that were present on the pattern. Because L_s was twice the bulk polymer spacing, however, some PMMA domains were forced to assemble above regions of the substrate that had no patterned line or spot. Due to the pinning of some of the PMMA domains to an ordered array of patterned regions on the substrate, the remaining PMMA domains were forced to assemble in an ordered array as well. As a result, directed assembly led to frequency doubling of the lamellar pattern, and frequency quadrupling in the case of the spotted pattern.

10.4
Conclusion

Directed assembly of block copolymer films provides a method for scientists to understand more thoroughly the formation of block copolymer domains, and is poised to provide benefits to a range of applications, such as semiconductor lithography and bit patterned media. Both the advancement of scientific understanding and the potential applications benefit from the fact that the final structures achieved from directed assembly are equilibrated structures. Thus, thermodynamic factors, as opposed to kinetic limitations, can be used to pre-determine features such as line width or pattern period, and can also improve the line edge roughness achieved with the initial lithographic exposure. Additionally, the essential set of semiconductor fabric architectures has been demonstrated with directed assembly, and the recent density multiplication results point to the possibility of achieving higher density features than can be provided by lithography. The combination of these features offers promise for the implementation of directed assembly of block copolymer films in a variety of applications, including advanced lithography.

References

1 Meier, D.J. (1969) *J. Polym. Sci. Pol. Sym.*, 81.
2 Helfand, E. (1975) *Macromolecules*, **8**, 552.
3 Leibler, L. (1980) *Macromolecules*, **13**, 1602.
4 Bates, F.S. and Fredrickson, G.H. (1990) *Annu. Rev. Phys. Chem.*, **41**, 525.
5 Bates, F.S. and Fredrickson, G.H. (1999) *Phys. Today*, **52**, 32.
6 International Technology Roadmap for Semiconductors - Emerging Research Materials. 2007 edn, International SEMATECH, Austin, TX.
7 Segalman, R.A., Yokoyama, H., and Kramer, E.J. (2001) *Adv. Mater.*, **13**, 1152.
8 Bita, I., Yang, J.K.W., Jung, Y.S., Ross, C.A., Thomas, E.L., and Berggren, K.K. (2008) *Science*, **321**, 939.
9 Kim, S.O., Solak, H.H., Stoykovich, M.P., Ferrier, N.J., de Pablo, J.J., and Nealey, P.F. (2003) *Nature*, **424**, 411.
10 Peters, R.D., Yang, X.M., Kim, T.K., Sohn, B.H., and Nealey, P.F. (2000) *Langmuir*, **16**, 4625.
11 Rockford, L., Liu, Y., Mansky, P., Russell, T.P., Yoon, M., and Mochrie, S.G. (1999) *J. Phys. Rev. Lett.*, **82**, 2602.
12 Shin, K., Leach, K.A., Goldbach, J.T., Kim, D.H., Jho, J.Y., Tuominen, M., Hawker, C.J., and Russell, T.P. (2002) *Nano Lett.*, **2**, 933.
13 Thurn-Albrecht, T., Steiner, R., DeRouchey, J., Stafford, C.M., Huang, E., Bal, M., Tuominen, M., Hawker, C.J.,

and Russell, T.P. (2000) *Adv. Mater.*, **12**, 1138.

14 Xiao, S.G., Yang, X.M., Edwards, E.W., La, Y.H., and Nealey, P.F. (2005) *Nanotechnology*, **16**, S324.

15 Yang, S.Y., Ryu, I., Kim, H.Y., Kim, J.K., Jang, S.K., and Russell, T.P. (2006) *Adv. Mater.*, **18**, 709.

16 Black, C.T., Guarini, K.W., Breyta, G., Colburn, M.C., Ruiz, R., Sandstrom, R.L., Sikorski, E.M., and Zhang, Y. (2006) *J. Vac. Sci. Technol. B*, **24**, 3188.

17 Chang, L.-W., and Wong, H.-S.P. (2006) *Proc. SPIE*, **6156**, 615611.

18 Zhang, Q.L., Xu, T., Butterfield, D., Misner, M.J., Ryu, D.Y., Emrick, T., and Russell, T.P. (2005) *Nano Lett.*, **5**, 357.

19 Black, C.T., Guarini, K.W., Milkove, K.R., Baker, S.M., Russell, T.P., and Tuominen, M.T. (2001) *Appl. Phys. Lett.*, **79**, 409.

20 Black, C.T., Guarini, K.W., Zhang, Y., Kim, H.J., Benedict, J., Sikorski, E., Babich, I.V., and Milkove, K.R. (2004) *IEEE Electron Device Letters*, **25**, 622.

21 Gratt, J.A. and Cohen, R.E. (2004) *J. Appl. Polym. Sci.*, **91**, 3362.

22 Yang, X., Xiao, S., Wu, W., Xu, Y., Mountfield, K., Rottmayer, R., Lee, K., Kuo, D., and Weller, D. (2007) *J. Vac. Sci. Technol. B*, **25**, 2202.

23 Xiao, S.G. and Yang, X.M. (2007) *J. Vac. Sci. Technol. B*, **25**, 1953.

24 Ross, C.A. and Cheng, J.Y. (2008) *MRS Bull.*, **33**, 838.

25 Stocker, W., Beckmann, J., Stadler, R., and Rabe, J.P. (1996) *Macromolecules*, **29**, 7502.

26 Rehse, N., Knoll, A., Magerle, R., and Krausch, G. (2003) *Macromolecules*, **36**, 3261.

27 Wang, Q., Nealey, P.F., and de Pablo, J. (2001) *J. Macromolecules*, **34**, 3458.

28 Tsarkova, L., Horvat, A., Krausch, G., Zvelindovsky, A.V., Sevink, G.J.A., and Magerle, R. (2006) *Langmuir*, **22**, 8089.

29 Konrad, M., Knoll, A., Krausch, G., and Magerle, R. (2000) *Macromolecules*, **33**, 5518.

30 Park, I., Park, S., Park, H.W., Chang, T., Yang, H.C., and Ryu, C.Y. (2006) *Macromolecules*, **39**, 315.

31 Knoll, A., Horvat, A., Lyakhova, K.S., Krausch, G., Sevink, G.J.A., Zvelindovsky, A.V., and Magerle, R. (2002) *Phys. Rev. Lett.*, **89**, 035501.

32 Knoll, A., Magerle, R., and Krausch, G.J. (2004) *Chem. Phys.*, **120**, 1105.

33 Edwards, E.W., Montague, M.F., Solak, H.H., Hawker, C.J., and Nealey, P.F. (2004) *Adv. Mater.*, **16**, 1315.

34 Daoulas, K.C., Muller, M., Stoykovich, M.P., Park, S.M., Papakonstantopoulos, Y.J., de Pablo, J.J., Nealey, P.F., and Solak, H.H. (2006) *Phys. Rev. Lett.*, **96**, 036104.

35 Matsen, M.W. and Bates, F.S. (1997) *J. Chem. Phys.*, **106**, 2436.

36 Wang, Q., Nath, S.K., Graham, M.D., Nealey, P.F., and de Pablo, J.J. (2000) *J. Chem. Phys.*, **112**, 9996.

37 Stoykovich, M.P., Muller, M., Kim, S.O., Solak, H.H., Edwards, E.W., de Pablo, J.J., and Nealey, P.F. (2005) *Science*, **308**, 1442.

38 Helfand, E. and Wasserman, Z.R. (1976) *Macromolecules*, **9**, 879.

39 Park, S.M., Stoykovich, M.P., Ruiz, R., Zhang, Y., Black, C.T., and Nealey, P.F. (2007) *Adv. Mater.*, **19**, 607.

40 Stoykovich, M.P., Edwards, E.W., Solak, H.H., and Nealey, P.F. (2006) *Phys. Rev. Lett.*, **97**, 147802.

41 Welander, A.M., Kang, H.M., Stuen, K.O., Solak, H.H., Muller, M., de Pablo, J.J., and Nealey, P.F. (2008) *Macromolecules*, **41**, 2759.

42 Huang, E., Pruzinsky, S., Russell, T.P., Mays, J., and Hawker, C.J. (1999) *Macromolecules*, **32**, 5299.

43 Huang, E., Rockford, L., Russell, T.P., and Hawker, C.J. (1998) *Nature*, **395**, 757.

44 Mansky, P., Liu, Y., Huang, E., Russell, T.P., and Hawker, C. (1997) *Science*, **275**, 1458.

45 Sauer, B.B. and Dee, G.T. (2002) *Macromolecules*, **35**, 7024.

46 Wu, S. (1970) *J. Phys. Chem.*, **74**, 632.

47 Russell, T.P., Hjelm, R.P., and Seeger, P.A. (1990) *Macromolecules*, **23**, 890.

48 Carriere, C.J., Biresaw, G., and Sammler, R.L. (2000) *Rheologica Acta*, **39**, 476.

49 Ellingson, P.C., Strand, D.A., and Cohen, A., Sammler, R.L., and Carriere, C.J. (1994) *Macromolecules*, **27**, 1643.

50 Dai, K.H. and Kramer, E.J. (1994) *Polymer*, **35**, 157.

51 Schulz, M.F., Khandpur, A.K., Bates, F.S., Almdal, K., Mortensen, K., Hajduk, D.A.,

and Gruner, S.M. (1996) *Macromolecules*, **29**, 2857.
52 Vitt, E. and Shull, K.R. (1995) *Macromolecules*, **28**, 6349.
53 Hu, W., Koberstein, J.T., Lingelser, J.P., and Gallot, Y. (1995) *Macromolecules*, **28**, 5209.
54 Ross, C.A., Jung, Y.S., Chuang, V.P., Ilievski, F., Yang, J.K.W., Bita, I., Thomas, E.L., Smith, H.I., Berggren, K.K., Vancso, G.J., and Cheng, J.Y. (2008) *J. Vac. Sci. Technol. B*, **26**, 2489.
55 Ji, S., Liu, C.C., Son, J.G., Gotrik, K., Craig, G.S.W., Gopalan, P., Himpsel, F.J., Char, K., and Nealey, P.F. (2008) *Macromolecules*, **41**, 9098.
56 Stoykovich, M.P. and Nealey, P.F. (2006) *Mater. Today*, **9**, 20.
57 Park, S.M., Craig, G.S.W., Liu, C.C., La, Y.H., Ferrier, N.J., and Nealey, P.F. (2008) *Macromolecules*, **41**, 9118.
58 Stoykovich, M.P., Kang, H., Daoulas, K.C., Liu, G., Liu, C.C., de Pablo, J.J., Mueller, M., and Nealey, P.F. (2007) *ACS Nano*, **1**, 168.
59 Park, C., Yoon, J., and Thomas, E.L. (2003) *Polymer*, **44**, 6725.
60 Ulbricht, M. (2006) *Polymer*, **47**, 2217.
61 Olson, D.A., Chen, L., and Hillmyer, M.A. (2008) *Chem. Mater.*, **20**, 869.
62 Edrington, A.C., Urbas, A.M., DeRege, P., Chen, C.X., Swager, T.M., Hadjichristidis, N., Xenidou, M., Fetters, L.J., Joannopoulos, J.D., Fink, Y., and Thomas, E.L. (2001) *Adv. Mater.*, **13**, 421.
63 Guarini, K.W., Black, C.T., Milkove, K.R., and Sandstrom, R.L. (2001) *J. Vac. Sci. Technol. B*, **19**, 2784.
64 Shibauchi, T., Krusin-Elbaum, L., Gignac, L., Black, C.T., Thurn-Albrecht, T., Russell, T.P., Schotter, J., Kastle, G.A., Emley, N., and Tuominen, M.T. (2001) *J. Magn. Magn. Mater.*, **226**, 1553.
65 Thurn-Albrecht, T., Schotter, J., Kastle, C.A., Emley, N., Shibauchi, T., Krusin-Elbaum, L., Guarini, K., Black, C.T., Tuominen, M.T., and Russell, T.P. (2000) *Science*, **290**, 2126.
66 Radzilowski, L.H., Carvalho, B.L., and Thomas, E.L. (1996) *J. Polym. Sci. Pol. Phys.*, **34**, 3081.

67 La, Y.H., Edwards, E.W., Park, S.M., and Nealey, P.F. (2005) *Nano Lett.*, **5**, 1379.
68 La, Y.H., Stoykovich, M.P., Park, S.M., and Nealey, P.F. (2007) *Chem. Mater.*, **19**, 4538.
69 Park, S.M., Craig, G.S.W., La, Y.H., and Nealey, P.F. (2008) *Macromolecules*, **41**, 9124.
70 Park, S.M., Craig, G.S.W., La, Y.H., Solak, H.H, and Nealey, P.F. (2007) *Macromolecules*, **40**, 5084.
71 Daoulas, K.C., Muller, M., Stoykovich, M.P., Papakonstantopoulos, Y.J., De Pablo, J.J., Nealey, P.F., Park, S.M., and Solak, H.H. (2006) *J. Polym. Sci. Pol. Phys.*, **44**, 2589.
72 Cheng, J.Y., Ross, C.A., Smith, H.I., and Thomas, E.L. (2006) *Adv. Mater.*, **18**, 2505.
73 Yang, X.M., Peters, R.D., Nealey, P.F., Solak, H.H., and Cerrina, F. (2000) *Macromolecules*, **33**, 9575.
74 International Technology Roadmap for Semiconductors - Lithography. 2007 edn, International SEMATECH, Austin, TX.
75 Patsis, G.P. and Gogolides, E. (2005) *J. Vac. Sci. Technol. B*, **23**, 1371.
76 Yamaguchi, T., Yamazaki, K., Nagase, M., and Namatsu, H. (2003) *Jpn. J. Appl. Phys. 1*, **42**, 3755.
77 Nakamura, J., Ban, H., and Tanaka, A. (1992) *Jpn. J. Appl. Phys. 1*, **31**, 4294.
78 Fedynyshyn, T.H., Thackeray, J.W., Georger, J.H., and Denison, M.D. (1994) *J. Vac. Sci. Technol. B*, **12**, 3888.
79 Reynolds, G.W. and Taylor, J.W. (1999) *J. Vac. Sci. Technol. B*, **17**, 334.
80 Edwards, E.W., Muller, M., Stoykovich, M.P., Solak, H.H., de Pablo, J.J., and Nealey, P.F. (2007) *Macromolecules*, **40**, 90.
81 Ruiz, R., Kang, H., Detcheverry, F.A., Dobisz, E., Kercher, D.S., Albrecht, T.R., de Pablo, J.J., and Nealey, P.F. (2008) *Science*, **321**, 936.
82 Cheng, J.Y., Rettner, C.T., Sanders, D.P., Kim, H.C., and Hinsberg, W.D. (2008) *Adv. Mater.*, **20**, 3155.

11
Surface Instability and Pattern Formation in Thin Polymer Films
Rabibrata Mukherjee, Ashutosh Sharma, and Ullrich Steiner

11.1
Introduction

The fabrication of large area polymer structures with feature sizes ranging from few microns down to the molecular level is key to various technologically important areas, examples of which include molecular electronics [1], flexible display screens [2], optical sensors [3], structural color [4], reusable super adhesives [5], super hydrophobic [6] and self cleaning surfaces [7], and scaffolds for tissue engineering [8]. The length scale of interest here, often referred to as the 'meso' scale, ranges from a few nanometers to hundreds of nanometers and interfaces the molecular and the macroscopic worlds. Thus, it becomes possible to observe simultaneous signatures of molecular interactions as well as macroscopic effects at these length scales, often giving rise to exciting new phenomena like quantum size effects (QSEs) [9], single electron tunneling (SET) [10], Coulomb blockade, and so on [11]. None of these effects can be realized either in purely macroscopic or molecular regimes. However, the success of the desired applications, harvesting the extraordinary scientific phenomena occurring at these length scales, depends strongly on the availability of suitable, easy to implement patterning techniques that can create defect-free structures over large areas. Present-day technologies offer three distinct approaches for mesoscale patterning. Two of the more popular ones are 'top-down' methods, that include various types of lithography and the serial writing methods [12–17] and 'bottom-up' self-assembly (SA) techniques, where individual components assemble in a desired fashion to form structures [18–20]. However, neither bottom-up nor top-down patterning techniques by themselves are capable of achieving structural control at the molecular and mesoscopic level combined with macroscopic addressability. Detailed discussion on either top-down lithography methods or bottom-up self-assembly methods are beyond the scope of this chapter, and an interested reader may refer to several excellent reviews and text books which are already available [12, 21–23].

A third approach which is gaining significant attention as a viable patterning technique, particularly in the context of liquid and soft solid thin films, is based on the

Generating Micro- and Nanopatterns on Polymeric Materials. Edited by A. del Campo and E. Arzt
Copyright © 2011 WILEY-VCH Verlag GmbH & Co. KGaA, Weinheim
ISBN: 978-3-527-32508-5

fact that small-scale or highly confined systems are usually unstable or can be made unstable by application of a suitable external field. Evolution of instability in the form of spatially periodic modes progressively alters the shape to a more intricate pattern which minimizes energy. The pattern morphology and length scale are governed by a competition among the destabilizing forces (e.g., van der Waals and electrostatic fields) and stabilizing forces (e.g., surface tension, elastic strain). The existence of such instabilities have been known for more than a century now [24, 25]. Thus, an understanding and control of instabilities can be employed to fabricate mesoscale patterns starting from a simple, but unstable shape such as a flat thin film. The approach, which we will refer to as self-organized patterning, is based on instability-mediated disintegration of thin films. Instabilities which are often considered undesirable because they tend to destroy the coating itself, are harnessed for the creation of meso- and nanoscale surface structures [40–281]. Due to several key factors (discussed later) the methods are particularly well-suited for polymeric films and surfaces [47–56]. Though there are several forms of instabilities, this chapter will focus primarily on morphological evolution and pattern formation caused by the spontaneous or externally stimulated growth of capillary waves, which are omnipresent on the surface of a thin visco-elastic film [26–28]. These surface capillary waves, resulting from molecular level thermal fluctuations have extremely low amplitudes (a few nm) [27] and are hardly of any significance in the macroscopic world. However, at the meso- and nanoscopic length scale, where intermolecular interactions are strong compared to the gravitational field, these fluctuations become important and significantly influence the stability of a thin film. These initial fluctuations can grow because of various forces such as the conjoining pressure resulting from the van der Waals forces [29, 30, 73–75], electrostatic forces [31], steric forces (molecular level entanglement, significant in polymers) [28, 110], and forces arising from the spatial confinement of fluctuation spectrum (like Casimir force) [32]. The growing fluctuations may lead to a progressive roughening of the surface of an initially flat film [33–170]. Eventually, either the growth stops to form a spatially periodic structure before the film ruptures, or the film break-up occurs leading to hole growth and coalesce to form larger scale patterns [33–39, 65]. These instabilities are mostly manifested in liquid or soft (elastic modulus $<$ 1 MPa) solid films, as the flow and deformation caused by the intermolecular interactions are rather weak. Dewetting is observed in ultrathin metal, polymer and solid films occurs only when they are molten, or, (above their glass transition) or have a small elastic modulus [40, 41]. The surface of a ruptured liquid film subsequently undergoes morphological evolution (like dewetting) that leads to the reduction of the free energy of the system [37, 38]. Understanding the science of instabilities is also of key importance in a variety of settings, including various industrial applications, such as enhancement of the stability of foams and emulsions, oil recovery, flotation, service life enhancement of coatings, vapor condensation on a tube wall, and detergent action in the cleaning of oil films [42]. Instabilities also play a critical role in host of biological and physiological phenomena particularly the small-scale deformations of membranes [43], such as the onset of microvilli in normal and neoplastic cells, the deformation and rhythmic movement (flickering) of red blood cells, adhesion, fusion, and attachment of cells and micro-organisms and phagocytosis, occurrence of tear film break up and dry eye syndrome [44].

While the science of dewetting of liquids on solid surfaces has been investigated for a long time [45], the spontaneous instability and film rupture continue to receive significant attention [47–171]. Energetically, the spontaneous break-up is only possible where the rupture and subsequent morphological rearrangement of the film results in the lowering of the system free energy as compared with an intact film [33–38]. A ruptured film may further evolve with morphological reorganization until thermodynamic equilibrium is attained [33–38, 56, 65, 84–88]. Interestingly, the morphology of the instability-induced structures in most cases bears the signature of the force that is responsible for the instability [56, 65]. Apart from spontaneous instabilities, the growth of the surface fluctuations can also be engendered by applying an external field (e.g., an electric field [230–264], or a thermal gradient) [265–270], or they can nucleate around heterogeneities or defects present in the film or on the substrate [117–139].

Irrespective of the precise mechanism leading to film instability, instability structures with lateral resolutions down to $\sim$100 nm can evolve in an unstable film [47–171]. Thus, based on thin film instabilies, it may in principle be possible to achieve structures having sub-micron lateral resolutions. It is this particular aspect which has led to a paradigm shift in the nature of research in the area of polymer thin film instability. Research, starting in the early 1990s, in the area focused more on the fundamental issues relating to the origin and the spatio–temporal growth of the instability due to various forces that are often antagonistic [65, 98, 123]. The basic intention was to enhance stability of thin films against dewetting based on the complete understanding of the physics leading to the destabilization process. In contrast, recent attention is significantly on exploring the possibility of using dewetting and other instability induced methods as viable surface patterning techniques for the fabrication of meso-and nanoscale structures, particularly in soft materials like polymers [172–281]. In this chapter, we will present a brief overview of some of the recent developments in the field of patterning, based on instabilities in thin polymer films [172–281].

Incidentally, most instability-mediated techniques result in structures with a well-defined mean length scale, lacking long-range order and desired detailed shapes [47–171]. Such isotropic and random structures have limited practical application, as most applications require regularly ordered and high-fidelity structures. Most soft lithography techniques can easily create such defect-free ordered structures over large areas [13]. Thus, strategies are necessary to impose long-range order on the final structures in order to develop instability-mediated morphologies as a viable surface patterning technique. This is often achieved by suitable templating strategies, such as the use of a patterned substrate or confining mask [174–229]. Templating, in conjugation with careful control of initial conditions such as the film thickness, surface and interfacial tensions and so on, offers a great deal of flexibility, allowing the tuning of the morphology, feature size, and pitch of the structures, thereby leading to the realization of novel concepts, such as *patterns on demand, reconfigurable structures,* and *patterning beyond the master.*

11.2
Origin of Surface Instability

The different mechanisms that may lead to instabilities in thin polymer films can be broadly subdivided into two major categories: instability resulting from the inherent dynamics of the system and instability triggered by an external field.

The first type of instability can be engendered by (i) spontaneous instability or spinodal dewetting, which can be caused by the amplification of surface fluctuations due to the action of intersurface forces, such as van der Waals, electrostatic and others [56–106]. It has very recently been shown that spontaneous dewetting can also be triggered due to release of residual stresses accumulated during film preparation, because of polymer chain entanglement and adsorption on the substrate, where many aspects of instability evolution as well the dewetted pattern morphology are similar to that observed in spinodal dewetting [106–116]; (ii) nucleated dewetting, which may be further subdivided into a heterogeneous nucleation mechanism relating to substrate defects, and homogeneous nucleation [117–139]. On a chemically heterogeneous surface, the gradient in wettability contrast is responsible for engendering an instability, leading to dewetting rather than the nonwettability of the substrate itself, as in the case of spinodal dewetting [129].

An external field-mediated instability can be engendered due to the application of a field like a thermal gradient or electrical field [236–281]. Some other causes, like density variations within the film, have also been proposed which may operate in special circumstances such as nanoparticle- or nanodefect-filled films [144, 282–284].

Interestingly, while spontaneous instabilities are possible only in thermodynamically unstable ultrathin films (<100 nm), nucleation can lead to the disintegration of thicker metastable films. Further, with the application of an external field it is even possible to destabilize a reasonably thick film (~500 nm) where the stabilizing influence of gravity is significant [123, 129, 137, 138]. Of all these mechanisms, only stress-induced instabilities caused by the deformation of polymer coils in the film are specific to high molecular weight polymers [107–116] and the rest are rather generic and independent of the precise nature of the liquid. As polymers are long chain molecules, the entanglement- and adsorption-mediated residual stresses generated, by non-equilibrium conformations of the individual molecules during spin-coating and solvent drying can influence the stability of the system significantly [107–116], an aspect that has been realized only very recently [108–116].

Irrespective of the precise origin of an instability, the amplification of the capillary waves leads to the deformation of the free surface (film–air interface) and results in localized flow of liquid from the thinner parts to the thicker parts of the film [38, 56–168]. This phenomenon eventually results in the rupture of the film with the formation of dry patches or holes, when the growing amplitude of the capillary wave spectrum equals the film thickness [65]. As the film ruptures, the two distinct interfaces (film–air and film–substrate) merge and a three phase contact line (film–air–substrate) is formed. Depending on the thermodynamics of the system, a ruptured hole may grow further with time, resulting in a subsequent morphological rearrangement of the film on the substrate [37, 38]. The role of surface tension during film instabilities needs a special mention. While it opposes the growth of the capillary waves, thereby playing a stabilizing role before a film has ruptured, it aids the growth of a hole after the rupture of the film [90].

11.3
Polymer Thin Film Dewetting

11.3.1
Spontaneous Instability or Spinodal Dewetting: Theoretical Aspects

In this section we review the theoretical framework that describes spontaneous instabilities or spinodal dewetting [33–39, 57, 64–67, 76–91]. A spontaneous instability is only possible in a thermodynamically unstable thin film. Thus, it is important to clearly define an unstable film and contrast it with other types of films, namely stable and metastable films [38, 65, 138]. The classification of films in terms of their stability is best understood in terms of excess free energy (ΔG), disjoining/conjoining pressure (Π), or effective interface potential (Φ) [65]. All three quantities manifest the excess energy in a thin film, as compared to the bulk, stemming from interactions between the two interfaces [38, 57, 64, 65, 138]. ΔG is a measure for the excess free energy per unit area required to bring two infinitely separated interfaces to a finite distance h. In the case of a thin film, the two interacting interfaces are the film–substrate and the film–air interfaces and h corresponds to the film thickness [65]. The disjoining pressure (Π) is the negative derivative of the excess free energy (per unit area) with respect to the film thickness ($\Pi = -\partial(\Delta G)/\partial h$) and is a measure of the excess pressure in a thin film arising from the interaction of the two interfaces [46, 65]. In a different notation, Φ refers to the effective intermolecular potential and is a measure of the excess intermolecular free energy per unit volume in the film ($\Phi = \partial(\Delta G)/\partial h$). It is obvious that for a thin film Φ is identical to the negative of the disjoining pressure, $-\Pi$ or conjoining pressure. In order to eliminate possible confusion we want to emphasize that some groups use the symbol Φ to represent so-called effective interface potential which is identical to the excess free energy (ΔG), in the notations used in this text [138].

The condition of instability vis-a-vis stability of a thin film in terms Φ or Π (or ΔG) as a function of h emerges from linear stability analysis (LSA) [33–38, 65]. In order to facilitate the discussion of LSA, a brief summary of the key theoretical results on thin film instability is presented here.

The general framework of the hydrodynamics of a free liquid surface have emerged after more than four decades of work by various researchers including Cahn [33], Vrij [34, 35], Ruckenstein and Jain [37], Williams and Davis [39], Derjaguin [36], Sharma [38, 64–67], and many others. The governing equation describing the two-dimensional (2D) dynamics and the shape of a thin film of Newtonian liquid on a flat substrate, neglecting the effect of evaporation/condensation and gravity is given as Equation 11.1:

$$3\mu \frac{\partial h}{\partial t} + \frac{\partial}{\partial x}\left[\gamma h^3 \frac{\partial^3 h}{\partial x^3}\right] + \frac{\partial}{\partial x}\left[-h^3 \frac{\partial \Phi}{\partial h}\frac{\partial h}{\partial x}\right] = 0, \tag{11.1}$$

where $h(x,t)$ is the local film thickness, x is the co-ordinate parallel to the substrate surface, t is time, μ is the viscosity, and γ is the surface tension of the liquid. The simplification of the x component of the Navier–Stokes equation in the limit of long wavelengths (or small slopes) combined with the continuity equation provides the individual components of velocity (u, w) along the x and z axes as functions of h, Φ and x [37, 38, 65, 67]. The small-slope approximation $(\partial h/\partial x \ll 0)$ originates from the assumption that the wavelengths of the surface fluctuations (λ) are much larger $(\lambda \gg h_0)$ compared with the mean film thickness (h_0), and is reasonably successful in capturing the essential physics of the system, unless the film thickness is of molecular dimensions [37, 38]. The velocity components are substituted in the 'kinematic boundary condition' $\partial h/\partial t + u_s \partial h/\partial x = w_s$, which is akin to the local mass balance at any point on the free liquid surface, and correlates the two components of velocity at the interface. The solution of Equation 11.1 gives the mean film thickness h as a function of the spatial coordinate (x) and time (t). When $h_t = 0$, the steady state film thickness is of the form $h(x)$, implying a spatial variation of film thickness even at a steady state. The consequent situation can be regarded as a dynamic steady state and can now be correlated to the preceding discussion on how surface tension opposes this spatial variation of film thickness engendered by molecular level fluctuations in favor of a flat surface, giving rise to surface capillary waves [38, 90].

Each of the constituent terms of Equation 11.1 represents a distinct force field. From left to right the terms represent the contributions of viscous forces, surface tension forces due to the curvature at the free interface (Laplace pressure), and the excess intermolecular forces (disjoining pressure) respectively [37, 38, 65, 67]. The viscous force in no way influences the stability as it merely controls the dynamics of the system. For tangentially immobile films, the prefactor of the viscous term **3** is replaced by **12** [38, 65]. The Laplace pressure arising from surface tension has a stabilizing influence, as already discussed. Thus, the only term that may induce an instability in the system is the one representing the excess intermolecular interactions [37, 38, 65].

The major practical use of Equation 11.1 lies in testing whether a film of uniform initial thickness h_0 remains stable or eventually become unstable with time. The solution of linearized equations of motion incorporating the effect of intermolecular forces, can be simplified by the lubrication approximation (non-inertial laminar flow in thin films) and admit space periodic solutions for the film thickness $h(x, t) = h_0 + \varepsilon \sin(kx) \exp(\omega t)$ with:

$$\omega = C[-\gamma_L k^4 - (\partial \Phi/\partial h)k^2] \tag{11.2}$$

Equation 11.2 represents the linear dispersion relation for the initial growth coefficient (ω) of an axis-symmetric perturbation on the free liquid surface with a small $(\varepsilon \ll h_0)$ initial amplitude having wavenumber of instability k [65]. The parameter

$C = (h_0^3/3\mu)$ depends on the film thickness and inversely on the film viscosity, but not on k. The necessary condition for the instability is a positive value of ω, which implies that the initial amplitude of the perturbation will grow with time. It can be seen from the expression of ω in the dispersion relation that both the term C and the first term with in the parenthesis (γk^4) are always positive and thus, growth of an instability ($\omega > 0$) is only possible for a negative spinodal parameter ($\partial\Phi/\partial h < 0$) [65]. In addition, for higher values of k, the k^4 term dominates over the term containing k^2 so that ω is negative even when $\partial\Phi/\partial h < 0$ for sufficiently high wavenumbers, signifying stability of the film. Thus, $\partial\Phi/\partial h < 0$ is the necessary condition for film instability. The sufficiency condition lies in values of k being in the range (0, k_C), where k_C is a critical wavenumber corresponding to $\omega = 0$ which is obtained by setting the dispersion equation to zero and solving for k. This observation implies that higher wavenumbers ($>k_C$) or low frequency oscillations fail to destabilize even intrinsically unstable films. It is also obvious that if $\partial\Phi/\partial h > 0$, the film is unconditionally (thermodynamically) stable. In such a film the conjoining pressure arising from the repulsive interaction between the two interfaces augments the Laplace pressure, imparting stability to the film [38, 65, 67]. Also, by setting $\partial\omega/\partial k = 0$, it is possible obtain an expression for the dominant wavenumber of the instability (k_m) and the fastest growth coefficient (ω_m). k_m is related to the length scale or periodicity of the fastest growing mode of instability as $\lambda_m = (2\pi/k_m)$ [38, 65, 67]. The dominant length scale of instability (λ_m) is an important parameter as it can be measured in dewetting experiments and can therefore be directly compared with theoretically predicted values. Physically, a negative value of $\partial\Phi/\partial h$ implies an attractive interaction between the film–air and film–substrate interfaces. This attraction, in the absence of any significant repulsive (stabilizing) forces, favors the growth of the deformations at the film–air interface which eventually results in rupture and subsequent dewetting of the film on the substrate, leading to the formation of a random collection of droplets [38, 57, 65]. During the growth of the deformation the flow of liquid is from the thinner zones of the film to the thicker parts, which is an example of negative diffusion.

It can be seen from the preceding discussion that the condition of instability of a thin film is purely thermodynamic and is independent of the film rheology. It critically depends on the sign of variation of Φ with the film thickness h. Thus, it becomes important to understand the forces that constitute the excess free energy (Φ) [65, 98]. For an unstable or a metastable film Φ is made up of antagonistic (attractive and repulsive) interactions with different distance variations (short and long range) [67, 98]. While surface instabilities require a negative disjoining pressure, contact line retraction upon hole formation is possible only with dominant short-range repulsion in the proximity of the substrate. In the absence of significant electrical double layer effects, which are unlikely to play a role for apolar polymer films in air, the key constituents of the excess intermolecular free energy (Φ) are the long-range apolar Lifshitz–van der Waals (LW) interaction and the much shorter ranged polar 'acid-base' (AB) interactions [31], which are often quantified as hydrophobic repulsion, hydration pressure, and so on [67]. The AB interactions can only occur between molecules that display conjugate polarities as measured by their electron (proton) donor and acceptor capabilities and thus, such interactions are also ruled out in most polymers such as PS, which are weakly electron-donor monopolar material and therefore cannot form hydrogen bonds.

Further, in situations where there is no physical origin of any short range repulsion (implying that $\Phi \rightarrow -\infty$ as $h \rightarrow 0$) a 'contact' repulsion of the form β/h^9 is included to remove the (nonphysical) singularity of Φ as $h \rightarrow 0$ [64, 65, 67]. Based on these conditions, a popular representation of the potential for a unstructured polymer thin film is given as Equation 11.3 [65]:

$$\Phi = (A_S/6\pi h^3) - \left(\frac{S_P}{l_P}\right)\exp(-h/l_P) - (\beta/h^9), \tag{11.3}$$

where, A_S denotes the effective Hamaker constant arising from the van der Waals interaction between the two interfaces of the film on a semi-infinite thick substrate. A positive or a negative value of A_S denotes a long-range attraction (the possibility of instability driven by van der Waals force) or repulsion (stabilizing role of van der Waals force), respectively. S_P in Equation 11.3 denotes the strength of a medium-range attraction ($S_P < 0$) or repulsion ($S_P > 0$) with l_P being the corresponding correlation (decay) length [65].

This term, though not significant for most polymer films, accounts for the AB-type interactions in polar liquids (for example, for water it accounts for the 'hydrophobic repulsion' with $S_P > 0$, $l_P \sim 0.2$–1 nm). The entropic confinement effects in polymer films due to adsorption/grafting at the film–substrate interface is also taken into account by this term and for this case $l_P \sim R_g$ (radius of gyration of the polymer).

For spinodally unstable films, LSA of Equation 11.1 gives the short-time initial length scale of instability, λ, the number density of features, N, and the time scale for the appearance of the amplified wave (τ) as:

$$\lambda = [-8\pi^2\gamma/(\partial\Phi/\partial h)]^{1/2} \tag{11.4a}$$

$$N = \lambda^{-2} = \frac{-(\partial\Phi/\partial h)}{8\pi^2\gamma} \tag{11.4b}$$

$$\tau = 12\gamma\mu[h^3(\partial\Phi/\partial h)^2]^{-1}\ln(h/\varepsilon) \tag{11.4c}$$

Equations 11.4 predict $\lambda \sim h^2$ and $\tau \sim h^5$, as well as $N \sim h^{-4}$ for instabilities driven by nonretard van der Waals interaction, and are widely used for the interpretation of thin film stability and determination of interaction potentials from measurements of λ, obtained from experimental results [65, 91].

Predictions based on LSA, although extremely useful and widely used in the analysis of instability results, is limited to the prediction the wavelength of the fastest growing mode of instability and the initial shape of two-dimensional axis-symmetric holes formed in the film [76–85], and cannot fully capture the morphological evolution during different stages of dewetting, and correlate the evolved patterns to the surface properties and intermolecular interactions, as surface evolution is inherently nonlinear. More detailed studies are based on full three-dimensional (3D) nonlinear simulations that provide a formalism for correlating the film morphology with the interfacial interactions and the film thickness, which makes it possible to directly compare theoretical predictions with experimental observations and therefore among other things, allow addressing the inverse problem of identifying the forces responsible for the instability based on the observed morphology [64].

11.3.2
Dewetting: Experiments

Polymer systems are preferred for dewetting experiments for several reasons: (i) the ability to switch between a liquid and a solid state by varying the system temperature above or below the glass transition temperature (T_g). This implies that while the dynamic and morphological reorganization of the film takes place in the liquid state, the structures can be made permanent by simply quenching below T_g; (ii) the low vapor pressure of polymers ensure that the geometry of the final structures do not change with time because of evaporation; (iii) the viscosity of the polymers can be tailored easily by adjusting the molecular weight, thereby controlling the dynamics of the system, enabling time-resolved experiments [47–56]. The films, which are solid at room temperature, are either heated beyond T_g or exposed to solvent vapor to decrease their T_g below the room temperature. It may be noted that instability-induced pattern formation is neither material specific (unlike techniques such as photolithography, which can directly pattern only optically-active photo resists) [12] nor limited to homopolymers. A variety of materials such as solid films [41], liquid crystals [40], functional polymers and polyelectrolytes [170, 171], conjugated polymers [172], blends, and block co-polymers [173], have been successfully patterned by this approach.

Very thin films (< 50 nm) of simple homopolymers such as polystyrene (PS), poly(methyl methacrylate) (PMMA), poly(acrylamide) and so on, coated on cleaned silicon wafers, quartz, cleaved mica or glass substrates by spin- or dip-coating from diluted solutions of the respective polymers dissolved in suitable solvents, are preferred model systems for studies on thin film instabilities [59–63, 65, 68–73, 93–98, 100–128, 137–139, 141–158]. Evaporation of the solvent during and after the coating leads to the deposition of thin layers of the polymer. The film thicknesses can be as low as few nm (~ 4 nm) and in most experiments involving dewetting, films thinner than ~ 100 nm are used. Films thicker than ~ 100 nm typically dewet only due to nucleation. On the other hand, it becomes progressively challenging and difficult to obtain smooth films of uniform thickness in films thinner than ~ 10 nm, as the inhomogeneities and defects present on the substrate adversely affect the quality of the films [57, 91]. For many polymers the glass transition temperatures T_g are around 100 °C or more and therefore, the films are solid-like at room temperature. Consequently, in order to engender dewetting, the samples are either heated beyond T_g or exposed to a solvent vapor. While heating beyond T_g liquefies the polymer (with several orders of magnitude reduction in the viscosity), exposure to solvent vapor also leads to the same effect, by reduction of the effective T_g below room temperature [172]. This happens by the penetration of solvent molecules into the polymer matrix, which results in a swelling of the film and reduction of the intermolecular cohesive interaction.

Experimentally, dewetting of an unstable polymer thin film on a defect-free smooth surface was first demonstrated by Reiter in 1992 [59, 60], where thermal dewetting of sub-100 nm-thick polystyrene films on non-wettable silicon wafer substrates was shown. The first signature of true spinodal dewetting in a polymer film was captured by Xie et al. [62]. and subsequently by several other groups [63, 138]. The scaling relations observed from the experiments agreed well with the theoretical predictions based on LSA [59, 60, 65]. The length scale of the dewetted features were seen to scale with the interfacial tension as $\lambda \sim \gamma^{0.5}$ [59, 60, 65]. However, it is important to point out that the scaling relations depend strongly on the precise decay behavior of intersurface forces, slippage at the interface [159–168], thermocapillarity, surfactant Margoni flow and heterogeneities of the substrate.

Dewetting experiments by Reiter [59–61] and subsequently by many others [62–66, 68–75, 92–128, 137–164] revealed the morphological evolution, self-organization, and pattern formation in thin polymer films. In most experiments the onset of instability is with the form of an undulation of the film surface (Figure 11.1a) [62] or the appearance of fairly equally sized but random collection of holes (Figure 11.1b) [59, 60]. The mean

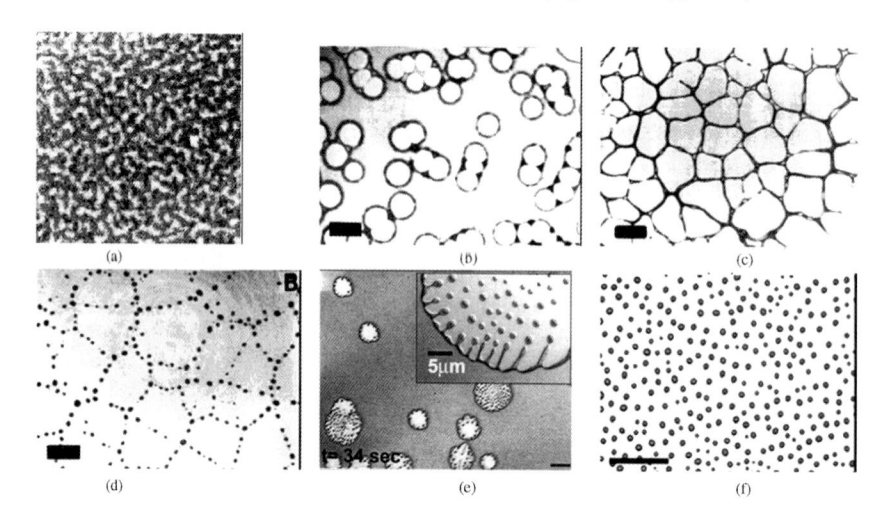

Figure 11.1 Examples of typical morphologies observed in different stages of dewetting of thin polymer films: (a) Undulations in the early stages of dewetting in a 4.5 nm-thick PS film annealed at 115 °C for 4 mins (Reproduced with permission from [62]. Copyright © (1998) American Physical Society); (b) formation and growth of holes in a 30 nm-thick PS film; (c) polygonal patterns observed in a 10 nm-thick PS film on a silanized silicon wafer; (d) final morphology of a dewetted film in the form of aligned polymer droplets where cellular patterns form, in a 25 nm-thick PS film. Such structures are due to the breakage of the polymer ribbons due to Rayleigh instability (b to d: Reproduced with permission from [60]. Copyright © (1993) American Chemical Society); (e) signatures of rim instability and detachment of droplets from the retracting three-phase contact line (inset) in a 35 nm-thick PS film on a silicon wafer substrate (Reproduced with permission from [155]. Copyright © (2004) American Institute of Physics); (f) final morphology of a dewetted film comprising a random collection of polymer droplets where rim instability is present, as observed after complete dewetting of a 24 nm-thick PS film on a crosslinked PDMS substrate. (Reproduced [207]. Copyright © (2008) The Royal Society of Chemistry.)

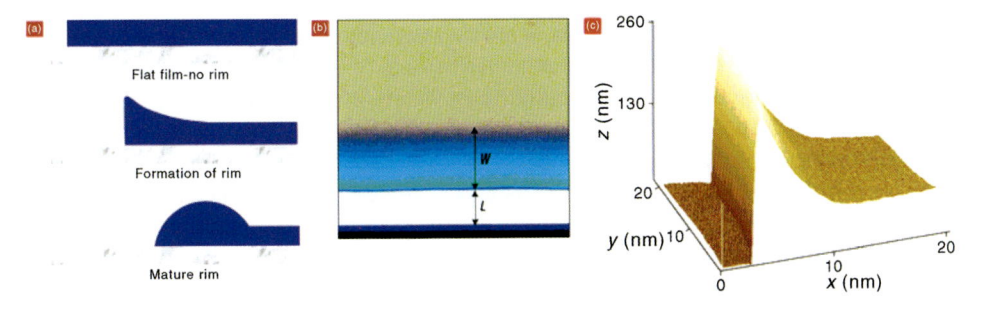

Figure 11.2 The shape of the rim during early stages of dewetting of thin visco elastic polymer films on nonadsorbing substrates. (a) The sequence of three fundamentally different shapes of the rim found experimentally; (b) a typical optical micrograph ($50 \times 50\,\mu m^2$) taken at the edge of the sample obtained after partial dewetting. The dewetted distance (l) and the width of the rim (w) are indicated. The colors can be directly related to the thicknesses, e.g, the brighter the blue, the thicker the film; (c) a typical rim profile as determined by AFM. (Reproduced with permission from [111]. Copyright © (2005) Macmillan Publishers Ltd.)

distance between the holes corresponds to the fastest growing mode of instability λ_m. Further morphological evolution of the film is associated with the spatial growth of the holes. The growth of holes in most dewetting experiments without polymer slippage on the substrate is associated with the formation of sharp rims or capillary ridges surrounding the holes (Figure 11.2) [111]. The rims result from a mismatch in the rates at which polymer retracts as the hole opens and subsequent redistribution of material to other thicker and intact parts of the film [59, 60, 65, 110, 111, 143, 147]. Further propagation of dewetting proceeds with the growth of the holes and retraction of the three phase contact line, until the rims of the adjacent growing holes touch each other, forming an interconnected ribbon structure of polymer, which is often referred to as the polygonal or cellular pattern (Figure 11.1c) [59]. The polymer ribbons subsequently breaks down due to a Rayleigh instability [25, 26], forming a polygonal array of isolated polymer droplets, as shown in Figure 11.1d [59, 60]. In some experiments, particularly on low-wettability substrates, the rims of the growing holes themselves become unstable, exhibiting a fingering instability (Figure 11.1e) [110, 143, 145, 153, 158, 168]. Droplets detach from the unstable rims of growing holes and the polygon formation is suppressed (Figure 11.1f), leading to the formation of a completely random collection of droplets [110, 143, 145, 153, 158, 168].

For a visco-elastic thin film with interfacial slippage, the shape of the rims progressively changes as dewetting proceeds and, based on the detailed geometry of the rim, three distinct regimes of dewetting have been identified [111, 146, 147]. At the onset of the instability, the energy resulting from capillary forces is primarily dissipated by means of viscous losses around the vicinity of the hole edges, which ensures that no rims form (Figure 11.2a) [111]. When the radius of the growing hole becomes larger than the decay length $\Delta = (h_0 b)^{0.5}$, where b is the slippage length ($\sim \eta/\zeta$; ζ is the coefficient of friction and η is the viscosity), the influence of friction at the substrate–film interface becomes significant and consequently, the velocity of the dewetting front reduces [111]. This eventually results in the appearance of a highly asymmetric rim (Figure 11.2b). While the steep side of the rim reaches up to a height H next to the three-phase contact line, an exponentially decaying profile of the rim with a decay length $\sim \Delta$ forms on the film side. At later stages of dewetting, because of the influence of surface tension, the sharp edges eventually become rounded, with further reduction in the dewetting velocity, resulting in a scaling of $v \sim t^{-1/3}$ at this stage [111].

A classic example of how the stability of a thin film is strongly influenced by the nature of interaction between the two interfaces (film–air and film–substrate) can be seen in the work by Reiter et al., where they were able to 'switch on' the instability in a stable thin film by inverting the role of dispersion forces from stabilizing to destabilizing by replacing the bounding media [66]. A thin liquid PDMS film on a silicon wafer in air is thermodynamically stable, as in a system comprising silicon–PDMS thin film–air the effective Hamaker constant is negative, signifying a long-range repulsion ensuring the stability of the film. However, when the same film was submerged under water the sign of the effective Hamaker constant became positive, which implies an attraction between the interfaces, resulting in immediate

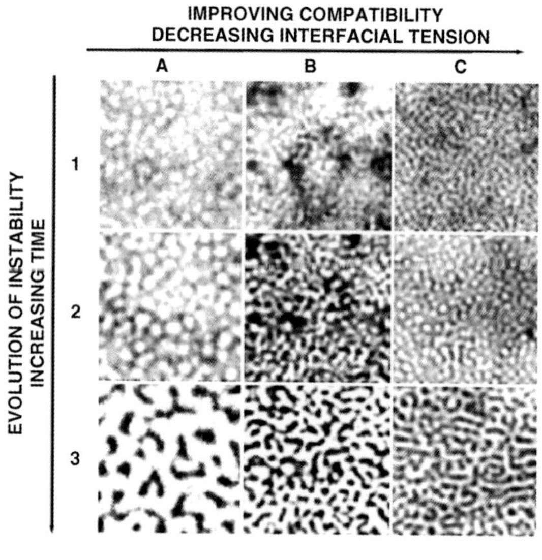

Figure 11.3 Changes in the length and time scales of break-up of a 60 nm-thick PDMS film under bounding liquids which have a varying level of compatibility with PDMS. The interfacial tensions change from $38.4 \pm 3\%$ mN m^{-1} in pure water (series A) to $8.1 \pm 8\%$ mN m^{-1} in series B (PP) and then finally to $0.3 \pm 33\%$ mN m^{-1} in series C (L77). The increasing levels of gray denote thicker portions. In each series three (1, 2, and 3) representative snapshots mark the different stages of evolution of instability. The actual times for series A, B, and C are (25, 65, and 200 s), (15, 45, and 100 s) and (5, 10, and 15 s), respectively. The area shown in frames of series A and B is $100 \times 100\,\mu m^2$. For clarity the results for series C are shown in an area which is 16 times smaller ($25 \times 25\,\mu m^2$). A rather smooth surface of the film evolves into a collection of circular depressions (frames A1, B1, and C1). These shallow depressions develop into circular holes (frames A2, B2, and C2) which grow laterally, coalesce with other growing holes and form long cylindrical ridges (frames A3, B3, and C3) that break up into small drops later. (Reproduced from [66]. Copyright © (2000) American Physical Society.)

rupture and dewetting of the film (Figure 11.3). The magnitude of the destabilizing attractive force could further be tailored by using dilute aqueous solutions of different surfactants resulting in changing the interfacial tension between the PDMS film and the bounding liquid [66].

11.3.3
Heterogeneous Dewetting

In reality most surfaces used for dewetting experiments are chemically heterogeneous on a submicron or nanometer length scale and therefore nucleated instabilities are present in most real settings [57, 91, 117–139]. It has been shown that the presence of chemical heterogeneities even at length scales significantly smaller than the spinodal length lead to a gradient of chemical potential along the solid–liquid interface [119, 129, 131, 133]. This in turn results in flow of liquid from lower wettability zones to higher wettability zones, resulting in the rupture of the film [119, 129, 131, 133]. Heterogeneous nucleation can lead to a rupture even in a spinodally stable film. For metastable films, once the local thickness at a depression is reduced to a point where the remnant film becomes spinodally unstable, both mechanisms of instability become active concurrently [121–123, 129, 133]. However, the time scale of heterogeneous dewetting is typically much faster than the spinodal time scale, especially near the critical thickness, and therefore signatures of nucleated dewetting become more pronounced [129]. The hallmarks of heterogeneous nucleation induced dewetting, particularly close to a heterogeneous patch are: (i) much faster evolution of the instability leading to film rupture; (ii) the absence of surface waves preceding the rupture of the film; and (iii) a local ordering of pattern around a heterogeneity. Interestingly, for nucleation over a chemically heterogeneous zone, it is not at all necessary that the heterogeneous zone has to be less wettable compared to its surroundings [129]. More-wettable patches are equally efficient in causing film rupture, as they lead to an inward flow of liquid from the patch periphery towards the center, thus causing a thinning of the film around the patch and lead to the formation of a annular holes. Thus, depending on the patch size (with respect to spinodal length scale), and film thickness extent of wettability contrast, and also on whether the heterogeneous patch is more or less wettable in contrast to the rest of the substrate, a variety of patterns can form, as can be seen in the different frames of Figure 11.4 [129]. In some experiments, satellite holes are seen to form around nucleated holes [121, 122]. It has been captured in experiments, and has been shown based on simulations, that satellite holes originate in the depression in the film adjacent to liquid rim, and in particular in the vicinity of regions where the rim is thicker. Over these regions, the Laplace pressure inside the depression acts as preferred nucleation center for the formation of the satellite holes [120–123].

The discussion up to this point may give the impression that polymer thin films dewetting experiments are rather simple, as they involve a few easy steps like film preparation, heating/solvent vapor exposure and finally, characterization of the evolved structures with an optical or atomic force microscope. However, extracting the exact information about the dewetting mechanism from a particular experiment

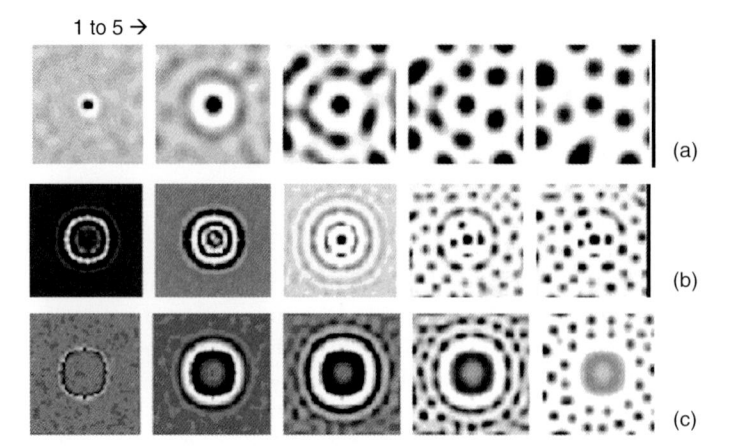

Figure 11.4 Morphological evolutions in a 30 nm-thick film on a heterogeneous substrate (a) with a less-wettable patch of diameter 31 μm. The figures from left to right correspond to nondimensional times 278, 3487, 7620, 13 598, and 24 710, respectively; (b) with a less-wettable patch of diameter 496 μm. The figures from left to right correspond to nondimensional times 1485, 2247, 8208, 17 269, and 20 838, respectively; (c) with a more-wettable patch of diameter 352 μm. The figures from left to right correspond to nondimensional times 63, 3164, 7608, 12 271, and 23 444, respectively. (Reproduced with permission from [130]. Copyright © (2001) American Institute of Physics.)

is a complicated exercise, primarily due to multiplicity of dewetting mechanisms like (spinodal, nucleation, release of residual stresses etc.) [57, 91, 111, 124, 133]. For thicker films, dewetting by heterogeneous nucleation of holes pre-empts the onset of the instability even for spinodally unstable films. Thus, even in films of the same polymer with same molecular weight, under identical film preparation and other experimental conditions but with slight variations in the initial film thickness, completely different evolution sequences and final morphologies may result, depending on which precise mode is dominant during the initial phases of the instability [91, 96, 137–139]. For example, Tsui et al. have shown that on oxide-coated silicon wafers, while dewetting of PS films having thickness $(h) > 13$ nm is dominated by heterogenous nucleation, films thinner than 13 nm dewetted by spinodal mechanism [139]. Seemann et al. have observed three distinct rupture mechanisms in the dewetting of low-molecular-weight PS thin films on silicon substrates: (i) spinodal dewetting in 3.9 nm-thick PS film on a silicon wafer; (ii) thermal nucleation in a 4.1 nm-thick film; and (iii) heterogenous nucleation in 6.6 nm-thick films [96, 138]. Even in unstable films that have initially ruptured by nucleation, rows of satellite holes are often seen forming around the rims of a nucleated hole once it (the hole) has grown to a certain size [121–123]. This type of hole-forming cascade, corresponding to a complex film rupture process, which is neither simply spinodal nor nucleated, has been captured in 3D numerical simulations [123, 131]. The coexistence of different modes in dewetting experiments has prompted researchers to define two distinct regimes of the spinodal instability itself: deep inside the spinodal territory (DIST) for the thinner films ($<\sim$5 nm-thickness) and defect-sensitive spinodal regime (DSSR) for slightly thicker films [57, 91].

11.3.4
Influence of Residual Stresses on Film Rupture and Dewetting

An unresolved issue in the area of polymer thin film dewetting that existed for a significant period of time, probably ever since research in the area started in the early 1990s, is the dewetting of thermodynamically stable films, and that too, resembling several features of spinodal dewetting! For example, there are reports on the dewetting of PS thin films on non-hydrophobized silicon wafer substrates. Bolline et al. have reported the dewetting of thermodynamically stable PS films on silicon substrates in which the behavior resembles spinodal dewetting though the process does not strictly follow the model for spinodal dewetting [142]. Another anomalous observation which has been reported is the spinodal-like dewetting of thicker films [59, 60, 65]. Ideally, in thicker films the influence of dispersion forces should be rather low and they are very unlikely to cause film instability. It may however be noted that most such conclusions on 'spinodal-like' dewetting are drawn from matching the length scale exponents to the scaling relations show in Equations 11.4a–11.4c. This approach of concluding about dewetting mechanism simply by observing the morphological evolution sequence or by checking how close the length scale exponents match the theoretically predicted values is itself questionable, as it has already been discussed how various destabilization modes arising from the interplay of different interactions are simultaneously present in typical dewetting experiments [91]. However, it has almost become customary to fit λ or N obtained experimentally with h on a double logarithmic plot, and if the exponents are close to 2 or -4, respectively, it is generally argued that dewetting is spinodal in nature [59, 60, 65, 91]. Thus, in cases where spinodal dewetting of thermodynamically stable films are reported, the essential logic for arriving at the conclusion is based on observed values of length scale exponents were close to 2 and -4 for λ and N with respect to h respectively [142, 143].

Several mechanisms, such as local density variation with film thickness [144], deformation of polymer molecules during film preparation [112], or even the influence of surface cleaning [169] have been proposed which are all aimed at explaining some of these anomalies, with limited success, as none of them could be convincingly proved in experiments. It was realized recently that the conditions of film preparation have a crucial role on the stability of the system. The rapid solvent evaporation during spin coating in all probability leads to a nonequilibrium conformation of the polymer molecules within the film, frozen-in during the vitrification of the material, resulting in accumulation of residual stresses within the films [108–116]. As the film is heated or exposed to solvent vapor for liquefaction of the polymer, these stresses are rapidly released, thereby providing additional driving force for dewetting that augments with the conjoining pressure [111]. Even at room temperature, these stresses are slowly released, and thus the level of residual stress in the film depends on the aging time of each sample (time between coating of the film and performance of the dewetting experiment) [111]. During aging, the chain-like molecules tend to adopt conformations closer to their equilibrium configuration, thereby reducing the extent of accumulated stresses. Further the probability of rupture strongly depends on the level of accumulated stresses [109, 111]. It seems that the magnitude of the force field arising from the release of the residual stresses is much stronger compared to the forces due to intermolecular interaction in thin films, and therefore irrespective of the thermodynamics of the system the release of the stresses dominate the dewetting scenario in some films [111]. This recent hypothesis in many ways explains the anomalous dewetting behavior of low energy spinodally stable films on high energy substrates [111]. However, the theoretical understanding of residual stress-induced dewetting is far from complete. It is also worth noting that there are several other issues involving the viscoelasticity of the polymers [100–106], extent of slippage at the interface [159–168], late stage coarsening [148–150], fingering instabilities around hole rims and so on [110, 143, 145, 151, 153, 158, 168], all of which significantly influence the dewetting process in thin polymer films and are therefore active topics. The details of these aspects of dewetting are beyond the scope of this chapter and the interested reader can refer to several good papers that are already available on some of these themes.

11.3.5
Pattern Formation in Polymer Thin Film Dewetting

While initial research on polymer thin film instabilities was focused more towards the basic issues such as the mechanisms of dewetting, correlating the length scale of the features with initial film properties and so on [59–86, 89–98, 117–126], recent emphasis is significantly on using instability-engendered morphological evolution as a viable and flexible meso patterning technique [87, 127–129]. This approach is based on the fact that, irrespective of the precise mechanism, in most dewetting experiments the final morphology comprises of a random collection of isolated but nearly equal sized droplets [59–168]. As mentioned already and shown in Figure 11.5, the periodicity and size of the droplets depend on initial film conditions such as the film thickness and surface and interfacial tensions of film and substrate [59, 65]. This provides a 'handle' to control the feature sizes by simply varying initial conditions without changing the generic morphology of the patterns. Further, a variety of distinct morphologies like holes, ribbons, droplets, or a combination of all of them can be formed by quenching or stopping the evolution sequence of the dewetting film at an intermediate stage [65, 87]. Additionally, in many nonlinear and complex systems, the equilibrium state is often not unique, as several metastable states of local energy minima exist, which enables a variety of different morphologies to be achieved, simply based on a judicious selection of the initial conditions and extent of dewetting. However, the lack of a long-range order of the structures formed, by dewetting or for that matter by most spontaneous self-organization processes, limits their use in practical application. In order to overcome this limitation, and to present dewetting as a truly effective patterning

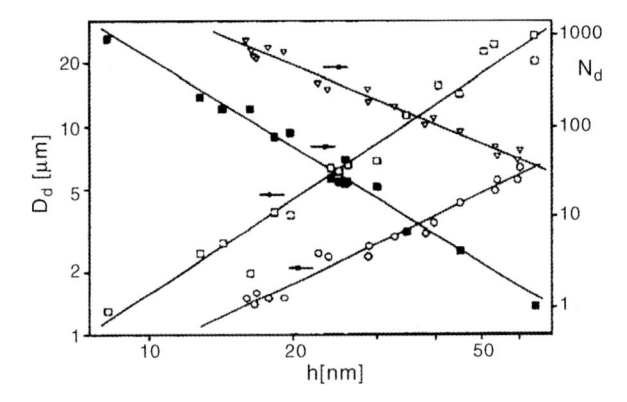

Figure 11.5 Double logarithmic plot of average diameter of droplets (D_d) and the average number of droplets per 10^4 μm² (N_d) as a function of film thickness (h). The symbols (■,▽) and (□, ○) refer to N_d and D_d respectively, for two different types of wafers, A and B. (Reproduced with permission from [59]. Copyright © (1992) American Physical Society.)

method, comparable to various soft lithography techniques, offering even more flexibility, it is important that a certain degree of long-range order is imposed on the resulting morphological structures [173–282].

11.4
Dewetting on Patterned Substrates

Dewetted structures have been successfully aligned by using either a chemically [176–190], physically [191–202], or physico-chemically patterned substrates, or even by simpler approaches such as rubbing [173–175]. The possibility of aligning the isotropic and randomly oriented dewetted structures was first demonstrated by Higgins and Jones by simply rubbing the substrate with a lens tissue before coating the film, resulting in the formation-aligned ridges on the substrate in the direction of rubbing [173]. The resulting directionality of the substrate influenced the dewetting pathway of the PMMA film and led to the formation of strongly anisotropic structures comprising a series of continuous lines running across the sample surface following the ridges [173]. Zhang et al. demonstrated a similar concept by brushing the surface of a coated PS films with a hard brush. This resulted in a subtle topographic alignment on the film surface, which influenced the dewetting pathway in a significant manner [174]. A variety of aligned structures such as nanogrooves, lines, and droplets formed during the intermediate and final stages of dewetting. More complex patterns were created when the film was coated on a photolithographically patterned substrate comprising a grating structure and then rubbed in a direction perpendicular or parallel to the stripes [174]. Dewetting of such films resulted in multi length scale structures, as shown in Figure 11.6, where the finer structures along the stripes are due to rubbing. These types of structure are useful in areas where strongly anisotropic patterns are necessary, for example involving liquid crystals for

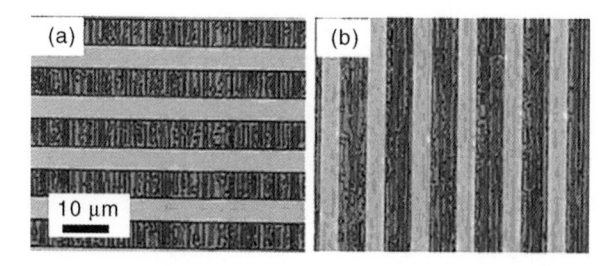

Figure 11.6 Structures formed by dewetting of a 6 nm-thick, rubbed PS film on a photolithographically patterned substrate, with the rubbing direction (a) orthogonal and (b) parallel to the direction of the patterns (chromium stripes). (Reproduced with permission from [174]. Copyright © (2005) Elsevier.)

various novel optical applications [176–178]. Müller-Buschbaum et al. have successfully created shallow (~5 nm high) nanochannels by dewetting of PDMS solution, by wiping a glass substrate with a lint-free paper pre-soaked with diluted PDMS solution, in a particular direction at very low (~few kPa) contact pressures [175]. The competition between oriented deposition of the solution and spontaneous dewetting resulted in the formation of the aligned structures with a periodicity of ~166 nm [175].

11.4.1
Dewetting on Chemically Patterned Substrates

While rubbing-based techniques are useful in demonstrating conceptually that isotropic dewetted structures can indeed be aligned [173–175], it is not possible to achieve patterns with a high degree of accuracy in terms of periodicity, fidelity, precise control of morphology, and long-range order by merely extending any of these methods. A systematic approach in this regard is to use physically (topographically) or chemically patterned substrates for dewetting a polymer thin film [176–202]. Such substrates are generally created by any of the existing lithography techniques, and thus pattern-directed ordering of dewetted structures is regarded as a hybrid approach, combining both top-down lithography and self-organization techniques.

The ordering of dewetted structures on a chemically patterned substrate, comprising microcontact printed patches of hydrophilic 11-mercaptoundecanoic acid (MUA) on a gold-coated glass substrate was first demonstrated by Meyer and Braun [179, 180]. The dewetting of a PS thin film on such a substrate was seen to initiate only over the MUA patches, which acted as preferred nucleation sites [179]. The dewetted structures aligned in commensuration with the chemical patterns on the substrate [179, 180]. It was subsequently shown by Sehgal et al. that even on a chemically patterned substrate, the extent of ordering of the dewetted structures depends strongly on the commensuration between the initial film thickness (h_0) and the periodicity of the patterned substrate (λ_S). They reported the dewetting of PS thin films on chemically patterned substrates comprising arrays of progressively narrower stripes (1–15 μm) of alternating –CH$_3$ (lower surface energy) and –COOH (higher surface energy) end-terminated self-assembled monolayers (SAMs) [181]. Using a library of patterns with varying feature sizes on the same sample allowed rapid comparative screening of the influence of the symmetry-breaking chemical field on dewetting behavior, while eliminating the possibilities of experimental errors involving different samples which invariably leads to slightly different initial conditions [181]. A variety of ordered, partially ordered and even disordered structures form on topographically smooth chemically patterned substrate, depending on the film thickness and the periodicity of the patterns. For ultrathin, spinodally unstable films (∼12 nm-thick) on the patterned areas, the intermediate stages exhibited strong anisotropy imposed by the underlying chemical patterns and superposition of spinodal-like structures. For a given film thickness, the final morphology of the dewetted structures was strongly influenced by λ_S, as shown in Figure 11.7 [181]. For the wider stripes ($\lambda_S = 12$–15 μm), two columns of droplets are seen to form per band. For narrower stripes ($\lambda_S = 9$ μm), the lateral confinement causes the coalescence of the two droplets and an array of aligned single droplets is seen. Further confinement of the substrate pattern to $\lambda_S = 6$ μm resulted in distorted oval shaped droplets. Subsequent reduction of λ_S leads to a loss of order and resulted in relatively distorted morphology. For the narrower patterns

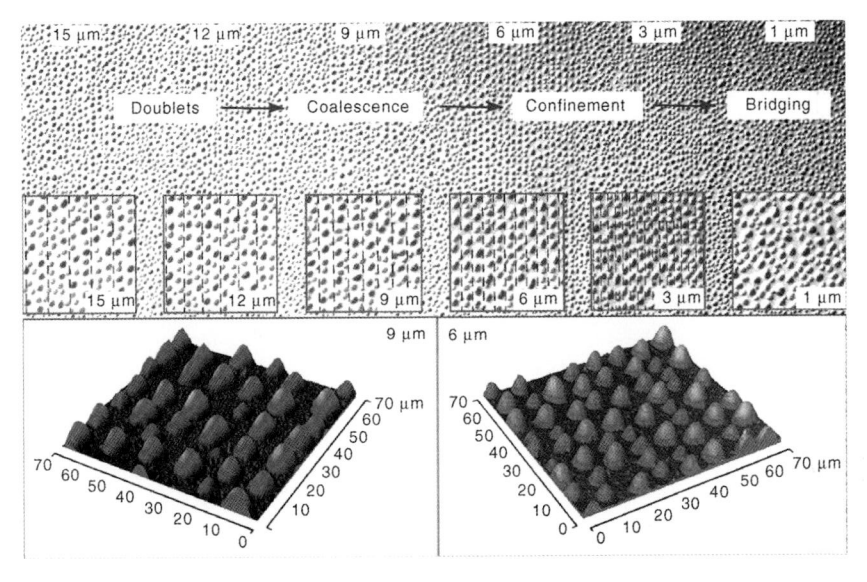

Figure 11.7 The influence of pattern dimension on the final morphology (aligned, partially aligned and random droplet arrays) of structures resulting from dewetting of a 12 nm-thick PS film. Optical micrographs and insets show transition from the doublet state (15–12 μm stripe width), to coalescence (9 μm), confinement (6 and 3 μm), and a heterogeneous morphology (1 μm) with bridging over multiple bands. Dashed lines (insets) indicate registry with the underlying chemical pattern period. AFM images (9 and 6 μm) highlight control of droplet size and spatial position. (Reproduced with permission from [181]. Copyright © (2002) American Chemical Society.)

($\lambda_S \sim 3\,\mu m$) the droplets and smaller and have distorted ellipsoidal shapes with the major axis oriented along the direction of the stripes. A late stage bridging transition leads to the coalescence of highly anisotropic (high surface energy) droplets across unfavorable –CH$_3$ bands to more rounded, lower energy structures spanning the very narrow stripes ($\lambda_S \sim 3\,\mu m$) [181]. Zhang et al. were successful in creating ordered 2D structures by using square-patterned substrates consisting of alternating high energy SiO$_x$ recessed squares in a low energy ocadecyltrichlorosilane (OTS) grid [182, 183]. In addition to wettability contrast the substrate also exhibited a distinct topography contrast arising from the presence of the OTS layer. Dewetting of a polymer film on such a substrate is triggered by the combined influences of topographic and wettability contrasts. The film ruptured at the edges of the squares, breaking-up into two isolated fragments on the silica and OTS patches [183]. Subsequent dewetting of the films over each fragment continued separately at a different pace, as shown in Figure 11.8. Dewetting of the detached film fragments on the silica squares lead to the formation of single isolated droplets in the center of each area. In contrast, dewetting of the remnant part of the film on the OTS grid surrounding the squares resulted in the formation of oval-shaped PS lines between two adjacent squares and larger spherical droplets at the corners of every square patch caused by the retraction of the contact line (Figure 11.8f) [183]. Several other groups have reported results on the ordering of dewetted structures by chemically patterned substrates for the creation of structures with a variety of

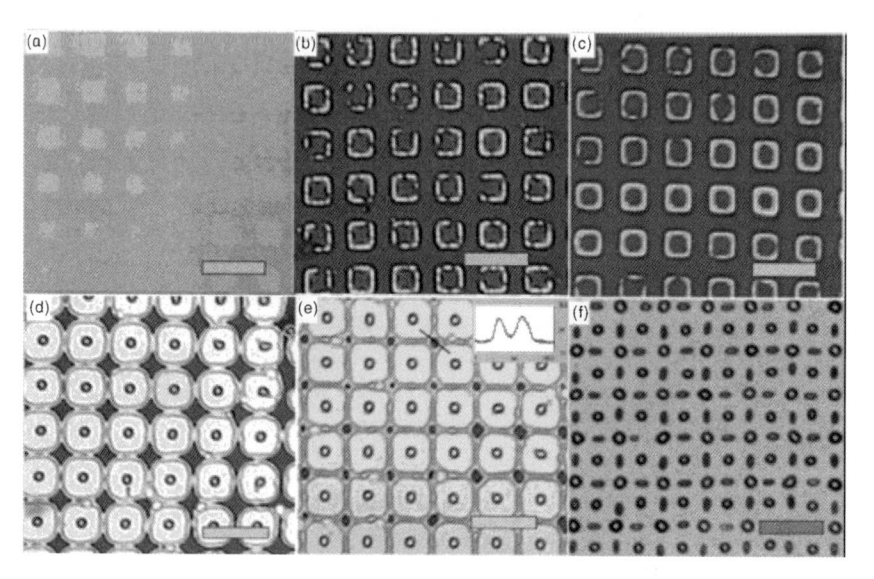

Figure 11.8 Morphological evolution sequence of a 47 nm-thin PS film on OTS squares patterned substrate. The substrate had alternating OTS and SiO$_x$ squares. The film was annealed at 170 °C for (a) 0 min; (b) 0.5 min; (c) 1 min; (d) 8 min; (e) 12 min and (f) 20 mins respectively. The scale bar is 30 mm. (Reproduced with permission from [183]. Copyright © (2003) Elsevier.)

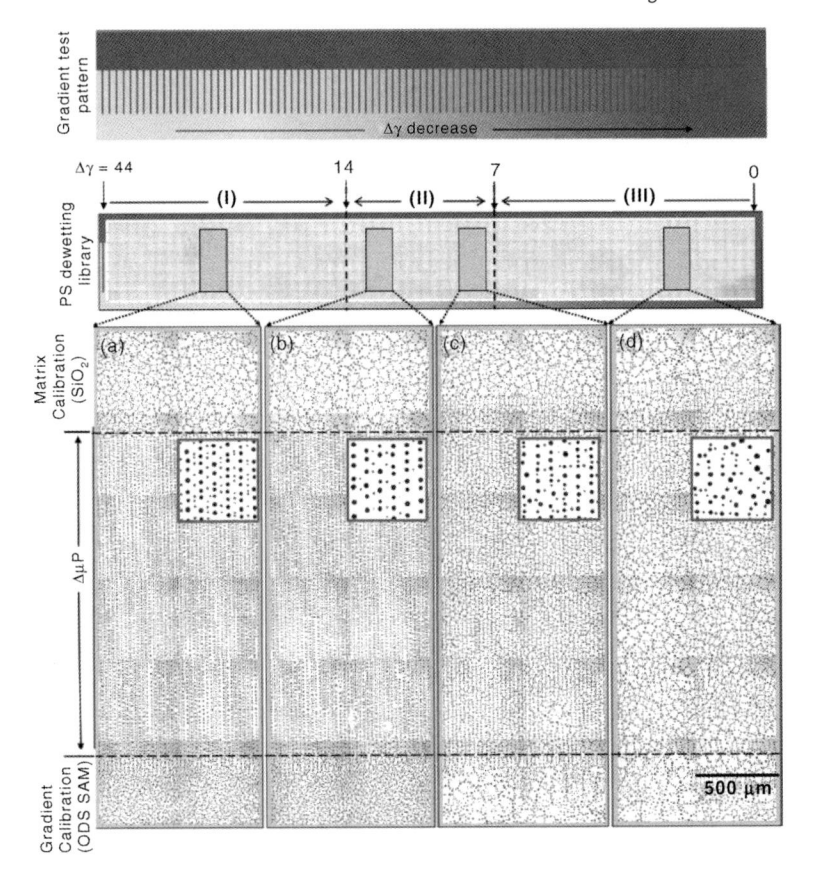

Figure 11.9 (Top) Schematic of the gradient test pattern illustrating pattern and gradient orientation in the experiments. (Middle) PS dewetting library compiled from 1900 contiguous optical micrographs over the entire specimen. Regions I, II and III, noted on the figure, are discussed in the text. (a–d) Strips of OM data extracted from the library showing the transition from pattern-directed to isotropic dewetting. Insets (250 μm × 250 μm) show magnified optical micrographs of representative dewetting morphologies. (Reproduced with permission from [190]. Copyright © (2007) The Royal Society of Chemistry.)

morphologies and geometries in many materials [184], including fluorescent and conducting polymers [185–189].

Julthongpiput *et al.* have recently shown how the magnitude of the wettability contrast (difference in surface energy, $\Delta\gamma$) in the substrate pattern affects the morphology and alignment of the dewetted polymer droplets [190]. For this purpose, a combinatorial test surface was used, with gradient micropatterns consisting of a series of 20 μm wide *n*-octyldimethylchlorosilane (ODS) lines with continuously changing surface energy (γ) with respect to a constant surface energy background. The lateral separation between the ODS stripes was constant at 4 μm. Figure 11.9 summarizes the variation of the dewetted pattern morphology on the gradient micropatterned surface as a function of $\Delta\gamma$ and shows the existence of three distinct regimes. Regime 1, corresponding to high $\Delta\gamma$ (~43 mJ m^{-2}), single aligned arrays of droplets distributed along the center of both the matrix and SAM portions of the patterns were observed. Regime II exhibits a crossover where the droplets gradually lose the extent of ordering in the direction of the stripes with decrease in the wettability contrast ($\Delta\gamma$), maintaining a subtle signature of directionality. In Regime III, the droplet arrangement was completely isotropic for $\Delta\gamma$ (~4 mJ m^{-2}) and the substrate patterns completely failed to impose any directionality on the dewetting process [190].

11.4.2
Dewetting on Physically Patterned Substrates

Dewetting on a physically or physico-chemically patterned substrate is in many ways more exciting and complex in comparison to dewetting on a substrate with chemical patterns only. The presence of the topographic contrast, in addition to influencing the dewetting pathway, also significantly affects the initial film preparation conditions, as direct spin coating on a topographically patterned substrate leads to nonuniformities in the initial film thickness over different areas [191]. In fact Bao et al. [192] and very recently, Ferrell and Hansford [193] have used dewetting during spin coating of a polymer solution on a topographically patterned substrate for fabricating micro- and nanoscale polymer structures. Bao et al. directly spun coated PMMA and polycarbonate (PC) solutions on topographically patterned silica substrates coated with a higher surface-energy silane on the raised portions and a lower-surface energy silane over the recessed zones. The coated polymer preferentially wetted the higher-surface energy zones, leading to the formation of well-aligned structures [192]. The final pattern morphology was strongly influenced by the geometry (height, lateral feature dimensions, and spacing) and polymer solution concentration [193].

Dewetting of a homopolymer film on a topographically patterned substrate was first reported by Rockfort et al., where striped substrates comprising periodically varying zones exhibiting polar (silicon oxide) and nonpolar (gold) interactions (stripe periodicity ~170 nm) were used to investigate dewetting of a 5 nm-thick PS film. It was seen that the PS film dewetted on the oxide stripes and segregated to the gold-coated domains, leading to the formation of strongly isotropic structures resembling the substrate pattern geometry [194]. Rehse et al. systematically investigated the stability of thin PS films of various molecular weights on regularly grooved silicon surfaces comprising triangular grooves or stripes with a mean stripe width of ~250 nm and a mean peak-to-valley height of ~5 nm without any surface energy contrast [195]. The films, which were directly spin coated on the patterned substrates showed a slight variation in thickness, in phase with the substrate geometry. The thickness of the coated films was slightly less over the raised parts of the stripes compared with the areas over the valleys. Rupture in such a films initiated over the thinner zones above the stripes, resulting in the break-up of the initially continuous film, forming isolated and long strips of polymer filling the grooves [195]. Thus, the profile of the dewetted film replicated the inverse morphology of the substrate. However, the instability was manifested only for films below a critical thickness as thicker films were found to be stable even after several days of annealing [195, 196]. The critical thickness was shown to be a linear function of the radius of gyration of the polymer molecules, and the effective dependence was experimentally found as t_{Peak} ~$0.55 R_g$ for PS, where t_{Peak} is the thickness of the thinnest part of the film [195]. Synchrotron X-ray reflection, in addition to atomic force microscopy (AFM) was used by some groups to understand how confinement affects liquid structure formation and spreading dynamics of a polymer thin film coated on a topographically patterned substrate. It was argued that in presence of an undulating substrate, the decay length of the surface modulation was much longer than that observed in case of simple liquids on a flat surface [196]. Several others have reported the strong influence of the substrate topography on the final dewetted morphology, in the form of aligned dewetted droplets [197–203]. For example, Yoon et al. have reported the influence of pattern geometry on the final morphology of the dewetted polymer films using both mesa and indent patterned substrates, where it was observed that dewetting initiated preferentially at the edges of individual pre-patterned mesas and subsequent morphological evolution lead to spherical cap domains located in the center of the mesas [199]. The domains were found to be much smaller than the individual mesas as a consequence of the significant pattern reduction. Arrays of 70 nm PS nanosphere caps were obtained from arrays of 200 nm square pre-patterned mesas. This method was also extended to other polymers such as poly(4-vinyl pyridine) (P4VP) containing Rhodamine 6G (Rh6G) dye on a pre-patterned PS substrate. The fluorescent polymer structures were found to be extremely stable [200]. Dewetting on topographically patterned substrates has been used by several groups to pattern functional photoluminescent materials as well as to fabricate organic polymeric thin-film transistors [204, 205].

The morphological evolution and pattern formation of polymer thin films with uniform thickness on topographically patterned substrates with different geometries has been investigated by Mukherjee et al. [206, 207]. In order to avoid thickness variations during direct spin coating, films of uniform thickness were initially spin coated on smooth surfaces and subsequently transferred onto topographically patterned substrates. It was shown that by adjusting the transfer conditions like, the rate and angle of lift-off, it is possible to control the initial morphology of the film with respect to the underlying substrate. For a faster pull rate with a horizontally held substrate, the overall morphology of the transferred films was nearly flat, with the films coming in contact only with the raised protrusions of the substrate pattern. This configuration has been termed as 'focal adhesion' or 'free hanging films' (Figure 11.10a) [206]. In contrast, a slow pull-off and vertical orientation of the capturing substrate results in the so-called 'conformal adhesion', where the film adheres closely to the contours of the substrate pattern (Figure 11.10c). With examples of dewetting on both striped as well as substrates comprising an array of square pillars, it was shown experimentally that the positioning of the dewetted droplets is strongly influenced by the nature of initial film adhesion. As shown in Figure 11.10, in the dewetting of a

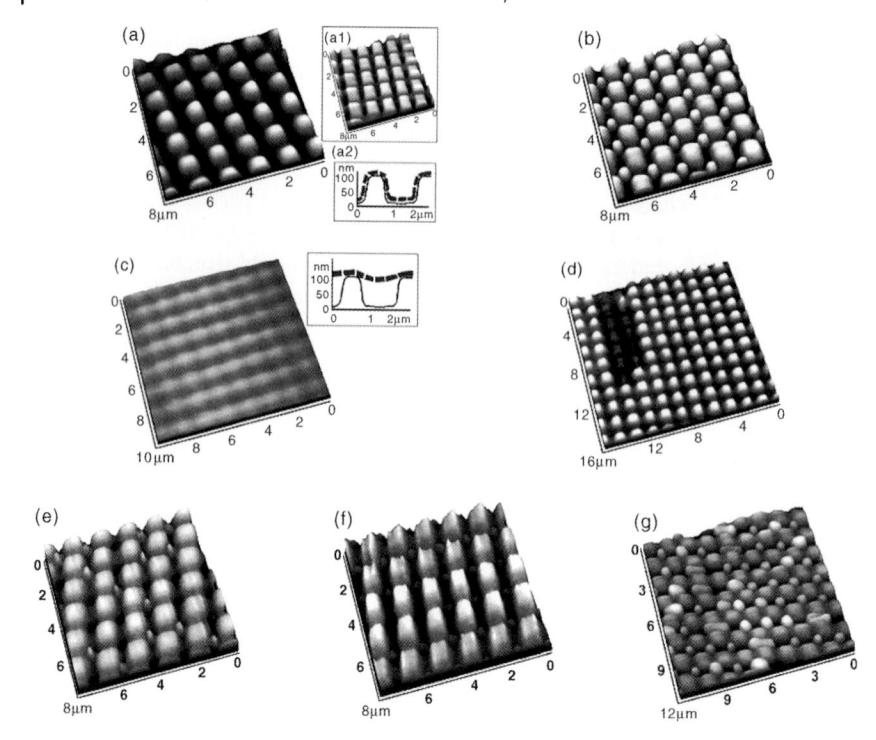

Figure 11.10 (a) A floated PS film conformally adhering to a topographically patterned substrate, comprising an array of square pillars (inset a1). Inset a2 shows the line profile of the substrate and the film. (b) A perfectly filled and ordered dewetted structure resulting from the dewetting of a 24 nm-thick film on the substrate shown in inset a1. (c) A 24 nm-thick floated PS film focally adhering to the same substrate as shown in a1. Inset shows the film (dashed line) in contact with the raised protrusions of the substrate only and freely hanging over other intervening areas. (d) The dewetted droplets are positioned on top of each pillar roof. A portion with a slight defect was deliberately used by the authors to distinguish between the dewetted droplets and the underlying pattern below, which can be seen at the place where the dewetted droplets are not placed on the pillar tops. (e–g) Variety of final dewetted morphologies as a function of initial film thickness for conformally adhering films: (e) Imperfect ordering with some missing droplets for an 11 nm film, (f) perfectly ordered but underfilled structure formation for a 17 nm-thick film, and (g) distorted, overfilled structure with occasional oversized or interconnected droplets for a 31 nm-thick PS film. (Reproduced from [207]. Copyright © (2008) The Royal Society of Chemistry.)

24 nm-thick PS film on a patterned substrate, the final dewetted morphology comprises an array of droplets, the positioning of which depends entirely on the initial conformation of the films (Figure 11.10a and c) with respect to the substrate. Figure 11.10b shows that the dewetted droplets accumulate at the intersections of the channels around the pillars for conformally adhering films [205]. In contrast, the droplets are seen forming on top of the pillars for films with same initial thickness but adhering focally on to the patterned substrate (Figure 11.10d) [207]. Even for conformally adhering films it was shown that perfectly filled and perfectly

ordered structures form only for a narrow range of film thickness for a specific geometry of a 2D patterned substrate (Figure 11.10d). For thicker (31 nm, Figure 11.10g) or much thinner (~11 nm, Figure 11.10e) films, either some of the droplets remain connected or some locations remain vacant, respectively, with both the states exhibiting deviation from the perfectly filled and ordered configuration obtained by dewelting of a 24 nm-thick film on a substrate with identical patterns (Figure 11.10b). Figure 11.10f shows another distinct morphology resulting from the dewetting of a 18 nm-thick film on the same substrate, which is best termed as a perfectly ordered but underfilled structure, as the droplet sizes are smaller than the largest possible size that can be accommodated at each channel intersection. Thus, there exists a critical thickness range for which perfect ordering is possible on a topographically patterned substrate with 2D structures. For substrates with 1D stripes, the droplet periodicity as well as their size changes with varying film thickness [207]. While the size of the droplets was seen to be strongly influenced by the width of the stripes, the droplet periodicity (λ_D) gradually decreased as the stripes became narrower. With gradual increase in film thickness, partially disordered structures in the form of droplets connected across stripes started to appear. The critical film thickness around which disordered structures started to appear was found to be a function of the stripe periodicity (λ_S) and channel width [207].

Even on substrates with identical geometry, such as those used for the experiments in Figure 11.10, a distinctly different dewetting pathway is observed for films thicker than ~40 nm (Figure 11.11). For these films, the onset of instability is through the formation of larger holes, which are completely uncorrelated to the substrate pattern (Figure 11.11a). The morphology in this stage is in many ways similar to that observed during dewetting of films on flat surfaces. However, as shown in Figure 11.11b, the periphery and the rims around the holes were aligned to some extent with the substrate patterns, resulting in the formation of noncircular, rectangular holes. The AFM scan of the dewetted portion inside a hole reveals tiny polymer droplets at each channel intersection on the patterned substrate (Figure 11.11c), left behind by the retracting contact line of a growing hole. The final morphology in Figure 11.11e shows the existence of polymer patterns at two distinct length scales, the larger polymer droplets resting on the patterned substrate due to bulk motion of the contact line of the growing holes and the finer perfectly filled and ordered structures (inset Figure 11.11e2) correlated with the substrate patterns, caused by the detachment of polymer drops from the retracting contact line [207].

11.4.3
Pattern Directed Dewetting: Theory and Simulation

It is evident from the preceding discussion that, like any other highly nonlinear process, pattern-directed dewetting is also influenced strongly by the initial conditions. Based on experimental observations, we have already seen that a slight change in initial conditions leads to a drastic change in the pattern morphology (Figure 11.10) or dewetting pathway (Figure 11.11) [207]. Such transitions are often extremely difficult to track experimentally, as will become clear from the following example.

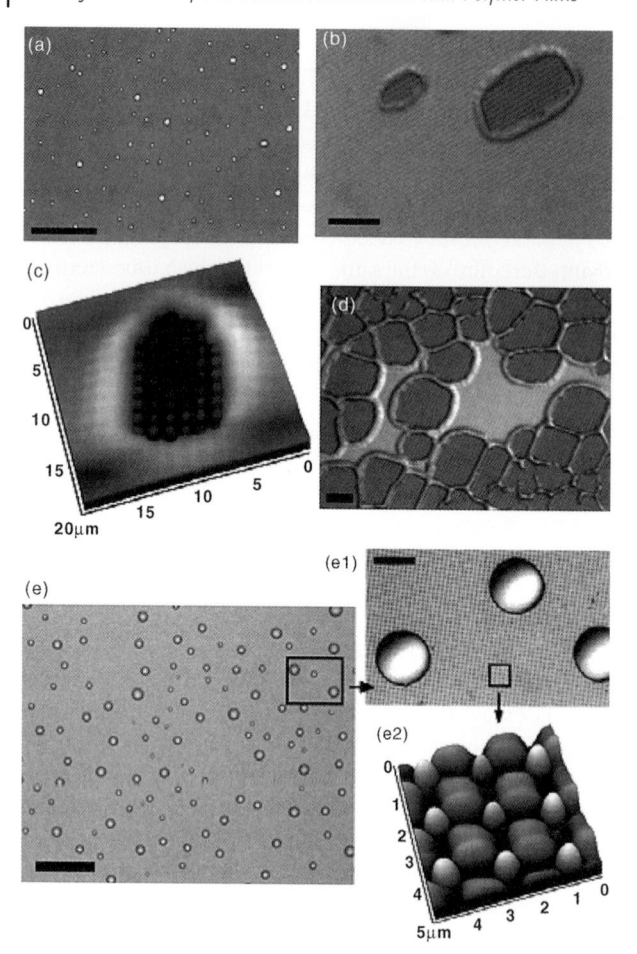

Figure 11.11 Dewetting pathway of a ~69 nm-thick PS film on a patterned surface. The initial stage of instability is manifested in the form of nucleation and growth of random, large holes not correlated to the substrate patterns.
(a) A large area optical micrograph showing the formation of random holes (scale bar 150 μm), (b) high magnification optical micrograph showing the exposed substrate patterns within a hole (scale bar 15 μm), and (c) an AFM image of a hole, revealing the presence of small polymer droplets at each interstitial position on the substrate within a hole. For images (a)–(c), exposure time is 4 minutes. (d) Formation of polymer ribbons after solvent vapor exposure for 11 minutes (scale bar 15 μm).
(e) Large area optical micrograph showing large polymer droplets, after solvent vapor exposure for 35 minutes (scale bar 100 μm). The inset e1 is a higher magnification optical micrograph, showing that the large polymer droplets are resting on a cross patterned substrate (scale bar 15 mm) and the AFM scan in inset e2 shows the presence of polymer droplets at the interstitial positions on the substrate patterns, forming a perfectly ordered and filled pattern. (Reproduced from [207]. Copyright © (2008) The Royal Society of Chemistry.)

Figure 11.10b and d–g and (11.11) shows the final dewetted structures of PS films having identical molecular weight but with thicknesses varying between 11 and 69 nm, respectively, on identical substrates. It is evident that not only the final morphology but also the dewetting pathway can be drastically different. As all experiments, however well-performed they may be, are associated with some degree of error, it is virtually impossible to determine the precise critical thickness at which the transformation in the dewetting pathway takes place! It is in situations like this where simulations are greatly useful, and thus form an important aspect of research on thin film instability on patterned substrates. Heterogeneous dewetting on patterned substrates, where the spatial variation of heterogeneity is well defined, is theoretically represented by Equation 11.5, which is obtained by slightly modifying Equation 11.1 with the incorporation of a parameter a, representing the roughness of the substrate ($z = a\,f(x,y)$). The local thickness of a film η is defined as $\eta = h(x, y, t) - a\,f(x, y)$ and the term h in Equation 11.1 is replaced by η [91, 208–217].

$$3\mu \frac{\partial \eta}{\partial t} + \frac{\partial}{\partial x}\left[\gamma_f (h-a)^3 \frac{\partial^3 \eta}{\partial x^3}\right] + \frac{\partial}{\partial x}\left[-(h-a)^3 \frac{\partial \phi}{\partial \eta}\frac{\partial \eta}{\partial x}\right] = 0 \tag{11.5}$$

On a chemically heterogeneous substrate, the interaction potential differs at different locations ($\Phi = \Phi(x,y)$) and the wettability gradient engenders the movement of fluid from less-wettable areas to the more-wettable ones, causing deformations of the free surface [206]. For topographically as well as chemically patterned substrates it has already been shown that there is a critical initial thickness range of the dewetting film (which to a large extent depends on the geometry and dimension of the substrate patterns) for which perfect ordering can be achieved. In addition, there can be initial thickness-dependent transitions between disordered to ordered states as well as morphological transitions between various ordered states. Although successfully captured experimentally with some degree of success by the groups of Karim [181] and Sharma [207] for some specific cases, such transitions are extremely difficult to capture experimentally. Kargupta and Sharma with their 3D nonlinear simulations have contributed significantly in this research topic. They have successfully predicted the morphology as well as dewetting pathways for films of various thicknesses on various types of chemically, topographically and physico-chemically patterned substrates, providing an understanding of the conditions under which the substrate patterns are faithfully reproduced into the film ('perfect templating'). In perfect templating, the final film morphology perfectly replicates the substrate surface energy map or the topographic pattern [206–214]. For example, the conditions for the evolution sequence of pattern replication on a substrate comprising alternating less and more-wettable stripes are as follows: (i) the periodicity of substrate pattern (λ_S) must be greater than the characteristic length scale of instability (λ_D), corresponding to the film thickness, but must be less than an upper cut-off scale ($\sim 2\lambda_D$); (ii) dewetting is initiated near the stripe boundary; (iii) the contact line eventually rests close to the stripe boundary; and (iv) the liquid cylinders that form on the more-wettable stripes remain stable [208, 209]. The relative combinations of film thickness (h_0), stripe width (l_P), and stripe periodicities (λ_S) are therefore crucial in determining the morphology of the final pattern. Using simulations, a variety of parametric studies are easily possible, predicting most of the final morphologies with near perfection. Many of the predictions of simulations have subsequently been validated by experiments. For example, the influence of variation of stripe width on the dewetting pathway and morphology with constant substrate pattern periodicity (λ_S) is shown in Figure 11.12 [208]. For small l_P ($\sim 0.4\lambda_S$, Figure 11.12a), rupture is initiated in the center of the less-wettable stripes by the formation of depressions that coalesce to form rectangular dewetted domains which are wider than the stripe width. In contrast, for a larger stripe width ($l_P \sim 0.8\lambda_S$), dewetting is initiated by formation of holes at the boundaries of the stripes (image 2 of Figure 11.12b) [208]. Coalescence of the two sets of holes leads to the dewetted regions comprising residual droplets (image 4 of Figure 11.12b). Upon further increase of the stripe width, the final pattern morphology changes to two rows of holes on each stripe separated by an elevated liquid cylinder (image 2 of Figure 11.12c). Interestingly, for this case at an intermediate stage of evolution, the number of cylindrical liquid ridges becomes twice the number of the more-wettable stripes on the substrate (image 3 of Figure 11.12c). The ordering is entirely lost with further increase in stripe width due to the formation and repeated coalescence of several rows of holes across stripes, eventually resulting in an array of isotropically arranged droplets

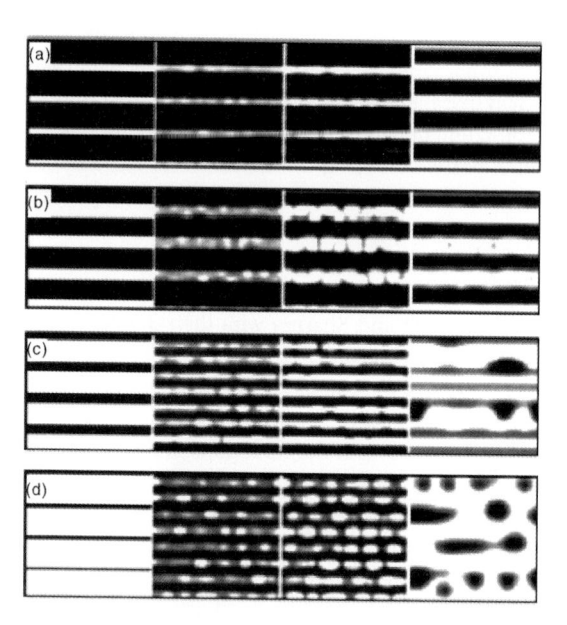

Figure 11.12 Morphological evolution in a 5 nm-thick film on a striped surface ($l_p = 3\,\mu m = 2\lambda$). $W = 0.6, 1.2, 2.1$, and $2.7\,\mu m$, respectively, for (a)–(d). The first image in this figure, as well as in the subsequent figures, represents the substrate surface energy pattern; black and white represent the more-wettable part and the less-wettable part, respectively. For other images describing the film morphology, a continuous linear gray scale between the minimum and the maximum thickness in each picture has been used. (Reproduced from [208]. Copyright © (2001) American Physical Society.)

(Figure 11.12d) [208]. The results suggest that the final morphology on a patterned substrate during dewetting of a thin polymer film can be greatly modulated by the competition between the different time scales (spinodal vs. heterogeneous dewetting) and length scales (spinodal, stripe width and periodicity, and film thickness) of the system.

Several other interesting case studies such as the dewetting on 2D patterned substrates with varying periodicity, patch dimension, topographic contrast, and dewetting on substrates with more complex patterns have been reported in great detail by the Sharma group. The morphology phase diagrams based on simulations, which correlates the nature of the final patterns to the initial conditions (film thickness, pattern dimensions, periodicity, etc.) for a specific pattern geometry are of significant importance. For example, the diagrams presented in Figure 11.13a and b summarize the major morphological transitions of the dewetted structures on a substrate comprising chemical patterns of more-wettable squares on a less-wettable substrate, and less-wettable squares on completely-wettable substrates, respectively, with respect to two normalized parameters: the dimensionless periodicity (l_P/λ_S) and the dimensionless width of the squares (W/λ_S). The dark shaded region in both the figures represents the parameter range for ideal templating. An interesting observation in Figure 11.13a includes the absence of heterogeneous rupture of the films (true dewetting) on substrates where $l_P < 0.7\lambda_S$, with only partial thickness deformations following the geometry of the substrate pattern. Also, for less-wettable area fractions greater than 0.5 or for square widths below a transition length (boundary 2, Figure 11.13a), liquid bridging across the less-wettable regions occurs for l_P less than $1.5\lambda_S$. An increase in l_P beyond boundary 3 eliminates the bridge formation at smaller W and produces trapped liquid domains on the less-wettable part, resulting in the

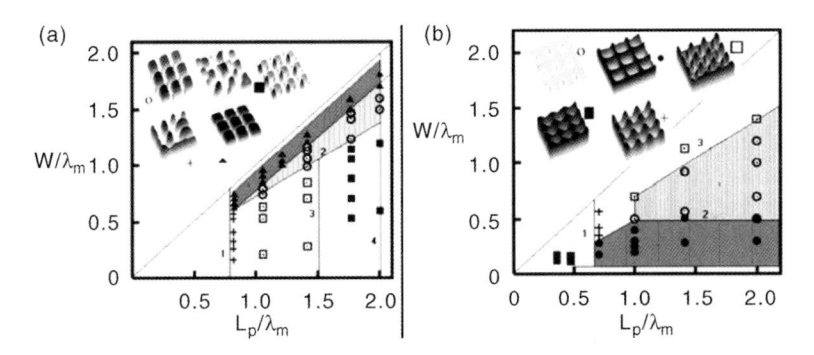

Figure 11.13 Morphological phase diagram for the isotropic substrate pattern consisting of (a) more-wettable blocks on a less-wettable substrate, lines 1, 2, 3, and 4 denote the boundaries between different regimes at the initial stage of dewetting. The dark-shaded region corresponds to the conditions for ideal templating; (b) less-wettable blocks on completely-wettable substrate. The lines 1, 2, and 3 denote the boundaries between different regimes at the initial stage of dewetting. The shaded region corresponds to the conditions for ideal templating. The 3D microstructures corresponding to the five symbols are also shown in respective insets. (Reproduced with permission from [210]. Copyright © (2003) American Chemical Society.)

'block mountain-rift valley' pattern. In contrast, on a completely-wettable substrate consisting of non-wettable squares, an increase in periodicity ensures ideal templating as long as the square width remains below a transition width (W_t, boundary 2, Figure 11.13b) [210]. For a pattern periodicity larger than λ, increasing the square width produces arrays of a 'castle moat' pattern (in parameter regime shown in a lighter shade), in which a liquid drop is trapped within a square-shaped region on each non-wettable area, while the liquid surface remains flat on the remaining completely-wettable areas. Many of the predicted morphologies on 1D and 2D patterned substrates by the Sharma group have been subsequently verified experimentally by other researchers. The availability of robust simulated phase diagrams is extremely useful in substantially reducing experimental efforts as they can be used to select the appropriate initial conditions (i.e., film thickness, pattern size and geometry etc.) to experimentally produce the desired final structures [210].

Pattern formation in films of block copolymers and blends are not covered in this chapter. A variety of interesting phenomena and mesoscale structures form in films of both types of materials on chemically and topographically patterned substrates, caused by the competition between the additional length scales of these systems. Several interesting papers in this field are available, some of which are cited here [194, 218–229]. For ultrathin polymer blend films, the morphological evolution reveals a coupling between the phase separation process and the surface deformation modes. The excitation of the surface modes occurs when the length scales of phase separation and the deformation modes are commensurate [219]. On the other hand, films of diblock copolymers can self-assemble into ordered periodic structures on the molecular scale [222]. However, in most cases perfect periodic domain ordering is achieved only over micron-sized regions separated by grain boundaries [223]. On a patterned substrate, it is possible to induce epitaxial self-assembly of the block copolymer domains, resulting in defect-free oriented patterns extending over large areas [224]. These structures are determined by the size and quality of the surface pattern rather than by the inherent limitations of the molecular level self-assembly process [224].

11.5
Instability due to Externally Imposed Fields

The discussion thus far has revealed how a thin polymer film becomes unstable and morphologically evolves as a result of the amplification of surface fluctuations engendered by a variety of causes, such as attractive van der Waals interactions [29, 30, 59–91], release of residual stresses [106–111], defects, and heterogeneities. In this section, we will discuss how a thin film can be destabilized by imposing an external field across the film. Examples of such external fields are electric fields [230–264], thermal gradients [265–270], or even a magnetic field [271]. Unlike spontaneous instabilities which occur only in very thin films (much below $\sim$100 nm), external field-mediated instabilities are possible all the way up to macroscopic length scales, because of the strength and long-range nature of electrostatic interactions, as compared to dispersion forces. They can therefore be used to destabilize thicker films by overcoming the stabilizing influence of the Laplace pressure (and possibly gravity for macroscopic liquid layers) and allow the control of structures on length scales which are difficult to manipulate in any other way. The influence of an electrical field on the surface of a viscous film was first investigated by Swan [230] as early as 1897 and was later proposed as an alternative to xerography [231–236]. However, the concept of using electro-hydrodynamic instabilities to pattern polymer thin films was demonstrated for the first time in 2000 by Schäffer et al. [237]; the technique has subsequently been christened as electro hydrodynamic (EHD) lithography. EHD surface instabilities are also important to a variety of other settings, for example in transformer oils, where instabilities lower the critical voltage and result in dielectric break-down [238, 239]. The existence of ripplons, a type of surface wave in liquid helium [240] and frosting in thermoplastics that lead to surface roughening, are also attributed to EHD instabilities [230].

A typical experimental set-up for electric field-induced pattern formation comprises sandwiching a liquid polymer film between two electrodes. The film is coated onto one of the electrodes, maintaining an air gap with the second electrode, as shown in Figure 11.14 [241–264]. A voltage is applied between the two electrodes, resulting in an electric field across the sandwiched film (Figure 11.14a). Upon application of the field, the capillary fluctuations at the liquid–air interface amplify in the direction of the field. This initially leads to an instability that is undulatory in nature (Figure 11.14b). With time the undulations amplify until the wave maxima touch the upper plate, thereby leading to the formation of polymer structures spanning the two electrodes [237, 241–248, 252–263]. The destabilization is engendered by the lowering of the total electrostatic energy in the capacitor for a liquid conformation that spans the two electrodes, compared with a layered structure of an intact film, which involves the build-up of energetically unfavorable displacement charges at the dielectric interface [237]. The final pattern morphology depends on several factors such as the field strength, the initial film thickness, the air-gap width, the surface tension, the structure of the film surface and so on [237, 241–243]. For example, as shown in Figure 11.14c, a random collection of columns spanning the two electrodes appears after annealing a 93 nm-thick PS film at 170 °C with an applied field strength of 50 V for 18 hours [237, 241]. Under similar conditions, a 193 nm-thick film shows much denser packing of the columns. This in turn results in a much enhanced electrostatic repulsion between the columns leading to formation of pillars with perfect hexagonal order (Figure 11.14d) [237, 241]. The ordering occurs because of the repulsion of equally charged peaks and valleys of the undulations. In most experiments, the two electrodes are never perfectly parallel which results in small lateral variation in the electric field. Consequently a variety of morphologies (undulations, columns) is generally observed on the same sample, as shown in Figure 11.14e [241]. In contrast to the hexagonally ordered structures shown in Figure 11.13d, a nucleated instability leads to a second-order effect as shown in Figure 11.14f [237, 241]. A locally higher value of film thickness at the point of nucleation leads to a higher strength of the local electric field, causing depletion of the nearest neighbor undulations. Amplification of the undulations occur again at the sites occupied by the next set of

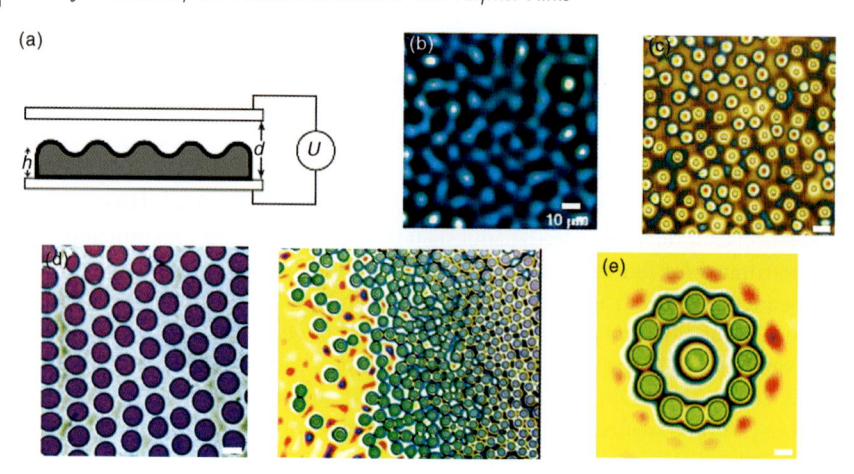

Figure 11.14 (a) Schematic representation of the capacitor device used to study capillary surface instabilities under an externally applied electric field. (b–f) Optical micrographs of polystyrene films exposed to an electric field; (b, c) film thickness 93 nm annealed for 18 h at 170 °C with an applied voltage U ~50 V. The gap spacing was larger in (c) than (b) (scale bar 10 μm); (d) thickness of PS film 193 nm, other conditions are same as (b), resulting into a denser packing of the polymer columns (scale bar 5 μm); (e) The figure shows a typical experimental scenario where the two plates are not perfectly parallel leading to a variation in pattern morphology. With decreasing plate spacing from left to right (color change from green to purple), the number density and extant of ordering of the columns is seen to increase. (Image width 145 μm); (f) a second-order effect is observed in a nucleated instability, in the form of a ring of 12 columns lies on a circle with a radius of 2λ. (scale bar 5 μm). The colors arise from the interference of light, and correspond to the local thickness of the polymer structures (for example, in f, yellow corresponds to a film thickness of ~200 nm, green ~450 nm). (a–d, f: Reproduced with permission from [237]. Copyright © (2000) Macmillan Publishers Ltd; e: Reproduced with permission from [243]. Copyright © (2001) Erik Schäffer.)

neighboring columns, resulting in a second-order ordering in the form of a rosette on a circle with a radius $r \sim 2\lambda_E$ and a circumference of $2\pi r < 12\lambda_E$, where λ_E is the characteristic wavelength of EHD instability. Beyond the circle of next-nearest neighbors, the instability decays with increasing distance [237]. In all the cases where EHD patterns are formed, the field strength is fairly high, on the order of $\sim 10^7 \, \text{Vm}^{-1}$. It has been reported by the Steiner group that no patterns form in the absence of an electric field, although pattern formation in the absence of electric fields had been claimed by the Chou group [272–277], which has been termed as lithographically assisted self-assembly (LISA).

In EHD instabilities, the fill factor, f (the ratio of the film thickness to the electrode spacing) is a parameter that significantly influences the pattern morphology, particularly during the later stages of pattern formation [237, 241, 251]. The initial phase of the film instability is qualitatively similar irrespective of the value of f. For low values of f (i.e., thin films compared to the electrode spacing) the wave maxima disconnect, giving rise to hexagonally-ordered columns with intercolumn distances that are commensurate with the wavelength of the initial instability. For higher values of f

(~0.5), the film stays partially interconnected when the wave maxima make contact with the top electrode. In this case, the late-stage evolution of the patterns is governed by a ripening process that leads to coalescence of some of the columns. This effect is even more pronounced for very high values of f, where the extensive coalescence of the initial wave pattern leads to a continuous polymer film with included voids leading to an inversion of the initial column structure. In a recent work, Heier et al. have looked at the response of a thin liquid polymer to a spatially modulated, heterogeneous electric field [248]. Salac et al. have studied the effect of an in-plane electric field on a polymer thin film and observed the disintegration of the film into polymer islands. The resulting structure morphologically resembles a dewetted film on a flat substrate and exhibits a narrow size distribution and spatial ordering [253].

While the topographical evolution of a polymer film occurs spontaneously with the application of the field, further control of the lateral structure can be achieved by laterally varying the electrode spacing d or effective field strength by using a topographically patterned top electrode. As the gap between the electrode and the film is smaller below every raised feature of the top electrode, the onset time for the formation of the instability structures is much shorter than for the recessed areas of the patterned top electrode [237]. The emerging instability in the film is therefore oriented towards the electrode protrusions and the final structure is a positive replica of the patterned electrode, as opposed to a negative replica achieved by conventional imprinting techniques (Figure 11.15a) [237]. The EHD patterning technique has also been extended to various other materials beyond simple homopolymers. For example, Dickey and coworkers have used EHD instability to pattern thin films of various photocurable materials, such as epoxy, vinyl ether, acrylate, thiolene, and so on demonstrating the material-independent nature of this technique [258]. Low molecular weight, low viscosity polymers, the EHD structures form at room temperature and are subsequently frozen by *in situ* photocuring. Lower viscosities also result in faster pattern formation dynamics (~seconds) [258]. Voicu et al. have patterned films of a metal oxide (titania) precursor stabilized in alcohol by EHD, which subsequently underwent a sol–gel condensation reaction to solidify, yielding meso-patterned titania films after high temperature annealing [247]. This is a rare combination where a 'soft' pattering technique has been successfully combined with sol–gel chemistry to form 'hard' metal oxide structures. Figure 11.15b shows a variety of patterns formed in titania precursor solution films using a patterned electrode [247]. However, since the films are in the liquid state during the pattern formation, the viscosity, surface tension, dielectric constant and overall volume change continuously during EHD instability because of solvent evaporation. This in turn influences the final pattern morphology. In addition, as the samples are pyrolyzed at high temperatures for the formation of structured anatase, the height of the formed structures shrank by a significant percentage (~85% in this case). The shrinkage is found not to significantly affect the lateral dimensions of the patterns. Using similar chemistries this method can be extended to pattern ferroelectric, piezoelectric, and ferromagnetic materials with high accuracy [247].

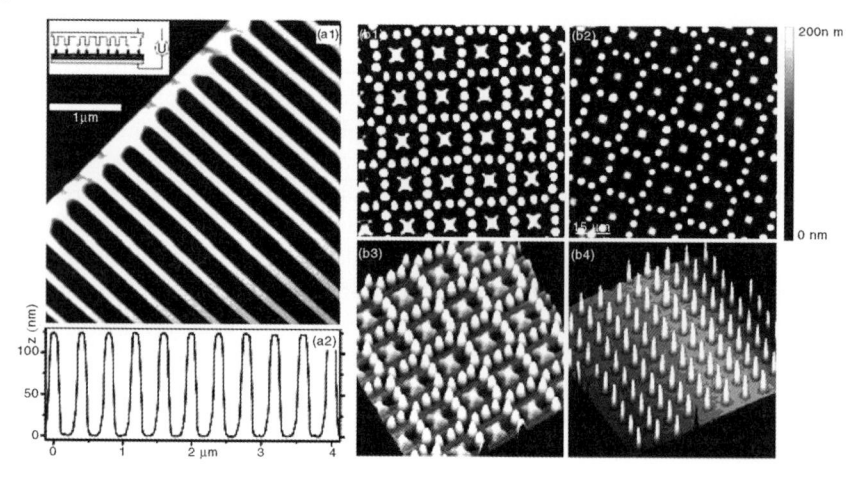

Figure 11.15 (a) Electrohydrodynamic lithography structures resulting from the use of a patterned top electrode (experimental arrangement shown in inset to a). The AFM image (a1; tapping mode at 360 kHz) shows 140-nm wide stripes (full-width at half-maximum) which replicate the silicon master electrode (200-nm stripes separated by 200-nm-wide and 170-nm-deep grooves). The cross-section (a2) reveals a step height of 125 nm; (b) Replication of patterns on a TiO$_2$ precursor solution film with patterned stamp: (b1), (b2) Conical pattern arranged in a square lattice. In addition to the replicated cones, a pattern in the center of the squares is observed. This pattern stems from the electric field-aided dewetting of square shaped patches of precursor film that remained once the cones have formed. Both the conical shape of the replicated pattern and the central shapes are a consequence of the solidification of the precursor material before the liquid can attain its equilibrium shape; (b3) Three dimensional representation of b2; (b4) Replication of a simple square pattern, showing cones with a base width of 2.5 μm and a height of 37.6 nm. (a: Reproduced with permission from [237]. Copyright © (2000) Macmillan Publishers Ltd; b: Reproduced from [247]. Copyright © (2007) The Royal Society of Chemistry.)

11.5.1
Electric Field-Induced Patterning: Theory

Theoretically, the origin of the film instability is best understood by considering the balance of forces acting at the polymer film–air interface. As discussed before, the film surface tension γ minimizes the area of the polymer–air interface and stabilizes the film. The electric field on the other hand polarizes the polymer resulting in an effective surface charge density. This result in an electrostatic pressure at the liquid–air interface. An expression for p_{EL} is obtained by minimization of the energy, F_{EL}, stored in the capacitor with a constant applied voltage, U, as [29]:

$$F_{EL} = \frac{1}{2}CU^2 \tag{11.6}$$

The total capacitance, C, is given in terms of a series of two capacitances, the film and the air gap, with dielectric constants ε_P and ε_{air} (~1) respectively. For two capacitors in series ($1/C = 1/C_P + 1/C_{Air}$), $1/C$ is given as

$$1/C = [\varepsilon_P d - (\varepsilon_P - 1)l]/\varepsilon_0 \varepsilon_P A \tag{11.7}$$

where l is the thickness of the polymer layer, d is the distance between the two electrodes, and A is the capacitor area. The dielectric displacement, D, is constant across the layers and can be used to calculate the electric field in the polymer film as:

$$D = \varepsilon_0 \varepsilon_P E_P = \varepsilon_0 \varepsilon_{Air} E_{Air} \tag{11.8}$$

where, ε_0 is the dielectric vacuum permittivity and E_{Air} is the electric field in air. Considering that the applied voltage is a summation of the voltage differences across the two layers [$U = E_P l + E_{Air}(d-l)$], and substituting the expressions of E_P and E_{Air} we get,

$$E_P = U/(\varepsilon_P d - (\varepsilon_P - 1)l) \tag{11.9}$$

The electrostatic pressure (p_{EL}) that destabilizes the film is opposed by the stabilizing Laplace pressure originating from surface tension. The total electrostatic pressure in the film is given by:

$$p_{EL} = -\varepsilon_0 \varepsilon_P (\varepsilon_P - 1) E_P^2 \tag{11.10}$$

The dispersion relation for a system dominated by electrostatic forces is given by [29, 243]:

$$1/\tau = -(l_0^3/3\eta)(\gamma q^4 + \partial_l p_{EL} q^2) \tag{11.11}$$

This is in contrast to the inviscid, gravity stabilized case with $\tau^{-1} \sim q$ [279]; the $\tau^{-1} \sim q^2$ variation is a signature of a typical dissipative system [280]. The necessary condition for the amplification of the fluctuations is $\tau > 0$. Since $\partial_l p_{EL} < 0$, all modes with $q = q_c = (-\partial_l p_{EL}/\gamma)^{1/2}$ are unstable and are spontaneously amplified. The wavelength of the fastest growing mode (λ), which eventually dominate the instability, is given as:

$$\lambda = 2\pi \sqrt{\frac{\gamma[\varepsilon_p d - (\varepsilon_p - 1)\ell]^3}{\varepsilon_0 \varepsilon_p (\varepsilon_p - 1)^2 U^2}}$$

$$= 2\pi \sqrt{\frac{\gamma U}{\varepsilon_0 \varepsilon_p (\varepsilon_p - 1)^2}} E_p^{-\frac{3}{2}} \tag{11.12}$$

However, the experimental results differ by a factor of 2 from the theoretical prediction, an effect which is not entirely understood.

Based on the above theoretical considerations, the dynamics of pattern formation in EHD instability has been modeled using LSA [236, 237, 241, 249, 250], and 3D nonlinear simulations [251, 252]. Linear stability analysis accounts for the forces acting on the film interface to determine the fastest mode of growth. The stability analysis of an interface between a dielectric and a conducting fluid by Herminghaus show that the electric fields can also arise naturally from the contact potentials of such an interface, giving rise to instabilities [236]. Pease and Russel have shown that for leaky dielectric liquids, finite conductivity leads to patterns of smaller wavelength and larger instability growth rates [249, 250]. Based on full 3D nonlinear simulations Verma et al. have simulated the electrostatic field-induced instability, morphology, and patterning of thin liquid films confined between two electrodes for both spatially homogenous and heterogenous fields [252]. In addition to the spinodal flow resulting from the lateral variation of the electric field arising from local film thickness changes, a laterally heterogenous field also causes the flow of liquid from the regions of low to high field strength, allowing precise control over the replicated pattern morphology [252]. For a patterned substrate, the simulations suggest that the electrode pattern is replicated into the film only when the pattern periodicity, L_P, exceeds the instability length scale calculated for the minimum inter-electrode separation distance. Thus, the formation of secondary structures (deviating from the template) can be suppressed by employing a patterned electrode with a high aspect ratio topography, giving rise to stronger field gradients. Simulation results by Verma et al. suggest that the number density of the electric field-induced structures can be controlled by tuning the mean film thickness, the fill ratio, periodicity and aspect ratio of the topography of the top electrode, and the applied voltage [252]. These observations have been substantiated experimentally by several groups [246].

11.5.2
Electric Field-Induced Patterning of Polymer Bilayers

When the electric field is applied across a polymer bilayer sandwiched between two electrodes (with or without an air gap, Figure 11.16a), it destabilizes both the air–polymer and the polymer–polymer interfaces [254–260], thereby resulting in the formation of more complex instability structures [257–260]. An electric field applied normal to an interface between two dielectric materials with different polarizabilities leads to an interfacial electrostatic pressure, which arises from the uncompensated displacement charges [260]. It is seen that there is a substantial amplification of the thermal fluctuations at an interface between two liquid films, owing to the reduced interfacial tension [254–256]. This leads to a significant reduction in feature size and periodicity, in addition to a much faster growth of instability compared to EHD instability in a single film [254, 255]. The field strength has a very strong influence on the evolution sequence in a polymer bilayer, as there are different hydrodynamic regimes based on low and high field strengths [255]. In a bilayer comprising a PS film coated on a PMMA thin film, heterogeneous nucleation can lead to the formation of holes in the films before the onset of electrostatically driven instability when the field strength is low. In this case, the dewetting kinetics is only marginally influenced by the applied electric field [254, 255]. Application of a stronger field leads to the build-up of electrostatically amplified surface waves which pre-empts heterogeneous nucleation and results in the formation of columns that span from the bottom layer right up to the top electrode [254–256]. The two regimes

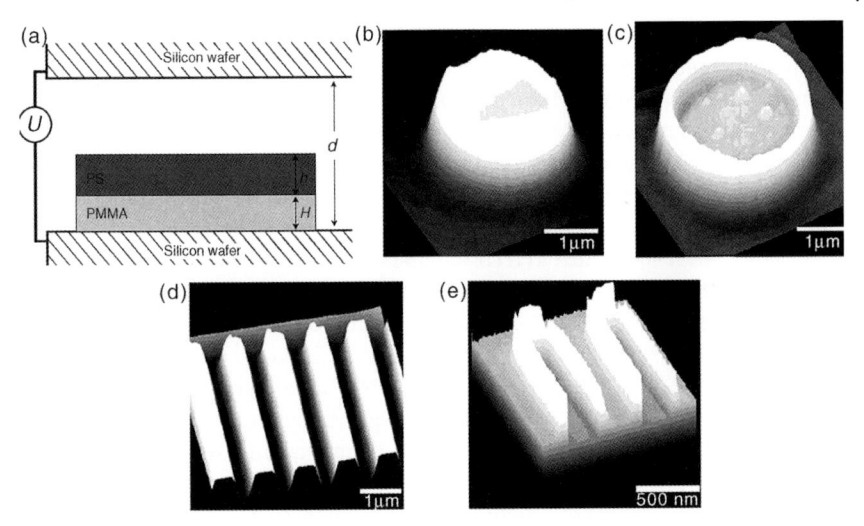

Figure 11.16 (a) Schematic representation of the capacitor device used to study capillary surface instabilities in a polymer bilayer comprising a PS top layer and a PMMA bottom layer with an externally applied electric field; (b, c) AFM images of a single column before and after removing the PS phase by washing the sample in cyclohexane, respectively. The PMMA phase in (b) forms a mantle around the PS column with a height of 170 nm and a width of ~200 nm; (d, e) EHD structures formed by using a stripe patterned electrode. Figure (d) shows the replicated stripes and (e) the secondary PMMA structures after washing the sample in Cyclohexane, removing PS preferentially. Feature height ~160 nm, width ~100 nm. (Reproduced with permission from [257]. Copyright © (2002) Macmillan Publishers Ltd.)

are drastically different in terms of hydrodynamic time scales [256]. While a slow-paced motion of the contact line during dewetting leads to only a small deformation of the lower liquid layer, a faster collective motion driven by the EHD instability in a stronger field leads to high shear stresses and therefore a more significant deformation of the PMMA contact line [256].

Electrohydrodynamic instabilities in a polymer bilayer have also been used to engineer hierarchical patterns exhibiting two independent and distinct characteristic length scales [257]. This was shown by Morariu et al. for a PS–PMMA bilayer. In their system, the time constant for destabilization of the air surface was smaller compared to the polymer–polymer interface [257]. This led to the destabilization of the top PS film on the PMMA underlayer [257]. The resulting lateral redistribution of PS induced a secondary instability in the PMMA underlayer, thereby creating a hierarchy of length scales of the formed structures. The distinct rim around the columns in Figure 11.16b is morphologically different from the columns formed in single layer films under similar conditions [237, 257]. The rims are the signature of a composite core–annular structure consisting of a cylindrical PS core surrounded by a PMMA shell, which is revealed when PS is selectively removed (Figure 11.16c) [257]. The composite patterns/hollow structures thus created can further be aligned using a patterned electrode, as shown in Figures 11.16d and 11.16e. While the overall EHD morphology induced by a stripe patterned stamp was indeed the positive replica of the electrode master (Figure 11.16d), each of these replicated stripes consisted of a PS core surrounded by a thin PMMA wall (Figure 11.16e) [257]. The height and width of the structures and their aspect ratio can be controlled by adjusting the electrode spacing, the lateral density of topographic features on the patterned electrode, and the initial film thickness of the layers [257].

It has also been shown that EHD instabilities in conjugation with dewetting at the polymer–polymer interface result in a variety of complex 3D structures [258–260]. The dielectric constant of the polymers in the two layers as well as their relative ordering are important parameters that strongly influence the morphology of the resulting pattern. In a PS–PMMA bilayer, PMMA has a higher dielectric constant compared to PS [257, 259]. The formation of cage type structures (Figure 11.17) was observed in a bilayer of PS on top of PMMA in a combination of dewetting and EHD. The top PS film initially dewetted on the PMMA surface before the electric field was switched on [259]. The PMMA meniscus showed a fingering instability around each PS column which originated from the transformation of the dewetted PS droplets into columns. This fingering instability is caused by the flow of PMMA along the outside PS columns driven by the electric field, eventually forming narrow strands surrounding each PS column. The PMMA, upon reaching the top electrode, which was chromium-coated, preferentially wetted the chromium surface displacing the PS layer thereby, detaching the PS columns from the top electrode and confining PS within a 'top-covered' PMMA cage. This can be clearly seen in Figure 11.17b where the PS has been selectively removed, revealing the cage-like PMMA structure. In contrast when the top electrode was coated with a fluoropolymer, the PMMA did not spread on the top electrode, the PS columns remained connected to the top electrode, with an open cage structures (Figure 11.17c) for an identical bilayer. In contrast, when PMMA was on top of PS, film the resulting morphology resembled closed-cell structures, confirming the importance of the layer sequence [259, 260]. While the size and spacing of the resulting structures are controlled by the electrostatic forces, it was

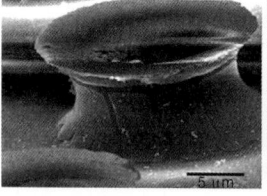

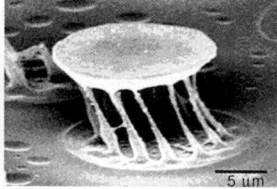

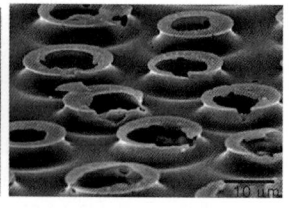

Figure 11.17 (a) SEM image of a single EHD column. The dark recesses along the height of the pillar are due to PMMA that has been degraded because of e-beam exposure in the SEM; (b) a single 'cage', after the sample was washed in cyclohexane to preferentially remove PS; (c) structures obtained with hydrophobic chromium layer coated on the top electrode after washing with cyclohexane. The structures are hollow and do not have ceiling. (Reproduced with permission from [258]. Copyright © (2006) American Chemical Society.)

the dewetting kinetics that dictated the order in which the two interfaces became unstable [259]. Further, a closed-cell structure formation is favored when a low molecular weight PS bottom layer is used [259].

11.5.3
Thermal Gradient-Induced Patterning

A polymer film can also be destabilized by an external thermal gradient [265–270], in a fashion similar to the application of an electric field [230–264]. This approach has been successfully used by several groups for patterning thin polymer films and to make meso- and nanoscale structures. The experimental set-up is similar to that of the electric field-mediated patterning, as the film is sandwiched between the substrate and a top plate, with an air gap separating them. Typically the two plates are maintained at two different temperatures [265–269]. The uniform temperature gradient across the air–polymer interface directs the interfacial instability in the direction of the temperature gradient, and consequently a variety of patterns, including columns and stripes, form, which span between the two plates. As shown in Figure 11.18, two distinct morphologies, columnar (Figure 11.18a) and stripes (Figure 11.18b) appear in experiments. The precise morphology depends on initial conditions such as the film thickness, the gap distance between the two plates, the magnitude of the thermal gradient, as well as the temperatures of the two plates. When the separation distance between the two plates (d) is large compared to the film thickness (h) the columnar morphology is predominant. Lower values of d/h lead to the formation of striped patterns [266]. Both the morphologies exhibit a single dominant length scale (λ) which depends inversely on the initial heat flux between the plates [266]. The formation of various other morphologies such as spirals patterns, or co-existence of several different

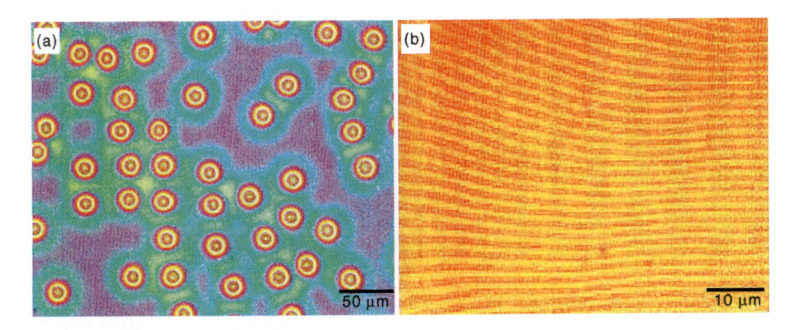

Figure 11.18 Optical micrographs of structures that have formed in polystyrene films sandwiched between two plates at different temperature. (a) Columnar patterns formed for film thickness ~309 nm, gap between plates ~755 nm and $\Delta T = 28\,°C$; and (b) stripe patterns formed for film thickness ~130 nm, gap between plates ~250 nm and $\Delta T = 19\,°C$. The colors stem from the constructive interference of the white microscope illumination and are an indication of the local film thickness. (Reproduced with permission from [266]. Copyright © (2003) American Chemical Society.)

morphologies on the same sample, have also been reported [275, 276]. Some experiments showed that the morphology of the patterns is also influenced by the proximity of the edges [276].

It is important to note that the nature of the instability and pattern formation in a thin film subjected to an externally imposed field is fundamentally different from a spontaneous process such as dewetting, where the morphological transformation is the transition from an unstable or metastable thermodynamic state to a stable state, corresponding to a lower free energy configuration [65]. In presence of a temperature gradient the system is intrinsically out of equilibrium and the definition of an equivalent of the Gibbs' free energy is not usually possible [281]. In contrast to the energy minimization for spontaneous instabilities, the morphological evolution maximizes the heat flux. Schäffer *et al.* found that the acoustically propagating phonons are reflected at the polymer–air interface, thereby exerting a destabilizing radiation pressure that eventually leads to an instability in the direction of the

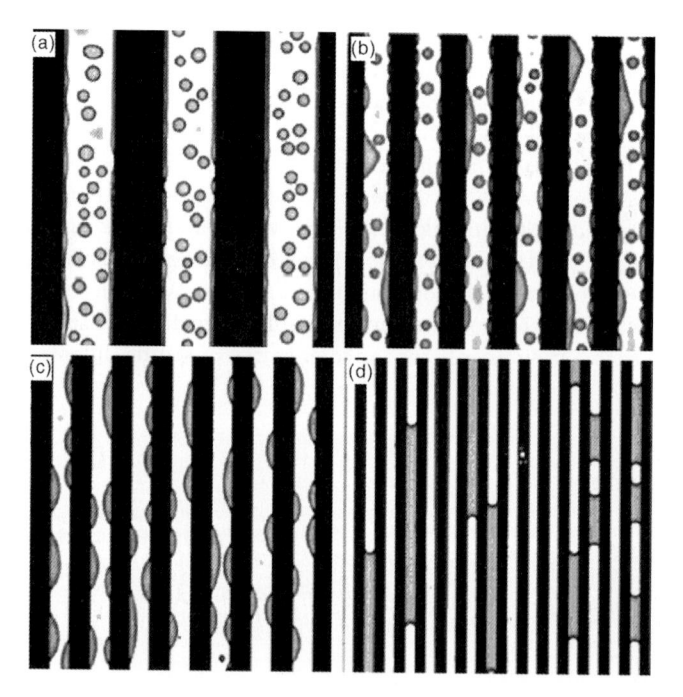

Figure 11.19 Temperature-gradient driven capillary instability in confined geometries. Topographic line-grid templates that were brought in contact with a liquid polymer film give rise to a quasi 1D confinement of width w of the free polymer surface. (a) $w = 12\,\mu m$; (b) $w = 6\,\mu m$: coexistence of free columns with circular cross-sections and wall-adsorbed columns with drop-shaped cross-sections; (c) $w = 4.5\,\mu m$: drop-shaped columns only; and (d) $w = 2\,\mu m$: plugs spanning the confining walls. The size of optical microscopy images are $70 \times 70\,\mu m^2$. The height of the polymer columns and plugs and is determined by the distance of the line grooves to the substrate (580–630 nm). (Reproduced from [281]. Copyright © (2005) The Royal Society of Chemistry.)

imposed field or gradient [265]. Nedelcu *et al.* have reported a pattern formation process by imposing a 1D confinement in the form of a 1D stripe patterned top plate under the influence of a temperature gradient. The presence of the patterns on the plate guides capillary instability along the stripe direction [281]. The final morphology of the patterns depend on the extent of confinement, which quantitatively is the ratio of the groove width of the confining stamp to the diameter of unconfined columns, that are formed in a laterally homogeneous temperature gradient with a flat top plate. Using a stripe-patterned stamp, two competing pattern formation mechanisms are coexistent: (i) amplification of capillary instabilities, leading to columns that span across the two plates; and (ii) thermocapillary driven flow that leads to the rise of the polymer meniscus along the confining walls. The final pattern morphology is a combination of both types of transport. Increasing the extent of lateral confinement leads to a transition from circular columns to wall-adsorbed columns with a drop-shaped cross section [281].

11.6 Conclusion

In this review we have summarized recent developments on unconventional surface patterning in thin, soft polymeric films, which are based on the morphological evolution arising from film instabilities. In this approach, pattern formation is engendered by a spontaneous or external field-mediated amplification of the capillary waves which are omnipresent on the free surface of a liquid film. This approach is thus distinctly different from the top-down replication based on standard soft lithography techniques. In this chapter we have primarily covered the instability-mediated pattern formation in homopolymer thin films in the liquid state (polymers heated beyond glass transition temperature or exposed to solvent vapor). We show that several distinct force fields such as the surface forces (in the form of surface and interfacial tensions of the substrate and polymer), intermolecular van der Waals forces, molecular level stresses, external force fields (electric field or a thermal gradient), and residual stresses in the film play a significant role in determining the evolution pathway, time and length scales, as well as the morphology of the final structures. Subsequent discussion on thin film instability has been classified into two parts. The first part covers the theoretical and experimental aspects of spontaneous dewetting in ultrathin films, and the second part deals with instability mediated by the application of an external field.

From the standpoint of patterning, both approaches are advantageous in terms of the flexible morphological control of the structures that can be created. For example, the feature size can be controlled by simply varying initial conditions, such as the film thickness, the magnitude of the applied external field, the gap between electrode and film, the fill ratio, and so on. This is in contrast to the necessity for a fresh stamp, master or a mask in any existing top-down technique whenever a change in feature size is required. In addition, it is also possible to control the morphology of the structure to a large extent by controlling the time evolution. For example, dewetting of a thin polymer film on a flat surface eventually leads to a random array of nearly equal-sized droplets. However if the dewetting experiment is allowed to proceed only up to an intermediate level, it is possible to obtain arrays of holes or ribbon like patterns, which appear as intermediate structures in the evolution sequence. This chapter cites many recent experimental and theoretical examples of how instability-mediated patterning enables flexibility and control over the resulting pattern morphology. Several key scientific issues are discussed, such as the role of surface tension during the various stages of dewetting, the formation and growth of rims, which are not only important in dictating the pattern morphology but are also of scientific interest, with many issues yet to be fully resolved.

Many instability-induced structures lack long-range order, and therefore find limited practical application. We have discussed several templating strategies which have been applied to overcome this limitation by using a chemically or topographically patterned substrate. The mere use of a template does not necessarily translate to perfect ordering, as the key issue is the commensuration between various length scales, mainly the intrinsic length scale of instability as well as the geometry and periodicity of the structures. With the help of several examples it is shown how a slight variation in the initial conditions can lead to a variety of evolution pathways and final pattern morphologies. This emphasizes the need for simulations that are able to accurately predict the final morphology.

The last topic of the chapter covers the external field-mediated instabilities and pattern formation. We have shown several examples of the suitability of EHD to become a novel and versatile patterning method. For example, it is a rare technique where it is possible to obtain structures with a reasonable degree of long-range order (hexagonally packed columns, for example) even when a flat, featureless top electrode is used. By using a polymer bilayer, several unique types of structure such as enclosed patterns, buried patterns and hierarchical structures have been successfully created by EHD, which are almost impossible to fabricate by existing lithography techniques.

The topics presented here are a subset of self-organization and pattern formation processes, which are also seen in various other settings such as evaporation, condensation, elastic instabilities in various materials including elastomers, block copolymers, blends, and liquid crystals. An interested reader is advised to consult some excellent review papers on each of the themes, some of which are cited in the reference section.

Acknowledgment

The authors would like to acknowledge the funding from a DST – UKIERI project.

References

1 Lloyd Carrol, R. and Gorman, C.B. (2002) *Angew. Chem. Int. Edn.*, **41**, 4378.
2 Kim, D.-H. and Rogers, J.A. (2008) *Adv. Mater.*, **20**, 4887.
3 Scott, B.J., Wirnsberger, G., and Stucky, G.D. (2001) *Chem. Mater.*, **13**, 3140.
4 Sato, O., Kubo, S., and Gu, Z.-Z. (2009) *Acc. Chem. Res.*, **42**, 1.

5. Geim, A.K., Dubonos, S.V., Grigorieva, I.V., Novoselov, K.S., Zhukov, A.A., and Shapoval, S.Y. (2003) *Nature Mater.*, **2**, 461.
6. Hassel, A.W., Milenkovic, S., Schurmann, U., Greve, H., Zaporojtchenko, V., Adelung, R., and Faupel, F. (2091) *Langmuir*, **23**, 2007.
7. Sun, M., Luo, C., Xu, L., Ji, H., Ouyang, Q., Yu, D., and Chen, Y. (2005) *Langmuir*, **21**, 8978.
8. Chin, V.I., Taupin, P., Sanga, S., Scheel, J., Gage, F.H., and Bhatia, S.N. (2004) *Biotech. Bioeng.*, **88**, 399.
9. Zhang, X., Jenekhe, S., and Peristein, J. (1996) *Chem. Mater.*, **8**, 1571.
10. Andress, R.P., Bein, T., Dorogi, M., Feng, S., Henderson, J.I., Kubiak, C.P., Mahoney, W., Osifchin, R.G., and Reifenberger, R. (1996) *Science*, **272**, 1323.
11. Drechsler, T., Chi, L.F., and Fuchs, H. (1998) *Scanning*, **20**, 297.
12. Madou, M.J. (2002) *Fundamentals of Microfabrication: The Science of Miniaturization*, 2nd edn, CRC Press, Boca Raton, FL.
13. Xia, Y. and Whitesides, G.M. (1998) *Angew. Chem. Int. Ed.*, **37**, 550.
14. Fischer, P.B. and Chou, S.Y. (2989) *Appl. Phys. Lett.*, **62**, 1993.
15. Tseng, A.A. (2005) *Small*, **1**, 924.
16. Ansari, K., van Kan, J.A., Bettiol, A.A., and Watt, F. (2004) *Appl. Phys. Lett.*, **85**, 476.
17. Piner, R.D., Zhu, J., Hong, S., and Mirkin, C.A. (1999) *Science*, **283**, 661.
18. Ozin, G.A. and Arsenault, A.C. (2005) *Nanochemistry*, RSC Publishing, London.
19. Whitesides, G.M., Mathias, J.P., and Seto, C.T. (1991) *Science*, **254**, 1312.
20. Whitesides, G.M. and Grzybowski, B. (2002) *Science*, **295**, 2418.
21. Innocenzi, P., Kidchob, T., Falcaro, P., and Takahashi, M. (2008) *Chem. Mater.*, **20**, 607.
22. del Campo, A. and Arzt, E. (2008) *Chem. Rev.*, **108**, 911.
23. Quake, S.R. and Scherer, A. (2000) *Science*, **290**, 1536.
24. Rayleigh, L. (1878) *Proc. London Math. Soc.*, **10**, 4.
25. Rayleigh, L. (1892) *Philos. Mag.*, **34**, 145.
26. Langbein, D. (2002) *Capillary Surfaces*, Springer-Verlag, Berlin.
27. Braslau, A. Pershan, P.S., Swislow, G., Ocko, B.M., Als-Nielsen, J. (1988) *Phys. Rev. A*, **38**, 2457.
28. Jiang, Z., Mukhopadhyay, M.K., Song, S., Narayanan, S., Lurio, L.B., Kim, H., and Sinha, S.K. (2008) *Phys. Rev. Lett.*, **101**, 246104.
29. Landau, L.D. and Lifshitz, E.M. (1960) *Electrodynamics of Continuous Media*, Pergamon, Oxford, pp. 368–376.
30. Dzyaloshinskii, I.E., Lifshitz, E.M., and Pitaevskii, L.P. (1961) *Adv. Phys.*, **10**, 165.
31. Van Oss, C.J., Chaudhury, M.K., and Good, R.J. (1988) *Chem. Rev.*, **88**, 927.
32. Casimir, H.B.G. (1948) *Proc. K. Ned. Akad. Wet.*, **B51**, 793.
33. Cahn, J.W. (1965) *J. Chem. Phys.*, **42**, 93.
34. Vrij, A. (1966) *Discuss. Faraday Soc.*, **42**, 23.
35. Vrij, A. and Overbeek, J. (1968) *J. Am. Chem. Soc.*, **90**, 3074.
36. Derjaguin, B.V. (1989) *Theory of Stability of Thin Films and Colloids*, Plenum, New York.
37. Jain, R.K. and Ruckenstein, E. (1976) *J. Coll. Interf. Sci.*, **54**, 108.
38. Sharma, A. and Ruckenstein, E. (1990) *J. Coll. Interf. Sci.*, **137**, 433.
39. Williams, M.B. and Davis, S.H. (1982) *J. Coll. Interf. Sci.*, **90**, 220.
40. Herminghaus, S., Jacobs, K., Mecke, K., Bischof, J., Fery, A., Ibn-Elhaj, M., and Schlagowski, S. (1998) *Science*, **282**, 916.
41. Pierre-Louis, O., Chame, A., and Saito, Y. (2009) *Phys. Rev. Lett.*, **103**, 195501.
42. Jain, R.K., Ivanov, I., Maldarelli, C., and Ruckenstein, E. (1979) Instability and Rupture of Thin Liquid Films, in *Dynamics and Instability of Fluid Interfaces* (ed T.S. Sorensen), Springer-Verlag, Berlin, pp. 140–167.
43. Poste, G. and Nicolson, G.L. (1978) *Membrane Fusion*, North-Holland, Amsterdam.
44. Sharma, A. and Ruckenstein, E. (1985) *J. Coll. Interf. Sci.*, **106**, 12.
45. de Gennes, P.G. (1985) *Rev. Mod. Phys.*, **57**, 827.
46. Israelachvili, J.N. (1995) *Intermolecular and surface forces*. 2nd edn Academic Press, New York.

47 Krausch, G. (1997) *J. Phys.: Cond. Matter,* **9**, 7741.

48 Müller-Buschbaum, P. (2003) *J. Phys.: Cond. Matter,* **15**, R1549.

49 Oron, A., Davis, S.H., and Bankoff, S.G. (1997) *Rev. Mod. Phys.,* **69**, 931.

50 Geoghegan, M. and Krausch, G. (2003) *Prog. Polym. Sci.,* **28**, 261.

51 Sharma, A. and Reiter, G. (2002) *Phase Trans.,* **75**, 377.

52 Jones, R.A.L. (1999) *Curr. Opin. Coll. Interf. Sci.,* **4**, 153.

53 Müller-Buschbaum, P., Bauer, E., Wunnicke, O., and Stamm, M. (2005) *J. Phys.: Cond. Matter,* **17**, S363.

54 Bucknall, D.G. (2004) *Prog. Mater. Sci.,* **49**, 713.

55 Craster, R.V. and Matar, O.K. (2009) *Rev. Mod. Phys.,* **81**, 1131.

56 Jacobs., K., Seeman, R., and Herminghaus, S. (2008) "Stability and dewetting of thin liquid films," in *Thin Liquid Films* (eds O.K.C. Tsui and T.P. Russel), World Scientific, ISBN: 978-9-812-81881-2.

57 Verma, R. and Sharma, A. (2007) *Ind. Eng. Chem. Res.,* **46**, 3108.

58 Brochard Wyart, F. and Daillant, J. (1084) *Can. J. Phys.,* **68**, 1990.

59 Reiter, G. (1992) *Phys. Rev. Lett.,* **68**, 75.

60 Reiter, G. (1993) *Langmuir,* **9**, 1344.

61 Reiter, G. (1994) *Macromolecules,* **27**, 3046.

62 Xie, R., Karim, A., Douglas, J.F., Han, C.C., and Weiss, R.A. (1998) *Phys. Rev. Lett.,* **81**, 1251.

63 Morariu, M.D., Schäffer, E., and Steiner, U. (2004) *Phys. Rev. Lett.,* **92**, 156102.

64 Sharma, A. and Khanna, R. (1998) *Phys. Rev. Lett.,* **81**, 3463.

65 Sharma, A. and Reiter, G. (1996) *J. Coll. Interf. Sci.,* **178**, 383.

66 Reiter, G., Khanna, R., and Sharma, A. (2000) *Phys. Rev. Lett.,* **85**, 1432.

67 Sharma, A. (1993) *Langmuir,* **9**, 861.

68 Reiter, G., Sharma, A., Casoli, A., David, M.O., Khanna, R., and Auroy, P. (1999) *Langmuir,* **15**, 2551.

69 Faldi, A., Composto, R.J., and Winey, K.I. (1998) *Langmuir,* **11**, 4855.

70 Müller-Buschbaum, P., Vanhoorne, P., Scheumann, V., and Stamm, M. (1997) *Europhys. Lett.,* **40**, 655.

71 Müller-Buschbaum, P. and Stamm, M. (1998) *Physica B,* **248**, 229.

72 Müller-Buschbaum, P. and Stamm, M. (1998) *Macromolecules,* **31**, 3686.

73 Reiter, G., Sharma, A., Khanna, R., Casoli, A., and David, M.-O. (1999) *J. Coll. Interf. Sci.,* **214**, 126.

74 Zhao, H., Wang, Y.J., and Tsui, O.K.C. (2005) *Langmuir,* **21**, 5817.

75 Zhao, H., Tsui, O.K.C., and Liu, Z. (2005) *Solid State Comm.,* **134**, 455.

76 Oron, A. (2000) *Phys. Rev. Lett.,* **85**, 2108.

77 Oron, A. and Bankoff, S.G. (1999) *J. Coll. Interf. Sci.,* **218**, 152.

78 Warner, M.R.E., Craster, R.V., and Matar, O.K. (2002) *Phys. Fluids,* **14**, 4040.

79 Warner, M.R.E., Craster, R.V., and Matar, O.K. (2003) *J. Coll. Interf. Sci.,* **268**, 448.

80 Mitlin, V.S. (1994) *Coll. Surf. A,* **89**, 97.

81 Mitlin, V.S. (1993) *J. Coll. Interf. Sci.,* **156**, 491.

82 Mitlin, V.S. and Sharma, M.M. (1993) *J. Coll. Interf. Sci.,* **157**, 447.

83 Saulnier, F., Raphael, E., and de Gennes, P.-G. (2002) *Phys. Rev. E,* **66**, 061607.

84 Sharma, A. and Jameel, A.T. (1993) *J. Coll. Interf. Sci.,* **161**, 190.

85 Sharma, A. and Jameel, A.T. (1994) *J. Chem. Soc. Faraday Trans.,* **90**, 625.

86 Khanna, R., Jameel, A.T., and Sharma, A. (1996) *Ind. Eng. Chem. Res.,* **35**, 3081.

87 Khanna, R. and Sharma, A. (1997) *J. Coll. Interf. Sci.,* **195**, 42.

88 Sharma, A. and Khanna, R. (1999) *J. Chem. Phys.,* **110**, 4929.

89 Sharma, A., Khanna, R., and Reiter, G. (1999) *Colloids Surf. B.,* **14**, 223.

90 Ghatak, A., Khanna, R., and Sharma, A. (1999) *J. Coll. Interf. Sci.,* **212**, 483.

91 Sharma, A. (2003) *Euro. Phys. J. E,* **12**, 397.

92 Sferrazza, M., Heppenstallbutler, M., Cubitt, R., Bucknall, D.G., Webster, J., and Jones, R.A.L. (1998) *Phys. Rev. Lett.,* **81**, 5173.

93 Ashley, K.M., Meredith, J.C., Amis, E., Raghavan, D., and Karim, A. (2003) *Polymer,* **44**, 769.

94 Herminghaus, S., Fery, A., Schlagowski, S., Jacobs, K., Seemann, R., Gau, H.,

Monch, W., and Pompe, T. (2000) *J. Phys.: Cond. Matter*, **12**, A57.

95 Seemann, R., Jacobs, K., and Blossey, R. (2001) *J. Phys.: Cond. Matter*, **13**, 4915.

96 Seemann, R., Herminghaus, S., and Jacobs, K. (2001) *J. Phys.: Cond. Matter*, **13**, 4925.

97 Seemann, R., Herminghaus, S., Neto, C., Schlagowski, S., Podzimek, D., Konrad, R., Mantz, H., and Jacobs, K. (2005) *J. Phys.: Cond. Matter*, **17**, S267.

98 Müller, M., MacDowell, L.G., Müller-Buschbaum, P., Wunnicke, O., and Stamm, M. (2001) *J. Chem. Phys.*, **115**, 9960.

99 Kim, H.I., Mate, C.M., Hannibal, K.A., and Perry, S.S. (1999) *Phys. Rev. Lett.*, **82**, 3496.

100 Al Akhrass, S., Ostaci, R.-V., Grohens, Y., Drockenmuller, E., and Reiter, G. (1884) *Langmuir*, **24**, 2008.

101 Damman, P., Baudelet, N., and Reiter, G. (2003) *Phys. Rev. Lett.*, **91**, 216101.

102 Shenoy, V. and Sharma, A. (2002) *Phys. Rev. Lett.*, **88**, 236101.

103 Herminghaus, S., Seemann, R., and Jacobs, K. (2002) *Phys. Rev. Lett.*, **89**, 056101.

104 Herminghaus, S., Jacobs, K., and Seemann, R. (2003) *Eur. Phys. J. E*, **12**, 101.

105 Gabriele, S., Damman, P., Sclavons, S., Desprez, S., Coppee, S., Reiter, G., Hamieh, M., Al Akhrass, S., Vilmin, T., and Raphael, E. (2006) *J. Polym. Sci. B*, **44**, 3022.

106 Reiter, G. (2001) *Phys. Rev. Lett.*, **87**, 186101.

107 Henn, G., Bucknall, D.G., Stamm, M., Vanhoorne, P., and Jerome, R. (1996) *Macromolecules*, **29**, 4305.

108 Reiter, G. and de Gennes, P.G. (2001) *Euro. Phys. J. E*, **6**, 25.

109 Richardson, H., Carelli, C., Keddie, J.K., and Sferrazza, M. (2003) *Euro. Phys. J. E*, **12**, 437.

110 Gabriele, S., Sclavons, S., Reiter, G., and Damman, P. (2005) *Phys. Rev. Lett.*, **96**, 156105.

111 Reiter, G., Hamieh, M., Damman, P., Sclavons, S., Gabriele, S., Vilmin, T., and Raphaël, E. (2005) *Nat. Mater.*, **4**, 754.

112 Yang, M.H., Hou, S.Y., Chang, Y.L., and Yang, A.C.-M. (2006) *Phys. Rev. Lett.*, **96**, 066105.

113 Hamieh, M., Al Akhrass, S., Hamieh, T., Damman, P., Gabriele, S., Vilmin, T., Raphaël, E., and Reiter, G. (2007) *J. Adhes.*, **83**, 367.

114 Damman, P., Gabriele, S., Coppée, S., Desprez, S., Villers, D., Vilmin, T., Raphaël, E., Hamieh, M., Al Akhrass, S., and Reiter, G. (2007) *Phys. Rev. Lett.*, **99**, 036101.

115 Vilmin, T. and Raphaël, E. (2006) *Eur. Phys. J. E*, **21**, 161.

116 Vilmin, T. and Raphaël, E. (2006) *Phys. Rev. Lett.*, **97**, 036105.

117 Stange, T.G., Evans, D.F., and Hendrickson, W.A. (1997) *Langmuir*, **13**, 4459.

118 Jacobs, K., Herminghaus, S., and Mecke, K.R. (1998) *Langmuir*, **14**, 965.

119 Redon, C., Brochard-Wyart, F., and Rondelez, F. (1991) *Phys. Rev. Lett.*, **66**, 715.

120 Jacobs, K., Seemann, R., Schatz, G., and Herminghaus, S. (1998) *Langmuir*, **14**, 4961.

121 Neto, C., Jacobs, K., Seemann, R., Blossey, R., Becker, J., and Grün, G. (2003) *J. Phys.: Cond. Matter*, **15**, S421.

122 Neto, C., Jacobs, K., Seemann, R., Blossey, R., Becker, J., and Grün, G. (2003) *J. Phys.: Cond. Matter*, **15**, 3355.

123 Becker, J., Grün, G., Seeman, R., Mantz, H., Jacobs, K., Mecke, K.R., and Blossey, R. (2003) *Nature Mater.*, **2**, 59.

124 Tsui, O.K.C., Wang, Y.J., Zhao, H., and Du, B. (2003) *Euro. Phys. J. E*, **12**, 417.

125 Lorenz-Haas, C., Müller-Buschbaum, P., Kraus, J., Bucknall, D.G., and Stamm, M. (2002) *Appl. Phys. A*, **74**, S383.

126 Karapanagiotis, I., Fennell Evans, D., and Gerberich, W.W. (2001) *Macromolecules*, **34**, 3741.

127 Karapanagiotis, I., Fennell Evans, D., and Gerberich, W.W. (2001) *Langmuir*, **17**, 3266.

128 Karapanageotis, I. and Gerberich, W.W. (2005) *Langmuir*, **21**, 9194.

129 Konnur, R., Kargupta, K., and Sharma, A. (2000) *Phys. Rev. Lett.*, **84**, 931.

130 Zope, M., Kargupta, K., and Sharma, A. (2001) *J. Chem. Phys.*, **114**, 7211.

131 Kargupta, K. and Sharma, A. (2002) *J. Coll. Interf. Sci.*, **245**, 99.

132 Thiele, U., Neuffer, K., Pomeau, Y., and Velarde, M.G. (2002) *Colloids Surf. A*, **206**, 135.

133 Thiele, U., Velarde, M.G., and Neuffer, K. (2001) *Phys. Rev. Lett.*, **87**, 016104.

134 Matar, O.K., Gkanis, V., and Kumar, S. (2005) *J. Coll. Interf. Sci.*, **286**, 319.

135 Brusch, L., Kühne, H., Thiele, U., and Bär, M. (2002) *Phys. Rev. E*, **66**, 011602.

136 Thiele, U. (2003) *Euro. Phys. J. E*, **12**, 409.

137 Meredith, J.C., Smith, A.P., Karim, A., and Amis, E.J. (2000) *Macromolecules*, **33**, 9747.

138 Seemann, R., Herminghaus, S., and Jacobs, K. (2001) *Phys. Rev. Lett.*, **86**, 5534.

139 Du, B., Xie, F., Wang, Y., Yang, Z., and Tsui, O.K.C. (2002) *Langmuir*, **18**, 8510.

140 Bischof, J., Scherer, D., Herminghaus, S., and Leiderer, P. (1996) *Phys. Rev. Lett.*, **77**, 1536.

141 Karapanageotis, I. and Gerberich, W.W. (2005) *Surf. Sci.*, **594**, 192.

142 Bollinne, C., Cuenot, S., Nysten, B., and Jonas, A.M. (2003) *Euro. Phys. J. E*, **12**, 389.

143 Seemann, R., Herminghaus, S., and Jacobs, K. (2001) *Phys. Rev. Lett.*, **87**, 196101.

144 Sharma, A. and Mittal, J. (2002) *Phys. Rev. Lett.*, **89**, 186101.

145 Leizerson, I., Lipson, S.G., and Lyushnin, A.V. (2004) *Langmuir*, **20**, 291.

146 Masson, J.-L. and Green, P.F. (2002) *Phys. Rev. Lett.*, **88**, 205504.

147 Brochard-Wyart, F., Debregeas, G., Fondecave, R., and Martin, P. (1997) *Macromolecules*, **30**, 1211.

148 Limary, R. and Green, P.F. (2003) *Langmuir*, **19**, 2419.

149 Glasner, K.B. and Witelski, T.P. (2003) *Phys. Rev. E*, **67**, 016302.

150 Green, P.F. and Ganesan, V. (2003) *Euro. Phys. J. E*, **12**, 449.

151 Saulnier, F., Raphaël, E., and de Gennes, P.-G. (2002) *Phys. Rev. E*, **66**, 061607.

152 Jeon, H.S., Dixit, P.S., and Yim, H. (2005) *J. Chem. Phys.*, **122**, 104707.

153 Karapanagiotis, I., Evans, D.F., and Gerberich, W.W. (2002) *Colloids Surf. A*, **207**, 59.

154 Oron, M., Kerle, T., Yarushalmi-Rozen, R., and Klein, J. (2004) *Phys. Rev. Lett.*, **92**, 236104.

155 Lee, S.H., Yoo, P.J., Kwo, S.J., and Lee, H.H. (2004) *J. Chem. Phys.*, **121**, 4346.

156 Besancon, B.M. and Green, P.F. (2004) *Phys. Rev. E*, **70**, 051808.

157 Leizerson, I., Lipson, S.G., and Lyushnin, A.V. (2004) *Langmuir*, **20**, 291.

158 Choi, S.-H. and Zhang Newby, B.-M. (2006) *J. Chem. Phys.*, **124**, 054702.

159 Bäumchen, O. and Jacobs, K. (2010) *J. Phys.: Cond. Matter*, **22**, 033102.

160 Redon, C., Brzoska, J.B., and Brochard-Wyart, F. (1994) *Macromolecules*, **27**, 468.

161 Brochard, F., de Gennes, P.G., Hervert, H., and Redon, C. (1994) *Langmuir*, **10**, 1566.

162 Fetzer, R., Jacobs, K., Munch, A., Wagner, B., and Witelski, T.P. (2005) *Phys. Rev. Lett.*, **95**, 127801.

163 Fetzer, R., Rauscher, M., Seemann, S., Jacobs, K., and Mecke, K. (2007) *Phys. Rev. Lett.*, **99**, 114503.

164 Ziebert, F. and Raphaël, E. (2009) *Europhys. Lett.*, **86**, 46001.

165 Sharma, A. and Kargupta, K. (2003) *Appl. Phys. Lett.*, **83**, 3549.

166 Kargupta, K., Sharma, A., and Khanna, R. (2004) *Langmuir*, **20**, 244.

167 Sharma, A. and Khanna, R. (1996) *Macromolecules*, **29**, 6959.

168 Reiter, G. and Sharma, A. (2001) *Phys. Rev. Lett.*, **87**, 166103.

169 Müller Buschbaum, P. (2003) *Euro. Phys. J. E*, **12**, 443.

170 Sheiko, S., Lermann, E., and Moller, M. (1996) *Langmuir*, **12**, 4015.

171 Erhardt, M.K. and Nuzzo, R.G. (2001) *J. Phys. Chem. B*, **105**, 8776.

172 Leibler, L. and Sekimoto, K. (1993) *Macromolecules*, **26**, 6937.

173 Higgins, M. and Jones, R.A.L. (2000) *Nature*, **404**, 476.

174 Zhang, X., Xie, F., and Tsui, O.K.C. (2005) *Polymer*, **46**, 8416.

175 Müller-Buschbaum, P., Bauer, E., Maurer, E., Schlögl, K., Roth, S.V., and Gehrke, R. (2006) *Appl. Phys. Lett.*, **88**, 083114.

176 Zhang, B., Lee, F.K., Tsui, O.K.C., and Sheng, P. (2003) *Phys. Rev. Lett.*, **91**, 215501.

177 Lee, F.K., Zhang, B., Sheng, P., Kwok, H.S., and Tsui, O.K.C. (2004) *Appl. Phys. Lett.*, **85**, 5556.
178 Ibn-Elhaj, M. and Schadt, M. (2001) *Nature*, **410**, 796.
179 Braun, H.-G. and Mayer, E. (1999) *Thin Solid Films*, **345**, 222.
180 Mayer, E. and Braun, H.-G. (2000) *Macromol. Mater. Eng.*, **276–277**, 44.
181 Sehgal, A., Ferreiro, V., Douglas, J.F., Amis, E.J., and Karim, A. (2002) *Langmuir*, **18**, 7041.
182 Zhang, Z., Wang, Z., Xing, R., and Han, Y. (2003) *Polymer*, **44**, 3737.
183 Zhang, Z., Wang, Z., Xing, R., and Han, Y. (2003) *Surf. Sci.*, **539**, 129.
184 Lu, G., Li, W., Yao, J., Zhang, G., Yang, B., and Shen, J. (1049) *Adv. Mater.*, **14**, 2002.
185 Wang, X., Tvingstedt, K., and Inganas, O. (2005) *Nanotechnology*, **16**, 437.
186 Wang, X., Ostblom, M., Johansson, T., and Inganas, O. (2004) *Thin Solid Films*, **449**, 125.
187 Granlund, T., Theander, T., Berggren, M., Andersson, M.R., Ruzeckas, A., Sundstrom, V., Bjork, G., Granstrom, M., and Inganas, O. (1998) *Chem. Phys. Lett.*, **288**, 879.
188 Wang, J.Z., Zheng, Z.H., Li, H.W., Huck, W.T.S., and Sirringhaus, H. (2004) *Nat. Mater.*, **3**, 171.
189 Cavallini, M., Facchini, M., Massi, M., and Biscarini, F. (2004) *Synth. Met.*, **146**, 283.
190 Julthongpiput, D., Zhang, W., Douglas, J.F., Karim, A., and Fasolka, M.J. (2007) *Soft Matter*, **3**, 613.
191 Li, Z., Tolan, M., Hohr, T., Kharas, D., Qu, S., Sokolov, J., and Rafailovich, M.H. (1998) *Macromolecules*, **31**, 1915.
192 Bao, L.-R., Tan, L., Huang, X.D., Kong, Y.P., Guo, L.J., Pang, S.J., and Yee, A.F. (2749) *J. Vac. Sci. Technol. B*, **21**, 2003.
193 Ferrell, N. and Hansford, D. (2007) *Macromol. Rapid Comm.*, **28**, 966.
194 Rockford, L., Liu, Y., Mansky, P., Russell, T.P., Yoon, M., and Mochrie, S.G.J. (1999) *Phys. Rev. Lett.*, **82**, 2602.
195 Rhese, N., Wang, C., Hund, M., Geoghegan, M., Magerle, R., and Krausch, G. (2001) *Euro. Phys. J. E*, **4**, 69.
196 Geoghegan, M., Wang, C., Rhese, N., Magerle, R., and Krausch, G. (2005) *J. Phys.: Cond. Matt.*, **17**, S389.
197 Luo, C., Xing, R., Zhang, Z., Fu, J., and Han, Y. (2004) *J. Coll. Interf. Sci.*, **269**, 158.
198 Xing, R., Luo, C., Wang, Z., and Han, Y. (2007) *Polymer*, **48**, 3574.
199 Yoon, B., Acharya, H., Lee, G., Kim, H.–C., Huhc, J., and Park, C. (2008) *Soft Matter*, **4**, 1467.
200 Khare, K., Brinkmann, M., Law, B.M., Gurevich, E.L., Herminghaus, S., and Seemann, R. (2007) *Langmuir*, **23**, 12138.
201 Volodin, P. and Kondyurin, A. (2008) *J. Phys. D: Appl. Phys.*, **41**, 065307.
202 Radha, B. and Kulkarni, G.U. (2009) *ACS Appl. Mater. Interf.*, **1**, 257.
203 Kuroda, A., Ishihara, T., Takeshige, H., and Asakura, K. (2008) *J. Phys. Chem. B*, **112**, 1163.
204 Chabinyc, M.L., Wong, W.S., Salleo, A., Paul, K.E., and Street, R.A. (2002) *Appl. Phys. Lett.*, **81**, 4260.
205 Luan, S., Cheng, Z., Xing, R., Wang, Z., Yu, X., and Han, Y. (2005) *J. Appl. Phys.*, **97**, 086102.
206 Mukherjee, R., Gonuguntla, M., and Sharma, A. (2007) *J. Nanosci. Nanotechnol.*, **7**, 2069.
207 Mukherjee, R., Bandyopadhyay, D., and Sharma, A. (2008) *Soft Matter*, **4**, 2086.
208 Kargupta, K. and Sharma, A. (2001) *Phys. Rev. Lett.*, **86**, 4536.
209 Kargupta, K. and Sharma, A. (2002) *J. Chem. Phys.*, **116**, 3042.
210 Kargupta, K. and Sharma, A. (2002) *Langmuir*, **19**, 5153.
211 Kargupta, K., Konnur, R., and Sharma, A. (2000) *Langmuir*, **16**, 10243.
212 Kargupta, K., Konnur, R., and Sharma, A. (2001) *Langmuir*, **17**, 1294.
213 Kargupta, K. and Sharma, A. (2002) *Langmuir*, **18**, 1893.
214 Suman, B. and Kumar, S. (2006) *J. Coll. Interf. Sci.*, **304**, 208.
215 Volodin, P. and Kondyurin1, A. (2008) *J. Phys. D: Appl. Phys.*, **41**, 065306.
216 Thiele, U., Brusch, L., Bestehorn, M., and Baer, M. (2003) *Euro. Phys. J. E*, **11**, 255.
217 Borcia, R. and Bestehorn, M. (2009) *Langmuir*, **25**, 1919.

218 Böltau, M., Walhelm, S., Mlynek, J., Krausch, G., and Steiner, U. (1998) *Nature*, **391**, 877.

219 Ermi, B.D., Nisato, G., Douglas, J.F., Rogers, J.A., and Karim, A. (1998) *Phys. Rev. Lett.*, **81**, 3900.

220 Nisato, G., Ermi, B.D., Douglas, J.F., and Karim, A. (1999) *Macromolecules*, **32**, 2356.

221 Karim, A., Douglas, J.F., Lee, B.P., Glotzer, S.C., Rogers, J.A., Jackman, R.J., Amis, E.J., and Whitesides, G.M. (1998) *Phys. Rev. E*, **57**, R6273.

222 Park, M., Harrison, C., Chaikin, P.M., Register, R.A., and Adamson, D.H. (1997) *Science*, **276**, 1401.

223 Huang, E., Rockford, L., Russell, T.P., and Hawker, C.J. (1998) *Nature*, **395**, 757.

224 Kim, S.O., Solak, H.H., Stoykovich, M.P., Ferrier, N.J., de Pablo, J.J., and Nealey, P.F. (2003) *Nature*, **424**, 411.

225 Walheim, S., Ramstein, M., and Steiner, U. (1999) *Langmuir*, **15**, 4848.

226 Karim, A., Slawecki, T.M., Kumar, S.K., Douglas, J.F., Satija, S.K., Han, C.C., and Russell, T.P. (1998) *Macromolecules*, **31**, 857.

227 Li, X., Wang, Z., Cui, L., Xing, R., Han, Y., and An, L. (2004) *Surface Sci.*, **571**, 12.

228 Sprenger, M., Walheim, S., Budkowski, A., and Steiner, U. (2003) *Interf. Sci.*, **11**, 225.

229 Cyganik, P., Budkowski, A., Steiner, U., Rysz, J., Bernasik, A., Walheim, S., Postawa, Z., and Raczkowska, J. (2003) *Europhys. Lett.*, **62**, 855.

230 Swan, J.W. (1897) *Proc. R. Soc. Lond.*, **62**, 38.

231 Glenn, W.E. (1870) *J. Appl. Phys.*, **30**, 1959.

232 Cressman, P.J. (1963) *J. Appl. Phys.*, **34**, 2327.

233 Taylor, G.I. and McEwan, A.D. (1965) *J. Fluid Mech.*, **22**, 1.

234 Killat, U. (1975) *J. Appl. Phys.*, **46**, 5169.

235 Saville, D.A. (1997) *Ann. Rev. Fluid Mech.*, **29**, 27.

236 Herminghaus, S. (1999) *Phys. Rev. Lett.*, **83**, 2359.

237 Schäffer, E., Thurn-Albrecht, T., Russell, T.P., and Steiner, U. (2000) *Nature*, **403**, 874.

238 Tonks, L.A. (1935) *Phys. Rev.*, **48**, 562.

239 Frenkel, J. (1935) *Phys. Z. Sowjetunion*, **8**, 675.

240 Leiderer, P. (1997) *Two-Dimensional Electron Systems*, Kluwer.

241 Schäffer, E., Thurn-Albrecht, T., Russell, T.P., and Steiner, U. (2001) *Europhys. Lett.*, **53**, 518.

242 Steiner, U. (2005) *Structure Formation in Polymer Films From Micrometer to the Sub-100nm Length Scales in Nanoscale Assembly*, Springer, New York.

243 Schäffer, E. (2001) Instabilities in Thin Polymer Films: Structure Formation and Pattern Transfer PhD Thesis, Konstanz, September.

244 Lei, X., Wu, L., Deshpande, P., Yu, Z., Wu, W., Ge, H., and Chou, S.Y. (2003) *Nanotechnology*, **14**, 786.

245 Dickey, M.D., Collister, E., Raines, A., Tsiartas, P., Holcombe, T., Sreenivasan, S.V., Bonnecaze, R.T., and Willson, C.G. (2043) *Chem. Mater.*, **18**, 2006.

246 Voicu, N.E., Harkema, S., and Steiner, U. (2006) *Adv. Funct. Mater.*, **16**, 926.

247 Voicu, N.E., Saifullah, M.S.M., Subramanian, K.R.V., Welland, M.E., and Steiner, U. (2007) *Soft Matter*, **3**, 554.

248 Heier, J., Groenewold, J., and Steiner, U. (2009) *Soft Matter*, **5**, 3997.

249 Pease, L.F. and Russel, W.B. (2002) *J. Non-Newtonian Fluid Mech.*, **102**, 233.

250 Pease, L.F. and Russel, W.B. (2003) *J. Chem. Phys.*, **118**, 3790.

251 Wu, N. and Russel, W.B. (2005) *Appl. Phys. Lett.*, **86**, 241912.

252 Verma, R., Sharma, A., Kargupta, K., and Bhaumik, J. (2005) *Langmuir*, **21**, 3710.

253 Salac, D., Lu, W., Wang, C.-W., and Sastry, A.M. (2004) *Appl. Phys. Lett.*, **85**, 1161.

254 Lin, Z., Kerle, T., Baker, M., Hoagland, D.A., Schäffer, E., Steiner, U., and Russell, T.P. (2001) *J. Chem. Phys.*, **114**, 2377.

255 Lin, Z., Kerle, T., Russell, T.P., Schäffer, E., and Steiner, U. (2002) *Macromolecules*, **35**, 3971.

256 Lin, Z., Kerle, T., Russell, T.P., Schäffer, E., and Steiner, U. (2002) *Macromolecules*, **35**, 6255.

257 Morariu, M.D., Voicu, N.E., Schäffer, E., Lin, Z., Russell, T.P., and Steiner, U. (2003) *Nat. Mater.*, **2**, 48.
258 Dickey, M.D., Gupta, S., Leach, K.A., Collister, E., Willson, C.G., and Russel, T.P. (2006) *Langmuir*, **22**, 4315.
259 Leach, K.A., Gupta, S., Dickey, M.D., Willson, C.G., and Russell, T.P. (2005) *Chaos*, **15**, 047506.
260 Leach, K.A., Lin, Z., and Russell, T.P. (2005) *Macromolecules*, **38**, 4868.
261 Shankar, V. and Sharma, A. (2004) *J. Coll. Interf. Sci.*, **274**, 294.
262 Craster, R.V. and Marter, O.K. (2005) *Phys. Fluids*, **17**, 032104.
263 Bandyopadhyay, D. and Sharma, A. (2007) *J. Coll. Interf. Sci.*, **311**, 595.
264 Wu, N., Kavousanakis, M.E., and Russel, W.B. (2010) *Phys. Rev. E*, **81**, 026306.
265 Schäffer, E., Harkema, S., Roerdink, M., Blossey, R., and Steiner, U. (2003) *Adv. Mater.*, **15**, 514.
266 Schäffer, E., Harkema, S., Roerdink, M., Blossey, R., and Steiner, U. (2003) *Macromolecules*, **36**, 1645.
267 Harkema, S., Schäffer, E., Morariu, M.D., and Steiner, U. (2003) *Langmuir*, **19**, 9714.
268 Verma, R., Sharma, A., Banerjee, I., and Kargupta, K. (2006) *J. Coll. Interf. Sci.*, **296**, 220.
269 Schäffer, E., Harkema, S., Blossey, R., and Steiner, U. (2002) *Europhys. Lett.*, **60**, 255.
270 Dietzel, M. and Troian, S.M. (2009) *Phys. Rev. Lett.*, **103**, 074501.
271 Cavallini, M., Gomez-Segura, J., Albonetti, C., Ruiz-Molina, D., Veciana, J., and Biscarini, F. (2006) *J. Phys. Chem. B*, **110**, 11607.
272 Deshpande, P. and Chou, S.Y. (2741) *J. Vac. Sci. Technol. B*, **19**, 2001.
273 Lei, C., Zhuang, L., Deshpande, P., and Chou, S.Y. (2005) *Langmuir*, **21**, 818.
274 Deshpande, P., Pease, L.F., Chen, L., Chou, S.Y., and Russel, W.B. (2004) *Phys. Rev. E*, **70**, 041601.
275 Peng, J., Han, Y., Yang, Y., and Li, B. (2003) *Polymer*, **44**, 2379.
276 Peng, J., Wang, H., Li, B., and Han, Y. (2004) *Polymer*, **45**, 8013.
277 Pease, L.F., Deshpande, P., Wang, Y., Russel, W.B., and Chou, S.Y. (2007) *Nat. Nanotech.*, **2**, 545.
278 Pease, L.F. and Russel, W.B. (2004) *Langmuir*, **20**, 795.
279 Melcher, J.R. (1963) *Field-Coupled Surface Waves*, MIT Press, Cambridge, Mass.
280 Cross, M.C. and Hohenberg, P.C. (1993) *Rev. Mod. Phys.*, **65**, 851.
281 Nedelcu, M., Morariu, M.D., Harkema, S., Voicu, N.E., and Steiner, U. (2005) *Soft Matter*, **1**, 62.
282 Barnes, K.A., Karim, A., Douglas, J.F., Nakatani, A.I., Gruell, H., and Amis, E.J. (2000) *Macromolecules*, **33**, 4177.
283 Krishnan, R.S., Mackay, M.E., Hawker, C.J., and Van Horn, B. (2005) *Langmuir*, **21**, 5770.
284 Mukherjee, R., Das, S., Das, A., Sharma, S.K., Raychaudhuri, A.K., and Sharma, A. (2010) *ACS Nano.* **4**, 3709.

Part Five
Applications

12
Cells on Patterns

Aldo Ferrari and Marco Cecchini

12.1
Introduction

Micro- and nano-patterning techniques for producing biomimetic scaffolds aim to tailor the interaction between artificial materials and target cells under specific experimental or physiological conditions [1]. In this frame, the viability and performance of biologically-oriented materials is assessed by testing the efficiency of supported cell proliferation, coverage, adhesion, and spreading [2]. Given that no comprehensive rule exists and that a large variability is provided by the cellular type under analysis, in this chapter we shall consider the general biological aspects involved in the establishment of cellular adhesions with engineered surfaces. And, to some extent, how this interaction can trigger a range of specific responses in mammalian cells (Figure 12.1).

The cellular interaction with the extracellular environment is mediated by specific (i.e., receptor dependent) and non-specific (i.e., governed by short- and long-range forces as van der Waals, electrostatic, steric, and solvation interactions) bonds. Although the latter are critical in determining early anchoring and proximity effects, *bona fide* biological adhesions are established only when membrane receptor molecules recognize and bind to functional groups or ligand deposited, absorbed or secreted over the surface [3].

Transmembrane adhesion receptors of the integrin family have a primary role in such recognition processes [4]. It is now evident that integrin-mediated adhesions, the focal adhesions (FAs), are cellular centers where all information regarding the biological, chemical, and physical properties of the extracellular environment are gathered [5]. The currently understood complexity of FAs (Figure 12.2) is sufficient to ensure that this information is integrated, processed, and ultimately translated into intracellular signals [6, 7]. Through this sensing activity the initial interaction with an engineered surface is readily converted into an educated cellular response which involves either spreading, migration, or proliferation. Yet, it can also lead to cell detachment and eventually apoptosis [8].

The induction of stereotyped cell activities through the control of the number, density, and maturation state of FAs is emerging as a key point in understanding the

Generating Micro- and Nanopatterns on Polymeric Materials. Edited by A. del Campo and E. Arzt
Copyright © 2011 WILEY-VCH Verlag GmbH & Co. KGaA, Weinheim
ISBN: 978-3-527-32508-5

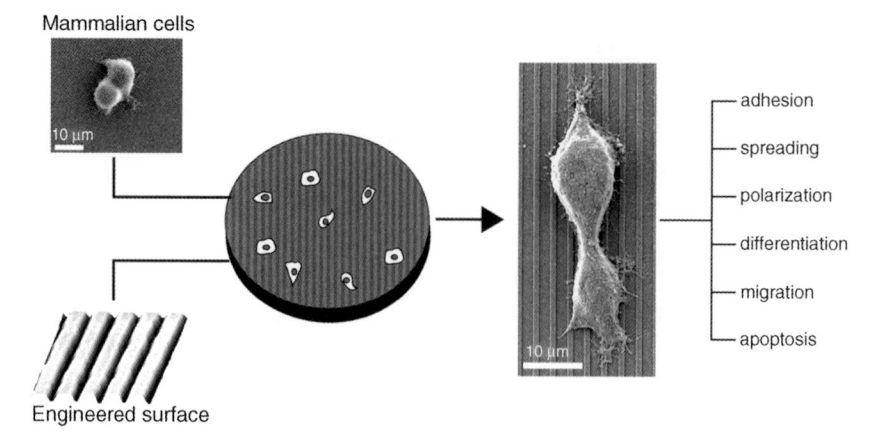

Figure 12.1 The fate and function of mammalian cells is modulated by the properties of micro- and nano-patterned polymeric surfaces.

impact of micro- and nanostructured polymeric surfaces on cellular behavior [9]. In this direction, the combination of unprecedented manufacturing techniques and a deeper molecular understanding will soon provide enough experimental boundaries to allow the rational design of substrates promoting desired cellular activities or inducing specific differentiation schemes [10].

In the next section we will define the substrate physicochemical properties which are read by cells and provide details on the parameter ranges that are of use in cellular studies. The final part of each sub-section will provide an overview of the techniques currently applied to independently control the defined substrate parameters.

12.2
Physicochemical Properties of the Substrate Read by Cells

Living cells grow and exert their activities while embedded in a dense and complex environment, the extracellular matrix (ECM). The ECM itself contains an array of structural and directional cues that ultimately guide and support both the polarization of individual cells and the morphogenesis of multicellular structures such as tissues and organs [11].

Individual cells sense these complex inputs through the FAs, which are generated and mature according to the local and global configuration of the ECM [3]. Thus, the cellular reaction to the ECM signals goes far beyond the simple chemical recognition of specific ligands and has a measurable effect on cell dynamics, function, and fate. In particular three independent ECM parameters play a major role in governing cell behavior:

1) **Density of adhesion points** – The density of ligands is a critical parameter that triggers the initial clustering and the subsequent enlargement of FAs.

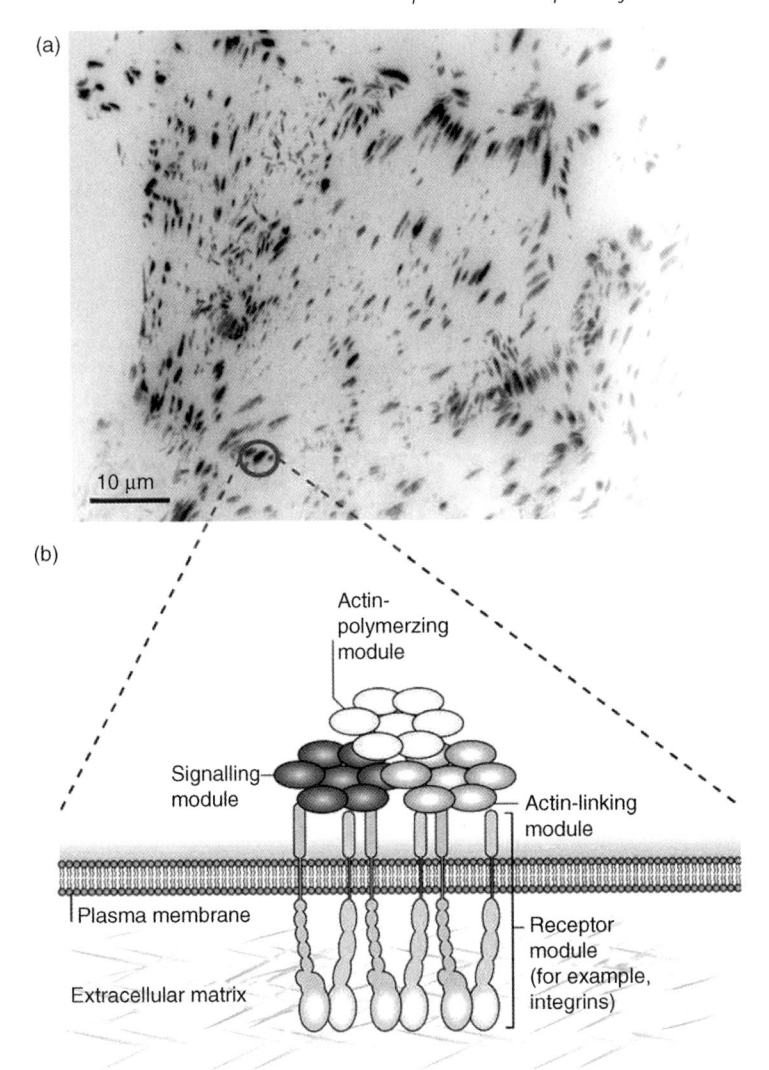

Figure 12.2 Integrin-mediated matrix adhesions. Rat embryo fibroblasts (REF52) expressing yellow fluorescent protein labeled paxillin display classical focal adhesion staining. (a) The inverted fluorescent signal at the cell-substrate interface (captured with a total internal reflection microscope) pinpoints the regions of paxillin accumulation (black spots). (b) A scheme of the active modules operating in a focal adhesion is reported. (Adapted with permission from [9]. Copyright © (2009) Macmillan Publishers Ltd.)

2) **Topography** – Topographical features in the micron and submicron range act as physical boundaries providing a local constraint to the formation and maturation of FAs.
3) **Rigidity of the substrate** – Cells apply force to the developing adhesions through acto-myosin contractility. The mechanical response of the matrix controls the

further maturation of the adhesion points in a molecularly regulated feedback loop.

In order to control the assembly and maturation of FAs, and thus to induce specific cellular responses, biomimetic scaffolds will provide independent modulation of the above listed parameters within the physiological ranges that are resolved by cells.

12.2.1
Density of Adhesion Points

The requirement of a minimal density of adhesion points for the establishment of stable interaction between a cell and a substrate is evidenced by a number of experimental observations. Nevertheless the technological challenges involved in uncoupling this ECM parameter from others (e.g., topography and rigidity of the substrate) and in achieving control of ligand spacing in the nanometer range (i.e., between 10 and 200 nm) over large areas (i.e., a few mm^2) limit these results to a qualitative level. Early approaches could assess the role of integrin clustering for adhesion reinforcement [12, 13]. In these studies, matrix proteins such as fibronectin or integrin binding short peptides (i.e., arginine-glycine-aspartate (RGD)) were homogenously absorbed at variable concentrations onto non adhesive surfaces. Yet the precise distance between individual ligands and thus the geometry of integrin binding could not be controlled.

The development of self-assembly patterning techniques offered an unprecedented combination of high spatial resolution, down to single molecule positioning, with rapid and cost effective fabrication of large functionalized surfaces [14]. Recently, block copolymer micelle nanolithography (BCMN) has emerged as the technique of choice for the study of assembly of FAs under controlled ligand spacing and density conditions. Figure 12.3 summarizes the general method for the preparation of spatially controlled nanopatterned surfaces biofunctionalized with RGD peptides.

The availability of large surfaces where the spacing between individual adhesion points, and thus between the engaged integrin receptors, can be modulated within the critical length scale of the involved molecules opened the way to a number of unprecedented investigations that quantitatively measured the relation between ligand spacing, integrin clustering and cell behavior.

Table 12.1 recapitulates these findings. Using a rat embryo fibroblast cell line (REF52) stably expressing a yellow fluorescent protein (YFP) labeled paxillin (a scaffold protein which gets enriched in FAs; [3, 22]) it was possible to reveal that the maximal ligand spacing supporting formation of FAs is 78 nm [18]. Larger ligand spacing ($d \geq 108$ nm; Table 12.1) resulted in delayed and inefficient formation of highly unstable FAs and finally in the detachment of the cell from the surface. These results demonstrated that integrin clustering over a typical distance of 50–70 nm is necessary for the establishment of a stable cell-to-substrate interaction (Figure 12.4).

In the same way, the minimal number of engaged integrin receptors required to assemble a biological adhesion ($n = 6$) was counted using patterned surfaces where gold nanoclusters were deposited on predefined geometries (Figure 12.4) [15]. These

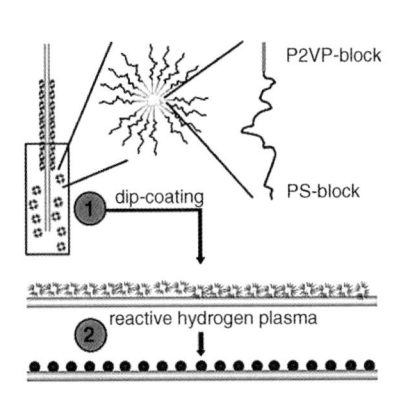

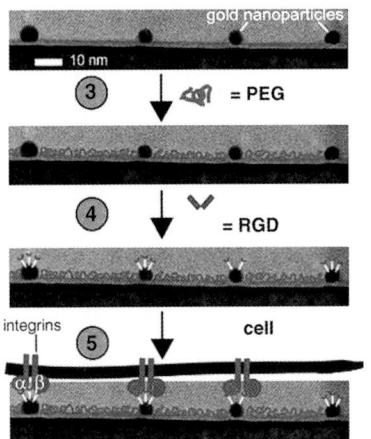

Figure 12.3 Block copolymer micelle nanolithography (BCMN) procedure to obtain large surfaces with controlled adhesion point density and ligand spacing. The procedure can be recapitulated in five steps: (1) Films of regular micelles are prepared by dipping a silicon wafer or a glass cover slip into a toluene solution of amphiphilic block copolymers of polystyrene (x)-*block*-poly(2-vinylpyridine)(y). PS(x)-*b*-P2VP(y) aggregate to uniform micelles where the PS blocks form a shell around the less soluble P2VP blocks. The diameter of the micelles is controlled by the molecular weight of the block copolymers and by the interactions between the polymer blocks and the solvent. The micellar core-shell structure allows for the selective dissolution of metal precursors salts (i.e., $HAuCl_4$) into the P2VP core. (2) After evaporation of the solvent the micellar films are treated with either H_2-, Ar- or O_2-plasma that entirely removes the polymer shell. Uniform Au nanoclusters of 1 to 15 nm are left to decorate the surface. The particle size is determined by the amount of metal precursor added to the micellar solution. The gold nanoclusters form a hexagonal lattice and their separation is tunable between 15 and 290 nm. (3) The space between the gold nanoclusters is selectively coated with a protein repellent layer of polyethylene glycol (PEG) to block cell adhesion and/or protein absorption. (4) The gold nanoclusters are functionalized with a RGD sequence serving as ligand for the integrin receptor. The final size of the resulting adhesive patch is approximately 8 nm. (5) Each RGD patch presents a binding site for a single integrin receptor due to steric hindrance when cells are plated on the nanopatterned surface. (Adapted with permission from [15, 16]. Copyright © (2009) RSC Publishing and John Wiley & Sons Ltd.).

were obtained by coupling BCMN with standard top-down lithography techniques (e.g., conventional e-beam lithography; [14]).

The possibility of independently controlling the density of adhesion points and the local geometry in which they are presented to the cell allowed for further investigation of the relevance of intermolecular spatial arrangement in adhesion formation. Experiments where the strength of initial (i.e., seconds to minutes) cell contacts was measured revealed that cooperative integrin binding governs the establishment of early cell contacts. And confirmed that, as described for other receptor-mediated cell responses such as T-cell activation [23], surface binding stability depends on the spatial arrangement of integrins rather than on the overall density of engaged receptors [19].

Table 12.1 Experimental boundaries for receptor clustering and cellular responses on patterned surfaces with controlled ligand density and spacing (s = nanocluster size; d = ligand spacing; n = number of nanoclusters; g = gradient of ligand spacing).

Cell type	Ligand	Nanocluster	Effect on FAs	Cell response	Ref.
fibroblast	RGD	s = 8 nm d = 58 nm	integrin clustering; maturation of FAs	adhesion & spreading	[17, 18]
fibroblast	RGD	s = 8 nm d = 108 nm	inefficient integrin clustering; high turnover of FAs	delayed adhesion & erratic migration	[17, 18]
fibroblast	RGD	s = 6 nm d = 58 nm n = 6–3000	no adhesion formation with less than 6 adhesion points		[15]
fibroblast	RGD	s = 6–8 nm d = 35/70/103 nm	rapid (seconds to minutes) adhesion reinforcement with d ≤ 55 nm	modulation of adhesion and spreading	[19]
osteoblast	RGD	i = 1–15 nm d = 15–250 nm g = 19–50 nm/mm	differential adhesion strength along the gradient	cell polarization, directed migration	[20]
sarcoma	TNF	s = 8 nm d = 58–290 nm	—	apoptosis with d ≤ 200 nm	[21]

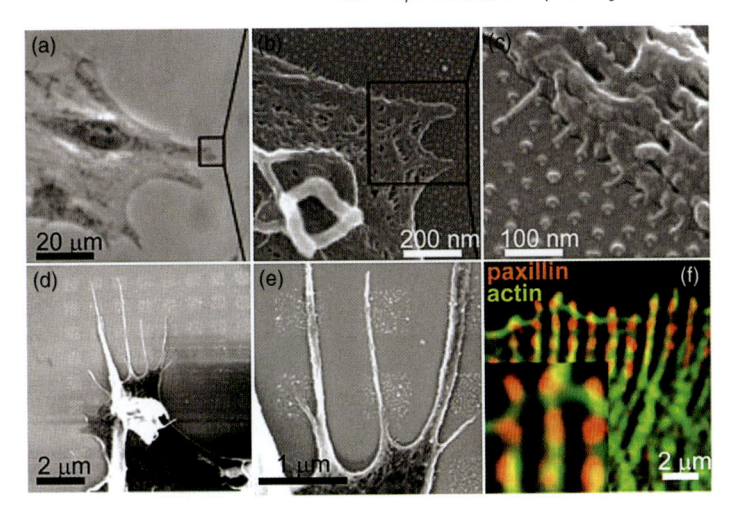

Figure 12.4 Hierarchically patterned substrates and their effect of FAs. (a) Transmission image of a REF52 fibroblast adhering to a glass surface presenting 6 nm sized nanoclusters; (b) scanning electron microscopy image; functionalized with cRGDfk peptides. (c) One-to-one adhesion of cellular protrusion to individual adhesion patches are revealed by SEM. (d, e) Patterned adhesive squares (side = 500 nm) locally control adhesion maturation as revealed by (f) fluorescence microscopy (green = actin, red = paxillin). (Adapted with permission from [24]. Copyright © (2009) RSC Publishing).

Recently, two-dimensional substrates presenting a gradient of ligand spacing ranging from 15 to 250 nm [20] were fabricated by means of BCMN. The gradient strength was modulated by controlling the length of the polymer blocks and the dipping velocity during micellar deposition. This allowed tailoring of surfaces where individual mouse osteoblasts (MC3T3) were contacting regions with different density of adhesion points at their opposing edges. This analysis demonstrated that cells can sense spacing differences as small as 1 nm and react by migrating in the direction of higher ligand density.

It is thus clear that patterning technologies to control the density and spacing of selective ligands offer a general approach to studying the link between receptor engagements and signal triggering in specific cell types. Recent examples are the application to the clustering of TNF receptors inducing apoptosis in human cancer cells (Kym-1) and to the role of DM-GRASP adhesion molecules in promoting attachment and differentiation of neurons [21, 25].

12.2.1.1 Topography

The description of contact guidance (i.e., the cell reaction with the topographical features in its local environment) in mammalian cells dates back to 1964 [26]. Since then, thanks to the development of various techniques for the topographical patterning of biocompatible materials, an increasing number of cell types have been shown to modulate their behavior when contacting flat or topographically-patterned surfaces [27].

Several techniques, mostly relying on lithographic protocols, have been exploited to create active artificial scaffolds presenting two- or three-dimensional topographies. The application of these methods yielded patterned surfaces with lateral resolution ranging from few microns (e.g., photolithography) down to tens of nanometers (e.g., electron beam lithography [28]). These techniques were generally applied to glass and quartz surfaces. Yet, the controlled realization of micro-patterns on biocompatible substrates proved a difficult task, because these materials are typically incompatible with standard fabrication techniques. Soft lithography and imprint lithography have mostly been used and shown that good versatility and reliability can be coupled with high resolution patterning (down to 100 nm) of polymeric materials [29–31]. The combination of these techniques with optically transparent polymers such as tissue culture polystyrene (TCPS) and cyclic olefin copolymer (COC) yielded patterned substrates where contact guidance could be observed by means of high resolution microscopy in living cells [29, 31–33].

Due to the vast difference among the cellular systems adopted (including endothelial and epithelial cells, neurons, osteoblasts, fibroblast, and many others [29, 30, 34–38]) and the range of topographical configurations tested (mainly gratings, but also pillars, pits and spikes; [39–42]) it is not possible to condense the available literature into a general scheme. However, the current understanding of contact guidance points to the modulation of the maturation of FAs as a key to interpreting the effect of topography on cell migration, polarity, and ultimately differentiation [43].

The physical parameters in the substrate topography which were reported to modulate contact guidance and formation of FAs in mammalian cells are the size, aspect ratio, and lateral spacing of the topographical features together with the isotropy and degree of disorder of the pattern. A concise summary of the quantitative results obtained is reported in Table 12.2.

Mammalian fibroblasts were mostly chosen in these investigations due to their rapid adhesion, constitutive migration, and fast proliferation rate. The response of these cells to the surface topography was generally measured by comparing the degree of cell anisotropy and alignment (i.e., the cell body circularity and orientation) and the direction of migration in cells contacting gratings or flat surfaces. Significant modulation of alignment and migration was observed in both the micron and submicron range, with clear contact guidance on gratings presenting topographical features as small as few tens of nanometers [34, 38, 44]. Similar results, with rare exceptions, were obtained in osteoblasts, mesenchymal, and nerve cells [34, 36, 46].

Thus a first general rule fixes the minimal topographical feature size detected by mammalian cells to be of the order of 10 to 100 nm. Below this threshold cell polarity and migration are indistinguishable from what is observed, in the same experimental conditions, on flat surfaces. A possible interpretation of this result will consider the dimension of features that cells encounter in the ECM *in vivo* (20 to 200 nm; [38]) and the typical size of actin-based cell membrane protrusions such as filopodia and microspikes (with diameter in the range of 100 nm [48]).

When regarding larger feature size, non-overlapping length scales appear to control adhesion formation and cell polarity. The observation of separate contact guidance regimes underpins diverse mechanisms interacting with the cell at

Table 12.2 Experimental boundaries for adhesion formation and cellular responses on patterned surfaces with controlled topography. Experiments on gratings (g = groove; r = ridge; lw = linewidth; dp = depth; a.r. = aspect ratio) and pits (diam. = diameter; d = distance; tol = tolerance) are reported.

Cell type	Pattern	Features	Effect on FAs	Cell response	Ref.
fibroblasts, endothelial cells, smooth muscle cells	grating	g = 2–10 µm dp = 50–200 nm	—	alignment & directed migration (dp ≥ 50 nm)	[34]
fibroblasts	grating	lw = 0.02–1 µm dp = 5–350 nm	—	alignment (lw ≥ 100 nm; dp ≥ 35 nm)	[44]
neurons	grating	lw = 0.5–5 µm dp = 200 nm	control of size, maturation & persistence of FAs	neurite alignment (lw ≤ 1 µm)	[29, 32]
fibroblasts	grating	lw = 0.1–10 µm dp = 0.1–1 µm a.r. = 0.01–1	—	alignment (a.r. ≥ 0.05)	[45]
osteoblasts	grating/pits	lw = 10–100 µm dp = 750 nm/ dp = 100 nm diam. = 120 nm d = 300 nm	modulation of formation of FAs	spreading & differentiation (lw ≥ 100 µm)	[46]
epithelial cells	grating	g = 0.33–2.1 µm r = 0.07–1.9 µm	modulation of width of FAs	alignment (r ≥ 70 nm)	[38]

(*Continued*)

Table 12.2 (Continued)

Cell type	Pattern	Features	Effect on FAs	Cell response	Ref.
mesenchymal stem cells	pits	dp = 100 nm diam. = 120 nm d = 300 nm tol = 0–50 nm	modulation of formation of FAs	differentiation (tol ≥ 50)	[39, 46]
fibroblasts	grating	r = 1 μm g = 1–9.1 μm dp = 400 nm	modulation of orientation of FAs	spreading & directed migration (g ≤ 3.8 μm)	[47]

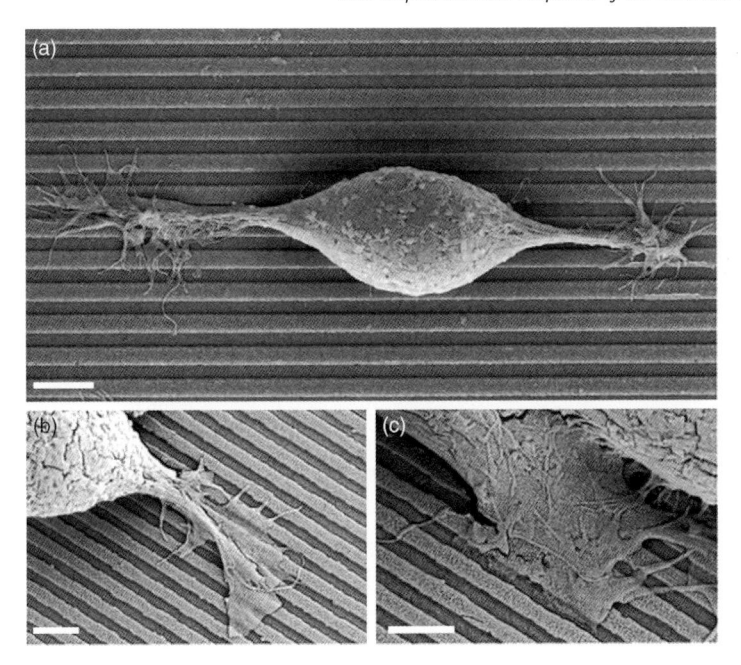

Figure 12.5 SEM images of a PC12 cell differentiating onto a submicron grating (a). Note that the cell body and developing neurites are standing on top of the ridges (b, c). Scale bars correspond to 2 μm. (Reproduced with permission from [30]. Copyright © (2008) IOP Publishing).

different levels. Gratings with lateral feature size in the submicron range (i.e., grooves and ridges $\leq 1\,\mu m$) were demonstrated to best induce cell alignment and directional migration of fibroblasts, smooth muscle, and endothelial cells [34, 44]. And, similarly, submicron gratings efficiently promoted neurite alignment, cell polarization and restricted migration in differentiating neurons (Figure 12.5) [29, 30, 42]. On such topographies the FAs, which are normally about 1 μm large and can be several microns long, are confined to ridges [32, 38, 49]. Hence, the width, size, maturation and persistence of FAs are controlled according to the size and geometry of the substrate, this yielding an overall reduction of adhesion formation (Figure 12.6; [32, 46]). Altogether these results support the counter-intuitive rule that efficient contact guidance correlates with reduced cellular adhesion strength [50].

Topographical features comparable to or larger than the cell body diameter (10–100 μm) confine the entire cell, or multiple cells, within a groove or on a single ridge. The lateral constraint provided by the groove walls or by the ridge edges was demonstrated to promote cell polarization and enhance differentiation in neurons [51] and osteoblasts [46] possibly by enhancing formation of FAs. In these conditions FAs tend to cluster around groove–ridge barriers as a potential response to the local surface curvature [52].

Finally we note the emerging role of disorder and symmetry in controlling cell behavior and, most importantly, cell differentiation and function. Human mesenchymal stem cells (MSC) are multipotent cells that, upon selective differentiation

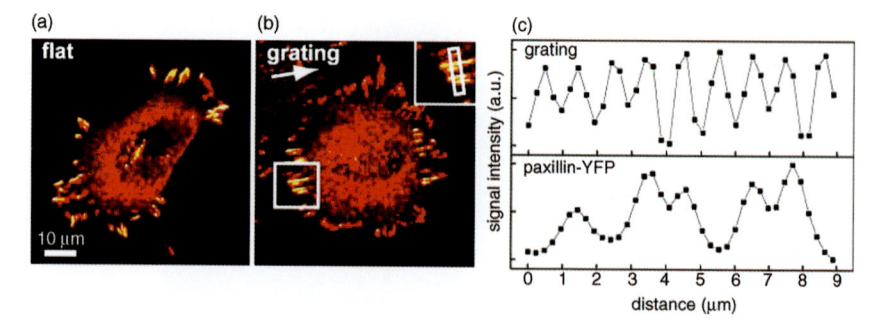

Figure 12.6 Focal adhesion confinement by submicron gratings in REF52. Paxillin-YFP signal from REF52 adhering to a flat substrate (a) and to a grating (b). The inset in (b) reports a zoomed view of the cell region enclosed in the white square. A white rectangle in the inset identifies the region encompassing the focal adhesions which was used for the graph of quantitative signal measurements (c). Signal intensity along the region containing the focal adhesions, measured in the transmission channel (grating) and in the fluorescence channel (paxillin-YFP) respectively. Note that the fluorescence peaks are confined to the ridges. (Adapted with permission from [32]. Copyright © (2010) Elsevier).

stimuli, can give rise to adipocytes, chondrocytes, osteoblasts, muscle cells, or fibroblasts [53]. Differentiation of MSC into bone cell populations was obtained, by means of pure topographical stimuli, using arrays of nanostructured pits with increasing degree of disorder (from an ordered square or hexagonal array to a fully random array; [39]). In particular, while fully ordered arrays proved inefficient (compared to the flat control) in inducing cell spreading, proliferation and differentiation, the best results were obtained using square arrays with a significant degree of disorder. These results further link formation of FAs with the cellular reaction to the underlying topography. And, suggest a model where substrate anisotropy contributes to cell polarization and directional migration by reducing the strength of cellular adhesion while, in the same frame, disorder acts ensuring that initial spreading and maturation of FAs are accomplished to the extent required for complete differentiation [39, 54].

12.2.2
Rigidity of the Substrate

During the morphogenesis of mammalian organs and tissues, developing cells adhere, contact and migrate within different extracellular environments, thus experiencing a wide range of matrix stiffness [55]. While, for example, differentiating neurons extend their protrusions (i.e., the neurites) within the soft brain matrix [56], other cell lineages, like muscle cells or osteoblasts, directly contact rigid collagen networks [57, 58]. It is now evident that the extent to which the local substrate stiffness (i.e., its mechanical response to traction) controls the cellular behavior, both in developing and adult tissues, goes beyond the modulation of early responses such as adhesion and migration. It dramatically affects cell commitment and fate [53].

How are the mechanical properties of the ECM probed by cells? In other words, what cellular machinery is used to sense and respond to the local substrate rigidity? This active molecular 'rheometer' is essentially composed of two modules: a probe and a transducer. The probe is required to generate traction forces and convey them to the substrate. The second module generates intracellular signals based on the mechanical response of the matrix.

Mammalian cells sense matrix rigidity through FAs [59, 60]. The scheme in Figure 12.2 shows that the cytoplasmic components of FAs (i.e., the adhesion plaque) are in direct contact with actin fibers [9]. These cortical fibers serve as scaffolds for local force generation through non-muscle myosin II-mediated contractility [61]. Through this link the force is transmitted first to the adhesion plaque and then to the ECM via the integrin contacts. FAs are thus the sites at which the cell-generated force is transmitted to the substrate [62]. Secondly, the adhesion plaque of a mature FA is enriched with several known signaling molecules, which get recruited and activated upon FA assembly and maturation [7]. The resulting signaling activity, modulated by the substrate rigidity, accounts for the role of FAs as on-the-spot mechanotransducers [63].

The complexity of this picture is further increased by the evidence that FA maturation itself requires tension generated by the cell and resisted by the substrate [3, 64]. Figure 12.7 schematically depicts the feedback loop regulating FA maturation. Individual FAs convey a constant stress of $5.5\,\text{nN}\,\mu\text{m}^{-2}$ to the substrate [3]. The local contraction requires, on the extracellular side, equivalent support from the matrix. In this case, the early adhesion matures anisotropically increasing in size along the direction of the exerted force [65, 66]. FA maturation is accompanied by the recruitment of adhesion plaque proteins and local actin polymerization [62]. This process eventually consolidates the cell adhesion and reshapes the cytoskeleton, and thus the cell polarity, accordingly [67]. Alternatively, if the substrate is too soft to resist the probing force, adhesion maturation fails and the contact is lost [68]. On the entire cell scale, this bidirectional flow of mechanical information encompassing FAs governs migration, spreading, matrix remodeling, cell differentiation, and tissue morphogenesis [53, 59, 68].

Experimentally different approaches have been used to produce substrates with controlled stiffness. A partial summary of the quantitative results obtained is reported in Table 12.3.

Two alternative solutions were generally adopted to produce substrates with variable rigidity. The first strategy exploits polyacrylamide (PAA) or poly(dimethyl siloxane) (PDMS) gels with variable monomer concentration and ratio between monomer and crosslinker to obtain flat substrates with tunable elastic modulus (i.e., Young's modulus E; [72, 74]). Alternatively PDMS pillar arrays (Figure 12.8) were obtained by casting the polymer on molding masks. In this case, control over the resulting structure size (the pillars' diameter and height) allows generation of bendable pillars with different spring constant [41, 69, 71, 75]. In all cases, the PDMS and PAA substrates were finally coated with cell adhesive proteins such as collagen or fibronectin to obtain surfaces presenting a homogenous ligand density (Figure 12.8).

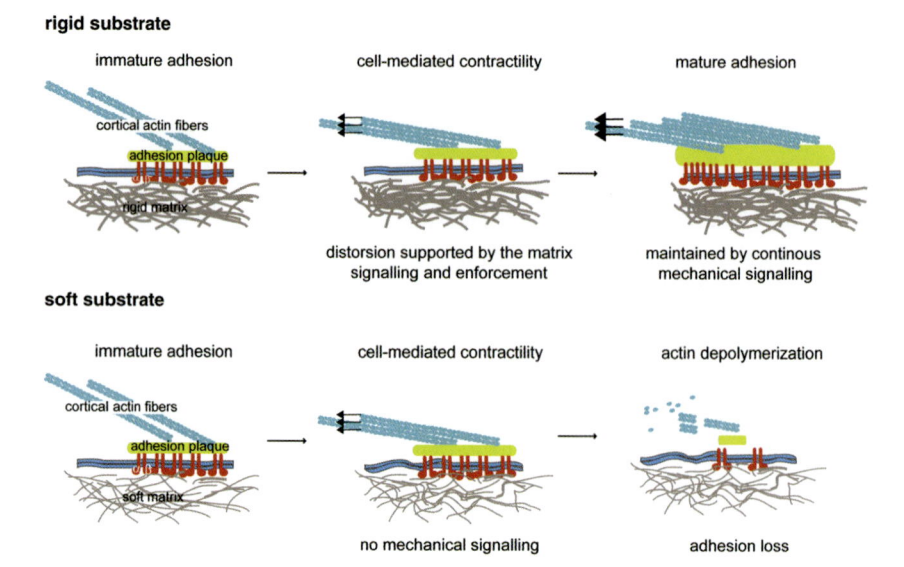

Figure 12.7 The influence of substrate elasticity on focal adhesions formed on rigid and soft substrates. When immature focal adhesions are formed (upper row) they do not necessarily occupy all available extracellular-matrix-binding. The distortions created in integrins (red) and proteins in the adhesion plaque (yellow) between the rigid extracellular binding sites (gray) and the contracting actin cytoskeleton (cyan) induces mechanical signaling. The subsequent recruitment of new integrins and adapter proteins together with actin polymerization leads to mature focal adhesions that, however, need to be continuously pulled to maintain mechanical signaling. When the cell establishes an immature adhesion on soft substrate (lower row), the subsequent acto-myosin dependent contraction is not able to induce signaling because of the lack of resistance of the substrate. The lack of focal adhesion reinforcement results in the dispersal of the focal adhesion site.

It was thus possible to demonstrate that both fibroblasts and endothelial cells naturally migrate to regions of the substrate with higher mechanical rigidity. This cellular behavior is defined as durotaxis [41, 73]. Similarly, using anisotropic PDMS pillar arrays (pillars with an ellipsoidal section where the spring constant changes depending on the bending direction) it was shown that kidney epithelial cells (MDCK) orient and move following the direction of higher rigidity [69, 75]. Using a similar approach MDCK cells were shown to enter alternative morphogenetic programs depending on the substrate rigidity in a molecularly regulated process that involves the regulation of cell contractility and FA maturation [61, 70, 71].

PDMS pillars have also been used as bidimensional on-site force sensor arrays [76]. In this application pillars (Figure 12.8) are designed and fabricated to have a spring constant yielding a resolvable deformation under cell-generated forces [71, 77, 78]. The correlation between pillar bending and the applied force is generally obtained using an atomic force microscope tip to apply a known force at

Table 12.3 Experimental boundaries for adhesion formation and cellular responses on substrates with controlled rigidity. Experiments using Polydimethylsiloxane (PDMS) or Polyacrilamide (PAA) are reported (E = Young's modulus; $k_{//}$ = spring constant; G = storage modulus).

Cell type	Material	Mechanical properties	Effect on FAs	Cell response	Ref.
fibroblasts	PDMS	$E = 15$ kPa	Maturation	spreading, migration	[3]
epithelial	PDMS	$E = 1800$ kPa $10 < k_{//} < 78$ nN μm^{-1}	asymmetric orientation	polarization, durotaxis	[69]
mesenchymal stem cells	PAA	$0.1 < E < 40$ kPa	differential maturation	lineage specification	[53]
epithelial cells	type I collagen/ PDMS	$G = 0.2$ Pa $k_{//} = 200$ N m^{-1}	signaling activation	differential morphogenesis	[70, 71]
epithelial cells/fibroblasts	PAA	$15 < E < 75$ Pa	differential maturation	spreading, migration	[72]
fibroblasts	PAA	$14 < E < 30$ kPa	—	durotaxis	[73]
fibroblasts	PAA	$10 < E < 20$ kPa	integrin engagement	spreading durotaxis	[59]
fibroblasts, endothelial cells	PAA/PDMS	$1.8 < E < 34$ kPa/ $2500 < E < 12\,000$ kPa	—	durotaxis	[74]
fibroblasts and cancer fibroblastic cells	PDMS	$500 < E < 2000$ kPa	—	polarization, migration and spreading	[41]

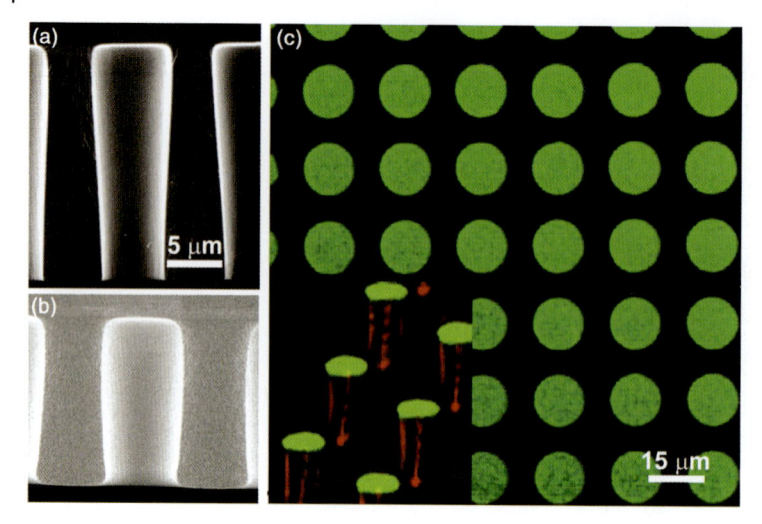

Figure 12.8 PDMS pillar arrays with different size and rigidity. Scanning electron microscopy images (a,b) of two different pillar configuration obtained by casting PDMS on a molding mask. The pillars in (a) have a diameter of 5.97 μm and height of 18.9 mm ($k_{//} = 50\,nN\,\mu m^{-1}$). The pillars in (b) have a diameter of 6.8 and a height of 11.7 μm ($k_{//} = 300\,nN\,\mu m^{-1}$). A collagen type I coating of the micropillars array is visualized by immunofluorescence labeling and confocal imaging (c). From the insert where bent pillars are imaged, it is clearly seen that the protein layer (shown in green) is restricted only to the top of the pillars, which were visualized in reflection (courtesy of Dr. Jens Ulmer).

the pillar tip. The out-of-focus signal recorded in a standard wide-field microscopy setup is used to directly calibrate the image with the force (Figure 12.9) yielding a local measure for its dynamics, direction and intensity. The relative transparency of the PDMS post array together with this automatic calibration permits direct live measurements of cell-generated forces during migration, contractility, and tissue morphogenesis [71].

Most importantly a publication by Engler and colleagues [53] clearly demonstrated that, in native human MSCs, the cellular commitment to a specific function can be specified by the mechanical properties of the substrate. The authors demonstrated that sparse cultures of MSCs, subjected to the same set of molecular stimuli in the culturing medium, developed alternatively neurons, myoblasts or osteoblasts when contacting soft ($E = 0.1–1$ kPa), intermediate ($E = 8–17$ kPa) and stiff ($E = 25–40$ kPa; Table 12.1) PAA scaffolds, respectively. This result provided a fundamental proof that cell differentiation (or re-differentiation) can be specified by means of pure material parameters. For this reason current tissue engineering approaches, aiming at producing biomimetic scaffolds for tissue regeneration or grafting, include a careful design of the visco-elastic properties of the material of choice. The ideal implant must provide an optimal biological support to the adhering cells. Yet, it should also closely match the mechanical properties of the tissue to be replaced or regenerated [79].

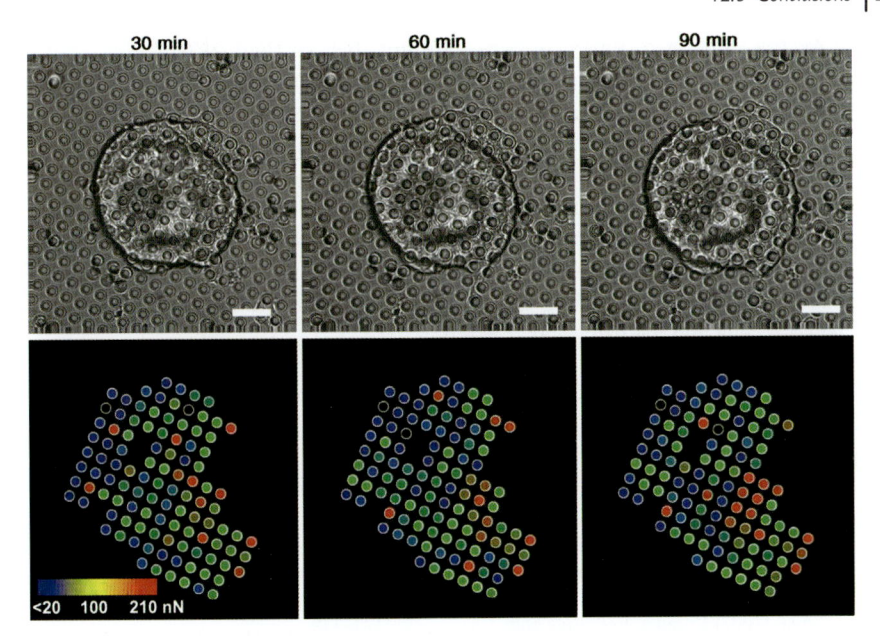

Figure 12.9 Epithelial cyst developing on top of an array of micro pillars. Transmission images of the cyst and pillars (diameter of 5 μm and height of 10 μm) at the cyst-pillar interface are shown for three time points (30, 60 and 90 minutes; upper row). The scale bar in each panel is 20 μm. In the lower row are reported the color coded contour plots of the measured forces on the pillars corresponding to the three time points shown in the upper row. A calibration bar (color to force) is reported in the lower left panel. (Adapted with permission from [71]. Copyright © (2010) Elsevier).

12.3
Conclusions

The set of new technological procedures described in this chapter has opened the possibility of controlling most of the material parameters which define the substrate identity toward interaction with mammalian cells. The generation of an optimal synthetic scaffold, bearing the ideal combination of physical, chemical and biological stimuli, is now conceivable and at the verge of implementation. Yet, the number of possible material combinations that are accessible has grown exponentially. Thus the final composition of a substrate is to be tailored among a large range of possibilities.

Current tissue engineering approaches demonstrated the control of two material parameters on one substrate. Chips with local independent variations of topography and surface energy, ligand density and rigidity, and topography and rigidity, were realized yielding bidimensional maps of cell adhesion, spreading and migration efficiency [41, 80, 81]. When applied to different cell types, these maps have the potential to individuate the best substrate composition to selectively promote the adhesion of desired cells (e.g., endothelial cells) while disfavoring others (e.g., smooth muscle cells or fibroblasts).

As this top-down approach moves further, the demand of high-throughput solutions and the degree of automation required, in the design, fabrication and characterization of materials, constantly increases. On the cellular side, efforts to unfold the structure and function of critical molecular machineries, such as the focal adhesions, have the potential to narrow the field by addressing the design of new materials to the most promising combinations.

Altogether a new area of research is shaping, which gathers expertise from computational modeling, polymer chemistry, biophysics, cellular biology and medicine, and aims at controlling the complex interaction standing at the interface between cells and the extracellular world.

References

1 Dalby, M.J. (2009) Nanostructured surfaces: cell engineering and cell biology. *Nanomed.*, **4**, 247–248.

2 Fink, J., Thery, M., Azioune, A., Dupont, R., Chatelain, F., Bornens, M. et al. (2007) Comparative study and improvement of current cell micropatterning techniques. *Lab on a chip*, **7**, 672–680.

3 Balaban, N.Q., Schwarz, U.S., Riveline, D., Goichberg, P., Tzur, G., Sabanay, I. et al. (2001) Force and focal adhesion assembly: a close relationship studied using elastic micropatterned substrates. *Nat. Cell Biol.*, **3**, 466–472.

4 Katsumi, A., Orr, A.W., Tzima, E., and Schwartz, M.A. (2004) Integrins in mechanotransduction. *J. Biol. Chem.*, **279**, 12001–12004.

5 Geiger, B., Bershadsky, A., Pankov, R., and Yamada, K.M. (2001) Transmembrane extracellular matrix-cytoskeleton crosstalk. *Nat. Rev. Mol. Cell Biol.*, **2**, 793–805.

6 Zaidel-Bar, R. (2009) Evolution of complexity in the integrin adhesome. *J. Cell Biol.*, **186**, 317–321.

7 Zaidel-Bar, R., Itzkovitz, S., Ma'ayan, A., Iyengar, R., and Geiger, B. (2007) Functional atlas of the integrin adhesome. *Nat. Cell Biol.*, **9**, 858–867.

8 Wozniak, M.A., Modzelewska, K., Kwong, L., and Keely, P.J. (2004) Focal adhesion regulation of cell behavior. *Biochim. Biophys. Acta*, **1692**, 103–119.

9 Geiger, B., Spatz, J.P., and Bershadsky, A.D. (2009) Environmental sensing through focal adhesions. *Nat. Rev. Mol. Cell Biol.*, **10**, 21–33.

10 Decuzzi, P. and Ferrari, M. (2009) Modulating cellular adhesion through nanotopography. *Biomaterials*, **31** (1), 173–179

11 Xu, R., Boudreau, A., and Bissell, M.J. (2009) Tissue architecture and function: dynamic reciprocity via extra- and intracellular matrices. *Cancer Metastasis Rev.*, **28**, 167–176.

12 Chen, C.S., Mrksich, M., Huang, S., Whitesides, G.M., and Ingber, D.E. (1997) Geometric control of cell life and death. *Science*, **276**, 1425–1428.

13 Massia, S.P. and Hubbell, J.A. (1990) Covalently attached GRGD on polymer surfaces promotes biospecific adhesion of mammalian cells. *Ann. N. Y. Acad. Sci.*, **589**, 261–270.

14 Glass, R., Moller, M., and Spatz, J.P. (2003) Block copolymer micelle nanolithography. *Nanotechnology.*, **14**, 1153–1160.

15 Arnold, M., Schwieder, M., Blummel, J., Cavalcanti-Adam, E.A., Lopez-Garcia, M., Kessler, H. et al. (2009) Cell interactions with hierarchically structured nano-patterned adhesive surfaces. *Soft Matter*, **5** (1), 72–77.

16 Aydin, D., Schwieder, M., Louban, I., Knoppe, S., Ulmer, J., Haas, T.L. et al. (2009) Micro-nanostructured protein arrays: a tool for geometrically controlled ligand presentation. *Small*, **5**, 1014–1018.

17 Cavalcanti-Adam, E.A., Micoulet, A., Blummel, J., Auernheimer, J., Kessler, H.,

and Spatz, J.P. (2006) Lateral spacing of integrin ligands influences cell spreading and focal adhesion assembly. *Eur. J. Cell Biol.*, **85**, 219–224.

18 Cavalcanti-Adam, E.A., Volberg, T., Micoulet, A., Kessler, H., Geiger, B., and Spatz, J.P. (2007) Cell spreading and focal adhesion dynamics are regulated by spacing of integrin ligands. *Biophys. J.*, **92**, 2964–2974.

19 Selhuber-Unkel, C., Lopez-Garcia, M., Kessler, H., and Spatz, J.P. (2008) Cooperativity in adhesion cluster formation during initial cell adhesion. *Biophys. J.*, **95**, 5424–5431.

20 Arnold, M., Hirschfeld-Warneken, V.C., Lohmuller, T., Heil, P., Blummel, J., Cavalcanti-Adam, E.A. et al. (2008) Induction of cell polarization and migration by a gradient of nanoscale variations in adhesive ligand spacing. *Nano Lett.*, **8**, 2063–2069.

21 Ranzinger, J., Krippner-Heidenreich, A., Haraszti, T., Bock, E., Tepperink, J., Spatz, J.P. et al. (2009) Nanoscale arrangement of apoptotic ligands reveals a demand for a minimal lateral distance for efficient death receptor activation. *Nano Lett.*, **9** (12), 4240–4245.

22 Zaidel-Bar, R., Cohen, M., Addadi, L., and Geiger, B. (2004) Hierarchical assembly of cell-matrix adhesion complexes. *Biochem. Soc. Trans.*, **32**, 416–420.

23 Mossman, K.D., Campi, G., Groves, J.T., and Dustin, M.L. (2005) Altered TCR signaling from geometrically repatterned immunological synapses. *Science*, **310**, 1191–1193.

24 Arnold, M., Schwieder, M., Blummel, J., Cavalcanti-Adam, E.A., Lopez-Garcia, M., Kessler, H. et al. (2009) Cell interactions with hierarchically structured nano-patterned adhesive surfaces. *Soft Matter*, **5**, 72–77.

25 Jaehrling, S., Thelen, K., Wolfram, T., and Pollerberg, G.E. (2009) Nanopatterns biofunctionalized with cell adhesion molecule DM-GRASP offered as cell substrate: spacing determines attachment and differentiation of neurons. *Nano Lett.*, **9** (12), 4115–4121.

26 Curtis, A.S. and Varde, M. (1964) Control of cell behavior: topological factors. *J. Natl. Cancer Inst.*, **33**, 15–26.

27 Martinez, E., Engel, E., Planell, J.A., and Samitier, J. (2009) Effects of artificial micro- and nano-structured surfaces on cell behaviour. *Ann. Anat.*, **191**, 126–135.

28 Shin, H. (2007) Fabrication methods of an engineered microenvironment for analysis of cell-biomaterial interactions. *Biomaterials*, **28**, 126–133.

29 Cecchini, M., Bumma, G., Serresi, M., and Beltram, F. (2007) PC12 differentiation on biopolymer nanostructures. *Nanotechnology*, **18**, 1–7.

30 Cecchini, M., Ferrari, A., and Beltram, F. (2008) PC12 polarity on biopolymer nanogratings. *J. Phys.: Conf. Ser.*, **100**, 012003.

31 Liliensiek, S.J., Wood, J.A., Yong, J., Auerbach, R., Nealey, P.F., and Murphy, C.J. (2010) Modulation of human vascular endothelial cell behaviors by nanotopographic cues. *Biomaterials*, **31** (20), 5418–26.

32 Ferrari, A., Cecchini, M., Serresi, M., Faraci, P., Pisignano, D., and Beltram, F. (2010) Neuronal polarity selection by topography-induced focal-adhesion control. *Biomaterials*, **31** (17), 4682–4694.

33 Ferrari, A., Faraci, P., Cecchini, M., and Beltram, F. (2010) The effect of alternative neuronal differentiation pathways on PC12 cell adhesion and neurite alignment to nanogratings. *Biomaterials*, **31** (17), 2565–2573.

34 Biela, S.A., Su, Y., Spatz, J.P., and Kemkemer, R. (2009) Different sensitivity of human endothelial cells, smooth muscle cells and fibroblasts to topography in the nano-micro range. *Acta Biomater.*, **5** (7), 2460–2466.

35 Dalby, M.J. (2005) Topographically induced direct cell mechanotransduction. *Med. Eng. Phys.*, **27**, 730–742.

36 Johansson, F., Carlberg, P., Danielsen, N., Montelius, L., and Kanje, M. (2006) Axonal outgrowth on nano-imprinted patterns. *Biomaterials*, **27**, 1251–1258.

37 Rajnicek, A., Britland, S., and McCaig, C. (1997) Contact guidance of CNS neurites on grooved quartz: influence of groove

dimensions, neuronal age and cell type. *J. Cell Sci.*, **110** Pt (23), 2905–2913.

38 Teixeira, A.I., Abrams, G.A., Bertics, P.J., Murphy, C.J., and Nealey, P.F. (2003) Epithelial contact guidance on well-defined micro- and nanostructured substrates. *J. Cell Sci.*, **116**, 1881–1892.

39 Dalby, M.J., Gadegaard, N., Tare, R., Andar, A., Riehle, M.O., Herzyk, P. *et al.* (2007) The control of human mesenchymal cell differentiation using nanoscale symmetry and disorder. *Nature Mater.*, **6**, 997–1003.

40 Francisco, H., Yellen, B.B., Halverson, D.S., Friedman, G., and Gallo, G. (2007) Regulation of axon guidance and extension by three-dimensional constraints. *Biomaterials*, **28**, 3398–3407.

41 Tzvetkova-Chevolleau, T., Stephanou, A., Fuard, D., Ohayon, J., Schiavone, P., and Tracqui, P. (2008) The motility of normal and cancer cells in response to the combined influence of the substrate rigidity and anisotropic microstructure. *Biomaterials*, **29**, 1541–1551.

42 Liliensiek, S.J., Wood, J.A., Yong, J., Auerbach, R., Nealey, P.F., and Murphy, C.J. (2010) Modulation of human vascular endothelial cell behaviors by nanotopographic cues. *Biomaterials*, **31** (20), 5418–26.

43 Kulangara, K. and Leong, K.W. (2009) Substrate topography shapes cell function. *Soft Matter*, **5**, 4072–4076.

44 Loesberg, W.A., te Riet, J., van Delft, F.C.M.J.M., Schon, P., Figdor, C.G., Speller, S. *et al.* (2007) The threshold at which substrate nanogroove dimensions may influence fibroblast alignment and adhesion. *Biomaterials*, **28**, 3944–3951.

45 Crouch, A.S., Miller, D., Luebke, K.J., and Hu, W. (2009) Correlation of anisotropic cell behaviors with topographic aspect ratio. *Biomaterials*, **30**, 1560–1567.

46 Biggs, M.J., Richards, R.G., Gadegaard, N., Wilkinson, C.D., Oreffo, R.O., and Dalby, M.J. (2009) The use of nanoscale topography to modulate the dynamics of adhesion formation in primary osteoblasts and ERK/MAPK signalling in STRO-1+ enriched skeletal stem cells. *Biomaterials*, **30**, 5094–5103.

47 Kim, D.H., Han, K., Gupta, K., Kwon, K.W., Suh, K.Y., and Levchenko, A. (2009) Mechanosensitivity of fibroblast cell shape and movement to anisotropic substratum topography gradients. *Biomaterials*, **30**, 5433–5444.

48 Mattila, P.K. and Lappalainen, P. (2008) Filopodia: molecular architecture and cellular functions. *Nat. Rev. Mol. Cell Biol.*, **9**, 446–454.

49 Biggs, M.J., Richards, R.G., Wilkinson, C.D., and Dalby, M.J. (2008) Focal adhesion interactions with topographical structures: a novel method for immuno-SEM labelling of focal adhesions in S-phase cells. *J Microsc.*, **231**, 28–37.

50 Dalby, M.J., Hart, A., and Yarwood, S.J. (2008) The effect of the RACK1 signalling protein on the regulation of cell adhesion and cell contact guidance on nanometric grooves. *Biomaterials*, **29**, 282–289.

51 Mahoney, M.J., Chen, R.R., Tan, J., and Saltzman, W.M. (2005) The influence of microchannels on neurite growth and architecture. *Biomaterials*, **26**, 771–778.

52 Walboomers, X.F., Croes, H.J., Ginsel, L.A., and Jansen, J.A. (1998) Growth behavior of fibroblasts on microgrooved polystyrene. *Biomaterials*, **19**, 1861–1868.

53 Engler, A.J., Sen, S., Sweeney, H.L., and Discher, D.E. (2006) Matrix elasticity directs stem cell lineage specification. *Cell*, **126**, 677–689.

54 Huang, J.H., Grater, S.V., Corbellinl, F., Rinck, S., Bock, E., Kemkerer, R. *et al.* (2009) Impact of order and disorder in RGD nanopatterns on cell adhesion. *Nano Lett.*, **9**, 1111–1116.

55 Cukierman, E., Pankov, R., Stevens, D.R., and Yamada, K.M. (2001) Taking cell-matrix adhesions to the third dimension. *Science*, **294**, 1708–1712.

56 Rauch, U. (2007) Brain matrix: structure, turnover and necessity. *Biochem. Soc. Trans.*, **35**, 656–660.

57 Engler, A.J., Griffin, M.A., Sen, S., Bonnemann, C.G., Sweeney, H.L., and Discher, D.E. (2004) Myotubes differentiate optimally on substrates with tissue-like stiffness: pathological implications for soft or stiff

microenvironments. *J. Cell Biol.*, **166**, 877–887.

58 Garcia, A.J. and Reyes, C.D. (2005) Bioadhesive surfaces to promote osteoblast differentiation and bone formation. *J. Dent. Res.*, **84**, 407–413.

59 Jiang, G., Huang, A.H., Cai, Y., Tanase, M., and Sheetz, M.P. (2006) Rigidity sensing at the leading edge through alphavbeta3 integrins and RPTPalpha. *Biophys. J.*, **90**, 1804–1809.

60 Riveline, D., Zamir, E., Balaban, N.Q., Schwarz, U.S., Ishizaki, T., Narumiya, S. et al. (2001) Focal contacts as mechanosensors: externally applied local mechanical force induces growth of focal contacts by an mDia1-dependent and ROCK-independent mechanism. *J. Cell Biol.*, **153**, 1175–1186.

61 Ferrari, A., Veligodskiy, A., Berge, U., Lucas, M.S., and Kroschewski, R. (2008) ROCK-mediated contractility, tight junctions and channels contribute to the conversion of a preapical patch into apical surface during isochoric lumen initiation. *J. Cell Sci.*, **121**, 3649–3663.

62 Bershadsky, A.D., Balaban, N.Q., and Geiger, B. (2003) Adhesion-dependent cell mechanosensitivity. *Annu. Rev. Cell Dev. Biol.*, **19**, 677–695.

63 Giancotti, F.G. and Ruoslahti, E. (1999) Transduction - Integrin signaling. *Science*, **285**, 1028–1032.

64 Butler, J.P., Tolic-Norrelykke, I.M., Fabry, B., and Fredberg, J.J. (2002) Traction fields, moments, and strain energy that cells exert on their surroundings. *Am. J. Physiol. Cell Physiol.*, **282**, C595–605.

65 Aroush, D.R.B., Zaidel-Bar, R., Bershadsky, A.D., and Wagner, H.D. (2008) Temporal evolution of cell focal adhesions: experimental observations and shear stress profiles. *Soft Matter*, **4**, 2410–2417.

66 Raz-Ben Aroush, D. and Wagner, H.D. (2006) Shear-stress profile along a cell focal adhesion. *Adv. Mater.*, **18**, 1537–.

67 Paul, R., Heil, P., Spatz, J.P., and Schwarz, U.S. (2008) Propagation of mechanical stress through the actin cytoskeleton toward focal adhesions: Model and experiment. *Biophys. J.*, **94**, 1470–1482.

68 Dembo, M. and Wang, Y.L. (1999) Stresses at the cell-to-substrate interface during locomotion of fibroblasts. *Biophys. J.*, **76**, 2307–2316.

69 Saez, A., Ghibaudo, M., Buguin, A., Silberzan, P., and Ladoux, B. (2007) Rigidity-driven growth and migration of epithelial cells on microstructured anisotropic substrates. *Proc. Natl. Acad. Sci. USA*, **104**, 8281–8286.

70 Wozniak, M.A., Desai, R., Solski, P.A., Der, C.J., and Keely, P.J. (2003) ROCK-generated contractility regulates breast epithelial cell differentiation in response to the physical properties of a three-dimensional collagen matrix. *J. Cell Biol.*, **163**, 583–595.

71 Zeng, D., Ferrari, A., Ulmer, J., Veligodskiy, A., Fischer, P., Spatz, J. et al. (2006) Three-dimensional modeling of mechanical forces in the extracellular matrix during epithelial lumen formation. *Biophys. J.*, **90**, 4380—4391.

72 Pelham, R.J. Jr. and Wang, Y. (1997) Cell locomotion and focal adhesions are regulated by substrate flexibility. *Proc. Natl. Acad. Sci. USA*, **94**, 13661–13665.

73 Lo, C.M., Wang, H.B., Dembo, M., and Wang, Y.L. (2000) Cell movement is guided by the rigidity of the substrate. *Biophys. J.*, **79**, 144–152.

74 Gray, D.S., Tien, J., and Chen, C.S. (2003) Repositioning of cells by mechanotaxis on surfaces with micropatterned Young's modulus. *J. Biomed. Mater. Res. Part A.*, **66**, 605–614.

75 du Roure, O., Saez, A., Buguin, A., Austin, R.H., Chavrier, P., Silberzan, P. et al. (2005) Force mapping in epithelial cell migration. *Proc. Natl. Acad. Sci. USA*, **102**, 2390–2395.

76 Tan, J.L., Tien, J., Pirone, D.M., Gray, D.S., Bhadriraju, K., and Chen, C.S. (2003) Cells lying on a bed of microneedles: an approach to isolate mechanical force. *Proc. Natl. Acad. Sci. USA*, **100**, 1484–1489.

77 Roos, W.H., Roth, A., Konle, J., Presting, H., Sackmann, E., and Spatz, J.P. (2003) Freely suspended actin cortex models on

arrays of microfabricated pillars. *Chemphyschem*, **4**, 872–877.

78 Roth, A., Roos, W., Spatz, J., and Sackmann, E. (2003) Elastic properties of freely suspended two-dimesional model actin cortex on microfabricated three-dimensional pillar substrates. *Biophys. J.*, **84**, 439a–440a.

79 Engelmayr, G.C., Cheng, M.Y., Bettinger, C.J., Borenstein, J.T., Langer, R., and Freed, L.E. (2008) Accordion-like honeycombs for tissue engineering of cardiac anisotropy. *Nature Mater.*, **7**, 1003–1010.

80 Roach, P., Eglin, D., Rohde, K., and Perry, C.C. (2007) Modern biomaterials: a review-bulk properties and implications of surface modifications. *J. Mater. Sci.-Mater. M.*, **18**, 1263–1277.

81 Sun, J.G., Graeter, S.V., Yu, L., Duan, S.F., Spatz, J.P., and Ding, J.D. (2008) Technique of surface modification of a cell-adhesion-resistant hydrogel by a cell-adhesion-available inorganic microarray. *Biomacromolecules*, **9**, 2569–2572.

13
Polymer Patterns and Scaffolds for Biomedical Applications and Tissue Engineering

Natália M. Alves, Iva Pashkuleva, Rui L. Reis, and João F. Mano

13.1
Introduction

Polymers have gained a remarkable place in the biomedical field as materials for the fabrication of various devices and tissue engineering applications. The versatility of their chemical composition and mechanical properties, the large pool of processing methodologies and the tailored biodegradability under physiological conditions have made them the obvious material of choice for many biomedical applications. Of course, the bulk properties are essential for the durability and the proper function of a biomaterial. However, the initial acceptance or rejection of an implantable device is dictated by the crosstalk of the material surface with the bioentities present in the physiological environment. Any medical intervention, including an introduction of a biodevice/biomaterial in the human body, induces changes in this environment. Controlled and predictable cellular response to these changes is highly desirable for tissue engineering and regenerative medicine. However, the regulation of cellular behavior in response to environmental changes is still one of the main obstacles in the design of biomedical materials and is associated with pathological states [1, 2], including blood clotting and wound healing defects as well as malignant tumor formation.

Living cells are extremely complex entities, presenting remarkable, inherent capacity to sense, integrate, and respond to environmental cues [3, 4]. Their native environment is a three-dimensional (3D) scaffold comprising an insoluble aggregate of several highly organized, multifunctional large proteins and glycosaminoglycans (GAGs), collectively known as extracellular matrix (ECM) [5, 6]. ECM provides a mechanical support for cells (most mammalian cells are anchorage-dependent, that is, they must adhere to a surface in order to survive) but also profoundly influences the fundamental cellular functions (e.g., migration, proliferation, differentiation, and apoptosis) of cells in contact with it (Figure 13.1). By interacting directly with the ECM, cells gather information about the chemical and physical nature of the environment, integrate and interpret it, and then generate an appropriate physiological response [4, 7]. Keeping in mind this extreme intelligence and sensitivity of

---

Generating Micro- and Nanopatterns on Polymeric Materials. Edited by A. del Campo and E. Arzt
Copyright © 2011 WILEY-VCH Verlag GmbH & Co. KGaA, Weinheim
ISBN: 978-3-527-32508-5

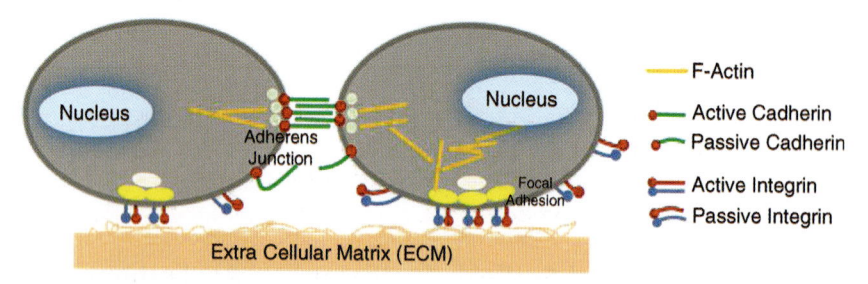

Figure 13.1 Schematic representation of cell–cell and cell–ECM interactions.

cells, it seems convenient that their behavior can be directed through a precisely designed environment.

It would be a naïve ambition to try to copycat the entire complex signaling cascade induced by the components of the extracellular environment. Nevertheless, advances in microfabrication and nanotechnology offer new tools for the investigation of this signaling and consequently for tailoring cellular responses, which mimic some elements of the signaling paths. In particular, the spatial regulation of cellular processes can be examined by engineering the chemical and physical micro- and nanoenvironments to which cells respond. Patterning methods and selective chemical modification schemes can provide biocompatible surfaces that control cellular interactions on the micron and submicron scales on which cells are organized. Although two-dimensional (2D) tailored surfaces are indispensable as models and give unique information about the components of ECM and to a certain extent about the interactions between these components and cells, they are still a long way from recreating the complexity and dynamics of the natural 3D environment of cells, their ECM. It is likely that cells require the full context of this 3D nanofibrous matrix to maintain their phenotypic shape and establish natural behavior patterns. Achieving effective temporal control over the signals that are presented to cells in 3D artificial matrices is still a key challenge [5] in the optimization of the outside-in signaling.

13.2
Cell Response to 2D Patterns

Cell behavior on surfaces presenting micrometer-sized features revealed that such topographies can induce cell polarization, direct cell migration and even regulate gene expression and cell signaling [8–10]. Many authors have analyzed the effects of ordered textures and, in particular microgrooves and ridges, on different cell types [11, 12]. Typically the cells become elongated and aligned in the direction of the grooves and migrate as guided by the grooves (see for example [12]). This phenomenon is known as contact guidance. It has also been found that the degree of alignment depends on the groove depth and width [12] and varies between different cells. Generally, the analyses of the effect of microgrooves on cell behavior have shown that the extent of orientation increases with groove depth up to 25 μm from topographies of ~1 µm. Based on these findings some authors suggested that these physical cues should not be greater than the actual size of the cell [12]. Ber *et al.* also found [13] that osteoblasts aligned and oriented on micropatterned collagen films within this size range, but the same did not occur when macropatterns are used. Very recently the differences in the morphological behavior between fibroblasts cultured on nanogrooved (groove depth: 5–350 nm, width: 20–1000 nm) and smooth polystyrene substrates were analyzed in detail in order to clarify to what extent cell guidance occurs on increasingly smaller topographies [14]. It was found that fibroblasts do not align for groove depths below 35 nm or ridge widths smaller than 100 nm. Besides cell alignment, other morphological change that has been observed in response to adhesion on microgrooves is the polarization of the cells [15]. The effect of other patterns such as pillars and pits, has also been analyzed [16], with cell proliferation thought to be increased with decreasing pit diameter [17] and cell attachment thought to be increased on microscale pillars compared with macroscale pits [18].

It has been demonstrated that cells do also respond to random surface topographies. Random patterns are easier to fabricate using low-cost techniques that do not require special equipment, such as lithography. Demixing of block copolymers, surface chemical etching, plasma treatments or electrospinning constitute significant examples that will be described in this section.

Human vascular endothelial cells were cultured on electrospun (randomly oriented nanofibers, that is, rough surface) and solvent-cast (smooth) poly(L-lactic acid) (PLLA) substrates. [19] Cell function, including cell phenotype, cell adhesion and proliferation was significantly improved on the rough surfaces. Moreover, the cell morphology was completely different on the electrospun substrates [19]. While anisotropic topographies such as ridges and grooves affect the individual cell behavior (cells align along the anisotropic direction), isotropic topographies, such as evenly or randomly distributed pits or protrusions such as the ones shown in Figure 13.2 obtained by submitting the substrates to plasma treatments performed at distinct conditions, affect collective cell behavior.

Roughness may also produce changes in the wettability of the surfaces that can affect cell behavior. In fact, PLLA surfaces combining nano- and micro-level roughness were developed using a phase separation method [20]. Such surfaces were found to be highly hydrophobic, exhibiting water contact angles that could be higher than 150°. The behavior of bone marrow derived cells in contact with these PLLA surfaces was found to be significantly affected by surface topography: such rough surfaces inhibit cell adhesion and proliferation, compared with a more flat one (Figure 13.3).

Random microtopographies have also been generated exploiting the crystallization process of polymer materials and the resulting surface morphologies. The crystallization kinetics of PLLA, for example, can be controlled and spherulitic development may be interrupted by different thermal treatments. [21–23]. Figure 13.4a shows the different spherulitic textures that were produced in a PLLA flat film upon different treatments. The adhesion and the cytoskeleton organization of primary human chondrocytes onto these substrates were influenced by such different topographies [24]. Cells adopted more elongated shapes onto bigger spherulites (L and M

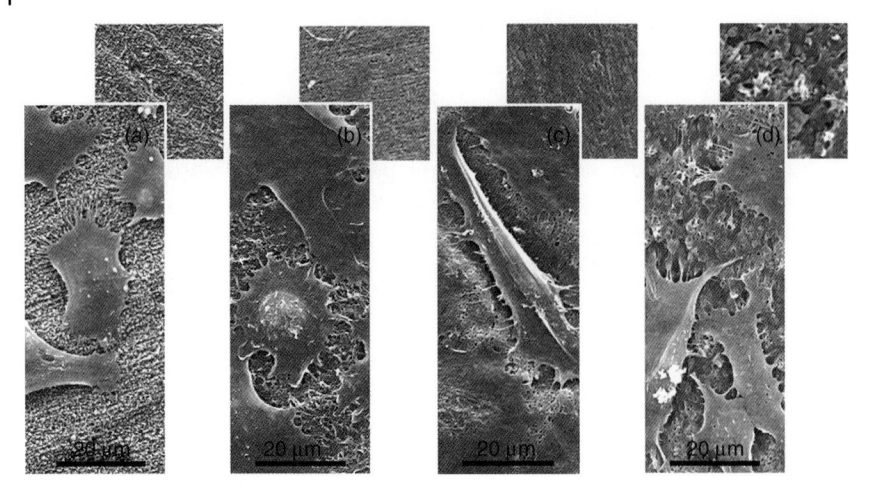

Figure 13.2 Substrates from blend of starch with ethylene vinyl alcohol (SEVA) (50/50 wt%) submitted to distinct plasma treatments (the small squares above) and cultured with a Saos-2 for 2 days: (a) 10 min in O_2 and applying a power of 30 W; (b) 30 min, O_2 and power of 80 W; (c) 5 min in argon and power of 30 W; and (d) untreated SEVA [25].

samples, that also show deeper microgrooves between the spherulites) and a more isotropic organization in the smoother surfaces (S and Am) – see Figure 13.4b. Note that such orientation observed in the cells is originated in textured surfaces without any special order, as spherulites are generated randomly throughout the volume of the material and in the end they are separated from each other by flat front lines. This study suggested that the observed differences could be assigned to the different conformation that proteins adopt upon adsorption to the surfaces showing distinct microtopographies. Such investigations were performed in 2D PLLA films, but similar different topographic features could be generated in the surface of 3D constructs.

Patterning has also been used in the so-called "cell sheet engineering". Such technology has been proposed to construct ideal transplantable tissues composed exclusively of cells [26, 27]. Okano and co-workers proposed temperature-responsive cell culture polystyrene substrates prepared by the grafting of thermo-responsive

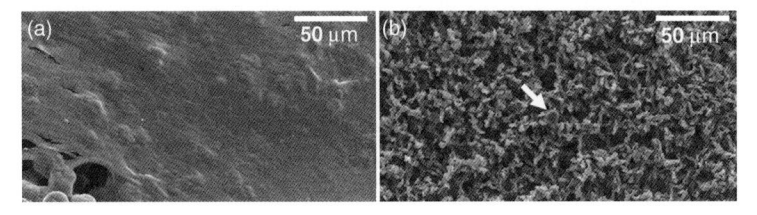

Figure 13.3 Bone marrow derived cells (RBMCs) on control films (a) and superhydrophobic substrates (b) after 21 days of culture. The white arrow identifies the only cell that was observed in the later surface. (Adapted with permission from [20]. Copyright © (2009) John Wiley & Sons Inc.)

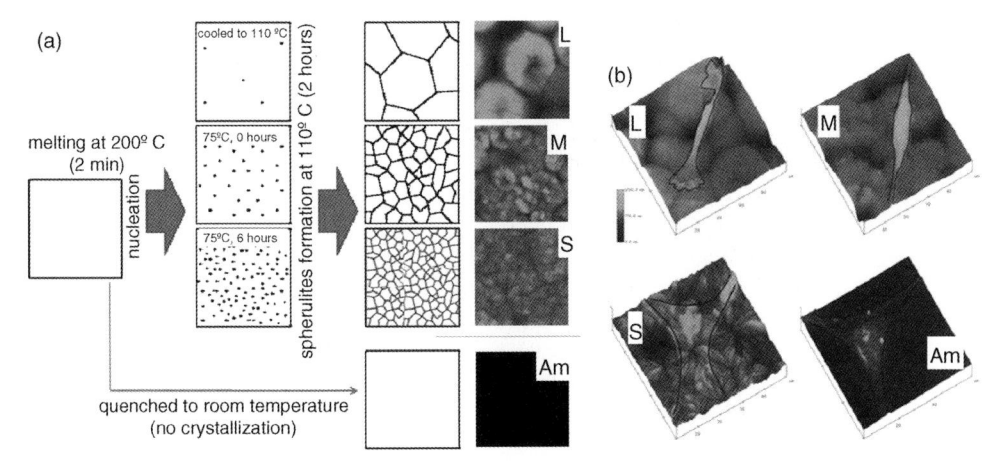

Figure 13.4 (a) Generation of different spherulitic textures on PLLA upon different thermal treatments. The amorphous material (Am) is produced by quenching the melt to room temperature; large spherulites (L) are obtained by directly crystallizing the melt to the crystallization temperature; medium (M) and small (S) spherulites are obtained by a previous nucleation induction by treating the material at 75 °C, for 0 and 6 hours, respectively, before the crystallization step. AFM images clearly show the different microtopographies obtained. (b) Morphology of chondrocytes on the different surfaces as seen by AFM; the lines follow the contour of the cells. (Data adapted with permission from the results in [24]. Copyright © (2008) Mary Ann Liebert, Inc.)

polymers, such as poly(*N*-isopropylacrylamide) (PNIPAAm), onto these surfaces, which allow the culture of confluent cell monolayers at 37 °C and their recovery as single cell sheets when the temperature is below the lower critical solubilization temperature (LCST) of PNIPAAm. An interesting approach was the grafting of PNIPAAm onto substrates that already presented micropatterns such as ridges and grooves [28]. Such topographic features influenced the orientation of endothelial cells seeded onto such surfaces, compared with the culture in flat substrates. After 2–3 weeks of culture at 37 °C the endothelial cells showed capillary-like tube formation on these thermo-responsive microtextured substrates. Such study provides an example where smart and patterned surfaces may be used to produce biological products with interesting therapeutical applications.

From the illustrative examples mentioned and from the extensive literature on this issue it is evident that topography affects cell adhesion, differentiation, proliferation, matrix production, gene expression, cell morphology, and orientation. Topography can provide directional growth for cells and may ultimately create tissue architecture at the cellular and subcellular level in a reproducible manner. So topographical cues could help shorten the tissue formation and the reorganization periods. Guidance is also of great importance when one deals with different kinds of cells and when we try to mimic the 3D structure and complexity of the activities in the tissue. The examples given here also point to the fact that the observed effects on cell behavior are extremely dependent not only on the substrate and the features of the micro/nano-texture that is created, but also on the cell type. It is known that surfaces acquire biological activity *in vivo* and *in vitro* through a layer of protein that is rapidly and largely irreversibly adsorbed from the surrounding medium [29]. Cells can then interact with the material in two ways: through the adsorbed proteins that will mediate cell adhesion and regulate multiple aspects of cell physiology [30, 31]; directly via pseudopodia that extend through the protein layer or by the consumption of adsorbed proteins [31, 32]. Consequently, the mechanism governing the varying response to different surface depends on many complex factors, namely, differences in protein adsorption or in the cytoskeleton arrangement could lead to different responses of distinct cell types when brought into contact with a given surface topography. So, for tissue engineering applications it would be a major breakthrough to discover the type of substrate topography that each cell type preferentially grows on, and create textured biomaterials in order to selectively propagate a particular cell type and optimize cellular adhesion, proliferation and other aspects of cell behavior but, as this section showed, this still constitutes a complex problem without a straightforward solution.

13.3
Cells onto 3D Objects and Scaffolds

So far, the majority of the research focused on the influence of the topography on the behavior of different cells has been carried out on planar surfaces. Those substrates represent poorly the *in vivo* environment which contain curved and porous objects such as blood vessels, bones, muscles, and so on. Hence, the ability to pattern curved walls would be very useful for controlling cell interaction with implanted stents and conduits. Recent efforts have succeeded in structuring tubular substrata. Using block copolymers, nylon tubes can be patterned with internal nanotopography [33, 34]. The process is similar to spin casting and leaves a thin layer of a suitable blend of polymers onto the substrate surface. These 3D nanotopographic features were found to decrease human osteoprogenitor cell spreading, reduce cytoskeletal organization and increase endocytotic activity [33]. This approach can be extended to 3D structures.

Three-dimensional scaffolds serve as temporary substrates for supporting and guiding tissue formation in various *in vitro* and *in vivo* tissue regeneration settings. Since the mid-1980s researchers have developed many novel techniques to shape polymers into complex architectures that exhibit the desired properties for specific tissue engineering applications. These fabrication techniques result in reproducible scaffolds with mechanical strength, interconnected porosity, varying surface chemistry and surface area, and unique geometries to direct specific tissue regeneration. Biodegradability and biocompatibility are other crucial issues in the design of 3D scaffolds; restoring the function of a tissue, or replacing an organ entirely, requires a porous scaffold that degrades on an appropriate time scale so that the new tissue replaces the resorbing scaffold. Therefore, the processing methodology must not affect both biocompatibility and biodegradability of the scaffolding material. The conventional techniques for scaffold fabrication include textile technologies, solvent casting, particulate leaching, and membrane lamination and melt molding [35].

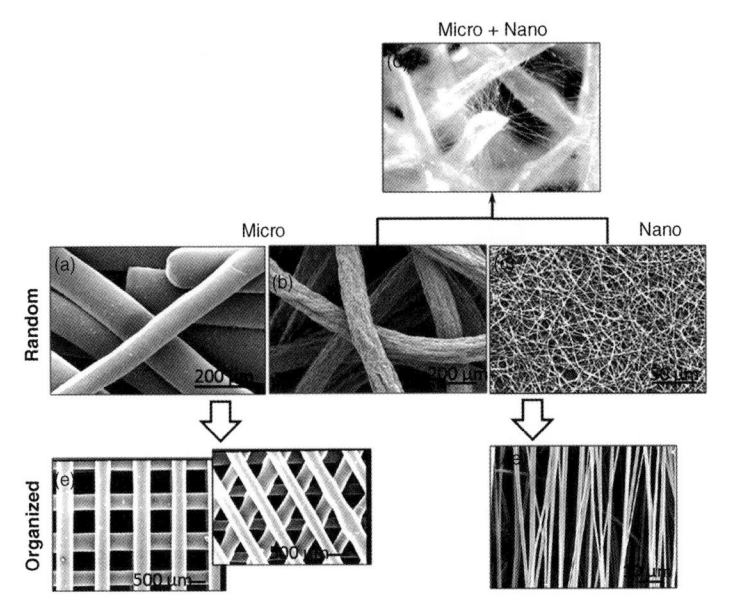

Figure 13.5 Typical morphology of microfiber mesh scaffolds processed by (a) melt spinning; (b) wet spinning; (e) rapid prototyping [37]; (c [38] and f [39]) nanofiber mesh scaffolds produced by electrospinning; and (d) combined micro- and nanofibrous scaffolds [40] obtained by electrospinning on wetspun microfibers. (Micrographs are reproduced with permission from [37].Copyright © (2002) Elsevier; [38] Copyright © (2008) Elsevier; [39] Copyright © (2005) Elsevier; [40] Copyright © (2005) Springer.)

Among them, the adapted textile technologies such as dry, wet and melt spinning, electrospinning and combinations of them, have shown remarkable versatility and potential to be applied in the design and fabrication of highly porous scaffolds (Figure 13.5) with tuned and controlled morphology. Fibers provide a large surface area-to-volume ratio and are therefore desirable as scaffold matrix material. Generally, nonwoven constructs have been used (Figure 13.5a–c) and promising results in tissue engineering of bone, cartilage heart valves and liver have been reported. Synthetic polyesters (PLA, PCL, PLLA, etc.) are mainly used for the fabrication of such constructs. As an example, scaffolds made of a blend of starch and PCL, which combines nano- and microfibrous topography, has been proposed for bone tissue engineering [36]. The nanotopography allowed endothelial cells to span between individual microfibers and influenced cell morphology. Such a structure was found to guide the 3D distribution of endothelial cells being, after implantation, easily available for blood vessel formation. Also, this structure does not compromise the structural requirements for bone regeneration. Furthermore, endothelial cells on nano- and microfibers maintained their structural integrity and their intercellular contacts and exhibited a marked angiogenic potential as shown by their ability to form extensive networks of capillary-like structures.

The upgrading of the conventional techniques with versatile computer-guided manufacturing systems has resulted in the development of so-called rapid

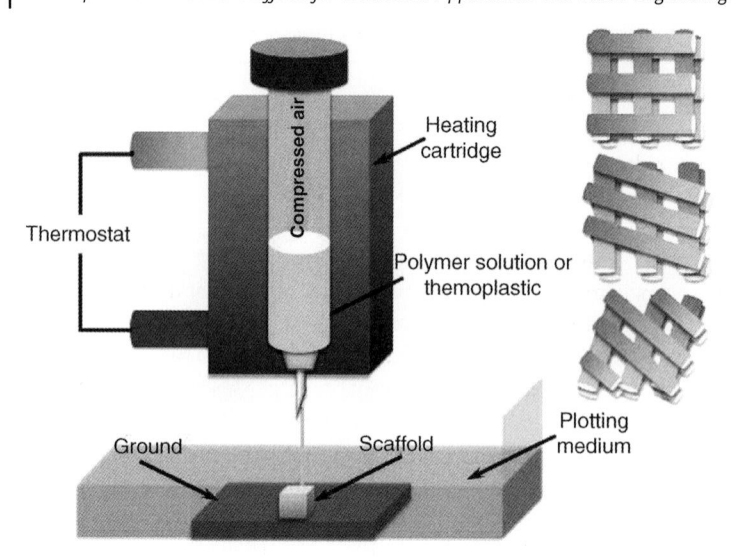

Figure 13.6 Schematic presentation of the plotting process and scaffolds with different morphologies which can be obtained by changing the angle of deposition of each layer. The thread of molten polymer or solution is solidified by the effect of the temperature or precipitation in a coagulation bath.

prototyping (RP). RP is a common name for a group of techniques such as fused deposition modeling, laminated object manufacturing, 3D printing, multiphase jet solidification and 3D plotting [41]. These techniques can be used to generate a physical model directly from computer aided design (CAD) data and hence they present superior control over design, manufacturing and reproducibility of tissue engineering scaffolds. It is an additive process in which each part is constructed in a layer-by-layer manner. RP technology allows production of complex 3D structures with controlled architecture including tuned pore size, porosity and pore distribution (e.g., Figures 13.5e and 13.6).

The fused deposition modeling uses computer tomography or magnetic resonance imaging to design the model of the scaffold. The image is imported into software, which mathematically slices the model into layers. The scaffold is created by extrusion of one layer at a time [37].

Three-dimensional printing was developed in MIT [42]. It is a solid fabrication technique which produces components by ink-jet printing a binder into sequential powder layers. Initially, a powder is spread over the surface of a powder bed and shapes the first layer using the CAD data. In the next step, a technology similar to that used in inkjet printing is used to deposit the binding material over the layer formed. The procedure is repeated until the model created by CAD is completed. The ink droplets cause formation of spherical aggregates from binder and powder which are then merged because of the capillary forces.

The key feature of 3D plotting is the 3D dispensing of a viscous plotting material into a liquid medium with a matching density (plotting medium, Figure 13.6).

In contrast to other mentioned techniques, 3D plotting can be applied at body temperature and in aqueous media, which makes it very important for the purposes of tissue engineering (e.g., living cells can be incorporated in the material). Another advantage of this technique is the large variability of the materials which can be processed namely polymer melts, polymer solutions, thermoset resins, cements and even bioactive polymers such as proteins [43].

Polyesters and recently natural polymers such as collagen and elastin are also used to fabricate synthetic extracellular matrices [44]. Natural ECM is composed of interconnected fibers with typical diameter between 10 to 300 nm. Its structure has inspired the development of the electrospinning technique for the fabrication of submicron and nanoscale fibrous scaffolds (Figure 13.5c, f), targeting closer mimics of the complex nanoscale details, that are observed in real organs at the level of cell-matrix interaction. Briefly, electrospinning involves the use of electric field to draw charged polymer solution from an orifice to a collector (Figure 13.7). As the electrical field increases, the shape of the drop at the end of the capillary tip is changed to a conical shape known as Taylor cone. When the critical value of electrical field is exceeded, a charged jet of the solution is ejected from the tip of Taylor cone. As the jet moves toward a collecting metal screen, solvent evaporates and solidified fibers are collected on the metal screen.

The morphology of this screen determines the orientation of the obtained fibers (e.g., Figure 13.5f). 3D scaffolds can easily be formed from the electrospun fibers during the deposition. The formed electrospun scaffolds are very useful for applications requiring minimal cellular infiltration such as vascular grafts. However, for

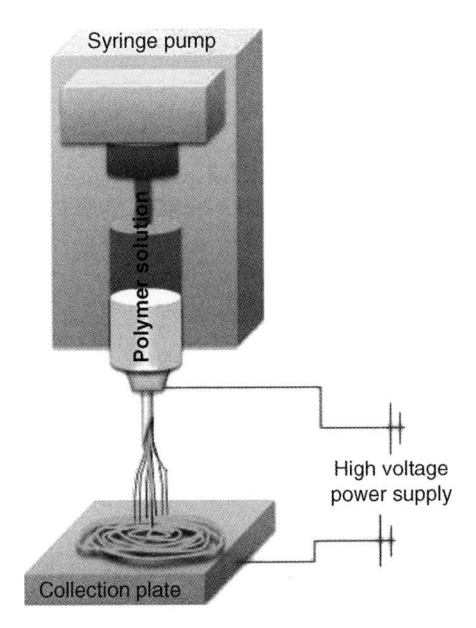

Figure 13.7 Schematic set-p of electrospinning apparatus.

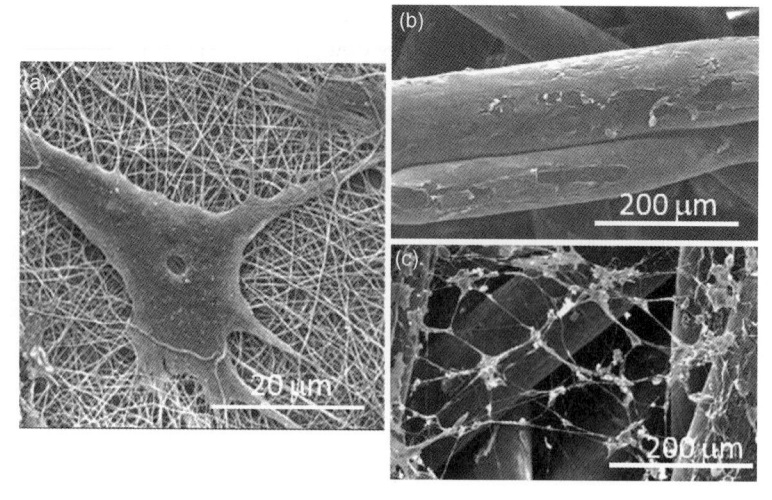

Figure 13.8 (a) Cells cultured on electrospun scaffolds [47]; (b) melt spun fiber mesh scaffolds and (c) combined micro- and nanofibrous scaffolds [36]. (Reproduced with permission from [47], Copyright © (2009) and [36] Copyright © (2008) Elsevier.)

other regeneration strategies where cell invasion of the scaffold is required, these structures are not appropriate because cell colonization is embarrassed by the size of the pores formed, which are usually smaller than average cell size [45]. The mean pore radius of electrospun matrices varies with fiber diameter. For example, a 100 nm fiber diameter yields a mean pore radius of less than 10 nm at a relative density of 80% [46]. The comparative size of a rounded cell (ranging from 5 to 20 mm) shows that such small pore sizes will obstruct cellular migration (Figure 13.8a).

As mentioned before, combined micro- and nanofibrous scaffolds (Figures 13.5d and 13.8c) have been proposed to overcome this problem [36, 40, 48, 49]. These scaffolds combine the nano-network aimed to promote cell adhesion with a microfiber mesh that provides the mechanical support.

13.4
Concluding Remarks

Distinct methods and strategies to tailor polymeric substrates with topographical signals have been discussed in this chapter. From the extensive work developed by many researchers and summarized here, it is evident that the majority of the studies has been performed onto 2D surfaces, mainly due to the limitation of the fabrication techniques. Further and continuous advances in micro- and nanofabrication, as well as other technologies are under development. It is expected that by applying these technologies it would be possible to translate the advanced 2D models into 3D structures with such a high structural and hierarchical complexity that could mimic more efficiently the *in vivo* environment.

Also, although this chapter has focused on patterning, that is, on topographical signals that influence cell behavior, it is well known that cell response is also affected by surfaces engineered with chemical and biomolecular cues. In fact, a way to control cell attachment and spreading is by altered surface chemistry, obtaining patterns which alternatively promote or prevent the adsorption of proteins. A major challenge will be to integrate the distinct signals, that is, topographical, chemical and biomolecular, in order to obtain highly-defined surfaces that will allow precisely controlled cell behavior.

Finally, it is expected that the continuous development of smart and biomimetic surfaces, for example, with cues that change by altering a given stimulus, will be of great importance as they will help to understand cell behavior in a dynamic environment, similar as the *in vivo* environment.

References

1. Tsuda, Y., Kikuchi, A., Yamato, M., Nakao, A., Sakurai, Y., Umezu, M., and Okano, T. (2005) *Biomaterials*, **26**, 1885.
2. Wehrle-Haller, B. and Imhof, B.A. (2003) *J. Pathol.*, **200**, 481.
3. Girard, P., Cavalcanti-Adam, E., Kemkemer, R., and Spatz, J. (2007) *Soft Matter*, **3**, 307.
4. Parent, C.A. and Devreotes, P.N. (1999) *Science*, **284**, 765.
5. Stevens, M.M. and George, J.H. (2005) *Science*, **310**, 1135.
6. Mrksich, M. (2002) *Curr. Opin. Chem. Biol.*, **6**, 794.
7. Spatz, J.P., and Geiger, B. (2007) *Methods Cell Biol.*, **83**, 89.
8. Dalby, M.J., Riehle, M.O., Johnstone, H., Affrossman, S., and Curtis, A.S.G. (2002) *Biomaterials*, **23**, 2945.
9. Dalby, M.J., Childs, S., Riehle, M.O., Johnstone, H.J.H., Affrossman, S., and Curtis, A.S.G. (2003) *Biomaterials*, **24**, 927.
10. Jungbauer, S., Kemkemer, R., Gruler, H., Kaufmann, D., and Spatz, J.P. (2004) *ChemPhysChem*, **5**, 85.
11. Milner, K.R. and Siedlecki, C.A. (2007) *J. Biomed. Mater. Res. A*, **82**, 80.
12. Wilkinson, C.D.W., Riehle, M., Wood, M., Gallagher, J., and Curtis, A.S.G. (2002) *Mater. Sci. Eng. C-Biomim.*, **19**, 263.
13. Ber, S., Kose, G.T., and Hasirci, V. (2005) *Biomaterials*, **26**, 1977.
14. Loesberg, W.A., te Riet, J., van Delft, F.C.M.J.M., Schon, P., Figdor, C.G., Speller, S., van Loon, J.J.W.A., Walboomers, X.F., and Jansen, J.A. (2007) *Biomaterials*, **28** 3944.
15. Oakley, C., and Brunette, D.M. (1995) *Biochem. Cell Biol.*, **73**, 473.
16. Wan, Y.Q., Wang, Y., Liu, Z.M., Qu, X., Han, B.X., Bei, J.Z., and Wang, S.G. (2005) *Biomaterials*, **26**, 4453.
17. Berry, C.C., Campbell, G., Spadiccino, A., Robertson, M., and Curtis, A.S.G. (2004) *Biomaterials*, **25**, 5781.
18. Turner, A.M.P., Dowell, N., Turner, S.W.P., Kam, L., Isaacson, M., Turner, J.N., Craighead, H.G., and Shain, W. (2000) *J. Biomed. Mater. Res.*, **51**, 430.
19. Xu, C.Y., Yang, F., Wang, S., and Ramakrishna, S. (2004) *J. Biomed. Mater. Res. A*, **71**, 154.
20. Alves, N.M., Shi, J., Oramas, E., Santos, J.L., Tomas, H., and Mano, J.F. (2009) *J. Biomed. Mater. Res. A*, **91**, 480.
21. Wang, Y., Ribelles, J.L.G., Sanchez, M.S., and Mano, J.F. (2005) *Macromolecules*, **38**, 4712.
22. Sanchez, M.S., Ribelles, J.L.G., Sanchez, F.H., and Mano, J.F. (2005) *Thermochim. Acta*, **430**, 201.
23. Sanchez, F.H., Mateo, J.M., Colomer, F.J.R., Sanchez, M.S., Ribelles, J.L.G., and Mano, J.F. (2005) *Biomacromolecules*, **6**, 3283.
24. Martinez, E.C., Hernandez, J.C.R., Machado, M., Mano, J.F., Ribelles, J.L.G.,

Pradas, M.M., and Sanchez, M.S. (2008) *Tissue Eng. A*, **14**, 1751.
25 Alves, N.M., Pashkuleva, I., Reis, R.L., and Mano, J.F. (2010) Small 6,2208.
26 Matsuda, N., Shimizu, T., Yamato, M., and Okano, T. (2007) *Adv. Mater.*, **19**, 3089.
27 Da Silva, R.M.P., Mano, J.F., and Reis, R.L. (2007) *Trends Biotechnol.*, **25**, 577.
28 Tsuda, Y., Yamato, M., Kikuchi, A., Watanabe, M., Chen, G.P., Takahashi, Y., and Okano, T. (2007) *Adv. Mater.*, **19**, 3633.
29 Horbett, T.A. (2004) The role of adsorbed proteins in tissue response to biomaterials, in *Biomaterials Science* (eds B.D. Ratner, A.E. Hoffman, F.J. Shoen, and J.E. Lemons), Elsevier Academic Press, San Diego, p. 237.
30 Saltzman, W.M. (2000) Cell interactions with polymers, in *Principles of Tissue Engineering* (eds R.P. Lanza, R. Langer, and J.P. Vacanti), Academic Press, New York, p. 221.
31 Schoen, F.J. and Mitchell, R.N. (2004) Tissues, the extracellular matrix, and cell-biomaterial interactions, in *Biomaterials Science* (eds B.D. Ratner, A.E. Hoffman, F.J. Shoen, and J.E. Lemons), Elsevier Academic Press, San Diego, p. 260.
32 Ingber, D.E. (2002) *Circ. Res.*, **91**, 877.
33 Berry, C.C., Dalby, M.J., Oreffo, R.O.C., McCloy, D., and Affrosman, S. (2006) *J. Biomed. Mater. Res. A*, **79**, 431.
34 Berry, C.C., Dalby, M.J., McCloy, D., and Affrossman, S. (2005) *Biomaterials*, **26**, 4985.
35 Hutmacher, D.W. (2001) *J. Biomat. Sci.-Polym. E*, **12**, 107.
36 Santos, M.I., Tuzlakoglu, K., Fuchs, S., Gomes, M.E., Peters, K., Unger, R.E., Piskin, E., Reis, R.L., and Kirkpatrick, C.J. (2008) *Biomaterials*, **29**, 4306.
37 Zein, I., Hutmacher, D.W., Tan, K.C., and Teoh, S.H. (2002) *Biomaterials*, **23**, 1169.
38 Cheng, M.L., Lin, C.C., Su, H.L., Chen, P.Y., and Sun, Y.M. (2008) *Polymer*, **49**, 546.
39 Lee, C.H., Shin, H.J., Cho, I.H., Kang, Y.M., Kim, I.A., Park, K.D., and Shin, J.W. (2005) *Biomaterials*, **26**, 1261.
40 Tuzlakoglu, K., Bolgen, N., Salgado, A.J., Gomes, M.E., Piskin, E., and Reis, R.L. (2005) *J. Mater. Sci. Mater. Med.*, **16**, 1099.
41 Yeong, W.Y., Chua, C.K., Leong, K.F., and Chandrasekaran, M. (2004) *Trends Biotechnol.*, **22**, 643.
42 Cima, L.G., Vacanti, J.P., Vacanti, C., Ingber, D., Mooney, D., and Langer, R. (1991) *J. Biomech. Eng.-T. ASME*, **113**, 143.
43 Pfister, A., Landers, R., Laib, A., Hubner, U., Schmelzeisen, R., and Mulhaupt, R. (2004) *J. Polym. Sci. Pol. Chem.*, **42**, 624.
44 Kim, B.S. and Mooney, D.J. (1998) *Trends Biotechnol.*, **16**, 224.
45 Tuzlakoglu, K. and Reis, R.L. (2009) *Tissue Eng. B*, **15**, 17.
46 Eichhorn, S.J. and Sampson, W.W. (2005) *J. Roy. Soc. Interface*, **2**, 309.
47 Prabhakaran, M.P., Venugopal, J.R., and Ramakrishna, S. (2009) *Biomaterials*, **30**, 4996.
48 Martins, A., Chung, S., Pedro, A.J., Sousa, R.A., Marques, A.P., Reis, R.L., and Neves, N.M. (2009) *J. Tissue Eng. Regen. Med.*, **3**, 37.
49 Park, S.H., Kim, T.G., Kim, H.C., Yang, D.Y., and Park, T.G. (2008) *Acta Biomater.*, **4**, 1198.

14
Nano- and Micro-Structured Polymer Surfaces for the Control of Marine Biofouling

James A. Callow and Maureen E. Callow

14.1
Introduction

14.1.1
The Fouling Interface

"Marine biofouling" is the colonization of submerged surfaces by unwanted marine organisms. It has detrimental effects on shipping and leisure vessels, heat exchangers, oceanographic sensors and aquaculture systems. A fouled hull can result in powering penalties of up to 86% at cruising speed [1]. Without effective antifouling measures, fuel consumption (and greenhouse gas emissions) would therefore increase significantly. "Green" alternatives to biocide-based technologies are urgently sought by the marine coatings industry, and there is considerable interest in developing coatings that rely on surface physico-chemical properties to either deter organisms from attaching in the first place, or which reduce the adhesion strength of those that do attach so that they are easily removed by the shear forces generated by ship movement or mild mechanical cleaning devices. Development of such coatings will be accelerated if there is a greater understanding of how the surface characteristics of a coating influence its resistance to fouling.

The colonization processes used by marine organisms represent the outcome of complex interactions at the substratum/water interface that occur during the initial recruitment and bonding processes. A major challenge in creating an effective fouling-resistant coating is that the diversity of fouling organisms is vast and the range of length scales that regulate attachment behaviors and adhesion mechanisms is correspondingly great. The colonizing ("recruitment") stages of fouling organisms range in size from microns (single-celled bacteria, spores of algae, diatom cells) to 100s of microns or even millimeters (larvae of invertebrates) (Figure 14.1). However, the critical length scale in determining settlement is not necessarily the size of the organism per se, but rather the size of the parts or structures involved in an organism's sensing apparatus, which determines whether or not the organism selects a surface for attachment. For example, cypris larvae of the barnacle,

---

Generating Micro- and Nanopatterns on Polymeric Materials. Edited by A. del Campo and E. Arzt
Copyright © 2011 WILEY-VCH Verlag GmbH & Co. KGaA, Weinheim
ISBN: 978-3-527-32508-5

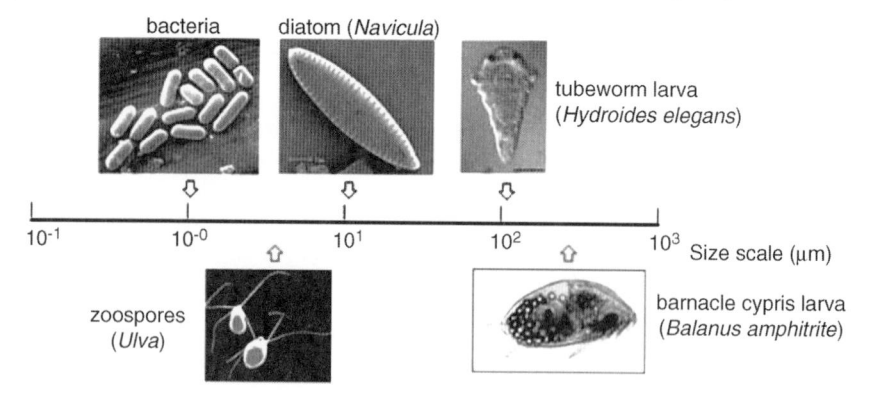

Figure 14.1 Length scales of the settlement stages of fouling organisms. The image of the barnacle cypris larva (kindly provided by Prof. A.S. Clare) shows the paired antennules (left) referred to in Figure 14.2. The image of the larva of *Hydroides elegans* was kindly provided by Dr B Nedved.

Balanus amphitrite, are c. 500 μm in length, but the substratum is explored (and probably "sensed") by a pair of specialized appendages called antennules. The antennules are approx. 200 μm long and are terminated by an oval attachment disk approx. 25–30 μm in the long axis and approx. 15 μm in width (Figure 14.2). The antennules bear a number of sensory setae and the attachment disks are covered by micro/nanoscale villi [2]. Is it the size of the antennules then that is relevant to considerations of the critical scale of surface features, the size and distribution of the sensory setae, or the size of the micro/nano villi? At an even smaller scale we also have

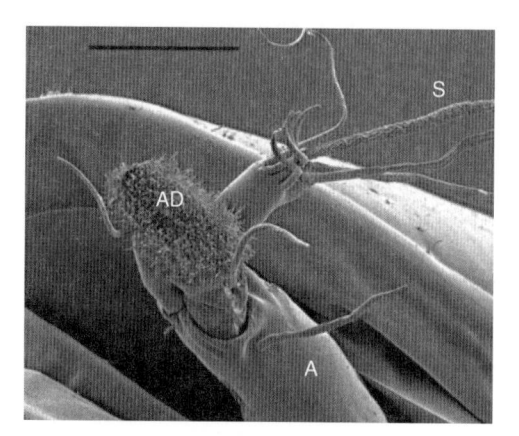

Figure 14.2 Scanning electron micrograph (SEM) of the distal region of an antennule (A) of the cypris larva of the barnacle *Balanus amphitrite*, showing the attachment disk (AD) covered in nano/microvilli and several sensory setae (S). Image kindly provided by Dr N. Aldred. Scale bar 20 μm. (Reproduced with permission from [2]. Copyright © (2008) Taylor & Francis.)

to consider structures at the cellular/macromolecular level. Perception of topological features in all types of cells is likely to depend on the distribution of small (nanometer sized) mechanotransducer proteins located within the plasma membrane of the cells. Then, assuming the organism has settled, the success of that organism in colonizing a surface and growing to a reproductive adult, within a turbulent marine environment, depends on how well the macromolecular adhesive polymers secreted by the settled organisms secure the adhesive interface. This is determined by the interfacial molecular interactions that are in turn influenced by the properties of a surface or coating at the molecular or nanoscale. The considerations outlined in this paragraph indicate the complexity of the situation facing the surface engineer attempting to design surfaces to repel fouling organisms in a rational way, and so far, such attempts have largely proceeded on an empirical basis.

14.1.2
Surface and Coating Designs to Investigate and Control Biofouling

The two main approaches to creating commercial fouling-resistant coatings are

- "antifouling coatings", that is, no settlement (attachment) of the colonizing larvae, spores or cells, and
- "fouling-release", that is, organism release under hydrodynamic forces because they are weakly adhered.

"Antifouling" coatings are typically based on the controlled release of biocides, to kill the colonizing organisms. With the recent, and increasing environmental restrictions placed on biocides (e.g., the ECs Biocidal Products Directive 98/8/CE), the emphasis in technological innovation is now on nontoxic coatings that do not release materials into the environment and which function through their physico-chemical surface properties. Since the critical events resulting in fouling are nano- or microscale in dimension, it follows that physico-chemical surface properties to control biofouling need to be on the same length scales. Figure 14.3 summarizes the range of interfacial properties that are considered to be relevant to the performance of fouling-resistant coatings against marine organisms [3].

An area of particular interest in recent years, and the focus of this chapter, is the impact of surface patterning (nano- and micro-) on fouling organisms. Surface patterns may be purely topographical, chemical, or combinations of the two since they are often inter-related, and can be generated by either "top-down" or "bottom-up" approaches. The former are most suitable for small scale surfaces that enable specific hypotheses to be tested in laboratory studies. The latter are more suited to practical coating designs through, for example phase-segregating polymer blends or self-assembling block copolymers. The importance of studying surfaces underwater is also becoming more important, especially for those surfaces that reconstruct in water, since it is the hydrated surface that the settling organisms/cells encounter. Not included in this chapter are examples of antifouling coatings where there is no specific, explicit reference to surface structuring.

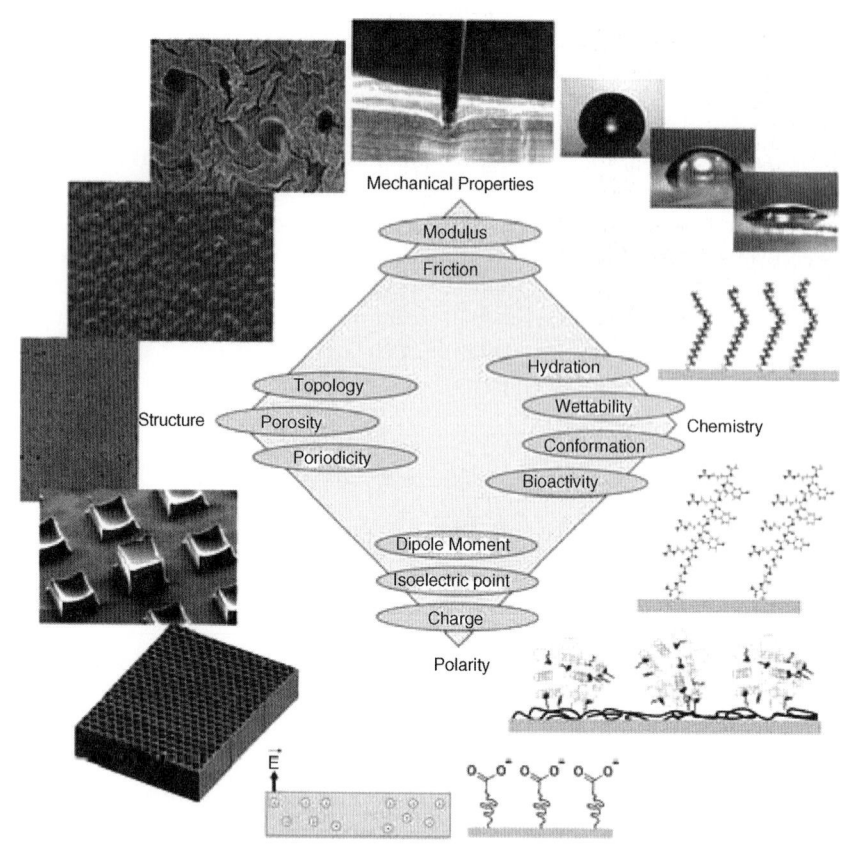

Figure 14.3 The range of physico-chemical coating properties that can influence settlement and adhesion of fouling organisms (Reproduced with permission from [3]. Copyright © (2008) Biointerphases.)

14.2
Replica Molding in PDMS and Other Polymers

Many studies have shown that topography of a substratum plays a significant role in governing the settlement of a number of species on both natural and man-made surfaces. In most studies, the scale of the features cover several orders of magnitude (microns-cm) and other factors such as environmental and biological factors make analysis of the data complex (e.g., [4, 5]). Other studies [6] showed that molded microtextured surfaces, with a range of textures from 1 to 100 μm in height, replicated in poly(methyl methacrylate) (PMMA) (from different grades of plankton netting) and polyvinylidene fluoride (PVDF) (riblet design) changed the exploratory behavior and inhibited settlement of cypris larvae of the barnacle, *Balanus improvisus*. The surfaces most effective at inhibiting settlement fell within a topographic range of 30–40 μm,

Figure 14.4 Scanning electron micrographs of engineered topographies on a PDMS surface: (a) 2 μm ribs of lengths 4, 8, 12, and 16 μm combined to create the Sharklet AF, (b) 10 μm equilateral triangles combined with 2 μm diameter circular pillars, (c) hexagonally packed 2 μm diameter circular pillars, (d) 2 μm wide ridges separated by 2 μm wide channels. Scale bars in all cases are 20 μm. (Reproduced with permission from [13]. Copyright © (2007) Taylor & Francis.)

recruitment being inhibited by 92% on these surfaces. It is interesting to note that this size range is similar to that of the cyprid antennular disc, which is 20–25 μm [6].

The development of techniques to produce discrete, highly ordered features at the micron scale has facilitated the investigation of specific feature geometries on the settlement of fouling organisms. Regular structures (channels, pillars) fabricated in polydimethylsiloxane (PDMS) at a range of scales from 5 to 20 μm (Figure 14.4) showed that spores of the green alga (seaweed) *Ulva* preferred to settle in or against structures that were of similar dimensions to the spore body (∼5 μm) [7], although lower numbers settled and more were removed by exposure to flow when the depth of channels was reduced from 5 to 1.5 μm [8].

The observation that many marine organisms do not become colonized by other species (e.g., [9]) and the identification of specific topographic components correlated with fouling resistance [10], has led to research on a range of both biomimetic and bioinspired surface designs. Biomimetic designs replicate nature whereas bioinspired designs employ biological concepts as a basis for improvement in order to develop novel technologies [11].

Topography inspired by the skin of fast moving sharks (Sharklet AF^{TM}) resulted in an 85% reduction in settlement of *Ulva* spores compared to smooth PDMS (Figure 14.5a; [12]). The Sharklet AF topography consists of 2 μm wide rectangular-like (ribs) periodic features (4, 8, 12, and 16 μm in length) spaced 2 μm apart. Further studies based on 2 μm features correlated spore settlement with an engineered roughness index (ERI_1) for various surface designs (Figure 14.5, [13]). ERI_1 is a dimensionless ratio based on Wenzel's roughness factor, depressed surface fraction and the degree of freedom of spore movement. Spore settlement decreased with an

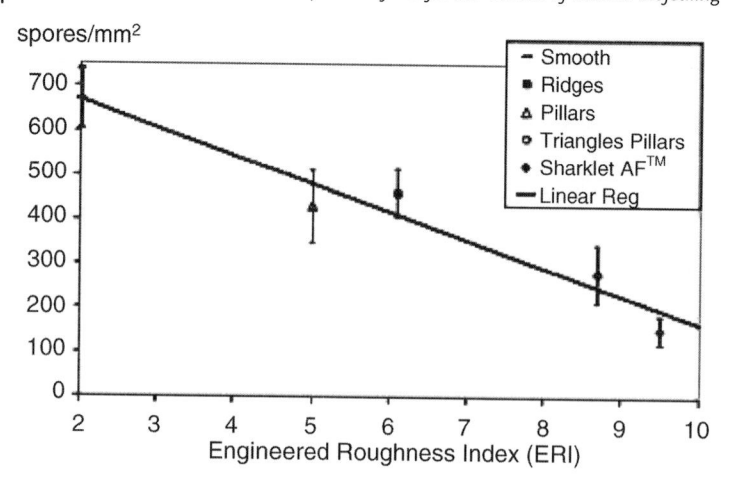

Figure 14.5 Correlation between settlement of zoospores of *Ulva* and the Engineered Roughness Index (ERI) at a fixed feature spacing of 2 μm. Plotted is the calculated ERI for the PDMS topographies shown in Figure 14.4, plus a smooth PDMS control, against the experimental mean settled spore density (spores/mm^2 ± standard error). (Reproduced with permission from [13]. Copyright © (2007) Taylor & Francis.)

increase in ERI_1: Sharklet AF with an ERI_1 of 9.5 showed a 77% reduction in settlement compared with the smooth surface.

Schumacher et al. [14] introduced the concept of nanoforce gradients to explain how different feature geometries might exert an influence on the settlement of spores of *Ulva*. It was hypothesized that nanoforce gradients caused by variations in topographical feature geometry will induce stress gradients within the lateral plane of the membrane (plasma membrane) of a settling cell during initial contact. Perception of such stress gradients via mechanotransducer proteins in the cell membrane could, through intracellular signal cascades, lead to modification of the settlement response. The generation of the nanoforce gradients was envisaged as a function of the bending moment or stiffness of the topographical features with which the cell is in contact. The geometric dimensions, including width, length, and height of the topographical feature as well as the modulus of the base material define its stiffness. By introducing geometric variations in features contained in the engineered topography, an effective force gradient between neighboring topographical features will be developed. To test the nanoforce gradient hypothesis, modifications were made to the design of Sharklet AF resulting in a range of nanoforce gradients [14]. The surfaces were then challenged with spores of *Ulva*. Surfaces with nanoforce gradients ranging from 125 to 374 nN all significantly reduced spore settlement relative to a smooth substrate with the highest reduction, 53%, measured on the 374 nN gradient surface. These results confirm that the designed nanoforce gradients may be an effective tool and predictive model for the design of unique, nontoxic, nonfouling surfaces for marine applications However, since different fouling organisms respond to topographies of different length scales, hierarchical patterning may be required. An initial study by Schumacher et al. [15] indicated that complex designs would be required to repel multiple settling species.

14.3
Stretched Topographies in PDMS

The expectation that no single topography could inhibit settlement of organisms the settling stages of which span several orders of magnitude (Figure 14.1) led Genzer to design surfaces with hierarchical corrugations from nanometers to millimeters [16]. The hierarchically wrinkled surfaces were fabricated in PDMS, which was stretched uniaxially. Surface modifications to alter for example, wettability, can be made to the wrinkled samples. Hierarchically wrinkled samples were not colonized by barnacles after 18 months exposure, but diatoms (unicellular algae) persisted in the grooves [17]. Furthermore, settlement of spores of *Ulva* was higher on most wrinkled surfaces compared to the smooth surface. The distribution of spores on biaxially stretched samples that have a random arrangement of wrinkles suggested that settlement was lower on areas where the distribution of wrinkles was most random [17]. The data from this preliminary investigation suggest that combining random hierarchical wrinkles with modifications to surface chemistry may produce surfaces capable of resisting the settlement of multiple species.

14.4
Structured Surfaces by Self-Assembly

14.4.1
Block Copolymers with Amphiphilic, Fluorinated and PEG-ylated Side Chains

The phase separation behavior of fluorinated copolymers and blends is well-suited to the generation of nanostructured, self-assembled, low surface energy materials (see Chapter 10 on diblock copolymer self-assembly). Amphiphilic block copolymers carrying hydrophilic oligoethylene glycol and hydrophobic perfluoroalkyl side chains were prepared by using side-chain grafting on a preformed poly(styrene-*block*-acrylic acid) copolymer [18]. Martinelli *et al*. [19, 20] developed similar amphiphilic copolymers using controlled atom transfer radical polymerization (ATRP) and a purely polystyrene backbone. The block copolymers, either alone or in a blend with commercial SEBS (poly(styrene-*block*-ethylene-*random*-butylene)-*block*-polystyrene), an elastomeric thermoplastic material, were spray-coated (~500 nm thick) on an SEBS underlayer (150–200 μm). In both studies, NEXAFS and angle-resolved X-ray photoelectron spectroscopy measurements demonstrated that the film surface underwent reconstruction when immersed in water, owing to its amphiphilic nature. This gave rise to a chemically heterogeneous nanostructure revealed by tapping mode atomic force microscopy (AFM) imaging [19]. In dry, annealed films, depending on the chemical composition of the block copolymers, spherical (about 20 nm diameter) and lying cylindrical (24–29 nm periodicity) nanodomains, with an rms roughness of about 1 nm, were observed (Figure 14.6) resulting from the thermodynamically induced phase segregation of the mutually incompatible components. After immersion in water, the underwater AFM patterns showed a mixed surface structure, in

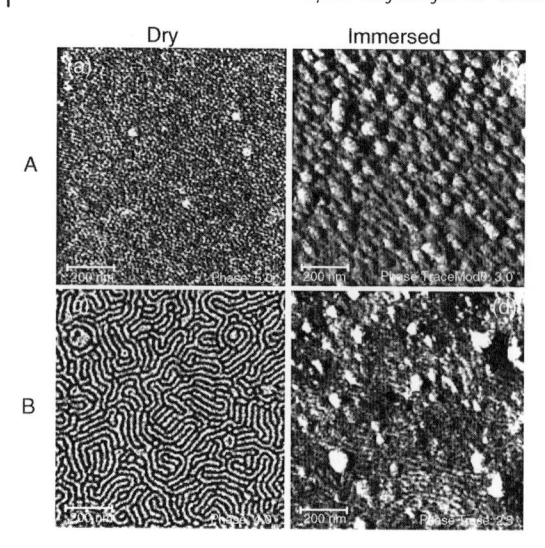

Figure 14.6 AFM tapping mode, phase images of amphiphilic, fluorinated/PEGylated styrene-based copolymers. The left hand column shows morphologies of dry films after annealing at 120°. The right hand column shows the same coatings after immersion for 7 days in artificial seawater. The two coatings (A) and (B) differ in the degree of polymerization of the polystyrene (S_n) and the relative lengths of the fluorinated-PEGylated polystyrene block (Sz_m). Coating (A) was $S_{26}Sz_{23}\_90$, and coating (B) was $S_{81}Sz_{19}\_90$. Both coatings illustrated contained 10% by weight of SEBS in the top layer and were deposited on a pure SEBS underlayer. The scale-bars are 200 nm. (Images reproduced with permission from [18]. Copyright © (2006) American Chemical Society.)

which the nanoscale heterogeneity and topography (rms 1–6 nm) were both increased to form larger nanoaggregates 50–75 nm in diameter overlying the original cylindrical or spherical morphologies (Figure 14.6).

In both studies the coatings were subjected to laboratory bioassays to explore their intrinsic ability to resist the settlement and reduce the adhesion strength of two marine algae, viz., the macroalga (seaweed) *Ulva linza* and the unicellular diatom *Navicula perminuta*. The amphiphilic nature of the copolymer coatings resulted in distinctly different performances against these two organisms [19]. *Ulva* adhered less strongly to the coatings richer in the amphiphilic polystyrene component, percentage removal being maximal at intermediate weight contents. In contrast, *Navicula* cells adhered less strongly to coatings with a lower weight percentage of the amphiphilic side chains.

In seeking a mechanistic explanation for these results, the general thesis behind this type of amphiphilic design is that it combines the nonpolar, low surface energy properties of the fluorinated block to reduce polar and hydrogen bonding interactions with the bioadhesives used by fouling organisms, with the well-known protein repellency properties of the hydrophilic, oligo(poly)ethylene glycols. The resulting chemical ambiguity may lower the entropic and enthalpic driving forces for the adsorption of the bioadhesives [18, 21]. Lau *et al.* [22] have speculated that the often-observed protein resistance of some nanopatterned, amphiphilic diblock copolymers is due to the intrinsic high density of surface interfacial boundaries.

14.4.2
Polystyrene-Based Diblock Copolymers

Diblock copolymers can self-assemble to produce a variety of ordered structures or domains at the nanoscale, such as spheres, cylinders, lamellae or double gyroids. Grozea et al. [23] produced a range of nanopatterned surfaces with hydrophobic and hydrophilic domains using the diblock copolymer polystyrene-*block*-poly(2-vinyl pyridine) (PS-*b*-P2VP) and polystyrene-*block*-poly(methyl methacrylate) (PS-*b*-PMMA). The PS-*b*-P2VP diblock copolymer, mixed with the crosslinker benzophenone and spin-coated onto silicon wafers, showed self-assembled cylindrical structures, which were retained after UV treatment for crosslinking. The thin films displayed stable cylindrical domains after immersion in water. The PS-*b*-PMMA diblock showed self-assembled cylindrical structures. PS-*b*-P2VP and PS*b*-PMMA cylindrical patterned surfaces showed reduced settlement of zoospores of the green alga *Ulva* compared with unpatterned surfaces.

14.4.3
Hyperbranched Amphiphilic Networks

Amphiphilic networks composed of hyperbranched fluoropolymers (HBFP) and poly (ethylene glycol)s (PEG) self-assemble on crosslinking to form complex surface topographies and chemical domains of both nanoscopic and microscopic dimensions [24]. The surface patterns are strongly influenced by immersion and by the relative proportions of the two polymers. The coating design anticipated that this surface complexity would either have a deterrent effect on the settling stages of fouling organisms, or be unfavorable for adsorption and unfolding of the proteinaceous biological adhesives. Gudipati et al. [25] subsequently showed that several macromolecules of biological origin (proteins, lipopolysaccharides) exhibited reduced adsorption to compositions with high concentrations of PEG (45–55% by wt.). These compositions were also effective against settlement of zoospores of the green alga *Ulva*, and small plants of *Ulva* adhered less well to the 45% PEG by wt. composition compared with PDMS. These are intriguing results, especially since other experimental amphiphilic coating designs (see Section 14.3.1) also show good biological performance. Intersleek 900®, the latest fouling-release coating produced by International Paint is marketed as an amphiphilic fluoropolymer [26].

14.4.4
Phase-Segregating Quaternized Siloxanes and Siloxane-Urethane Nanohybrids

Commercially available fouling-release (FR) coatings for marine use are based on polysiloxanes, which provide the combination of surface energy, low modulus and lubricity that favors interfacial fracture of adherands. Investigators have explored the potential for nanohybrid coatings that combine PDMS with other polymers or functionalities.

Majumdar et al. [27] investigated the antifouling and FR properties of PDMS with tethered quaternary ammonium salts (QAS). The initial concept was to combine the biocidal properties of QAS with the FR properties of PDMS. Surface analysis of the resultant coatings revealed a heterogeneous, two-phase morphology with increased water contact angle (increased hydrophobicity). Depending on the QAS alkyl chain length, surface protrusions were either isolated, 0.7–2.6 µm in size (C14), or interconnected, 0.60–2.9 µm in size (C18). The C18 coatings also showed higher levels of nanoroughness. In laboratory tests with a range of fouling organisms, the rougher C18 coatings showed the best FR performance against the macroalga, *Ulva*, exceeding that of a commercially-available FR coating. Fouling-release performance against *Ulva* is unlikely to be related to the tethered biocidal functionality *per se*, since the coatings did not affect growth of the alga, but rather seems to be related to the nanoroughness. On the other hand, it was shown that the QAS coatings inhibited microbial biofilm formation of the bacterium *Cellulophaga lytica*, and the diatom *Navicula incerta*, which is most likely due to the biocidal functionality.

In an attempt to improve the mechanical properties of PDMS, Majumdar et al. [28–30] synthesized thermosetting, crosslinked PDMS–polyurethane nanohybrid copolymer coatings, certain compositions of which (those containing only 10% PDMS) spontaneously phase-separated to form microtopographic domains composed of PDMS surrounded by the polyurethane matrix. The authors suggested that this was a consequence of the restricted ability of the PDMS to fully cover the surface of the coating [28]. The domains appeared to increase in size after 2 weeks immersion: stability depended on the casting solvent [28] and mixing time [30] used. Compositions with a higher proportion of PDMS did not form domains but rather formed a smooth stratified coating of low surface energy PDMS over a sublayer of higher surface energy polyurethane. In detachment studies with "pseudobarnacles" (an epoxy-bonded stud used as a proxy for live barnacles) and with reattached live barnacles, it was shown that the pull-off force in both cases was lower on compositions with surface domains than those without [30].

14.5
Nanocomposites

Commercial fouling-release coatings can be considered as nanocomposites since all contain fillers such as TiO_2 and Fe_3O_4 in the case of Intersleek (International Paint) and calcium carbonate in the case of RTV11™ (General Electric) [31]. Arce et al. [32] showed by AFM and SEM that exposure of RTV11 to seawater resulted in a modification of its morphology and mechanical properties. More recently two novel types of nanoparticle-containing polysiloxane coatings with interesting antifouling and/or FR properties have been reported.

14.5.1
Carbon Nanotube-Filled Polysiloxanes

Hydrosilation-cured silicone elastomers, reinforced through the incorporation of small quantities of multiwall carbon nanotubes (MWCNTs) and sepiolite nanoclays showed enhanced FR performance in laboratory assays with algae and barnacles compared with unfilled PDMS controls as well as in short-term field tests [33]. The use of MWCNTs is particularly significant since the improved performance was obtained at low loadings of the nanofiller (0.05–0.2% by weight). At this level of loading the bulk properties of the coatings (tensile modulus and crosslink density) appear to be unchanged but the coatings became slightly more hydrophobic and there was a significant change in shear-thinning behavior [34]. The changed rheological properties were attributed, on both theoretical and experimental grounds, to a strong molecular affinity between the siloxane chains and the MWCNTs, via CH-π interactions involving methyl groups of the PDMS and aromatic rings of the MWCNTs [34]. This affinity is also important in reducing the likelihood of MWCNTs being released into the marine (or any other) environment, which is relevant to the current debate on potential issues of nanoparticle toxicity. The improved FR performance appears to be correlated with the effect of MWCNTs in inducing time- and immersion-dependent changes in surface nanotopography [35]. Under tapping-mode AFM both unfilled and MWCNT-filled PDMS showed a smooth morphology (rms roughness $\sim$1 nm). After immersion in water there was a complex time-dependent surface restructuring of both filled and unfilled coatings, but those containing 0.1% MWCNTs exhibited a significant restructuring at the nanoscale (Figure 14.7). It is not yet clear how such nanostructuring contributes towards FR performance; some indications will be given in Section 14.5.2.

14.5.2
Nanostructured Superhydrophobic Surfaces

Superhydrophobic surfaces are water-repellent, having a high water contact angle (typically >150°) and a very low roll-off angle that is, the inclination angle at which a

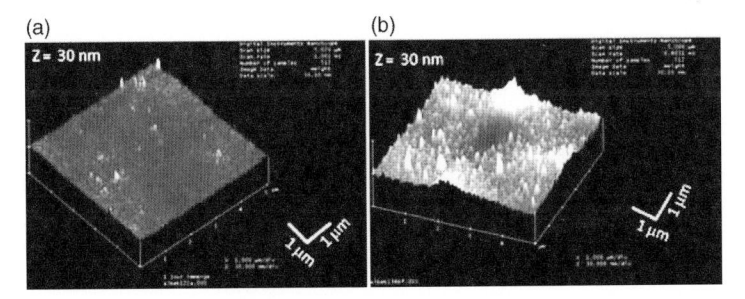

Figure 14.7 AFM tapping mode images of cured, unfilled PDMS (a) and PDMS filled with 0.1% multiwall carbon nanotubes (b), 1 day after immersion. (Images reproduced with permission from [35]. Copyright © (2009) Journal of Nanostructured Polymers and Nanocomposites.)

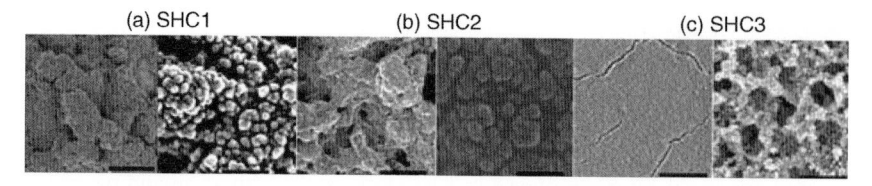

Figure 14.8 Scanning Electron Microscopy images for three superhydrophobic polysiloxane coatings. Each pair of coatings was imaged at two magnifications. Scale bars are 10 μm (left hand image of each pair) and 600 nm (right hand image of each pair). (Reproduced with permission from [10]. Copyright © (2009) Taylor & Francis.)

water drop rolls off the surfaces [36]. Although a number of the coating technologies discussed in previous sections are hydrophobic in character, they do not fully comply with this definition. The superhydrophobic effect relies on the trapping of air and is exemplified by a number of natural surfaces for example, the Lotus leaf and insect wings [37]. Marmur [38] discussed the ways in which the "Lotus" effect could be mirrored in the aquatic environment through the design of surfaces that minimize the air-wetted area when submerged, by retaining an air film between the water and the surfaces.

Recently, Scardino et al. [10] have reported three superhydrophobic coatings (SHC) that inhibit the settlement/attachment of five fouling organisms viz cells of the diatom *Amphora* sp., spores of the green alga *Ulva rigida*, spores of the red alga *Polysiphonia sphaerocarpa*, larvae of the bryozoan *Bugula neritina* and cypris larvae of the barnacle *Amphibalanus amphitrite*. The coatings were made by spraying fumed silica-filled siloxanes. All three coatings were superhydrophobic (SHC 1 contact angle $\theta_A = 169°$, SHC 2 $\theta_A = 155°$, SHC 3 $\theta_A = 169°$) and exhibited either micro/nanoroughness (SHC 1 and 2) or only nanoroughness (SHC 3) (Figure 14.8). All five organisms avoided settling on the SHC 3 coating, which had large pores around 350 nm in diameter and small pores in the range 10–50 nm. small angle X-ray scattering (SAXS) was used to characterize the partial wetting of the superhydrophobic surfaces *in situ* under immersion conditions. The broad spectrum antifouling effect was attributed to the larger amount of unwetted interface on SHC 3 compared to SHC 1 and 2 when immersed in seawater. Whether the unwetted interface will resist wetting (and hence the accumulation of fouling) following long term immersion needs to be determined.

14.6
Nanostructured Polymer Surfaces by Vapor Deposition Methods

Vapor deposition methods can also be employed to pattern polymer surfaces. SiO_x-like coatings were deposited on glass slides from a hexamethylsiloxane monomeric precursor by plasma-assisted CVD [39]. Surface energies (23.1–45.7 mJ m^{-2}) were correlated with the degree of surface oxidation and hydrocarbon contents. Tapping mode AFM revealed a range of surface topographies with R_a values 1.55–3.16 nm and rms roughness 1.96–4.11 nm. Settlement of spores of the green alga *Ulva* was significantly less, and detachment under shear significantly more on the lowest surface energy coatings. Removal of young plants (sporelings) of *Ulva* under shear was positively correlated with reducing the surface energy of the coatings. The most hydrophobic coatings also showed good performance against a freshwater bacterium, *Pseudomonas fluorescens*, significantly reducing initial attachment and biofilm formation, and reducing the adhesion strength of attached bacteria under shear. In a subsequent paper [40], this relationship was explored in more detail by varying the deposition parameters even further, to either increase or reduce the degree of oxidation at the two extremes of the coating spectrum. Significantly improved antifouling performance was detected with the most reduced coatings deposited using extended ion-cleaning to increase the deposition temperature, which may result in higher decomposition rates and hence an increased crosslinking of the polymer. Films with improved performance have potential for use as coatings in the control of biofouling in applications such as heat exchangers, where thin films are important for effective thermal transfer, or optical windows where transparency is important.

14.7
Conclusions

Studies on the influence of polymeric nanostructured coatings and surfaces on marine fouling organisms are in their infancy. A number of nano- and microscale surface designs discussed within this article show clearly that fouling organisms exhibiting a wide range of length scales, are indeed influenced in either or both their initial surface colonization behavior, and the subsequent development of adhesion strength. In some cases the effective coatings are real candidates for practical application, after further development work. But the outstanding issue concerns the mechanism(s) by which nanostructured coatings exert their effects. At best we have correlations between specific types of morphological or chemical surface structuring and biological performance. However, correlation does not imply causality and there are real knowledge barriers to understanding how these coatings work. Advances are needed to understand the underlying biological mechanisms. For example, there is almost no information available for the significant macrofouling organisms on what the critical length scale is of the structures involved in surface sensing. It is presumed that unicellular zoospores of fouling algae, like other cell types, possess mechanotransducing proteins and stretch-sensitive ion channels in their surface membranes that can sense and initiate responses to surface structures, but there are no actual reports of such proteins. This is partly due to the fact that fouling organisms do not provide the best models for the sort of intensive cell biological studies using patch clamping or other investigative procedures. Apart from a dearth of knowledge on biological aspects, more understanding is needed of coating surface properties in the immersed condition, rather than dry or *in vacuo*, since immersion changes the dynamics of self-assembly and surface organization. Concerning coatings based on amphiphilic PEGylated and fluorinated copolymers, the hypothesis is that these are structured to present a chemically complex surface to fouling organisms. But, is the reduced adhesion observed (see Section 14.4.1 above) solely due to the reduced ability of adhesive proteins to adsorb to the PEGylated domains? The development of novel, nanorough, FR polysiloxanes containing carbon nanotubes (Section 14.5.1) is intriguing, but at present there is no understanding as to the mechanistic basis of this effect: does the nanoroughness influence the cells to adhere less strongly (i.e., a "biological" basis) or is a "physical" mechanism more likely whereby nano- and micro-air incursions, such as those postulated by [10] for superhydrophobic coatings, reduce the interfacial contact between adhesive and surface, thereby reducing adhesion strength? Interdisciplinary studies combining chemistry, biology and advanced physical techniques for interfacial characterization [10] will be crucial in advancing this field.

Acknowledgements

The authors acknowledge support from the AMBIO project (NMP-CT-2005-011827) funded by the European Commission's 6th Framework Programme, and support from the US Office of Naval Research (#N00014-08-1-0010).

References

1 Schultz, M.P. (2007) Effects of coating roughness and biofouling on ship resistance and powering. *Biofouling*, **23**, 331–341.
2 Aldred, N. and Clare, A.S. (2008) Mini-review: the adhesive strategies of cyprids and development of barnacle-resistant marine coatings. *Biofouling*, **24**, 351–363.
3 Rosenhahn, A., Ederth, T., and Pettitt, M.E. (2008) Advanced nanostructures for the control of biofouling: the FP6 integrated project AMBIO. *Biointerphases*, **3**, IR1–IR5.
4 Hills, J.M. and Thomason, J.C. (1998) The effect of scales of surface roughness on the settlement of barnacle (*Semibalanus balanoides*) cyprids. *Biofouling*, **12**, 57–69.
5 Thomason, J.C., Letissier, M.D.A.A., Ocampo-Thomason, P., and Field, S.N. (2002) Optimising settlement tiles: the effects of surface texture and energy, orientation and deployment duration upon the fouling community. *Biofouling*, **18**, 293–304.
6 Berntsson, K.M., Jonsson, P.R., Lejhall, M., and Gatenholm, P. (2000) Analysis of behavioural rejection of micro-textured surfaces and implications for recruitment by the barnacle *Balanus improvisus*. *J. Exp. Mar. Bio. Ecol.*, **251**, 59–83.
7 Callow, M.E., Jennings, A.R., Brennan, A.B., Seegert, C.E., Gibson, A., Wilson, L., Feinberg, A., Baney, R., and Callow, J.A. (2002) Microtopographic cues for settlement of zoospores of the green fouling alga *Enteromorpha*. *Biofouling*, **18**, 237–245.
8 Hoipkemeier-Wilson, L., Schumacher, J.F., Carman, M.L., Gibson, A.L., Feinberg, A.W., Callow, M.E., Finlay, J.A., Callow, J.A., and Brennan, A.B. (2004) Antifouling potential of lubricious, micro-engineered, PDMS elastomers against zoospores of the green fouling alga *Ulva* (*Enteromorpha*). *Biofouling*, **20**, 53–63.

9 Bers, A.V., D'Souza, F., Klijnstra, J.W., Willemsen, P.R., and Wahl, M. (2006) Chemical defence in mussels: antifouling effect of crude extracts of the periostracum of the blue mussel *Mytilus edulis*. *Biofouling*, **22**, 251–259.

10 Scardino, A.J., Zhang, H., Cookson, D.J., Lamb, R.N., and de Nys, R. (2009) The role of nano-roughness in antifouling. *Biofouling*, **25** (8), 757–767

11 Ralston, E. and Swain, G. (2009) Bioinspiration- the solution for biofouling control? *Bioinspiration and Biomimetics*, **4**, 1–9.

12 Carman, M.L., Estes, T.G., Feinberg, A.W., Schumacher, J.F., Wilkerson, W., Wilson, L.H., Callow, M.E., Callow, J.A., and Brennan, A.B. (2006) Engineered antifouling microtopographies – correlating wettability with cell attachment. *Biofouling*, **22**, 11–21.

13 Schumacher, J.F., Carman, M.L., Estes, T.G., Feinberg, A.W., Wilson, L.H., Callow, M.E., Callow, J.A., and Brennan, A.B. (2007) Engineered antifouling microtopographies-effect of feature size, geometry and roughness on settlement of zoospores of the green alga *Ulva*. *Biofouling*, **23**, 55–62.

14 Schumacher, J.F., Long, C.J., Callow, M.E., Finlay, J.A., Callow, J.A., and Brennan, A.B. (2007) Engineered nanoforce gradients for inhibition of settlement (attachment) of swimming algal spores. *Langmuir*, **24**, 4931–4937.

15 Schumacher, J.F., Aldred, N., Callow, M.E., Finlay, J.A., Callow, J.A., Clare, A.S., and Brennan, A.B. (2007) Species-specific engineered antifouling topographies: correlations between the settlement of algal zoospores and barnacle cyprids. *Biofouling*, **23**, 307–317.

16 Efimenko, K., Rackaitis, M., Manias, E., Vaziri, A., Mahadevan, L., and Genzer, J. (2005) Nested self-similar wrinkling patterns in skins. *J. Nat. Mater.*, **4**, 293–297.

17 Efimenko, K., Finlay, J.A., Callow, M.E., Callow, J.A., and Genzer, J. (2009) Development and testing of hierarchically wrinkled coatings for marine antifouling. *Appl. Mater. Interface*, **5**, 1031–1040.

18 Krishnan, S., Ayothi, R., Hexemer, A., Finlay, J.A., Sohn, K.E., Perry, R., Ober, C.K., Kramer, E.J., Callow, M.E., Callow, J.A., and Fischer, D.A. (2006) Antibiofouling properties of comb-like block copolymer with amphiphilic side-chains. *Langmuir*, **22**, 5075–5086.

19 Martinelli, E., Agostini, S., Galli, G., Chiellini, E., Glisenti, A., Pettitt, M.E., Callow, M.E., Callow, J.A., Graf, K., and Bartels, F.W. (2008) Nanostructured films of amphiphilic fluorinated block copolymers for fouling release application. *Langmuir*, **24**, 13138–13147.

20 Martinelli, E., Menghetti, S., Galli, G., Glisenti, A., Krishnan, S., Paik, M.Y., Ober, C.K., Smilgies, D.-M., and Fischer, D.A. (2009) Surface engineering of styrene/PEGylated-fluoroalkyl styrene block copolymer thin films. *J. Polym. Sci. Polym. Chem.*, **47**, 267–284.

21 Lin, F.-Y., Chen, W.-Y., and Hearn, M.T.W. (2002) Thermodynamic analysis of the interaction between proteins and solid surfaces: application to liquid chromatography. *J. Mol. Recognition*, **15**, 55–93.

22 Lau, K.H.A., Bang, J., Hawker, C.J., Kim, D.H., and Knoll, W. (2009) Modulation of protein-surface interactions on nanopatterned polymer films. *Biomacromolecules*, **10**, 1061–1066.

23 Grozea, C.M., Gunari, N., Finlay, J.A., Grozea, D., Callow, M.E., Callow, J.A., Lu, Z.-H., and Walker, G.C. (2009) Water-stable diblock polystyrene-block-poly(2-vinyl pyridine) and diblock polystyrene-block-poly(methyl methacrylate) cylindrical patterned surfaces inhibit settlement of zoospores of the green alga *Ulva*. *Biomacromolecules*, **10**, 1004–1012.

24 Gudipati, C.S., Greenleaf, C.M., Johnson, J.A., Pryoncpan, P., and Wooley, K.L. (2004) Hyperbranched fluoropolymer and linear poly(ethylene glycol) based amphiphilic crosslinked networks as efficient antifouling coatings: An insight into the surface compositions, topographies, and morphologies. *J. Polym. Sci. Polym. Chem.*, **42**, 6193–6208.

25 Gudipati, C.S., Finlay, J.A., Callow, M.E., Callow, J.A., and Wooley, K.L. (2005) The anti-fouling and fouling-release

performance of unique hyperbranched fluoropolymer (HBFP)-poly(ethylene glycol) (PEG) composite coatings evaluated by protein adsorption and the settlement of zoospores of the green fouling alga *Ulva* (syn. *Enteromorpha*). *Langmuir*, **21**, 3044–3053.

26 International Paint http://www.international-marine.com/products/productsearch/pages/Intersleek_970_3782.aspx, (accessed 23 August 2010).

27 Majumdar, P., Lee, E., Patel, N., Ward, K., Stafslien, S.J., Daniels, J., Chisholm, B.J., Boudjouk, P., Callow, M.E., Callow, J.A., and Thompson, S.E.M. (2008) Combinatorial materials research applied to the development of new surface coatings IX: An investigation of novel antifouling/fouling-release coatings containing quaternary ammonium salt groups. *Biofouling*, **24**, 185–200.

28 Majumdar, P. and Webster, D.C. (2005) Preparation of siloxane-urethane coatings having spontaneously formed stable biphasic microtopograpical surfaces. *Macromolecules*, **38**, 5857–5859.

29 Majumdar, P. and Webster, D.C. (2006) Influence of solvent composition and degree of reaction on the formation of surface microtopography in a thermoset siloxane–urethane system. *Polymer*, **47**, 4172–4181.

30 Majumdar, P., Stafslien, S., Daniels, J., and Webster, D.C. (2007) High throughput combinatorial characterization of thermosetting siloxane–urethane coatings having spontaneously formed microtopographical surfaces. *J. Coat. Technol. Res.*, **4**, 131–138.

31 Arce, F.T., Avci, R., Beech, I.B., Cooksey, K.E., and Wigglesworth-Cooksey, B. (2003) Microelastic properties of minimally adhesive surfaces: a comparative study of RTV11™ and Intersleek elastomers™. *J. Chem. Phys.*, **119**, 1671–1682.

32 Arce, T.F., Avci, R., Beech, I.B., Cooksey, K.E., and Wigglesworth-Cooksey, B. (2006) Modification of surface properties of a poly(dimethylsiloxane)-based elastomer, RTV11, upon exposure to seawater. *J. Chem. Phys.*, **119**, 671–1682.

33 Beigbeder, A., Degee, P., Conlan, S.L., Mutton, R.J., Clare, A.S., Pettitt, M.E., Callow, M.E., Callow, J.A., and Dubois, P. (2008) Preparation and characterisation of silicone-based coatings filled with carbon nanotubes and natural sepiolite, and their application as marine fouling-release coatings. *Biofouling*, **24**, 291–302.

34 Begbeider, A., Linares, M., Devalckenaere, M., Degee, P., Claes, M., Beljonne, D., Lazzaroni, R., and Dubois, P. (2008) CH-π interactions as the driving force for silicone-based nanocomposites with exceptional properties. *Adv. Mater.*, **20**, 1003–1007.

35 Beigbeder, A., Jeusette, M., Mincheva, R., Claes, M., Brocorens, P., Lazzaroni, R., and Dubois, P. (2009) On the effect of carbon nanotubes on the wettability and surface morphology of hydrosilylation-curing silicone coatings. *J. Nanostruct. Polym. Nanocomp.*, **5** (2), 37–43.

36 Marmur, A. (2004) Adhesion and wetting in an aqueous environment: theoretical assessment of sensitivity to the solid surface energy. *Langmuir*, **20**, 1317–1320.

37 Genzer, J. and Marmur, A. (2008) Biological and synthetic self-cleaning surfaces. *MRS Bulletin*, **33**, 742–746.

38 Marmur, A. (2006) Superhydrophobicity fundamentals: implications for biofouling prevention. *Biofouling*, **22**, 107–115.

39 Akesso, L., Pettitt, M.E., Callow, J.A., Callow, M.E., Stallard, J., Teer, D., Liu, C., Wang, S., Zhao, Q., D'Souza, F.D., Willemsen, P.R., Donnelly, G.T., Donik, C., Kocijan, A., Jenko, M., Jones, L.A., and Guinaldo, P.C. (2009) The potential of nanostructured silicon oxide type coatings deposited by PACVD for control of aquatic biofouling. *Biofouling*, **25**, 55–67.

40 Akesso, L., Navabpour, P., Teer, D., Pettitt, M.E., Callow, M.E., Liu, C., Su, X., Wang, S., Zhao, Q., Donik, C., Kocijan, A., Jenko, M., and Callow, J.A. (2009) Deposition parameters to improve the fouling-release properties of thin siloxane coatings prepared by PACVD. *Appl. Surf. Sci.*, **255**, 6508–6514.

15
Bioinspired Patterned Adhesives
Marleen Kamperman, Eduard Arzt, and Aránzazu del Campo

15.1
Introduction

Researchers have pondered the ability of some insects and geckos to firmly attach to and rapidly detach from varied types of surfaces in diverse conditions for many centuries [1–6]. Microscopic examination of their attachment pads has revealed a complex surface hyperstructure consisting of long hairs, sometimes organized in a hierarchical arrangement spanning the millimeter to nanometer range (Figure 15.1). These hairs have characteristic sizes and geometries depending on the animal, and their small dimension seems to be the key factor enabling their strong but reversible adhesion [7]. In the particular case of the gecko foot, which shows the best adhesion performance, the keratinous hairs ("setae") are typically 30 to 130 µm long and contain hundreds of projections terminating in 200 to 500 nm spatula-shaped structures. The gecko uses noncovalent surface forces to achieve adhesion, and research suggests that they rely primarily on van der Waals and capillary forces [8, 9]. The architecture is therefore the primary design variable in natural adhesive systems, not the chemistry.

Recent innovations in the area of micro- and nanofabrication with polymeric materials have created a unique opportunity for mimicking natural attachment systems as a new generation of moderately strong but reversible adhesives [10–12]. Increasing efforts are being made to establish fabrication technologies which allow the production of polymeric structured surfaces with greater geometrical complexity and finer structures, closer to the natural systems in design and performance. These include patterns made of polymer materials possessing elongated fibers (high aspect ratio), ordered in a tilted arrangement, coated fibers, exhibiting several hierarchy levels or three-dimensional (3D) spatula-like terminals. This chapter describes a number of micro- and nanofabrication approaches that have been applied to generate such structures. The methods are classified according to the complexity of the structures produced (vertical, tilted, coated, hierarchical, 3D). At the end, recent developments in the area of switchable adhesives based on structured surfaces and responsive materials will be described.

Generating Micro- and Nanopatterns on Polymeric Materials. Edited by A. del Campo and E. Arzt
Copyright © 2011 WILEY-VCH Verlag GmbH & Co. KGaA, Weinheim
ISBN: 978-3-527-32508-5

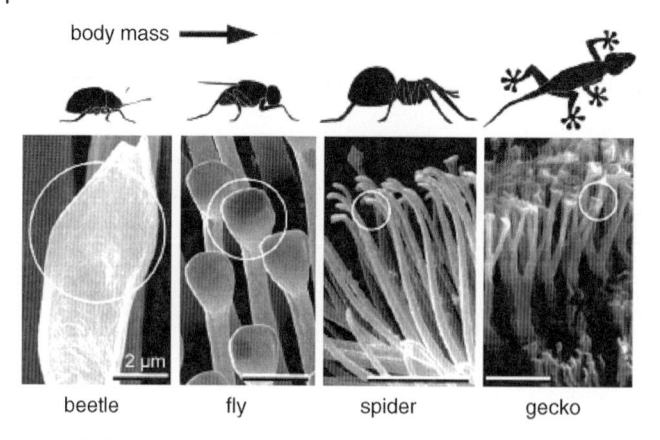

Figure 15.1 Terminal elements (circles) in animals with hairy design of attachment pads. Note that heavier animals exhibit finer adhesion structures. Scale bar represents 2 μm. (Reproduced with permission from [7]. Copyright © (2003) National Academy of Sciences, USA.)

15.2
Vertical Structures

15.2.1
E-Beam Lithography

Nanofabrication of fibrillar polyimide (PI) structures without the use of a template was realized using electron-beam lithography (Figure 15.2) followed by plasma etching and pattern transfer [13]. The etching step is necessary in order to obtain high aspect ratio structures, since the maximum penetration depth of low energy electrons is about 100 nm. Fiber diameters ranged between 0.2 and 4 μm, heights from 0.15 to 2 μm and spacings from 0.4 to 4.5 μm. Electron-beam lithography is appropriate to obtain model structures with small dimensions. The main limitations of this method are the costs associated with the processes and the small patterning areas [14].

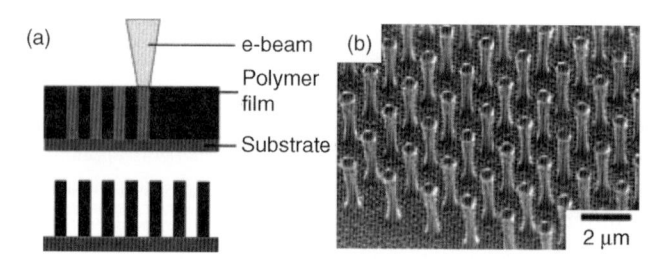

Figure 15.2 (a) Direct nanofabrication by e-beam lithography. (b) SEM micrograph of PI pillars array. (Reproduced with permission from [13]. Copyright © (2003) Macmillan Publishers Ltd.)

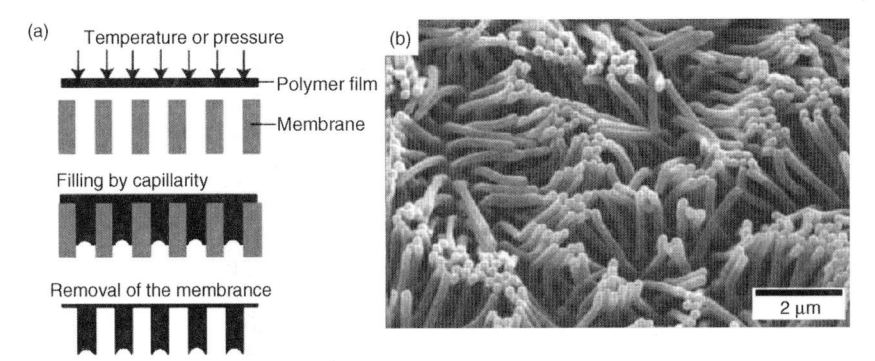

Figure 15.3 (a) Fabrication of high aspect ratio fibers by filling porous AAO membranes. (b) Fibers collapse after removal of the membrane (dissolution) because of the large capillary forces acting during the drying process. (Reproduced with permission from [18]. Copyright © (2007) Springer.)

15.2.2
Filling Porous Membranes

Commercially available track-etched polycarbonate (PC) and anodic alumina (AAO) membranes, with pore sizes ranging from a few nm to a few μm and different spacings and thicknesses, were used to produce arrays of long, nanosized fibers (Figure 15.3) [15–21]. The pores of the membrane can be filled with polymer precursors, solutions or melts through casting or electrodeposition [17, 21] to obtain cylinders with dimensions reproducing those of the pores. PC membranes containing randomly distributed cylindrical pores of 0.6 μm diameter and spacings <5 μm have been filled with PI solutions [15]. Since the PC membrane is flexible, it can be peeled off from the solidified PI film to release dense arrays of high aspect ratio nanofibers. Due to their large aspect ratio, these fibers were shown to collapse laterally and therefore reduced adhesion is expected. AAO membranes with pores between 200 and 400 nm have been also filled with two-parts epoxy resin [15], with a polystyrene (PS) solution [20], and with UV curable precursors [18]. Because of their rigidity, AAO membranes cannot be removed from the polymer by peeling off but need to be selectively dissolved in NaOH solution. This represents a strong disadvantage for fabrication, since the template is destroyed and dissolution takes long times and may cause polymer swelling. In addition, wet etching is followed by a drying step during which capillary forces usually may cause fibers lateral collapse and, consequently, reduced adhesion (Figure 15.3b) [18].

15.2.3
Photolithographic Templating

Photolithographic templates might be able to overcome these shortcomings, since the template can be peeled off and re-used. In addition, photolithographic templates allow precise control over the geometric parameters, which can be exploited to obtain well-defined fibrillar arrays and to perform model studies [22–25]. Photolithography using SU-8 photoresist has been proven to be particularly suited for obtaining regular templates [24]. This resist material has been specially formulated to obtain high-aspect-ratio features, like holes which are required for obtaining long fibers in the replication process [26]. The negative replica of a micro- or nanostructured hard master is prepared by casting and thermal curing a liquid prepolymer (mostly poly (dimethyl siloxane) (PDMS)) on the master (Figure 15.4a). The elastomeric character allows the PDMS stamp to be released easily from the master (or molded polymer), even in the presence of complex and fragile structures, like high-aspect ratio fibers. Moreover, its low interfacial free energy and chemically inertness reduce mold sticking. Arrays of PDMS fibers with radii between 1 and 25 μm and lengths between 5 and 80 μm were reported over 25 cm^2 areas (Figure 15.4d) [24]. These systems have been used to analyze the influence of different geometrical parameters in adhesion, such as contact radius and aspect ratio of the fibers [24]. The mechanical stability of the fibers is limited by the compliance of the material, and this limits the aspect ratio

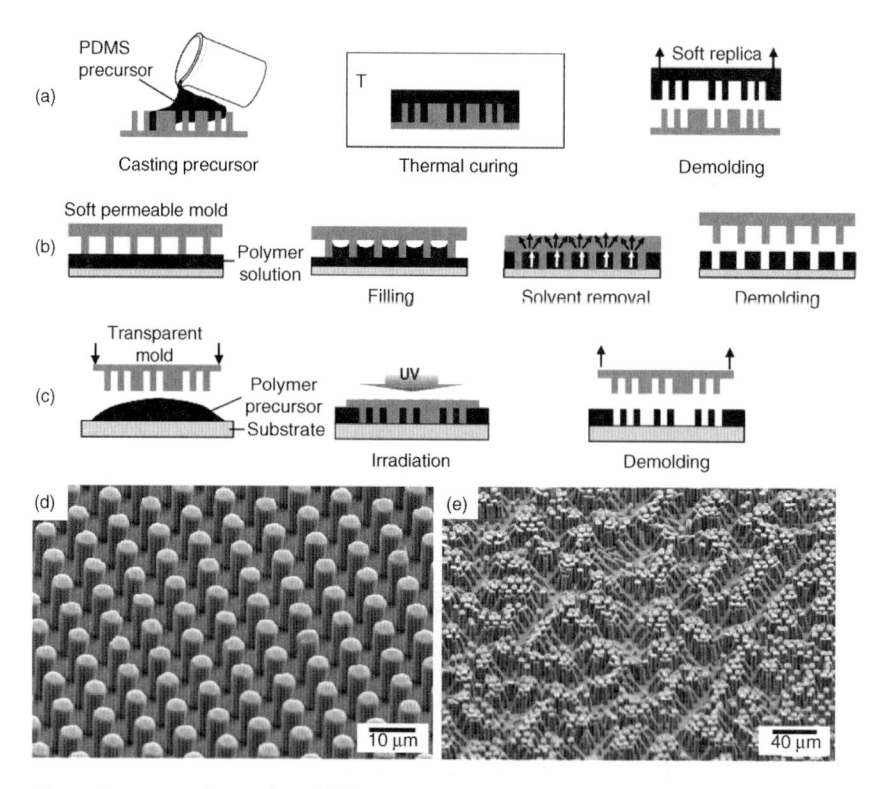

Figure 15.4 (a) Replication by soft lithography. (b) Soft molding polymer solutions using PDMS stamps. (c) Soft-molding UV curable prepolymers. (d) SEM micrograph showing an array of fibers made by soft-molding PDMS onto SU-8 photolithographic templates. Fibers have a radius of 2.5 μm and a height of about 20 μm. The minimum interpillar distance is 5 μm. (e) PDMS pillar with higher aspect ratios collapse after demolding due to the low mechanical stability of PDMS. (Reproduced with permission from [24]. Copyright © (2007) American Chemical Society).

of the structures that are possible to obtain by this method. For PDMS microfibers collapse typically occurs with aspect ratios exceeding four (Figure 15.4e) [12, 27]. This fact is more critical in nanosized fibers and their fabrication requires the use of stiffer materials, such as PS and poly(methyl methacrylate) (PMMA) [17].

The PDMS replica can also be used as mold (or stamp) for patterning other polymeric materials in subsequent replication processes (Figure 15.4b and c). For example, PDMS stamps were used to soft-mold liquid polyurethane (PUR) precursors and fibers with 20 μm diameter, 40 to 100 μm length and 40 μm spacing were obtained [28]. In addition, PDMS is transparent and can be used to mold UV-curable prepolymers (Figure 15.4c). Using this method, hard PUR patterns with fibrils of 0.5 to 4 μm height and diameters between 1 and 4 μm have been generated [25]. Alternative ways of producing templates include indenting a wax surface with an AFM tip [16], laser ablation of a metallic surface [29] and interference lithography [30]. Another example of template fabrication is by etching of colloidal patterns, creating nanofibers of parylene with ~250 nm diameter and aspect ratio 10 [31].

15.2.4
Hot Embossing

Hot-embossing involves shaping a polymer melt by conformal contact of a micro- or nanostructured mold using heat and pressure and can be used to pattern thermoplastic materials (Figure 15.5). The polymer melt is able to flow and fill the mold cavities in the processing conditions. The filling depends on the viscosity, wetting properties and pattern geometry, as well as on applied pressure or vacuum. Solidification of the polymer after filling is achieved by cooling below the crystallization temperature in semicrystalline polymers, or below the glass transition temperature in amorphous polymers. Removal of the mold releases a structured polymer with

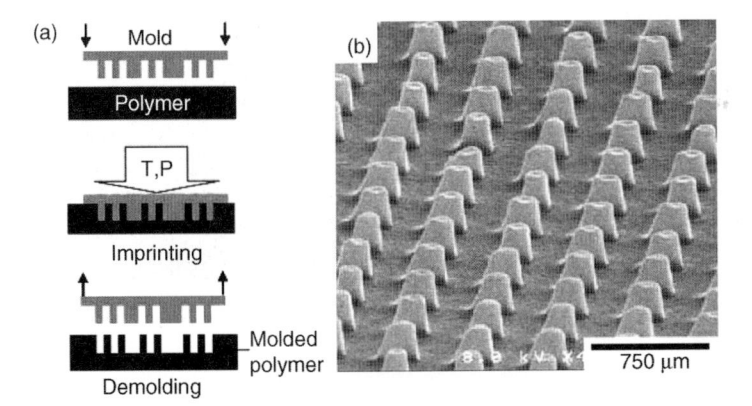

Figure 15.5 (a) Hot-embossing method and (b) example of fabricated PMMA fibers. The mold had pillar features with a lateral dimension of 150 nm at base end, 50 nm at top, and 500 nm height. (Reproduced with permission from [33]. Copyright © (2006) Springer.)

features reproducing its particular geometry. This can be done by peeling off (demolding) or by selective dissolution of the template. Demolding is preferred since it permits using the same mold for additional molding processes. An example of embossed PMMA fibrillar surface is given in Figure 15.5b. The embossing process was performed at temperatures above the glass transition of PMMA ($\sim$120 °C) using a poly(urethane acrylate) (PUA) mold. Arrays of fibers with 150 nm diameter and up to 500 nm height were obtained [32, 33].

The fabrication of high-quality moulds is one of the most important requirements in performing successful embossing. These are typically made of silicon or silicon dioxide by dry etching technologies, or by deposition of nickel or other metals on patterned resist substrates (LIGA process). Mold fabrication is the most time and cost consuming step involved in these patterning techniques, and is likely to constitute the biggest limitation in potential industrial applications.

15.3
Tilted Structures

15.3.1
Filling Nanoporous or Photolithographic Templates

To mimic the directionality that is typically observed in natural attachment systems, tilted fibrillar structures have been fabricated. The tilted disposition of the fibers allows peeling-off the adhesive and, therefore, easy removal [34]. Tilted structures were obtained using nanoporous PC membranes by processing the patterned film through two heated rollers [35]. To obtain tilted pillars using photolithographic templates, templates were fabricated by exposing the resist layer at well-defined angles (between 0 and 50° with respect to the substrate), as illustrated in Figure 15.6 [28]. Soft-molding PUR precursors with a PDMS negative replica of the SU-8 master yielded arrays of PUR microfibers with a tilting angle of 25° [28]. The directional nature of these angled systems was shown by measuring the shear performance in both directions [36].

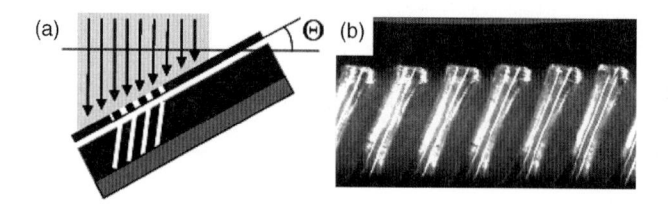

Figure 15.6 (a) Fabrication of tilted fibers by inclined lithography. (b) The light microscopy image shows tilted polyurethane pillars obtained after soft-molding with the photolithograhic SU-8 masters. (Reproduced with permission from [28]. Copyright © (2007) American Chemical Society.)

15.3.2
Drawing Polymer Fibers

Slanted and straight nanofiber arrays have been obtained by drawing fibers from polymer drops on nonwettable surfaces by contacting them with a hot plate and then moving both surfaces apart [37]. The process is based on surface tension and capillary forces and is available in a number of versions. For example, films of thermoplastic polymers can be brought in contact with a hot structured master possessing pillars. When pulling the master apart, fibers will be drawn from the contacting points (Figure 15.7). Alternatively, structured polymer films with large features can be brought in contact with a hot surface and then removed to obtain elongated fibers on the top of the features [37]. Other authors have miniaturized this method by using an AFM tip to draw nanosized fibers from a melt polymer film [38]. These fabrication methods seem to be more suitable for large area patterning, and are therefore more likely to be applied in manufacture. However, this requires severe alignment between hot plate (or a roll) and the polymer film across large areas, if nanosized fibers need to be obtained.

A creative alternative has been reported recently by combining molding and fiber drawing processes to obtain high aspect ratio nanofibers from thin PS and PMMA films [39]. This was achieved by preforming the demolding step at temperatures above the glass transitions of the materials. Capillary forces induced deformation of the polymer melt into the void spaces of the mold, and the filled nanofibers were elongated upon removal of the mold due to tailored adhesive force at the mold/polymer interface (Figure 15.7c). PS and PMMA fibrils with 80 nm diameter and AR > 20 have been reported [40].

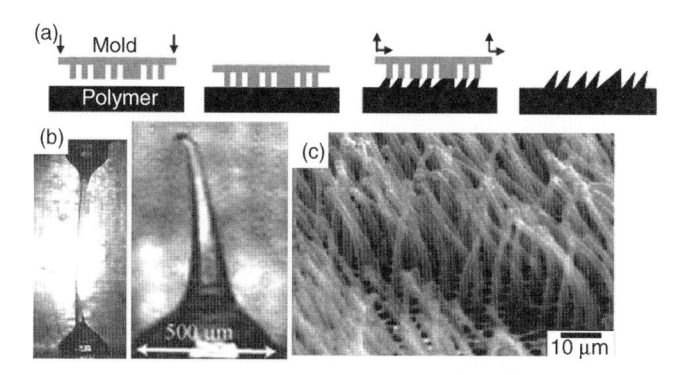

Figure 15.7 (a) Fabrication of structured surfaces by fiber drawing method. (b) Photograph of nylon hook fabrication and (c) SEM picture of a PMMA fibrillar surface. ((b): Reproduced with permission from [37]. Copyright © (2005) Institute of Physics; (c): Reproduced with permission from [39]. Copyright © (2006) American Chemical Society.)

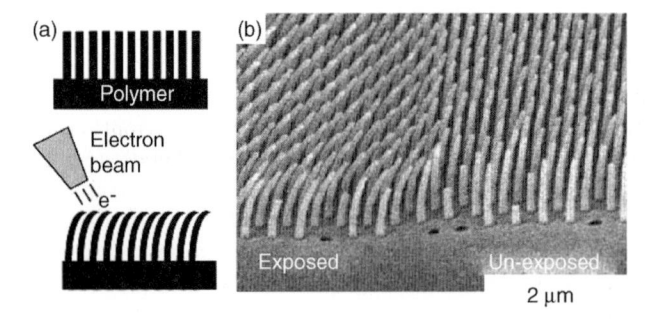

Figure 15.8 (a) Fabrication of slanted nanofibrils by e-beam exposure. (b) SEM image of exposed and un-exposed PUA fibrils. (Reproduced with permission from [41]. Copyright © (2009) Wiley-VCH Verlag GmbH & Co. KGaA.)

15.3.3
Electron-Beam Irradiation

Tilted structures were also prepared by a post-molding electron-beam irradiation step. The irradiated fibril surface shrinks more than the opposite surface, resulting in bending of soft-molded fibrils [41, 42]. With this method, PUA nanopillars (100 nm diameter, and aspect ratio of 10) were fabricated with tilting angles between 30° and 80° (Figure 15.8).

15.4
Coated Structures

Recently, the interest on adhesive fibrillar surfaces has turned to include adhesion against soft substrates or under wet environments. This is achieved by coating the micropatterned substrates with wet adhesives (Figure 15.9) [43–45] or viscous oils [46]. Patterned PDMS [44] or PUR [45] surfaces were coated with poly(dopamine methacrylate-co-2-methoxyethyl acrylate), a copolymer containing high amounts of cathecol, and a key component of wet adhesive proteins found in mussels. Such

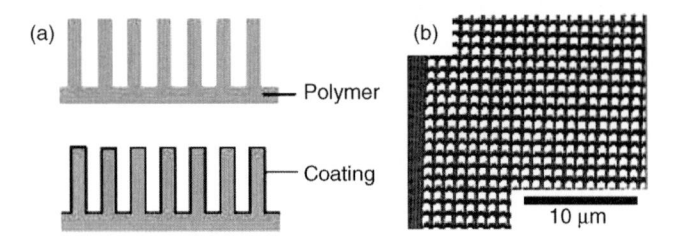

Figure 15.9 (a) Modification of fibrillar arrays after molding by coating to modify the chemistry of the pillars. (b) SEM image of patterned PDMS coated with poly(dopamine methacrylate-co-2-methoxyethyl acrylate). (Reproduced with permission from [44]. Copyright © (2007) Macmillan Publishers Ltd.)

proteins were shown to interact with different surfaces, such as metals, organic polymers and oxides, by a variety of interactions [44, 45]. Both single fibers and film-coated pillar arrays showed dramatic increase (up to 23-fold) in adhesion forces in wet conditions over noncoated controls [44, 45]. In a different approach, micropatterned substrates were coated with an aldehyde-functionalized polysaccharide and the ability to adhere to tissue was evaluated [43].

Alternatively, thin metal coatings were applied either in combination with electron-beam irradiation or applied anisotropically to create tilted structures. The irradiated or coated pillar surfaces shrank more than the opposite surface, resulting in bending of soft-molded pillars, and subsequent anisotropic adhesion behavior [41, 47, 48].

15.5
Hierarchical Structures

Gecko setae have a hierarchical organization, which enables adaptability and adhesion to rough surfaces of any kind. To mimic the multiple levels of compliance present in the gecko setae, structures ordered over multiple length scales have been proposed.

15.5.1
Filling Stacked Membranes

Hierarchical fibrillar structures were obtained by filling stacked micro- and nanoporous AAO membranes with different pore diameters [49]. The nanoporous membrane (pore diameter ≈60 nm and interpore distance ≈100 nm) was generated by the anodization of an aluminum film in an oxalic acid solution. The microporous alumina was produced by conventional lithography and anisotropic chemical etching of the thick film of anodic alumina pores. The micro- and nanoporous alumina membranes were subsequently brought into intimate contact and filled with PMMA solution. Hierarchical polymeric microfibrils (fibril diameter ≈10 μm; fibril length ≈70 μm) with nanofibril arrays at their tips were obtained after removal of the solvent by heating and selective etching of the AAO membranes (Figure 15.10a and b) [49]. The nanofibril has a lateral dimension of approximately 60 nm with length-to-diameter aspect ratios as high as 100:1. In a recent study a combination of direct laser writing and membrane templating resulted in hierarchical systems composed of compliant lamellar flaps covered with high aspect ratio nanofiber arrays (Figure 15.10c, d and e) [50].

15.5.2
Multistep Exposure in Photolithography

Multistep photolithography in combination with molding was used to create hierarchical structures with stacked fibers of different dimensions [51, 52]. Figure 15.11a and b show the fabrication procedure and an example of micrometric adhesive

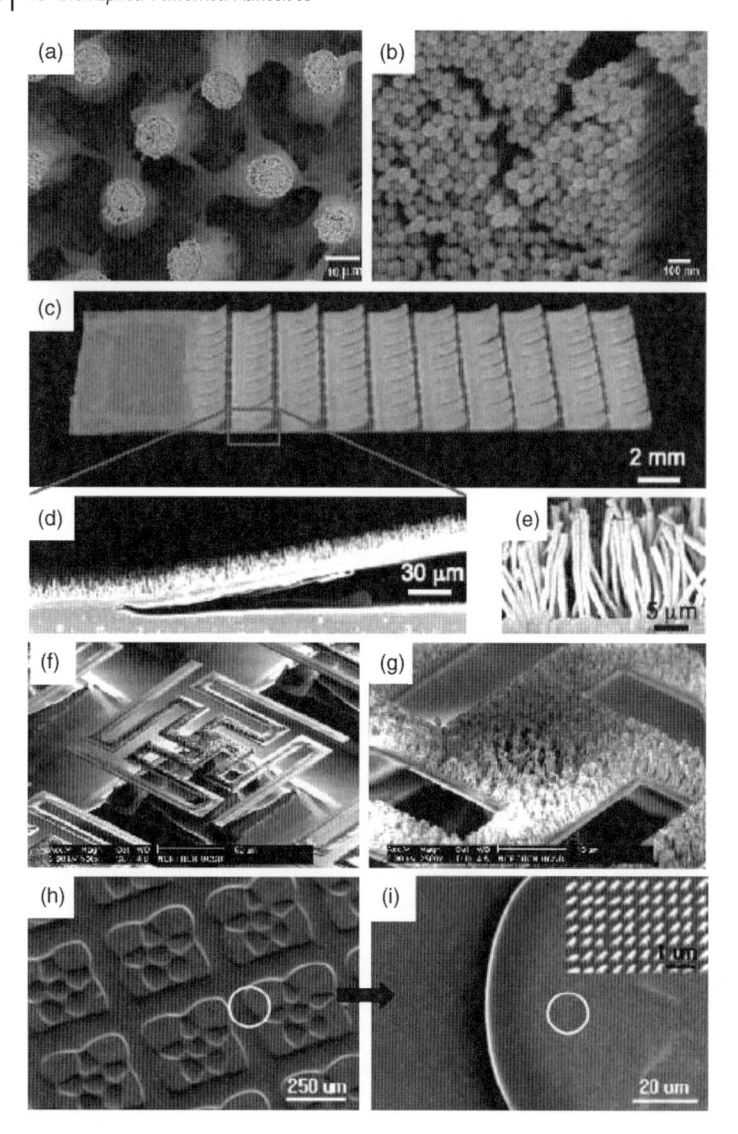

Figure 15.10 (a) Hierarchical microfibril array fabricated by filling porous AAO; (b) microfibers are 10 µm wide and 70 µm long, and each branches into nanofibrils about 60 nm wide and 0.5 µm long. (c) Optical and (d, e) SEM images of lamellar flaps covered with nanofiber arrays. (f) Multiscale microfabricated structures consisting of single crystal silicon pillars supporting a silicon dioxide platform and (g) coated by nanometer-sized polymer fibers. Hierarchical PMMA pattern fabricated by superposition of two embossing steps (h, i). ((a,b): Reproduced with permission from [49]. Copyright © (2007) Institute of Physics; (c–e): Reproduced with permission from [50]. Copyright © (2009) American Chemical Society; (f, g): Reproduced with permission from [57]. Copyright © (2006) Elsevier; (h, i): Reproduced with permission from [32]. Copyright © (2006) American Chemical Society.)

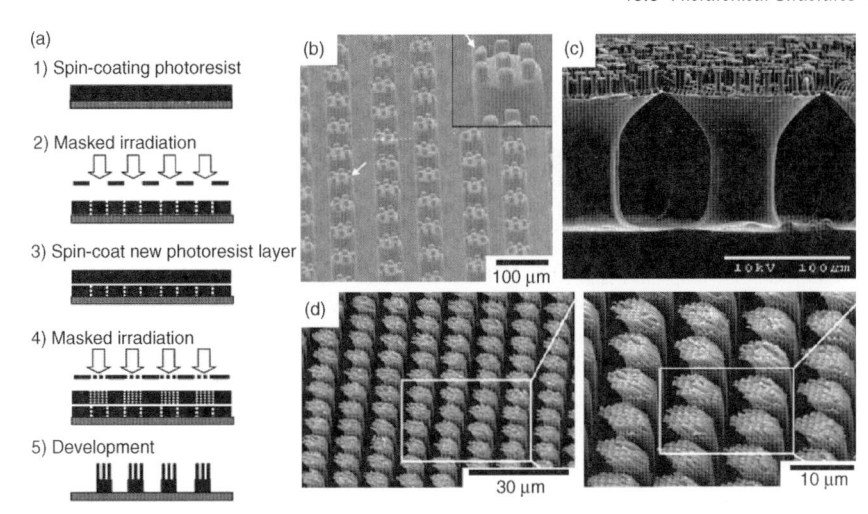

Figure 15.11 (a) Schematic of hierarchical structures by double exposure photolithography and (b) SEM image of PDMS hierarchical pillar array. (c) PU structures with a complex tip geometry prepared by soft-lithography and capillary molding. (d) PUA hierarchical structures obtained by reactive ion etching and capillary molding. ((a, b) Reproduced with permission from [51]. Copyright © (2009) Wiley-VCH Verlag GmbH & Co. KGaA; (c): Reproduced with permission from [54]. Copyright © (2009) American Chemical Society; (d): Reproduced with permission from [55]. Copyright © (2009) National Academy of Sciences, USA.)

structure with two hierarchical levels obtained by SU-8, respectively [53]. Superposition of coating and irradiation steps enables fabrication of structures with several organization levels using traditional 2D setups and alignment markers on the mask for guiding superposition. Additional levels are possible by just increasing the number of coating and irradiation steps. Geometries are not restricted to cylindrical fibers and depend only on the mask used for irradiation. This can also be replicated using soft-molding methods and transferred to other materials.

Two-step molding has been also used to fabricate PUR [54] (Figure 15.11c) and PUA [55] (Figure 15.11d) hierarchical structures: the base fibers were formed by soft-lithography, while the top ones were formed by capillary molding and a second curing step (first level of fibrils 5 to 50 μm in diameter, 25 to 100 μm high; second level with fibers 350 nm to 3 μm in diameter, 2.8 to 20 μm high) [54, 55]. This method also yielded a three-level structure, by using macrofibers (1.2 mm in height, 300 μm in diameter) as the first hierarchical level [54].

15.5.3
Microfabrication Technologies

Hierarchical structured surfaces have also been obtained by combining photolithography and dry etching methods, as typically used in MEMS fabrication [56, 57]. Single crystal silicon posts (1 μm in diameter and height up to 50 μm) supporting microsized silicon dioxide platforms (2 μm thick and 100 to 150 μm on a side) were coated by photoresist fibers with ~250 nm diameter and ~4 μm height in average (Figure 15.10f and g).

15.5.4
Two-Step Embossing

Sequential embossing steps have also been shown to result in hierarchical patterns with micrometric fibers decorated with nanosized fibers of various sizes and spacings (Figure 15.10h and i) [32]. A PMMA film was patterned with microfibers (20 to 120 μm diameter and low aspect ratio) using a PDMS stamp. Subsequently, nanofibers (100–150 nm width and 600 nm height) were patterned on the top of the preformed microfibers using a hard PUA mold. The PUA mold replaces the PDMS mold for sub-100 nm lithography since high-aspect ratio sub-100 nm features in a PDMS mold do not retain dimensional stability due to the low Young's modulus of PDMS. The resulting micro/nanoscale combined structures were robust and demonstrated enhanced water-repellent properties as a consequence of the hierarchical arrangement.

15.6
3D Structures

Theoretical and experimental reports have stressed the significance of contact shapes and have been found to play a decisive role for the final adhesion performance in artificial systems [58, 59]. Spherical, conical, filament-like, band-like, sucker-like, spatula-like, flat and toroidal tip shapes have all been observed in different organisms [60].

15.6.1
Reactive Ion Etching

3D templates can be produced through an etching process (Figure 15.12). For example, a Faraday cage was adapted to a plasma etching chamber, allowing direct etching of angled cavities from a substrate [55]. By using an initial isotropic etching of the substrate and a silica etch-stop layer, pillars could be fabricated with thicker bottoms and spatular tips of different sizes. Two-step molding resulted in hierarchical structures: the base pillars were formed by soft-lithography, while the top ones were formed by capillary molding and a second curing step [55]. Three-dimensional geometries have also been obtained combining photolithography and dry etching methods [56, 61–63]. Arrays of spatula-like fibers have been fabricated by filling templates containing 3D holes with PUA [62] or PDMS [63]. Demolding the 3D profile from the template was only possible using PDMS [63] and required etching of the template in the case of PUA. Alternatively, 3D structures have been produced by micromachined templates [64, 65].

(a) Masked irradiation

(b)

Reactive ion etching

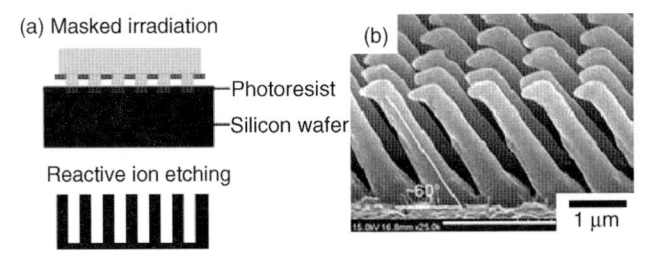

Figure 15.12 (a) Complex mold fabrication by a combination of photolithography and etching techniques. (b) SEM image of PUA pillar array with 3D structure. (Reproduced with permission from [55]. Copyright © (2009) National Academy of Sciences, USA.)

15.6.2
Post-Molding Inking

Using a modification of the soft-molding method, arrays of PDMS [26, 53, 66] or PUR [36, 67] fibers with spherical and mushroom-like tips were reported. These surfaces usually showed higher adhesion strength than planar surfaces, and in some cases even surpassed the gecko adhesion [53]. Arrays of microfibrils with controlled tip-shapes were obtained by inking a fibrillar PDMS surface in a thin film of PDMS precursor. Figure 15.13 shows the fabrication methods. For example, pressing an inked stamp against a flat substrate followed by curing resulted in pillars with a flexible and flat mushroom-shaped tip. Other curing schemes resulted in different

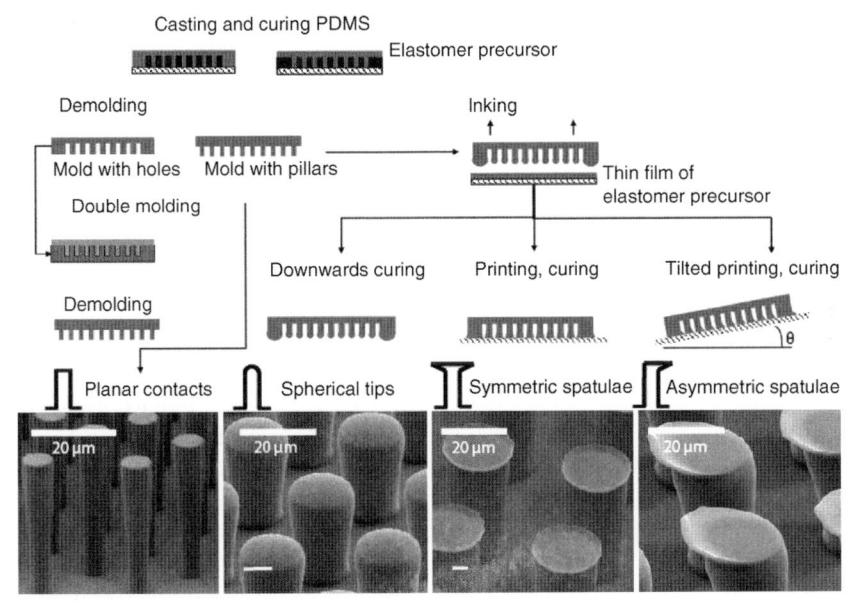

Figure 15.13 Structuring methods for 3D tip shape fabrication. (Reproduced with permission from [52]. Copyright © (2007) Wiley-VCH Verlag GmbH & Co. KGaA.)

geometries. A similar method was applied to obtain mushroom tips onto tilted (0 to 33°) PUR microfibers [36], and dangling chain functionalized structures [68]. Tilted PUR fibers were modified with mushroom tips placed at a controlled angle with respect to the fiber by applying a certain load during curing, causing bending of the fibrils [67]. This system is the first demonstration in which fiber and spatula are tilted independently, thereby representing the most advanced gecko-inspired system to date. Other authors have used this method to obtain a continuous thin film of PDMS on the top of a fibrillar surface, which is also advantageous for adhesion purposes [69].

15.7
Switchable Adhesion

The great challenge to mimic nature's specific mechanism of "on demand" strong adhesion and easy release is being explored by several research groups using topographical changes upon applying an external stimulus. In a first example, an array of shape memory polymer (SMP) microfibers was mechanically deformed above its shape-memory transition temperature, followed by cooling to room temperature in the deformed position. This yielded a temporary nonadhesive surface consisting of pillars in a tilted position. By reheating above the transition temperature, the patterned surface switched from the temporary nonadhesive state to a permanent adhesive surface, increasing the adhesion force by a factor of 200 [70].

A fully reversible adhesive based on magnetic switching of nickel cantilevers (10 μm × 130 μm) coated with vertically aligned polymeric nanorods was demonstrated by Northen *et al.* [71]. The cantilevers were able to reorient under a magnetic field, such that the tips rotated away from the counter-surface, decreasing the adhesion force [71].

Recently, switchability was also demonstrated for nonresponsive material systems, that is, PDMS fibrillar arrays. Mechanical stretching as external stimulus was used to switch reversibly from adhesion to nonadhesion of a PDMS fibrillar array. The array was giving a wrinkled configuration through UV treatment of the PDMS. Adhesion could be turned on by orientating the fibrils normal to the surface upon stretching [72]. PDMS fibrillar arrays were also shown to switch adhesion with applied pressure (preload) as external stimulus. The increased preload caused the pillars to lose the interfacial adhesive contact. The elasticity of PDMS aided the fibril recovery to the upright position upon removal of preload enabling repeatability of the switch [73].

15.8
Outlook

Biomimetic fibrillar adhesives with increasing complexity have been generated by different micro- and nanofabrication methods. The achieved geometries still represent a coarse simplification of natural systems and this may limit their adhesion performance. Before large scale manufacturing can be demonstrated and fibrillar adhesives commercialized several issues have to be addressed:

Finer contact elements will result in stronger adhesion and will also enhance the adaptability to rough surfaces. Reduction of structural features of the fibrillar adhesives, however, requires materials systems that are strong, tough and durable, since small structures are prone to lateral collapse and break easily. Therefore, more efforts are required to develop materials systems that accommodate these requirements.

Up till now combinations of different (microfabrication) techniques are required for geometrical designs that span over three dimensions and in different length scales (from mm to nm). Microfabrication techniques strongly restrict material's selection to a few resists, all of them quite expensive. In addition, patterning can only be performed in specialized laboratories and requires costly equipment and long processing times. For this reason, scientists have started to look for alternative patterning techniques which do not require a template (e.g., the mold in soft-lithography or the mask in photolithography). An interesting, largely unexplored fabrication technique for biomimetic fibrillar arrays that may enable complex 3D and hierarchical adhesives in one-step processes is self-assembly.

Other prerequisites for the successful development of useful products lie in so far unexplored properties: How do gecko surfaces behave under repeated contact formation and breakage? How do they respond to changes in temperature and humidity? What is their long-term reliability in specific environments? And, most important, can they ever be fabricated cost-effectively over large areas? Only if these and similar problems can be successfully overcome, will gecko-inspired adhesives realize their potential in applications.

References

1 Hiller, U. (1968) *Z. Morph. Tiere*, **62**, 307.
2 Gorb, S. and Scherge, M. (2000) *Proc. R. Soc. B.*, **267**, 1239.
3 Gorb, S., Gorb, E., and Kastner, V. (2001) *J. Exp. Biol.*, **204**, 1421.
4 Gorb, S.N., Beutel, R.G., Gorb, E.V., Jiao, Y.K., Kastner, V., Niederegger, S., Popov, V.L., Scherge, M., Schwarz, U., and Votsch, W. (2002) *Integr. Comp. Biol.*, **42**, 1127.
5 Autumn, K. (2006) *Am. Sci.*, **94**, 124.
6 Autumn, K. and Peattie, A.M. (2002) *Integr. Comp. Biol.*, **42**, 1081.
7 Arzt, E., Gorb, S., and Spolenak, R. (2003) *Proc. Natl. Acad. Sci. USA*, **100**, 10603.
8 Autumn, K., Sitti, M., Liang, Y.C.A., Peattie, A.M., Hansen, W.R., Sponberg, S., Kenny, T.W., Fearing, R., Israelachvili, J.N., and Full, R.J. (2002) *Proc. Natl. Acad. Sci. USA*, **99**, 12252.
9 Huber, G., Mantz, H., Spolenak, R., Mecke, K., Jacobs, K., Gorb, S.N., and Arzt, E. (2005) *Proc. Natl. Acad. Sci. USA*, **102**, 16293.
10 Gates, B.D., Xu, Q.B., Stewart, M., Ryan, D., Willson, C.G., and Whitesides, G.M. (2005) *Chem. Rev.*, **105**, 1171.
11 Geissler, M. and Xia, Y.N. (2004) *Adv. Mater.*, **16**, 1249.
12 del Campo, A. and Arzt, E. (2008) *Chem. Rev.*, **108**, 911.
13 Geim, A.K., Dubonos, S.V., Grigorieva, I.V., Novoselov, K.S., Zhukov, A.A., and Shapoval, S.Y. (2003) *Nature Mater.*, **2**, 461.
14 Jeong, H.E. and Suh, K.Y. (2009) *Nano Today*, **4**, 335.

15 Majidi, C., Groff, R., and Fearing, R. (2004) Clumping and packing of hair arrays manufactured by nanocasting. Presented at 2004 ASME International Mechanical Engineering Congress & Exposition, Anaheim, CA, USA.
16 Sitti, M. and Fearing, R.S. (2003) *J. Adhes. Sci. Technol.*, **17**, 1055.
17 Schubert, B., Majidi, C., Groff, R.E., Baek, S., Bush, B., Maboudian, R., and Fearing, R.S. (2007) *J. Adhes. Sci. Technol.*, **21**, 1297.
18 Kim, D.S., Lee, H.S., Lee, J., Kim, S., Lee, K.H., Moon, W., and Kwon, T.H. (2007) *Microsyst. Technol.*, **13**, 601.
19 Cho, W.K., and Choi, I.S. (2008) *Adv. Func. Mater.*, **18**, 1089.
20 Jin, M.H., Feng, X.J., Feng, L., Sun, T.L., Zhai, J., Li, T.J., and Jiang, L. (2005) *Adv. Mater.*, **17**, 1977.
21 Lu, G.W., Hong, W.J., Tong, L., Bai, H., Wei, Y., and Shi, G.Q. (2008) *ACS Nano*, **2**, 2342.
22 Crosby, A.J., Hageman, M., and Duncan, A. (2005) *Langmuir*, **21**, 11738.
23 Glassmaker, N.J. and Hui, C.Y. (2004) *J. Appl. Phys.*, **96**, 3429.
24 Greiner, C., del Campo, A., and Arzt, E. (2007) *Langmuir*, **23**, 3495.
25 Lamblet, M., Verneuil, E., Vilmin, T., Buguin, A., Solberzan, P., and Léger, L. (2007) *Langmuir*, **23**, 6966.
26 Del Campo, A. and Greiner, C. (2007) *J. Micromech. Microeng.*, **17**, R81.
27 Xia, Y.N. and Whitesides, G.M. (1998) *Angew. Chem. Int. Edit.*, **37**, 551.
28 Aksak, B., Murphy, M.P., and Sitti, M. (2007) *Langmuir*, **23**, 3322.
29 Peressadko, A. and Gorb, S.N. (2004) *J. Adhesion*, **80**, 247.
30 Kim, S., Sitti, M., Jang, J.-H., and Thomas, E.L. (2008) Fabrication of bio-inspired elastomer nanofiber arrays with spatulate tips using notching effect. Presented at 2008 8th IEEE Conference on Nanotechnology, Arlington, Texas, USA.
31 Kustandi, T.S., Samper, V.D., Yi, D.K., Ng, W.S., Neuzil, P., and Sun, W. (2007) *Adv. Func. Mater.*, **17**, 2211.
32 Jeong, H.E., Lee, S.H., Kim, J.K., and Suh, K.Y. (2006) *Langmuir*, **22**, 1640.
33 Yoon, E.S., Singh, R.A., Kong, H., Kim, B., Kim, D.H., Jeong, H.E., and Suh, K.Y. (2006) *Tribol. Lett.*, **21**, 31.
34 Federle, W. (2006) *J. Exp. Biol.*, **209**, 2611.
35 Lee, J., Fearing, R.S., and Komvopoulos, K. (2008) *Appl. Phys. Lett.*, **93**, 191910.
36 Murphy, M.P., Aksak, B., and Sitti, M. (2007) *J. Adhes. Sci. Technol.*, **21**, 1281.
37 La Spina, G., Stefanini, C., Menciassi, A., and Dario, P. (2005) *J. Micromech. Microeng.*, **15**, 1576.
38 Harfenist, S.A., Cambron, S.D., Nelson, E.W., Berry, S.M., Isham, A.W., Crain, M.M., Walsh, K.M., Keynton, R.S., and Cohn, R.W. (2004) *Nano Lett.*, **4**, 1931.
39 Jeong, H.E., Lee, S.H., Kim, P., and Suh, K.Y. (2006) *Nano Lett.*, **6**, 1508.
40 Jeong, H.E., Lee, S.H., Kim, P., and Suh, K.Y. (2008) *Colloid. Surfaces A*, **313–314**, 359.
41 Kim, T.-i., Jeong, H.E., Suh, K.Y., and Lee, H.H. (2009) *Adv. Mater.*, **21**, 2276.
42 Kim, T.-i., Pang, C., and Suh, K.Y. (2009) *Langmuir*, **25** (16), 8879.
43 Mahdavi, A., Ferreira, L., Sundback, C., Nichol, J.W., Chan, E.P., Carter, D.J.D., Bettinger, C.J., Patanavanich, S., Chignozha, L., Ben-Joseph, E., Galakatos, A., Pryor, H., Pomerantseva, I., Masiakos, P.T., Faquin, W., Zumbuehl, A., Hong, S., Borenstein, J., Vacanti, J., Langer, R., and Karp, J.M. (2008) *Proc. Natl. Acad. Sci. USA*, **105** 2307.
44 Lee, H., Lee, B.P., and Messersmith, P.B. (2007) *Nature*, **448**, 338.
45 Glass, P., Chung, H., Washburn, N.R., and Sitti, M. (2009) *Langmuir*, **25**, 6607.
46 Cheung, E. and Sitti, M. (2008) *J. Adhes. Sci. Technol.*, **22**, 569.
47 Yoon, H., Jeong, H.E., Kim, T.I., Kang, T.J., Tahk, D., Char, K., and Suh, K.Y. (2009) *Nano Today*, **4**, 385.
48 Yoon, H., Woo, H., Choi, M.K., Suh, K.Y., and Char, K. (2010) *Langmuir*, **26**, 9198.
49 Kustandi, T.S., Samper, V.D., Ng, W.S., Chong, A.S., and Gao, H. (2007) *J. Micromech. Microeng.*, **17**, N75.
50 Lee, J., Bush, B., Maboudian, R., and Fearing, R.S. (2009) *Langmuir*, **25**, 12449.
51 Greiner, C., Arzt, E., and del Campo, A. (2009) *Adv. Mater.*, **21**, 479.
52 del Campo, A., Greiner, C., Álvarez, I., and Arzt, E. (2007) *Adv. Mater.*, **19**, 1973.
53 del Campo, A., Greiner, C., and Arzt, E. (2007) *Langmuir*, **23**, 10235.

54 Murphy, M.P., Kim, S., and Sitti, M. (2009) *Appl. Mater. Interfaces*, **1**, 849.
55 Jeong, H.E., Lee, J.K., Kim, H.N., Moon, S.H., and Suh, K.Y. (2009) *Proc. Natl. Acad. Sci. USA*, **106**, 5639.
56 Northen, M.T. and Turner, K.L. (2005) *Nanotechnology*, **16**, 1159.
57 Northen, M.T. and Turner, K.L. (2006) *Sensor. Actuat. A-Phys.*, **130–131**, 583.
58 Spuskanyuk, A.V., McMeeking, R.M., Deshpande, V.S., and Arzt, E. (2008) *Acta Biomaterialia*, **4**, 1669.
59 del Campo, A. and Arzt, E. (2007) *Macromol. Biosci.*, **7**, 118.
60 Spolenak, R., Gorb, S., and Arzt, E. (2005) *Acta Biomaterialia*, **1**, 5.
61 Northen, M.T., Greiner, C., Arzt, E., and Turner, K.L. (2006) A hierarchical gecko-inspired switchable adhesive. Presented at Solid-State Sensors, Actuators, and Microsystems Workshop, Hilton Head Island, SC, USA.
62 Kim, D.S., Lee, S.H., Ahn, C.H., Lee, J.Y., and Kwon, T.H. (2006) *Lab on a Chip*, **6**, 794.
63 Davies, J., Haq, S., Hawke, T., and Sargent, J.P. (2009) *Int. J. Adhes. Adhes.*, **29**, 380.
64 Gorb, E., Kastner, V., Peressadko, A., Arzt, E., Gaume, L., Rowe, N., and Gorb, S. (2004) *J. Exp. Biol.*, **207**, 2947.
65 Santos, D., Spenko, M., Parness, A., Kim, S., and Cutkosky, M. (2007) *J. Adhes. Sci. Technol.*, **21**, 1317.
66 del Campo, A., Álvarez, I., Filipe, S., and Wilhelm, M. (2007) *Adv. Func. Mater.*, **17**, 3590.
67 Murphy, M.P., Aksak, B., and Sitti, M. (2009) *Small*, **5**, 170.
68 Sitti, M., Cusick, B., Aksak, B., Nese, A., Lee, H.I., Dong, H.C., Kowalewski, T., and Matyjaszewski, K. (2009) *Appl. Mater. Interfaces*, **1**, 2277.
69 Glassmaker, N.J., Jagota, A., Hui, C.Y., Noderer, W.L., and Chaudhury, M.K. (2007) *Proc. Natl. Acad. Sci. USA*, **104**, 10786.
70 Reddy, S., Arzt, E., and Del Campo, A. (2007) *Adv. Mater.*, **19**, 3833.
71 Northen, M.T., Greiner, C., Arzt, E., and Turner, K.L. (2008) *Adv. Mater.*, **20**, 3905.
72 Jeong, H.E., Kwak, M.K., and Suh, K.Y. (2010) *Langmuir*, **26**, 2223.
73 Paretkar, D., Kamperman, M., Schneider, A.S., Martina, D., Creton, C., and Arzt, E. (2010) *Mater. Sci. Eng. C*, doi: 10.1016/j.msec.2010.10.004

16
Patterned Materials and Surfaces for Optical Applications
Peter W. de Oliveira, P. Rogin, M. Quilitz, and Eduard Arzt

16.1
Introduction

16.1.1
Optical Materials, Light and Structures

Visible light is an electromagnetic radiation ranging in wave length from about 380 (violet) to 780 nm (red). Materials can interact with light in specific manners. The simplest kind of interaction is the *transmission* of light, as through an ordinary window pane. Even though it seems that light passes unchanged, a closer look reveals that every nonperpendicular ray of light is slightly displaced from its original path due to *refraction* at the air–glass interfaces. Using a transmitting object with nonparallel surfaces, this effect can be exploited to redirect light. Since refraction depends on the wavelength of light, it is possible to separate a light beam into different colors by using a prism. *Absorption* of light is another, ubiquitous mode of interaction between light and matter. It is strongly dependent on the wavelength, thus giving color to our everyday life. A third mode of interaction is the *emission* of light, for example, from the hot wire of a light bulb, from the coating inside a fluorescent tube or from a light-emitting diode. Transmission, refraction, absorption and emission are phenomena which can be described by a few simple, although wavelength-dependent, material properties.

Now, what is the role of structure in optical materials? A lens can be used as a magnifying glass due to its specific shape. It is the exact distribution of black, white or colors that turns a sheet of paper into a tax declaration form, a love letter, or any other document. On this scale, each structural element such as a printed letter is typically perceived individually and has its own clearly defined significance. This can apply to smaller scales, too: in optical data storage media, the information of one bit is represented by the reflection properties of a well-defined area of typically sub-micrometer size.

Anyone who ever inspected a CD or DVD more closely will have noticed the iridescent colors that appear, depending on how the disk is oriented relative to the

Generating Micro- and Nanopatterns on Polymeric Materials. Edited by A. del Campo and E. Arzt
Copyright © 2011 WILEY-VCH Verlag GmbH & Co. KGaA, Weinheim
ISBN: 978-3-527-32508-5

eye and to the light source. This is an example of an effect generated by the collective action of many structural elements together: the individual bit areas are aligned in a long track spiraling outwards, with a well-defined spacing between two consecutive turns. This overall structure acts as a grating which diffracts light, depending on its wavelength.

This transition from individual to collective perception usually happens at the scale of the wavelength of the light, that is, in the submicrometer region. It is not possible to attribute optical effects to individual structural elements if they are separated by less than approximately half the wavelength: this is known as the *resolution limit*. An optical material structured at a scale significantly below this limit appears to be homogeneous in the sense that its appearance does not depend on the specific position. In this case, the material properties can be changed by bulk or surface structuring. Small nanoparticles dispersed in a transparent matrix can lead to drastic changes of the absorption and refraction properties, and slightly larger particles can introduce scattering.

In this chapter, we will first review the basics of refraction, diffraction and reflection which are used in understanding optical effects caused by patterns or microstructures. Examples from nature will then be given that illustrate these principles. The major part of this chapter will deal with technical structures developed for optical purposes, with a discussion of their fabrication methods and of typical applications.

16.1.2
Interaction of Light and Matter – Basic Considerations

This section will provide a basis for understanding the optical effects to a reader who is not familiar with optics. Derivations will not be given. For a deeper, accurate and quantitative understanding the reader is referred to textbooks on electrodynamics and optics [1, 2].

Light is a form of electromagnetic radiation, that is, it can be described as an electromagnetic wave [1, 2]. Neglecting lateral boundaries, a collimated beam of monochromatic light can be described as a planar wave. The electric field vector of this wave as a function of position and time is Equation 16.1:

$$\vec{E}(\vec{x}, t) = \mathrm{Re}[\vec{E}_0 \cdot \exp(i \cdot \vec{k} \cdot \vec{x} - i \cdot \omega \cdot t)] \qquad (16.1)$$

where $\vec{x}$ is the position in space, t is the time, and $\omega = 2\pi f$ is the angular frequency of oscillation. The quantity $\vec{k}$ is the so-called wave vector, which points in the direction of propagation of the wave fronts. Its length k is related to the angular frequency ω and to the wavelength λ (Equation 16.2):

$$k = \frac{2 \cdot \pi}{\lambda} = \frac{n \cdot \omega}{c} \qquad (16.2)$$

where c is the velocity of light (in vacuum) and n is the ratio between the velocities of light in vacuum and the respective medium, which is also known as the refractive index of that medium.

$\vec{E}_0$ is the amplitude vector of the wave. The direction of this amplitude vector characterizes the polarization of the light, and the square of its length is proportional to the light intensity. In an isotropic medium, $\vec{k}$ and $\vec{E}_0$ are perpendicular, and $\vec{k}$ also describes the direction of the propagation of energy.

An important phenomenon that can occur with electromagnetic waves is *interference*. If several waves are present, the total electrical field at any place and time is the sum of the electrical fields of each of the waves. If the field vectors point in the same direction and are appropriately in phase, they add up to a stronger field, which is called constructive interference; if they point in opposite directions, they cancel each other, which is referred to as destructive interference. If the phase relationship between the different waves does not change over time, constructive and destructive interference will lead to areas of high and low intensity, respectively. In that case, the waves are said to be coherent.

How does an electromagnetic wave interact with matter? Even a material that looks homogeneous, at the scale of the wavelength of light is composed of polarizable units – atoms, molecules, crystalline unit cells – that respond to the electric field. As a result of field-induced polarization, each of these units acts as a small dipole antenna and emits radiation (with spherical wave fronts). The result of their superposition is a radiation that is coherent with the incident wave, but has a phase shift and an amplitude depending on the specific material properties. All the secondary waves emitted from the units interfere with each other and with the incident wave. This interference is destructive for all directions except for the original direction of incidence where it causes a retardation of the phase, relative to the original incident wave. So, the overall effect is that the velocity of the light is reduced by a factor given by n.

Next, consider an interface between two materials as shown in Figure 16.1.

Light passes from medium A into medium B, with different refractive indices n_A and n_B. In Equation 16.2, both the frequency and the vacuum velocity of light are invariant. In order to fulfill Equation 16.2 in both media, the wave vector $\vec{k}$ must change upon going from A to B. As the relation between the electrical fields on the

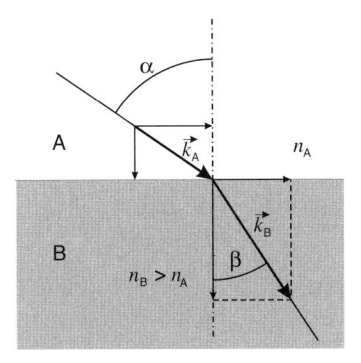

Figure 16.1 Illustration of refraction at the interface between two media A and B with different refractive indices n_A and n_B.

two sides of the interface must be independent of position, the component of $\vec{k}$ parallel to the interface must remain invariant. Thus, the change is restricted to the perpendicular component. This results in a well-known relationship, called Snell's law (Equation 16.3):

$$n_A \cdot \sin\alpha = n_B \cdot \sin\beta. \tag{16.3}$$

This is the fundamental equation for describing the path of light through macroscopic transparent objects such as lenses or prisms. Now it can happen that, when trying to solve Equation 16.3 for the exit angle β, it is found that $\sin\beta > 1$, which makes no sense mathematically. Physically, all the light is then reflected from that surface, which behaves like a perfect mirror. Translated to the description of light as an electromagnetic wave, the perpendicular component of the wave vector is imaginary. This means that there is still some electric field directly at the interface but this so-called evanescent wave decays rapidly with increasing distance from the interface.

Total reflection at an interface is a special case of reflection, which is described by Fresnel's equations. Physically, they originate from the necessity to fulfill certain boundary conditions of the electromagnetic field. The surface reflection depends on the refractive indices, on the angle of incidence and on the polarization of the light. For normal incidence, the *reflectivity*, that is, the ratio of field amplitudes, is given by (Equation 16.4):

$$\frac{E_{\text{reflected}}}{E_{\text{incident}}} = \left|\frac{n_B - n_A}{n_B + n_A}\right|. \tag{16.4}$$

Since the intensity of the light is proportional to the square of the amplitude, the *reflectance*, i.e. the fraction of reflected power, is given by the square of this value. Taking $n_A = 1$ and $n_B = 1.5$, we thus find a surface reflectance of 4% for a typical air–glass interface.

The superposition principle can also be applied to larger units on the scale of the wavelength of light. The reaction of a single unit then has to be described as a multipole antenna instead of a dipole. If these individual units do not form a homogeneous medium, the secondary waves no longer interfere destructively for any direction except for the original "forward" direction. For indiscriminately distributed inhomogeneities, a randomization of the direction of the light is obtained. This is known as *scattering*. For ordered structures, there are certain directions typically different from the forward direction where constructive interference is obtained as well. This is then referred to as *diffraction*.

A special case of a diffracting object is a line grating, as shown in Figure 16.2.

The condition for constructive interference is that the path difference between waves diffracted at neighboring lines (highlighted in Figure 16.2) be an integral multiple of the wavelength. This leads to (Equation 16.5):

$$\Delta \cdot (\sin\alpha + \sin\beta) = m \cdot \lambda, \tag{16.5}$$

where Δ is the line spacing and m is an integer.

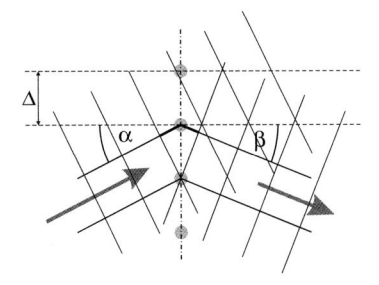

Figure 16.2 Illustration of diffraction caused by a grating consisting of parallel lines (assumed to extend infinitely in the viewing direction) with spacing Δ.

A variation of the above case is a grating composed of parallel planes (Figure 16.3).

In this case an additional condition must be met: the angle of incidence of the incoming wave and of the outgoing wave must be identical. Otherwise, waves diffracted from different lateral positions would interfere destructively. This latter condition implies that such a grating behaves like a mirror for special combinations of angle and wavelength. This is why this mode of diffraction is often called "Bragg reflection", after W. H. Bragg and W. L. Bragg [3], who first used this principle to determine distances between crystal lattice plane using X-rays.

In the case of Bragg reflection, Equation 16.5 changes to its well-known form (Equation 16.6):

$$2\Delta \cdot \sin \alpha = m \cdot \lambda \tag{16.6}$$

Referring to Figure 16.3, an interesting fact is that if the two waves are present without the diffracting structure, their interference will form an intensity pattern corresponding exactly to the diffracting structure. By using a suitable recording medium, the diffracting structure can thus be created by interference. If subsequently only the incident wave is provided from an external source, the outgoing wave is reconstructed by the diffraction pattern. This is the basis for *holography* [4]. A grating produced by interference of two coherent beams is referred to as a "holographic grating".

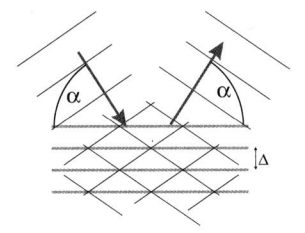

Figure 16.3 Diffraction from an array of parallel planes spaced at a distance Δ ("Bragg reflection").

16.1.3
Optical Microstructures in Nature

Many visible weather phenomena involve naturally occurring microstructures on various scales. A prominent and beautiful example is a rainbow. A combination of refraction and total internal reflection in spherical water droplets causes sunlight to be scattered backward at a well-defined angle relative to the original direction of incidence. Looking into the falling rain, we see this light coming only from a direction where this specific angular relationship is fulfilled, giving rise to the beautiful arc shape in the sky. Due to the dispersion of the refractive index of water, this angle depends on the wavelength, which results in the splitting of the arc into different colors, as shown in Figure 16.4.

This is an example where a collective effect is generated by comparatively large structures: although the raindrops are typically millimeter-sized, they cannot be resolved individually because of the large viewing distance. In some rare cases, when a rainbow is caused by small droplets (less than 100 µm), the color spectrum is seen to repeat itself, although weakly, on both sides of the main rainbow. This is an effect of diffraction, and the angular spread allows an estimate of the droplet size [5]. If droplets become smaller, the angular spread by diffraction increases. Because of the larger number of droplets the probability of multiple scattering events is increased. This results in random, white scattering, commonly known as a cloud.

In minerals, nature provides optical structures with various degrees of ordering. A well-known example is opal, an amorphous form of silica ($SiO_2 \cdot nH_2O$). It is made up of submicron spheres of silica with a small amount of water (typically between 3 and 20%) and air trapped in the interstices. In precious opal, the silica spheres are almost monodispersed and arranged in a regular 3D packing [6], which reflects specific colors, like a Bragg grating, as a function of incident angle. The resulting play of color qualifies opal as a valuable gemstone.

A similar structure of biogenic origin is nacre, which forms the inner shells of some molluscs. Nacre is a composite material of aragonite, a crystalline modification

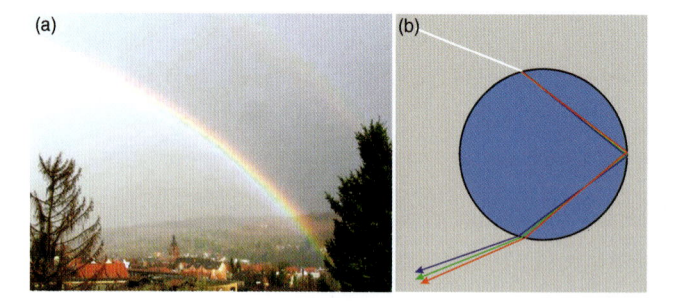

Figure 16.4 (a) A rainbow over Sulzbach near Saarbrücken, Germany; (b) schematic of refraction and total reflection inside a water droplet.

Figure 16.5 Tip of a peacock feather showing colors produced by combined interference and pigmenting.

of calcium carbonate, and various proteins. The aragonite is present in the form of platelets of about 0.5 μm thickness. These platelets are stacked in layers separated by 30 to 50 nm of organic matrix. Here, too, the highly regular structure acts as a Bragg grating, which is perceived as the well-known luster of pearls. The formation and structure of nacre has been the subject of intense research [7, 8].

Nature seems to have developed nacre mainly for its mechanical strength, whereas the attractive optical appearance is only a side effect. In other cases it is really the look that matters. Brilliant color effects can be seen, for example, on butterfly wings or peacock feathers, as shown in Figure 16.5. In these cases, optical interference is often combined with pigments to maximize the effect [9]. Strength of peacock feathers may appear to be a side effect here, but is also important [10]

Yet, another example of a natural optical microstructure is the surface of some insect eyes. The cornea surface of a moth eye has microscopic bumps about 200 nm in height and ordered in a hexagonal pattern with a period of 300 nm. Due to their size, the individual features cannot be resolved in visible light, but instead produce a meta-material, where the index of refraction changes gradually from that of air to that of the bulk material. As can be surmised from Equation 16.4, this gives rise to a very low reflectivity [11]. As a result, the moth eye is totally black and absorbs light from any direction. Several technological approaches are currently being investigated to mimic this principle to obtain better antireflective surfaces.

16.2
Optical Micro- and Nanostructures for Applications

Knowledge of fundamental optics, sometimes combined with lessons from nature, has led to the design of micro- and nanostructures for specific applications. They may be based on reflective, refractive or diffractive effects. The most important designs are now presented.

16.2.1
Effects of Reflection and Refraction

There are many applications where it is desirable to eliminate reflections from optical surfaces. Examples are ophthalmic or photographic lenses, mobile phone or flat panel display screens, or picture frames with a protective glass. Several approaches are known to achieve this goal, all of them involving structures on a small scale.

16.2.1.1 Anti-Glare

One of the simplest ways to minimize reflections is to randomize reflected light by means of a slightly uneven surface ("anti-glare structure"). Such surfaces have randomly distributed hills and valleys with a typical lateral feature size of up to 100 µm and typical elevations of 1 µm or less. The relatively smooth unevenness ensures that light is scattered only by a small angle of less than ten degrees. This is sufficient to blur sharp reflections into diffuse ones, without inducing excessive haze. On the other hand, the deflection of transmitted light is much smaller; an object located closely (a few millimeters) behind the structured material still appears in focus. This is illustrated in Figure 16.6.

This principle has been used for picture frame glasses for many decades and has gained significant importance for flat panel displays. Manufacturing methods include surface etching of glass or microreplication techniques, such as mold casting and embossing; the randomized structure in the mold can be obtained by sandblasting or etching. Several papers and patents describe coatings with structures formed spontaneously during the manufacturing process, for example, due to incomplete leveling of a spray coating [12], agglomeration of fine particles [13], or the separation of two immiscible phases upon evaporation of the common solvent [14].

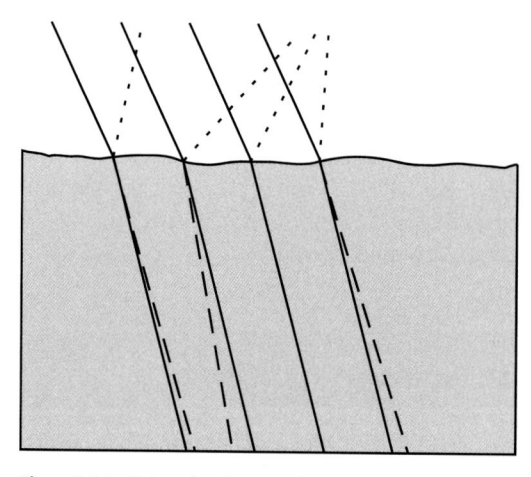

Figure 16.6 Principle of an anti-glare structure: The reflected light rays (dotted) are scattered by the surface roughness, whereas the deviation of the transmitted rays (dashed) from the apparent ones (solid) is much smaller.

Anti-glare films for displays have a disadvantage: they scatter the light emitted by the display, resulting in a less crisp image for the viewer. The matte finish mutes clarity, color and contrast. For such anti-reflective components the total transmission, integrated over the visible range, is between 75 and 85%, which is not sufficient for many applications. In the next sections, two possibilities for making highly efficient antireflection coating by structuring a polymer surface are described.

16.2.1.2 Dielectric Antireflective Coatings

Dielectric coatings are thin transparent films or stacks of several thin films coated on a substrate so that there are abrupt changes of the refractive index at the interfaces. When a ray of light passes through this system, a partial wave is reflected at each interface. The refractive indices of the materials and the layer thicknesses are tuned so that destructive interference results at the wavelengths of interest. Such coatings suffer less from a reduction in contrast than anti-glare surfaces.

The simplest case is a single film having a low refractive index (such as MgF_2, $n = 1.38$) on a substrate with a higher refractive index (such as glass, $n = 1.52$). The effective optical thickness of the film (i.e., the geometrical thickness times the refractive index) is then chosen to be a quarter of the design wavelength. A first partial wave is then reflected at the air–MgF_2 interface and a second one at the MgF_2–glass interface. The second partial wave experiences a phase shift of π, leading to destructive interference (Figure 16.7).

The reflectance does not vanish completely even for the design wavelength. The reason for this is that the refractive index of MgF_2 is still too high, even though it is one of the lowest values found for compact solid materials. The optimum refractive index would be the geometric average (square root of the product) of the two bulk media – about 1.23 in this case. Furthermore, there is still a significant wavelength dependence. It is possible to deal with both difficulties with more complex stack designs [15, 16]. With three layers, it is possible to achieve a low reflectance over a broad wavelength range (Figure 16.8).

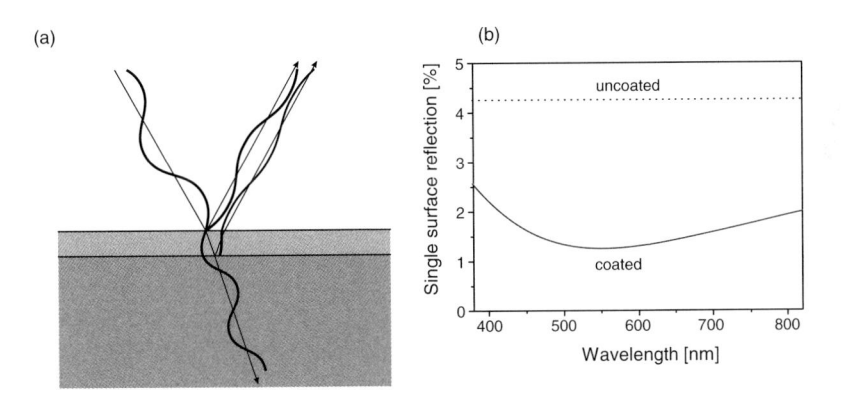

Figure 16.7 Single layer antireflection coating. (a) Schematic structure and working principle; (b) simulated reflectance spectrum with refractive indices of 1.52 and 1.38 for the substrate and the film, respectively (no dispersion assumed).

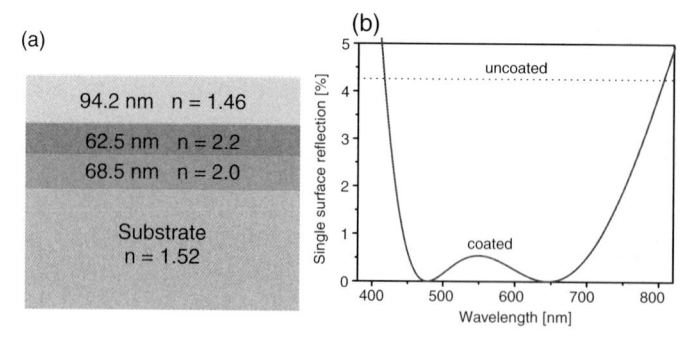

Figure 16.8 Example of a triple layer antireflection coating. (a) Schematic structure; (b) simulated reflectance spectrum (no dispersion assumed).

Complex commercial antireflection coatings for advanced applications are usually produced by gas phase deposition techniques, such as evaporation or sputtering. These processes produce high quality film systems, albeit on relatively small substrates and at comparatively high cost [16]. Sol–gel and related techniques can be an alternative route to produce dielectric films of generally lower performance, but with higher throughput and at lower cost [17].

Dielectric films from sol–gel processes are typically structured on two scales. On the 100 nm scale, the vertical structure is the same as that of the corresponding evaporated films. On the 10 nm scale, the films frequently contain nanoparticles to influence the refractive index. Colloidal particles can impart both antireflection and high reflection properties to dielectric films [18]. Films composed of colloidal particles alone are typically very porous, which reduces the refractive index significantly below the bulk value. This principle can be used to create single layer antireflective coatings with virtually zero reflectance at the design wavelength. An example from our laboratory is shown in Figure 16.9.

A disadvantage of such films is their low mechanical strength due to the fragile interparticle bridges. Therefore, they are only useful for internal surfaces in closed lens systems or in certain laboratory applications. By embedding nanoparticles in a polymeric or hybrid inorganic–organic matrix, composite materials with a mechanical strength comparable to that of vacuum-coated films can be obtained [19]. Refractive indices between 1.49 and 1.93 can be achieved using embedded SiO_2 and TiO_2 nanoparticles. This range is not as wide as for vacuum-coated films, but still sufficient for many applications.

16.2.1.3 Moth Eyes

Dielectric antireflective films are usually not completely antireflective over a broad wavelength range. The residual reflection can be lowered by designs of higher complexity, but this is at the expense of higher cost and greater risk of inducing defects. As an alternative, several research groups are trying to exploit the moth eye principle. Randomly structured graded index surfaces obtained by etching were described as early as 1976 [20]. More recent approaches include microreplication

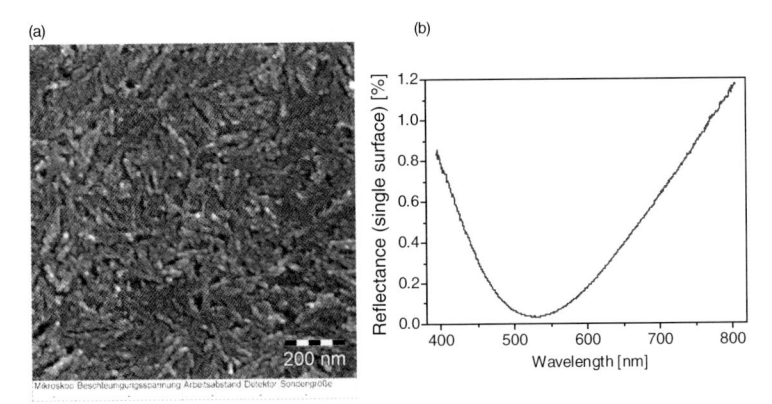

16.2 Optical Micro- and Nanostructures for Applications

Figure 16.9 Porous coating acting as an antireflection layer with low refractive index. (a) SEM image; (b) measured reflection spectrum. The material is hydroxy apatite (HAP) on a glass slide. Although the refractive index of bulk HAP is 1.63, the reflection spectrum is consistent with a refractive index close to the optimum value of 1.23 due to the porosity.

by embossing from a holographically produced master structure [21] and plasma etching of near-periodic structures using a self-aligned etch mask pattern [22]. Typical structures are shown in Figure 16.10.

16.2.1.4 Holograms

Holograms are, in the most general definition, optical devices with patterns that modify light by interfering with its phase composition. In last 20 years, holograms have increasingly found new applications. Contemporary holographic techniques cover a wide range of imaging and production formats, all of which have their own

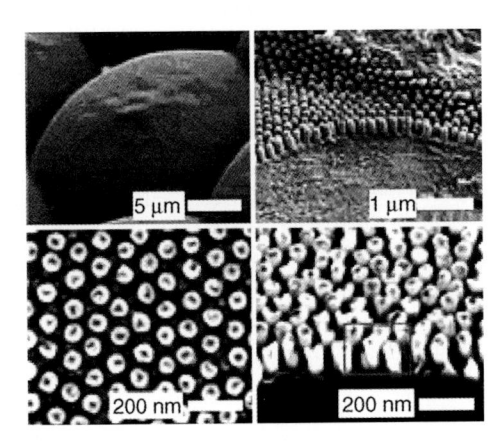

Figure 16.10 Natural moth eye structure (upper row) and technically produced biomimetic structure (lower row). (Images courtesy of Max-Planck-Institut für Metallforschung, cf. ref. [22]).

specific qualities, advantages and limitations. The most significant limitation to applying holograms for design, security and displays is the choice of material. Five characteristics need to be considered to obtain holographic optical elements with high quality: lighting, depth, color, and viewing angle are characteristics relevant to all applications; for some applications, the ability to store multiple images (multi-channel properties) is also essential. All of these parameters are interrelated and are essential to design phase or relief structures with high brilliance or high total transmittance.

To overcome the lighting limitation, techniques like pixelgram or dot matrix holograms have been developed. These types of two-dimensional holograms consist of several small diffraction gratings, each corresponding to a pixel color value of an image file. For high depth in a hologram display, a light sensitive material with highly linear optical behavior is required. The pattern can be printed or embossed using a photopolymer, dichromated gelatin (DCG) or silver halide. Most forms of holography and materials will give monochromatic images, with blue being a notoriously difficult color to achieve with a high degree of brightness. Scientists at DuPont have developed photopolymers for reflection holograms designed for true color, which can be exposed using three color laser beams [23]. Holograms recorded by an interference pattern have a viewing range which extends over 45° horizontally and vertically from the central axis of the hologram. Outside of this viewing angle the image will disappear. The angle of view in a photopolymer depends on the maximum of the refractive index gradient possible in the photopolymer.

Holograms which exhibit different functions or images at different angles are called multichannel holograms. The holographic material needs to have the ability to store different images at different angles. High quality multichannel holograms can be obtained as volume phase holograms for example, in photopolymers. The angle of view of each image needs to be very precise in order to avoid the appearance of so-called ghost images, that is, the presence of traces of images from the other channels when one particular channel is viewed.

16.2.1.4.1 **Volume Holograms: Recording Mechanism in Photopolymers** Colburn and Haines recorded some of the first volume holograms in organic photopolymers in 1971 [24]. Photopolymers for holographic recording developed by DuPont were first demonstrated by Booth (DuPont OnmiDex® and DuPont HRF®). Higher demands must be met by holograms as safety features on articles prone to counterfeiting, such as bank notes or credit cards. In the holographic material from the type of photopolymers, such as Du Pont's HRF, phase gratings are recorded through an interdiffusion of monomers during the photopolymerization mechanism. The photopolymer materials include four main components: a polymerized binder material, vinyl monomers or variations thereof, a photosensitive dye and a polymerization initiator. Before the photo polymerization process, the material is in a gelatinous or liquid state, where the monomers are free to diffuse through the volume of the material. When the material is irradiated by an interference pattern, three processes take place: photo polymerization, diffusion and fixing by a subsequent irradiation (flood exposure). Monomers, driven by chemical potential differences, tend to diffuse from regions of low optical density to regions of higher optical density. At the same time, the absorption of light by the photosensitive dye in the regions of higher optical intensity starts a chemical reaction, which allows the polymerization initiator to link monomer units together. More monomer units diffuse into the irradiated regions and polymerize until either the supply of monomers is depleted or the photosensitive dye is bleached. A subsequent flood exposure polymerizes any remaining monomer units.

As the monomers in the material polymerize, the volume of the material shrinks by 7 to 12% in volume due to the formation of covalent bonds. In an alternative material, ULSH-500® developed by Polaroid, the volume occupied by each monomer unit expands during the polymerization process, reducing the dimensional change of the material by as much as an order of magnitude.

Holographic materials can also be used for optical elements such as diffusers and light management devices. Optical diffuser films are key elements in liquid crystal displays; they spread the incident light over a wide angle to prevent the light source from being perceived directly by the viewer and to keep the brightness uniform over the display area. Diffusers can be classified into two types: the particle-diffusing type, which relies on transparent beads inside the plastic film to scatter light [25–27], and the surface-relief type. The latter scatter light by its surface microstructure, e. g. in microlens diffusers [28, 29], random phase diffusers [30], deterministic diffractive diffusers [31] and holographic diffusers [32–38]. Holographic diffusers may also be designed to act as volume gratings. Much research has been performed on such diffusers due to their unique properties, such as controllable diffusion angle, directional property, volume refractive index variation and high transmittance. Hologram materials such as silver halide sensitized gelatin [39], DCG, photopolymer and azobenzene polymer have been used for this purpose.

Ionic liquids are attracting considerable attention due to their unique physicochemical properties: extremely low volatility, wide liquid temperature range, good thermal stability, good dissolving ability, excellent microwave absorbing ability, designable structure, high ionic conductivity, wide electrochemical window, and so on [40]. Recently, the Leibniz INM has reported that ionic liquids, based on dialkylimidazolium, as additives in the photopolymerizable hologram materials can increase the sensitivity, the resolution and the diffraction efficiency [41]. A micrograph of the cross-section of a typical optical diffuser produced by this method is shown in Figure 16.11; here the light pipes were formed by interdiffusion of functionalized nanoparticles during the UV irradiation step.

Using the diffusion of functionalized nanoparticles, volume structures with a periodicity from 1 to 10 µm can be formed. The ionic liquid based composite provides phase gratings with diffraction efficiency up to the theoretical maximum due to the strong light induced diffusion during the polymerization process. An important reason for choosing a hologram as an optical diffuser is its ability to perform more than one function, this ability arises from the fact that in a hologram more than one type of information can be stored and retrieved independently. It is generally assumed that the superposition of different gratings for such purposes will not result in "cross-talk" during reconstruction.

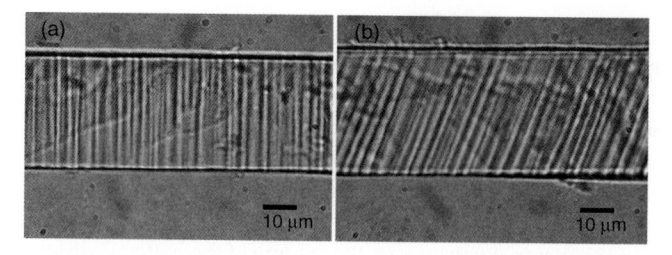

Figure 16.11 Micrographs of a diffuser cross section in photonanomer (phase microscopy technique): the light pipes formed by functionalized nanoparticles are clearly distinguishable. (a) This diffuser produces diffuse light when transmitted at a perpendicular angle, while being transparent at an angle of about 30°, (b) asymmetric diffuser with 25° of irradiation (Reproduced with permission from [41]. Copyright © (2007) American Institute of Physics.)

16.2.1.4.2 Relief Holograms and Diffractive Elements by Embossing

Various embossing techniques have been employed in roll-to-roll processes [42, 43] with high throughput. Some materials can be embossed in the cold state (e.g., aluminum foil), others such as the thermoplastic polyethylene terephthalate (PET) require higher temperature. The most common process is reactive embossing, where a UV-curable resin, coated onto a transparent foil, is wound around an embossing roll and irradiated while still in contact with the roll. In order to speed up such processes, a thixotropic embossing process was developed at INM which separates the embossing from the curing steps, thus leaving more time for curing [44]. The rheological properties of the coated film must then permit separation from the master before curing.

The master used for embossing is produced by holographic or lithographic techniques. Such a process consists of optically writing a structure into a photoresist and, after developing the resist, replicating the structure into a nickel shim by (electro)chemical plating. An interesting option is holographic dot matrix printing, where many of such gratings, each confined to a small area, are placed next to each other to fill the whole image area. [45] By varying the line spacing and orientation, interesting effects can be achieved, for example simple rainbow effects, three-dimensional images, hidden images and angle-dependent appearance of different images [46].

Diffractive optical elements are also used in optical communication, for example, in micro Fresnel lenses for fiber coupling. A Fresnel lens is obtained by segmenting the convex surface of a plano-convex lens into rings and displacing all the rings to the planar surface, as shown in Figure 16.12a; a large-area scan in an atomic force microscope (AFM) is shown in Figure 16.12b. Such a thin lens has similar characteristics as the much thicker original lens but with significantly less material.

The action of a classical Fresnel lens is purely refractive. On scaling down of the lens structure, diffractive effects appear. If the structure height is tuned to the design wavelength, both effects add up and result in a highly efficient focusing element. Such lenses can be produced by embossing from a master structure obtained by precision micromachining.

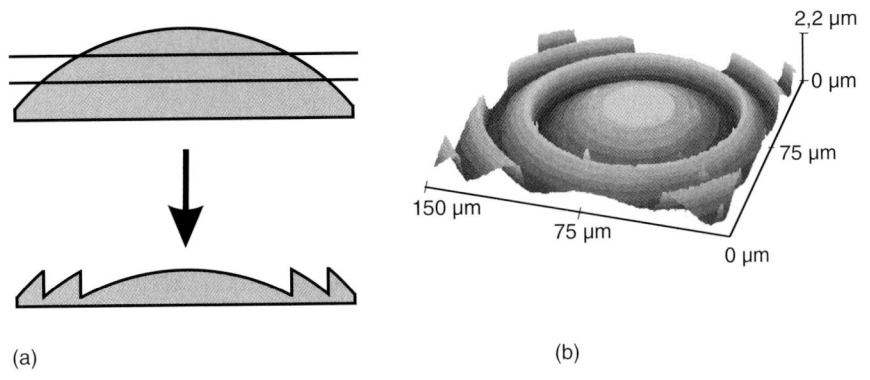

Figure 16.12 Fresnel lens: (a) Principle of creating a refractive Fresnel lens by segmenting a normal plano-convex lens. (b) Large-area AFM image of the center of an embossed micro Fresnel lens (side length is 150 µm).

Purely diffractive micro Fresnel lenses can also be obtained holographically, by the interference of a convergent or divergent laser beam with a planar wave in a Mach–Zehnder interferometer, as shown in Figure 16.13. Figure 16.14 shows the structure of a lens produced by this technique from a photosensitive nanocomposite material.

The optical construction as shown in Figure 16.13 produces axially aligned Fresnel lenses. However, it can easily be modified to permit the production of Fresnel lenses or zone plates with focus point outside of the geometrical axis of the optical elements. Generally, holography is a fabrication technique for optical elements allowing for great design flexibility. It is even possible to integrate more than one function in

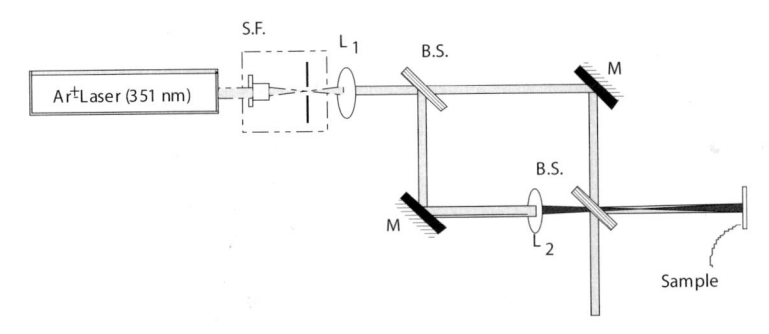

Figure 16.13 Setup for holographic writing of a micro Fresnel lens. A linearly polarized collimated beam from an Ar + -Laser (wavelength 351 nm) is divided by a beam splitter into two beams; the object beam illuminates the test object through a lens L2. The reference beam recombines with the object beam at the second beam splitter (B.S.). A photopolymerizable coating of composite material, set at a certain distance of from the object, records the hologram resulting from the interference of the two waves. (M = mirror, S.F. = spatial filter).

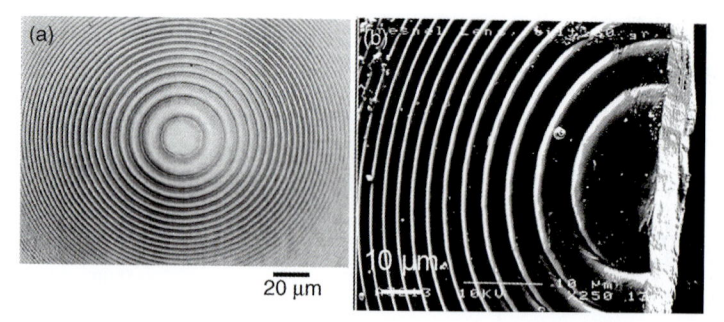

Figure 16.14 Holographic micro Fresnel lens as seen by (a) optical microscopy; (b) and scanning electron microscopy (Reproduced with permission from [47]. Copyright © (2007) SPIE.)

a single optical element, because in a hologram multiple information can be stored and retrieved independently. Care must be taken that the superposition of different wave fronts for such purposes will not result in cross-talk during reconstruction.

A pilot scale embossing station for replicating microstructures in thixotropic coatings is shown in Figure 16.15. This set-up is integrated into a pilot scale continuous roll-to-roll coating line. As an example of a microstructure produced by this setup, Figure 16.16 shows the atomic force micrograph of an embossed hologram with multichannel properties.

Figure 16.15 Continuous embossing setup at INM for the scaling up of the thixotrophic material to produce storage information optical elements. Up to three different master structures can be mounted on the upper embossing roller. The maximum area per field is about 160 mm long and 260 mm wide. Subsequently, the thixotropic material is cured by a UV lamp at intensities of up to $2\,\mathrm{W\,cm^{-2}}$.

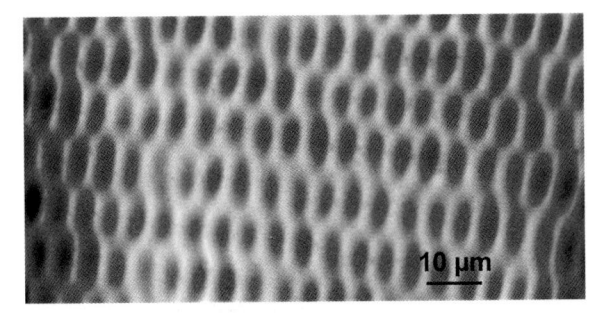

Figure 16.16 Atomic force micrograph of an embossed hologram on thixotropic composite based on ZrO_2. The surface of the hologram is not a perfect grating because of the superposition of different interference patterns.

16.2.2
Waveguides

Waveguides are structures that confine the propagation of light inside a medium, with a comparatively high refractive index, by total reflection at the boundary to the surrounding media, which have a lower refractive index. The simplest case is a planar waveguide as shown in Figure 16.17. By lateral structuring, light can be further confined to propagate in only one direction. The dimensions of a waveguide are typically macroscopic in the direction of propagation (millimeters to centimeters on planar substrates, up to tens of kilometers in the case of telecommunication fibers), whereas the vertical and transversal dimensions are typically a few micrometers maximum.

Structuring of either the guiding layer or one of the surrounding layers can be used to confine the light further to only a line or a stripe. The same principles governing the vertical confinement of light are also applicable to the additional transversal confinement, which is achieved by raising the effective refractive index of the planar waveguide modes along the stripe. This can be done by structuring the guiding medium itself or the surrounding media, as illustrated in Figure 16.18.

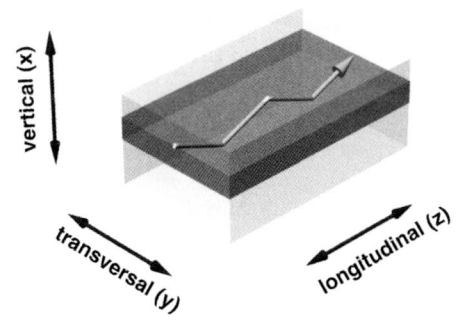

Figure 16.17 Guiding of light in a planar waveguide by total reflection, together with a designation of directions in the coordinate system, described in the text.

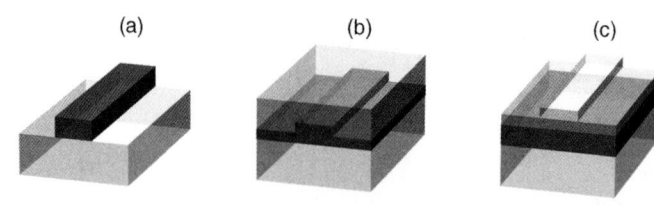

Figure 16.18 Some possible geometries for stripe waveguides. The simplest case (a) is a free ridge of guiding material in air. Transversal guiding is also achieved if the guiding material is thicker in the waveguide region than outside (b). Furthermore, the waveguide can be clad by some material of lower refractive index (also b). It is even possible to achieve guiding by producing a ridge on a very thin cladding alone (c), since the evanescent wave is sensitive to structures that can be up to a few 100 nm away from the guiding layer.

16.2.2.1 Fabrication of Waveguides

Planar polymer-based waveguides can be fabricated from thin-film coatings by a variety of methods [48]. Structuring is performed by optical means, either by direct laser writing in photosensitive polymers or sol–gel materials, or by using mask-based lithographic processes. In direct structuring, the irradiated material becomes, after a suitable development process, the high-n material used for guiding. This approach has been taken with positive or negative photoresists [49–51] and with materials that change their refractive index upon irradiation, for example, based on gelatin [52], other polymers [53] or sol–gel materials [54]. In more complex processes, grooves are first formed in a substrate layer and subsequently filled with the high-n material [55]; alternatively, a metal layer on a substrate is patterned and is then used as a mask for an ion exchange process [56]. Other sources of irradiation are also possible, for example, by means of electron beam writing [57]. Alternatively, waveguide structures can be manufactured by microreplication techniques such as mold casting [58, 59] or embossing [60].

16.2.2.2 Waveguide Devices

Glass fibers are the most usual form of waveguides used to transmit light over long distances, as a carrier of either energy or information. To manipulate the light, the fibers are usually coupled to waveguide devices located on planar substrates. Such manipulations can either be passive or active. Typical passive manipulations are filtering of specific wavelengths out of a broader spectrum and combining light of different wavelengths. This is used in wavelength division multiplexing (WDM). This technique, which relies on transmitting data over a single optical fiber in several channels using slightly different wavelengths, is the basis for the enormous data rates achieved today in long range optical fiber communication. Active manipulation of light refers to changing the transmission of light through a device in response to an external stimulus such as an electrical field, temperature or the chemical environment. Minute changes of the waveguide properties due to such influences can be translated into large changes of the optical response by exploiting principles of optical interference, which allows for the construction of highly sensitive sensors as well as fast electro-optical modulators. In the following, we present a few basic designs of such devices.

16.2.2.2.1 Gratings Waveguide gratings generally consist of a periodic structure superimposed on the waveguide structure, typically in an orientation perpendicular to the direction of propagation. Such structures can be produced by lithographic [61], holographic [62] or microreplication [63] techniques. Depending on the periodicity of the gratings, they can be used in two different ways.

One possibility refers to gratings of a periodicity comparable to or longer than the wavelength of the guided light. These can be used as coupling elements to inject light into the waveguide or extract light from them. For a given wavelength, this coupling occurs at a very specific incident or emergent angle, which depends on the waveguide properties defining k. If the propagation constant is made to depend for example, on the presence of absorbed humidity or adsorbed biomolecules, such a device can be used as a sensor [64].

The second possibility refers to gratings having a periodicity which is equal to half the (effective) wavelength of the guided light. Such a grating acts as a retroreflector for this specific wavelength in analogy to a Bragg grating in bulk media and is therefore often referred to as a (distributed) Bragg reflector. If the grating modulation is weak, a highly selective wavelength filter useful for demultiplexing of WDM signal streams is obtained. Such filters can be made tunable by changing the refractive index of the waveguide medium through changing the temperature [65]. It is also possible to design a sensor, for example, a strain sensor which reacts to variations of the periodic length of the grating [66].

16.2.2.2.2 Mach–Zehnder Interferometers In a classical two-beam interferometer a light beam is divided between two branches, where the partial beams may experience different phase shifts before being coherently combined again. The light intensity at the output is a sensitive function of any phase difference between the two branches. This can be exploited for either measuring external parameters through their influence on the phase difference of for modulating the light intensity by controlling such parameters.

This principle can be transferred to waveguides. A scheme that has found particularly widespread use is that of an integrated Mach–Zehnder interferometer, as shown in Figure 16.19. It works by splitting and recombining the light in Y-shaped junctions with two parallel waveguides in between.

Following the path of the light, the waveguide mode is first stretched in the taper section of the Y splitter and then transforms adiabatically into two independent modes confined to the two branches of the interferometer. In the beam combiner at the output end, this process is reversed if the electromagnetic waves are in phase. In this case the interferometer is highly transmitting. If, in contrast, the two waves are in phase opposition, the field distribution in the combiner corresponds to a first order lateral mode. Since the exit waveguide is too narrow to guide this mode, all the light is radiated into the substrate, and the transmission function of the interferometer is zero. This has been used for sensors [67, 68] and modulators [69, 70].

(a)

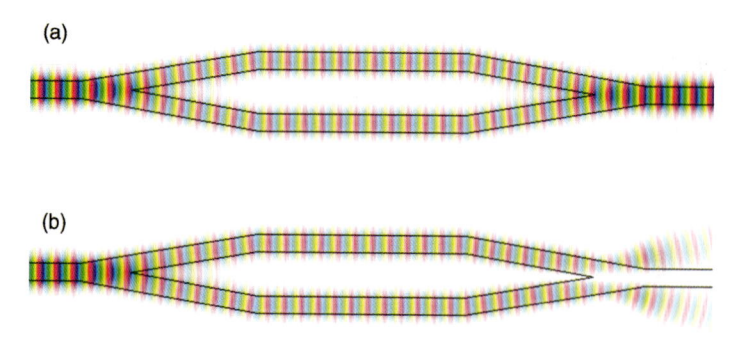

(b)

Figure 16.19 Layout of an integrated Mach–Zehnder interferometer and illustration of the field distributions in the case of constructive (a) and destructive (b) interference. In the bottom figure, an increased refractive index in the lower branch causes the waves to be out of phase.

16.2.2.2.3 **Microring Resonators** The concept of microring resonators is more recent than gratings and Mach–Zehnder interferometers, although it is fairly simple. The basic idea is to bend a stripe waveguide such that it forms a ring. Light can circulate inside this ring provided that the phase change after one round-trip is an integral multiple of 2π; if there is even a very small deviation from this phase condition, a light wave inside the ring will interfere destructively with itself and decay rapidly. In theory, light having the right wavelength to match this phase condition will circulate forever in the ring, giving rise to an infinitely sharp resonance. In practice, the width of the resonance will be finite due to absorption and scattering losses, although a bandwidth in the range of tenths of a nanometer at a wavelength of 1550 nm is not unusual.

Light is coupled into ring resonators from an adjacent linear waveguide through the field of the evanescent wave, as shown in Figure 16.20. If the ring resonator is in resonance with the particular wavelength, the coupling becomes very efficient, which

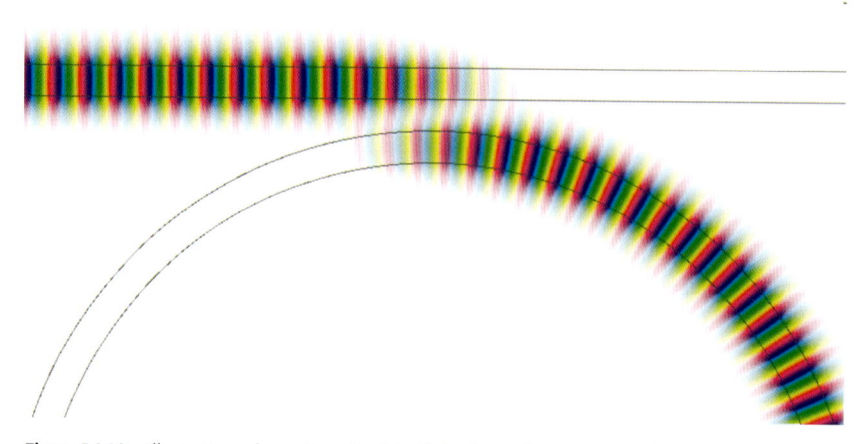

Figure 16.20 Illustration of coupling of guided light from a linear waveguide into an adjacent ring waveguide through the evanescent wave. The picture shows optimized coupling; in general, the coupling efficiency depends on the geometrical parameters.

results in a sharp dip in the transmission of the linear waveguide. This behavior is similar to that of a Bragg grating. Simple microring resonators therefore can be used for wavelength filtering. Any external parameter changing the refractive index of the waveguide material or the surrounding medium will also shift the resonance frequency, which is the basis for sensors and modulators. Furthermore, light coupled into a resonator can also be coupled back into a different linear waveguide, which results in a so-called add-drop filter [71] transferring light between waveguides with high wavelength selectivity. Arrays of such add-drop filters operating at slightly different wavelenghts can form fully integrated multiplexers/demultiplexers for WDM [72, 73].

Multiple-coupled ring resonators offer further possibilities. Recent developments succeeded in demonstrating compact optical delay elements [74, 75] and optical gyroscopes based on the principle of a Sagnac interferometer [76]. Additional interesting effects can be expected from future research.

16.3
Conclusion

In this article, we have reviewed the development of novel optical and polymeric materials designed for surface structuring. The combination between material development and structuring technology can open perspectives for multiple or parallel application domains and create enabling technology platforms, as has been described in this chapter. Optical techniques based on nano- and microsurface structuring have experienced significant growth over the last two decades [77–79]. This growth stems largely from the developing infrastructure for telecommunication [80, 81] and computer sectors. In telecommunications, optical components such as new light sources, detectors, and optical fibers have appeared. In the computer industry, new light sources have been developed to "read" decreasing feature sizes in optical storage media, such as DVDs and blu-ray disk. Although these optical components were developed for completely different purposes, they have spilled over into applications in the optical sensing arena as well as in nanomedicine.

Future components will enable smaller, low-powered, integrated devices to be designed and used for a variety of optical applications in different environments [82] which have not previously been possible due to size or power limitations. Further improvements and new capabilities must be expected that will advance developments in the telecommunications and computer industries and in optical sensing [83]. Optical materials will very likely see a "bright" future.

References

1 Lipson, S.G., Lipson, H.S., and Tannhauser, D.S. (1995) *Optical Physics*, 3rd edn, Cambridge University Press.

2 Jackson, J.D. (1998) *Classical Electrodynamics*, 3rd edn, John Wiley & Sons, Inc., New York, USA.

3 Bragg, W.H. (1912) *Nature*, **90**, 219; Bragg, W.L. (1912) *Nature*, **90**, 410.
4 Caulfield, H.J. (1979) *Handbook of Optical Holography*, Academic Press, New York.
5 Scott Barr, E. (1963) *Appl. Opt.*, **2**, 639–647.
6 Klein, C., Hurlbut, C.S., and Dana, J.D. (1998) *Manual of Mineralogy*, 21th edn, John Wiley & Sons, Inc., New York, USA.
7 Weiss, I.M., Tuross, N., Addadi, L., and Weiner, S. (2002) *J. Exp. Zool.*, **293**, 478–491.
8 Weiss, I.M., Kaufmann, S., Heiland, B., and Tanaka, M. (2009) *J. Struct. Biol.*, **167**, 68–75.
9 Zi, J., Yu, X., Li, Y., Hu, X., Xu, C., Wang, X., Liu, X., and Fu, R. (2003) *PNAS*, **100**, 12576–12578.
10 Weiss, I.M. and Kirchner, H.O.K. (2010) *Adv. Eng. Mater.*, **12** (5), 412–416.
11 Wilson, S.J. and Hutley, M.C. (1982) *Opt. Acta*, **29**, 993–1009.
12 Al-Dahoudi, N., Bisht, H., Göbbert, C., Krajewski, T., and Aegerter, M.A. (2001) *Thin Solid Films*, **392**, 299–304.
13 Haga, Y., Matsumura, S., Watanabe, H., and Nagahama, T., (2008) Anti-glare film, method of manufacturing the same, and display device. European patent EP 1962111 A1 (JP 2003-100671)/JP 2007-341462).
14 Takahashi, H. and Ushida, H. (2006) Anti-glare film. United States Patent US 7128428 B2 (JP 2003-100671).
15 Thelen, A. (1989) *Design of Optical Interference Coatings*, 1st edn, McGraw-Hill Book Company, New York.
16 Flory, F.R. (1995) *Thin Films for Optical Systems*, Marcel Dekker Inc., New York, pp. 41–248.
17 Dislich, H. (1988) in *Sol-Gel Technology for Thin Films, Fibers, Preforms, Electronics and Specialty Shapes* (ed. L.C. Klein,), Noyes Publications, Park Ridge, N.J., pp. 50–79.
18 Floch, H.G., Priotton, J.-J., and Thomas, I.M. (1989) *Thin Solid Films*, **175**, 173–178.
19 Mennig, M., Oliveira, P.W., and Schmidt, H. (1999) *Thin Solid Films*, **351**, 99–102.
20 Minot, M.J. (1976) *J. Opt. Soc. Am.*, **66**, 515–519.
21 Gombert, A., Glaubitt, W., Rose, K., Dreibholz, J., Bläsi, B., Heinzel, A., Sporn, D., Döll, W., and Wittwer, V. (2000) *Sol. Energy*, **68**, 357–360.

22 Lohmüller, T., Helgert, M., Sundermann, M., Brunner, R., and Spatz, J.P. (2008) *Nano Lett.*, **8**, 1429–1433.
23 Bjelkhagen, H.I. (2005) Color holography, in *Encyclopedia of Modern Optics*, (eds B.D. Guenther & D.G. Steel) Elsevier, Amsterdam, pp. 64–72.
24 Haines, K., Colburn, W., Arends, T., and Kurtzner, E. (1971) *IEEE J. Quantum. Elect.*, **7**, 321–322.
25 Kim, G.H. (2005) *Eur. Polym. J.*, **41**, 1729–1737.
26 Kim, G.H., Kim, W.J., Kim, S.M., and Son, J.G. (2005) *Display*, **26**, 37–43.
27 Kim, G.H. and Park, J.H. (2007) *Appl. Phys. A*, **86**, 347–351.
28 Chang, S.I. and Yoon, J.B. (2004) *Opt. Express*, **12**, 6366–6371.
29 Chang, S.I., Yoon, J.B., Kim, H., Kim, J.J., Lee, B.K., and Shin, D.H. (2006) *Opt. Lett.*, **31**, 3016–3018.
30 Garcia-Guerrero, E.E., Mendez, E.R., Escamilla, H.M., Leskova, T.A., and Maradudin, A.A. (2007) *Opt. Express*, **15**, 910–923.
31 Parikka, M., Kaikuranta, T., Laakkonen, P., Lautanen, J., Tervo, J., Honkanen, M., Kuittinen, M., and Turuner, J. (2001) *Appl. Opt.*, **40**, 2239–2246.
32 Gu, C., Lien, J.R., Dai, F., and Hong, J. (1996) *J. Opt. Soc. Am. A*, **13**, 1704–1711.
33 Kim, S.I., Choi, Y.S., Ham, Y.N., Park, C.Y., and Kim, J.M. (2003) *Appl. Opt.*, **42**, 2482–2491.
34 Wadle, S., Wuest, D., Cantalupo, J., and Lakes, R.S. (1994) *Opt. Eng.*, **33**, 213–218.
35 Wadle, S. and Lakes, R.S. (1994) *Opt. Eng.*, **33**, 1084–1088.
36 Joubert, C., Loiseaux, B., Delboulbe, A., and Huignard, J.P. (1997) *Appl. Opt.*, **36**, 4761–4764.
37 Honma, H., Maekawa, Y., Takano, M., Murillo-Mora, L.M., Sato, A., Hirose, K., and Iwata, F. (2004) *Proc. SPIE*, **5290**, 74–80.
38 Sakai, D., Harada, K., Kamemaru, S.I., El-Morsy, M.A., Itoh, M., and Yatagai, T. (2005) *Opt. Rev.*, **12**, 383–386.
39 Wasserscheid, P. and Welton, T. (2003) *Ionic Liquids in Synthesis*, Wiley-VCH Verlag GmbH, Weinheim, Germany.
40 Parvulescu, V.I. and Hardacre, C. (2007) *Chem. Rev.*, **107**, 2615.

41 Lin, H., Oliveira, P.W., and Veith, M. (2008) *Appl. Phys. Lett.*, **93**, 141101.
42 Ahn, S.H. and Guo, L.J. (2009) *ACS Nano*, **3**, 2304–2310.
43 Ahn, S. and Guo, L.J. (2008) *Adv. Mater.*, **20**, 2044–2049.
44 Gier, A., Kunze, N., Mennig, M., de Oliveira, P.W., Sepeur, S., Schäfer, B., and Schmidt, H., (2004) European patent EP 1248685 B1.
45 Davis, F. (1998) System for making a hologram of an image by manipulating object beam characteristics to reflect image data. United States patent US 5822092.
46 Psaltis, D., Barbastathis, G., and Levene, M. (1996) in *Trends in Optics: Research, Developments, and Applications* (ed. A. Consortini), Academic Press, London, pp. 189–206.
47 Oliveira, P.W., Krug, H., and Schmidt, H. (1997) *Proc. SPIE*, **3136**, 442–451.
48 Eldada, L. and Shacklette, L.W. (2000) *IEEE J. Sel. Top. Quant.*, **6**, 54–68.
49 Mukherjee, A., Eapen, B.J., and Baral, S.K. (1994) *Appl. Phys. Lett.*, **65**, 3179–3181.
50 Tung, K.K., Wong, W.H., and Pun, E.Y.B. (2005) *Appl. Phys. A.*, **80**, 621–626.
51 Hikita, M., Yoshimura, R., Usui, M., Tomaru, S., and Imamura, S. (1998) *Thin Solid Films*, **331**, 303–308.
52 Chen, R.T. (1992) *Appl. Phys. Lett.*, **61**, 2278–2280.
53 Eldada, L., Xu, C., Stengel, K.M.T., Shacklette, L.W., and Yardley, J.T. (1996) *J. Lightwave Technol.*, **14**, 1704–1712.
54 Krug, H., Tiefensee, F., de Oliveira, P.W., and Schmidt, H. (1992) *Proc. SPIE*, **1758**, 448–455.
55 Ashley, P.R. and Tumolillo Jr., T.A. (1991) *Appl. Phys. Lett.*, **58**, 884–886.
56 Schlotter, N.E., Jackel, J.L., Townsend, P.D., and Baker, G.L. (1990) *Appl. Phys. Lett.*, **56**, 13–16.
57 Wong, W.H., Zhou, J., and Pun, E.Y.B. (2001) *Appl. Phys. Lett.*, **78**, 2110–2112.
58 Huang, Y., Paloczi, G.T., Scheuer, J., and Yariv, A. (2003) *Opt. Express*, **11**, 2452–2458.
59 Fujii, A., Suzuki, T., Shimizu, K., Yatsuda, K., Igusa, M., Ohtsu, S., and Akutsu, E. (2007) *Proc. SPIE*, **6775**, 677506.
60 Chao, C.Y. and Guo, L.J. (2002) *J. Vac. Sci. Technol. B*, **20**, 2862–2866.
61 Oh, M.-C., Lee, M.-H., Ahn, J.-H., Lee, H.-J., and Han, S.G. (1998) *Appl. Phys. Lett.*, **72**, 1559–1561.
62 Mukherjee, N., Eapen, B.J., Keicher, D.M., Luong, S.Q., and Mukherjee, A. (1995) *Appl. Phys. Lett.*, **67**, 3715–3717.
63 Kim, D.-H., Chin, W.-J., Lee, S.-S., Ahn, S.-W., and Lee, K.-D. (2006) *Appl. Phys. Lett.*, **88**, 071120.
64 Lukosz, W. (1995) *Sensor. Actuat. B*, **29**, 37–50.
65 Oh, M.-C., Lee, H.-J., Lee, M.-H., Ahn, J.-H., Han, S.G., and Kim, H.-G. (1998) *Appl. Phys. Lett.*, **73**, 2543–2545.
66 Kersey, A.D., Davis, M.A., Patrick, H.J., LeBlanc, M., Koo, K.P., Askins, C.G., Putnam, M.A., and Friebele, E.J. (1997) *J. Lightwave Technol.*, **15**, 1442–1462.
67 Heidemann, R.G., Kooyman, R.P.H., and Greve, J. (1993) *Sensor. Actuat. B*, **10**, 209–217.
68 Luff, B.J., Wilkinson, J.S., Piehler, J., Hollenbach, U., Ingenhoff, J., and Fabricius, N. (1998) *J. Lightwave Technol.*, **16**, 583–592.
69 de Ridder, R.M., Driessen, A., Rikkers, E., Lambeck, P.V., and Diemeer, M.B.J. (1999) *Opt. Mater.*, **12**, 205–214.
70 Enami, Y., Mathine, D., DeRose, C.T., Norwood, R.A., Luo, J., Jen, A.K.-Y., and Peyghambarian, N. (2007) *Appl. Phys. Lett.*, **91**, 093505.
71 Little, B.E., Chu, S.T., Haus, H.A., Foresi, J., and Laine, J.-P. (1997) *J. Lightwave Technol.*, **15**, 998–1005.
72 Little, B.E., Chu, S.T., Absil, P.P., Hryniewicz, J.V., Johnson, F.G., Seiferth, F., Gill, D., Van, V., King, O., and Trakalo, M. (2004) *IEEE Photonics Technol. Lett.*, **16**, 2263–2265.
73 Deng, W.-Y., Sun, D.-G., E, S.-L., and Xu, W. (2009) *Optik*, **120**, 188–194.
74 Xia, F., Sekaric, L., and Vlasov, Y. (2007) *Nature Photonics*, **1**, 65–71.
75 Yanik, M.F. and Fan, S. (2004) *Phys. Rev. Lett.*, **92**, 083901.
76 Scheuer, J. and Yariv, A. (2006) *Phys. Rev. Lett*, **96**, 053901.

77 Ehrlich, H., Janussen, D., Simon, P., Bazhenov, V., Shapkin, N., Erler, C., Mertig, M., Born, R., Heinemann, S., Hanke, T., Worch, H., and Vournakis, J. (2008) *J. Nanomater.*, **2008**, Article ID 670235, 8 pages.
78 Koller, D.M., Galler, N., Ditlbacher, H., Hohenau, A., Leitner, A., Aussenegg, F.R., and Krenn, J.R. (2009) *Microelec. Eng.*, **86**, 1314.
79 Atwater, H.A., Tsai, H.A., and Krenn, J.R. (2008) *IEEE Sel. Top. Quantum Electron.*, **14**, 1393.
80 Szameit, A., Garanovich, I.L., Heinrich, M., Sukhorukov, A.A., Dreisow, F., Pertsch, T., Nolte, S., Tünnermann, A., and Kivshar, Y.S. (2009) *Nature Physics*, **5**, 271–275.
81 Szameit, A., Kartashov, Y.V., Dreisow, F., Heinrich, M., Pertsch, T., Nolte, S., Tünnermann, A., Vysloukh, V.A., Lederer, F., and Torner, L. (2009) *Phys. Rev. Lett.*, **102**, 153901.
82 Lo, S.-S., Chen, C.-C., Garwe, F., and Pertsch, T. (2007) *J. Phys. D: Appl. Phys.*, **40**, 754–758.
83 Kagan, D., Calvo-Marzal, P., Balasubramanian, S., Sattayasamitsathit, S., Manesh, K.M., Flechsig, G.-U., and Wang, J. (2009) *J. Am. Chem. Soc.*, **131**, 12082–12083.

Index

a

acrylate monomer for UV-NIL resists   12, 20
adhesion
– biological   269, 320f
– cellular   269, 279f, 296
– fibrillar   319ff
– of fouling organisms   306
– between mold and resist   18, 20
– promoter   33, 35
– reinforcement   274
adhesive
– adhesive structures of animals   320f
– bioinspired patterned structures   319ff
air-bridged structure   78f
air entrapment in NIL processes   10
alignment accuracy   4
alignment of dewetted structures   233
angle-resolved colloidal lithography (ARCL)   187
antiadhesion layer   18
antifouling coating   305ff
antiglare structure   343
antireflection coating
– dielectric   345f
– hologram   347
– moth eye   346
antisticking property   19, 31, 34
antisticking strategy   17f
application
– of block copolymer micelle nanolithography (BCMN)   275
– of colloidal patterned polymers   188f
– of decal transfer microlithography (DTM)   89
– of dewetting of polymer films on patterned structures   234f
– of directed assembly of block copolymers   208
– of direct ink writing   97
– of dynamic nano-inscription (DNI)   39
– of embossing techniques   351f
– of excimer ablation of polymers   147
– of femtolaser ablation   156f
– of holographic techniques   350f
– of laser ablation in microstructure fabrication   141
– of laser direct writing (LDW), see application of TPP
– of microcontact printing (μCP)   59
– of micromolding in capillaries (MIMIC)   63
– of microtransfer molding (μTM)   88f
– of nanoimprint lithography (NIL)   10, 29f
– of optical micro- and nanostructures   343ff
– of reversal nanoimprinting (RNi)   88
– of soft molding in 3D patterning   48
– of solvent-assisted molding   53f
– of two-photon polymerization (TPP)   120f, 123f, 128ff

b

barnacle settlement   304, 309
BCMN, see block copolymer micelle nanolithography
bilayer transfer process   30
binary supercrystal   184f
biomedical material   291
biomimetic application
– of direct ink writing   97
– polyelectrolyte structures as templates for   97ff
biomimetic scaffold   269, 272, 284
biomimetric patterned polymer adhesive   319ff
biomolecular surface patterning   57, 59
block copolymer   199

– with amphiphilic side chains   309f
– with fluorinated PEG-ylated side chains   309f
– free energy of a microphase separated   200
– polystyrene-based diblock copolymer   311
– polystyrene-*block*-polydimethylsiloxane   205f
– polystyrene-*block*-poly(methyl methacrylate)   202ff
– polystyrene-*block*-poly(2-vinylpyridene)   205f
– polystyrene-*block*-(*t*-butylacrylate)   207
block copolymer micelle nanolithography (BCMN)   272f
Bragg diffraction   189
Bragg reflection   341
Brownian motion   173f
bubble dissolution in NIL processes   10
bubble evacuation in NIL processes   10
buckling   60

c

capacitor device for studying capillary surface instabilities   248, 253
capillarity   52f
capillary assembly of polymer particles   182
capillary depression   50
capillary force   7, 63
capillary force lithography (CFL)   53
capillary rise   50
capillary surface instability   248
capillary wicking   86
carbon nanotube-filled polysiloxane   313
CARS microscopy, *see* coherent anti-Stokes Raman scattering microscopy
cationic photopolymerization   113ff
cell-matrix interaction, *see* cell-substrate interaction
cells on patterns   269ff
cell-substrate interaction   272
cellular interaction with patterned surfaces   269ff
– contact guidance in   275ff
– effect of microgrooves on   292ff
– interaction with hierarchically patterned substrates   275
– interaction with three-dimensional scaffolds   296f
– interaction with topographically patterned substrates   275f
– sensing of the matrix rigidity by the cells   281f
chemically directed self-assembly   199
coagulating reservoir   94

coalescence of polymer droplets   234f
coating of polymer microstructures   130
coherent anti-Stokes Raman scattering (CARS) microscopy   125ff
colloidal lithography   185ff
colloidal monolayer production   185f
colloidal polymer patterning   171, 177
– fabrication of three-dimensional arrays   178f
– fabrication of two-dimensional arrays   179f
– pre-patterning   178
– by sedimentation   179f
– using an dieletrophoretic assembly   181
– using capillary interaction   181
colloidal system
– phase diagram for colloidal spheres   176
– structural transitions in   176f
concentric nanograting pattern   39
confinement of dewetting polymer films   235f, 239
curing speed   11

d

decal transfer microlithography (DTM)   76f
demolding   4, 10, 12, 63, 71
deposition nozzle   93
dewetted structure   236
– induced by EHD instabilities   250
– ordering by chemically patterned substrates   236
dewetting   50, 218
– on chemically patterned substrates   234ff
– heterogeneous   229ff, 252
– method for residual layer removal   84
– nucleated   220, 229
– pathway   242ff, 249
– pattern formation caused by   226f
– on physically (topographically) patterned substrates   234, 238ff
– rim formation   227
– during spin-coating   238
– spinodal   220ff, 231
– theory of pattern-directed   241f
dewetting experiment   225ff
– on combinatorial test surfaces with gradient micropatterns   237f
– influence of aging time on   231
– influence of residual stress on   231f
– simulation of pattern-directed   241ff
dip-coating   178, 180, 225
dip-pen lithography   183
directed assembly system
– energetics of   200
– free surface energy of   202

– interfacial energy contrast of   203
– non-bulk morphologies   207
directed self-assembly of block copolymer films   199ff
– on chemical patterns   211
– fabrication of one-dimensional nanoparticle arrays   208
– lithographic application of   210
– process steps of   201
direct ink writing   95
direct NIL patterning of soft material   28f
direct-write assembly   93ff
– of 3D hydrogel scaffolds   102f
disorder-to-order transition of hard spheres   176
dispersed particle behaviour   173ff
distortion of printed patterns   4, 13
DLVO theory   176
DMI, see duo-mold imprinting
DNI, see dynamic nano-inscription
drop-casting process, see colloidal monolayer production
drop-dispensing   6
drop-on-demand™ 6, see also drop-dispensing
DTM, see decal transfer microlithography
duo-mold imprinting (DMI)   77f
dynamic nano-inscription (DNI)   36ff
dynamics of two-dimensional surfaces   220f

e

effective absorption coefficient of laser radiation   146
EHD, see electro hydrodynamic lithography
elastomeric mold   44
electrohydrodynamic-base particle assembly   186
electro hydrodynamic lithography (EHD)   247ff
electron beam lithography   3, 16, 57, 180, 320
embossing   44, 47, 75, 350
emulsion polymerization   172f
epoxy monomer for UV-NIL resists   12
epoxysilicone nanograting   33f
epoxysilicone resist layer   32f
equilibrium bulk lamellar period   200, 203
etch resistance   11f
etch selectivity   199
ethylene tetrafluoroethylene (ETFE) mold   31ff

f

fabrication
– of biodegradable microdevices   150, 154
– of biomimetic adhesives   89, 320ff
– of diffractive optical elements   350ff
– of fibrillar polyimide nanostructures   321f
– of holograms   348
– of light-driven machines   130
– of membranes   150
– of microchannels in polymers   152
– of microelectronic devices   157
– of microfluidic systems   147
– of micropattern comprising different materials   123f, 160
– of nanofibers   321ff
– of organic thin film transistor via Rni   88
– of semiconductors by directed self-assembly of polymers   212
– of three-dimensional biocompatible scaffolds   296f
– of waveguides   88, 157, 354
fabrication of three-dimensional microstructures   97, 110ff
– with freely movable parts   123, 130f
– free-standing microstructures   112, 120f, 123, 160
– photonic crystals   129f
– photonic devices   153
– polymeric channel structures   152f
fabrication of three-dimensional nanostructures   330f
fabrication of two-dimensional nanoparticle arrays   186f
femtosecond laser   153f
film-air interface   220, 228
film rupture   219, 229
– influence of residual stress on   213
– mechanism of   230
film-substrate interface   220, 228
fingering instability   232
flat panel display   344
flourinated silane as release agent   19
fluorinated mold treatment-resist interaction   20f
fluorinated self-assembled monolayer (F-SAM)   19f
focused ion beam lithography   57
fouling interface   303f
fouling organism   303
– impact of surface patterning on   305
– settlement stages of   304
fouling release coating   304
– based on nanocomposites   312
– superhydrophobic   313f
free-form array pattern   36
Fresnel lens   350f

fully integrated multiplexer/demultiplexer   357
fused silica mold   16

g
generative laser process   158ff
glass transition temperature   27, 72, 75, 225
gold nanograting   38

h
Hamaker constant   228
hardening   4
hierarchical microfibril array   327ff
high aspect ratio feature   7, 10, 33, 159, 321
high resolution nanoscale patterning   3ff, 27, 109
high-throughput nanoscale patterning   3f, 27
hollow-woodpile polymer structure   98
holographic material   348f
holography   341, 347ff
HSQ, see hydrogen silsesquioxane
hybrid organic-inorganic monomer for UV-NIL resists   14
hydrogel ink   101f

i
imaging techniques for microstructure characterization   125
imprinting force   7, see also imprinting pressure
imprinting material   11
imprinting of a wire-grid polarizer   35
imprinting of thin grating structures   34f
imprinting on topographies   79f
imprinting pressure   3ff, 7, 10
imprinting step   4
imprinting time   7
imprint lithography, see also nanoimprint lithography
– limitations of   43
initial resist thickness   8f
ink coagulation   96
ink deposition process   95
inkpad   75, 80
ink viscosity   93
instability-mediated disintegration   218
integral microstereolithography   158
interaction of light and matter   337ff
– phenomena in nature   342ff
interfacial energy   205
interfacial interaction   228
interfacial tension   226, 228
interference   339

intersurface force   220, 229
inverted microcontact printing (i-μCP)   61
ion implantation   3

j
"jet and flash" imprint lithography™ (J-FIL)   6

l
lamellar morphology period   200, 203
Laplace pressure   222, 247, 251
laser ablation   141ff
– application in bioengineering   150, 154f
– choice of wavelength   144
– excimer   147ff
– heat affected zone (HAZ)   145
– incubation effect   143
– material damage caused by   145
– mechanism of   144
– morphology of the ablated region   149
– photochemical   145f, 148
– of polymers   143f, 147ff
– thermal   145
laser beam quality   143
laser direct writing (LDW)   109ff, 158ff
– experimental TPP setup   116ff
– limitations of   131
– modes of   117f
– resolution in   119f
laser fluence   142, 145, 148
laser-induced forward transfer (LIFT)   146
laser-induced roughening   155f
laser micromachining of polymers   141ff, 147ff
– application of   149f
– fabrication of microcapillaries by   152
– of fluorinated polymers   153
– by melting   144f
– parallel mode of   143
– by photolytic decomposition   145f
– serial mode of   143
– using a femtosecond laser   153
– using an excimer laser   147ff
– by vaporizing   144f
laser pulse duration   142
laser pulse energy   142
laser repitition rate   142
laser source   109, 111, 142, 145
layer-by-layer fabrication of three-dimensional structures   79f, 85f, 158ff
LDW, see laser direct writing
lift-off process   3, 8, 28
linear grating mold   38
line edge roughness (LER)   10, 16, 212
lithography techniques

– colloidal lithography   185ff
– dip-pen lithography   183
– focused ion beam lithography   57
– microstereolithography   158ff
– nanoimprinting lithography   4ff
– scopes and limitations   57
– top-down method   217ff
lotus effect   155

m
marine biofouling   303
material deposition   3
maximum peak intensity of laser radiation   142
mesopatterning   232f, 255ff, *see also* dewetting experiment
mesoscale patterning effect   217
microcontact printing (µCP)   43, 49, 57, 183
– inverted (i-µCP)   61
– in liquid media   61
– of self-assembled monolayers   59f
microfluidic device   58
microlens fabrication   147
micromolding in capillaries (MIMIC)   43, 53, 57, 63
micromolding of hydrophilic biopolymers   49
micromolding of hydrophobic polymers   49
micromolding of PEG   49
micropatterning   36, 43f
– of biocompatible substrates   276
– of filamentary features   93
microphase separation   199f
microreplication   348ff, 354
microstereolithography (µSL)   158ff
microtransfer molding (µTM)   43, 57, 64, 72ff
modulation of cell behaviour by polymer surfaces   270f
mold   3, *see also* stamp
mold alignment   79
mold characterization   16f
mold copy   21f
mold degradation   20
mold design   11
mold fabrication   16f
mold feature geometry   10
mold filling   71ff
mold flatness   10
mold flexibility   30
mold inspection   11
mold material   16
– polymeric   21
mold repair   11
mold-resist sticking   17f
mold surface treatment   82f

mold treatment   18f
monodispersed polymer lattice   171
multichannel hologram   348
multilayer printing   79ff, *see also* imprinting on topographies

n
nanofiber
– coated structures   326f
– hierarchical structures   327ff
– with three-dimensional structured tips   331
– tilted structures   324ff
– vertical structures   320ff
nanofiber fabrication
– by combined photolithography-dry etching method   329
– by drawing from polymer drops on nonwettable surfaces   325
– by filling porous membranes   321, 327
– by hot embossing   323f
– by inclined litography   324
– by multistep exposure in photolithography   327ff
– by photolithographic templating   321f
– by two-step embossing   330
– using electron beam litography   320f, 326
nanoimprinting with soft stamps   62f
nanoimprint lithography (NIL)
– continous roll-to-roll   30
– continous thermal   30
– relevant process parameters for application of   10
– roll-to-plate   32
– roll-to-roll   29ff
– schematic of   4, 27f
– thermal   3, 7, 27
– UV-assisted   3ff, *see* UV-NIL
nanopatterning
– of ceramic materials   63
– of flexible surfaces   79f
– of functional polymers   36ff
– of metals   36
– over curved surfaces   36
– of polymer surfaces by CVD   314f
– thermal gradient-induced   255ff
– of topographies   79f
next generation lithography (NGL)   3
NIL, *see* nanoimprint lithography
nucleation mechanism
– film defect-related homogeneous   220
– substrate defect-related heterogeneous   220, 229, 252

o

one-photon excitation 112
optical microstructure
– application of 343ff
– in nature 342f
optical projection lithography 3
optical storage media 337, 357
organic light emitting diode (OLED) 147
organic thin film transistor 88
oxygen plasma treatment 34, 83, 186

p

particle-particle interaction 174f
particle-surface interaction 174f, 183
pattern deformation in μCP 61
pattern density multiplication 213
pattern formation 3, 176ff
– bottom-up self-assembly technique 217
– electric field-induced 247ff, 252ff
– engineering of hierachical patterns 253
– film thickness-dependent phase transitions 243
– growth of holes in dewetting experiments 227
– on photocurable polymer films 249
– on polymer bilayers 252ff
– polymers for instability-induced 225
– in polymer thin film dewetting 232ff
– theory of electric-field induced 250f
– thermal gradient-induced 255ff
– top-down method 217
pattern morphology
– biomimetic 307
– 'block mountain-rift valley' pattern 246
– 'castle moat' pattern 246
– cellular pattern 227
– columnar polymer pattern 255
– conformal adhesion 239f
– control of, in dewetting experiments 232ff
– free hanging film 239
– influence of template morphology on 235
– of instability-induced structures 219, 226
– parameters influencing EHD instabilities 248
– phase diagram of dewetted structures 245
– polygonal pattern 227
– spiral polymer pattern 256
– stretched topographies in PDMS 309
– of thin polymer films in different stages of dewetting 226
pattern rectification 213
pattern transfer 50, 59, 63, 65, 73f, 77f, 80, 111f, 186
PDMS, see poly(dimethylsiloxane)

PDMS fiber 322
PDMS mold 44, 58ff, 72, 306
PEG, see polyethylene glycol
PEM, see polyelectrolyte multilayer
phase contact line
– retraction of 227
photo-ablation 148ff
photocurable ink 93
photo-initiator 113ff, 122
photolithography
– limitations of 43
photonic band gap (PBG) 97, 128
photonic crystal 128
– application of 128, 188
– fabrication of 97, 128, 188
– optical properties of 188f
photopolymer 348f
photopolymerizable precursor 109
photopolymerizable resin composition 113
photopolymerization 110f
placement accuracy 4, 6
planarization layer 5, 7
plasma composition 14
plasma enhanced vapor deposition (PECVD) 19
plasma etching 14f
– anisotropic "breakthrough" 3, 14
plastic polymer deformation 37
PMMA, see poly(methyl methacrylate)
poly(dimethylsiloxane) (PDMS) 22, 30, 58f, 75, 306
polyelectrolyte ink 95
polyelectrolyte multilayer (PEM) 54
polyethylene glycol (PEG) 49
polymer bilayer 252
polymer dispersion
– capillary interactions in 175, 177
– depletion interaction in 175
– electrostatic interaction in 174
– mean particle displacement in 173
– sedimentation of 174
– steric interaction in 175
– van der Waals interaction in 174
– viscous drag of 174
polymer distributed feedback (DFB) laser 54
polymer droplet 232
polymer film thickness 221
– critical dewetting 221, 239
– local 243
polymer film transfer 75f
polymeric ink formulation 93
polymer inking 74
polymer inverse opal 188
polymerization intensity threshold 122

polymer light emitting device 54
polymer material for SLS 161
polymer mold 31
polymer-mold interaction 83
polymer opal 187f
polymer particle assembly 182
– based on specific adsorption 183
– based on vertical deposition of two-dimensional films 187
– based on wetting phenomena 183
– of functional and biopolymers 183f
– high precision techniques 184
– of polymer opals 187
– using sedimentation 187
– using templates 184
polymer-polymer interface 252
polymer substrate for DNI 37
polymer substrate-laser interaction 109
polymer superstructure 183
poly(methyl methacrylate) (PMMA) 151, 203, 225
protrusion density of a mold 8, 11

q
quartz template 18

r
radical photopolymerization 113ff
radical polymerization 172
Raleigh instability 226
random sequential adsorption (RSA) 183
raster-scan mode 117f
reactive-ion-etching (RIE) 27f, 49, 84, 186, 330
reflectance 340, 345
refractive index 97
relief hologram 350
replica molding (REM) 43ff, 306f
repulsive iinterfacial interaction 223
residual layer 79, 83f
resist adherence
– to mold 12
– to substrate 12
resist contamination 7
resist depensing 4
resist displacement 7
resist flow 4, 7f
resist layer
– application on wafers 5ff
– examples of imprinted 8f
– low-viscosity 4
– residual 4f
resist mask 3, 7, 11
resist material, see also imprinting material

– composition of UV-NIL 12ff
– liquid low-viscosity 5
– liquid UV-curable epoxysilicone 30
– organic 6
– perfluorinated 15f
– requirements on 11f
– silicon-containing 14
– silicon-rich plasma-etch-resistant 6, 15
resist-oriented antisticking solution 21
resist removal, see resist stripping
resist shrinkage 13f
resist stripping 15f
resolution limit 338
resolution loss in µCP 60
reversal nanoimprinting (RNi) 71ff
– with soft inkpads (RNsi) 75f
Reynolds number 174
rheology of direct writing inks 96, 99, 102
RNi 88, see reversal nanoimprinting
roller imprinter 31
roof collapse 60
room-temperature imprinting 36
rubbing 233f

s
SADEM, see solvent-assisted dewetting molding
SAMIM, see solvent-assisted microcontact molding
sedimentation of colloidal polymer suspensions 178
selective laser sintering (SLS) 160
self-assembling polymer with structured surface
– amphiphilic block copolymers 309ff
– hyperbranched amphiphilic networks 311
– phase-segregating quaternized siloxanes 311f
– phase-segregating siloxane-urethane nanohybrids 311f
– polystyrene-based diblock copolymers 311
self-assembly of nanoparticles 186f
self-organized patterning 218
self-removal of residual layers 84
self-supporting filaments 93, 96
S-FIL, see "step and flash" imprint lithography™
S-FIL reverse nanoimprint process (S-FIL/R) 6
shape memory polymer (SMP) 332
silane coupling agent treatment 83
silicone elastomer mold 43
silicone mold 36, 38
silk fibroin ink 98f

silsesquioxane
– fluorinated photocurable 31
– hydrogen silsesquioxane (HSQ) 16
SLS, see selective laser sintering
soft lithography 43, 57
– application of 43, 58, 322
– forced 65
– principles of 58f
soft molding, see solvent-assisted molding
sol-gel process 63
solvent absorption rate 45f
solvent-assisted dewetting molding (SADEM) 49f
solvent-assisted microcontact molding (SAMIM) 43ff, 47ff, see also solvent-assisted molding or soft molding
solvent-assisted molding 64
– etching in 46
– with a good solvent 47ff, see also SAMIM
– with a poor solvent 49f, see also SADEM
– principles of 44f
– properties of solvents for 47f
solvent evaporation rate 46
solvent for soft-molding 47f
spider silk 95
spin-coating 5, 72, 225
– dewetting during 238
spinning dope 95
spinodal instability, see also dewetting and surface instability
– deep inside the spinodal territory (DIST) 230f
– defect-sensitive spinodal regime (DSSR) 231
spinodal length scale 229
spinodal parameter 223
"spot-by-spot" patterning 109
stage bridging transition in dewetting of polymer films 235f
stamp 3, see also mold
– fused silica 16
– for microcontact printing 49
– PDMS 49, 53, 59, 323, 330
– poly(methyl methyacrylate) (PMMA) 61
– preparation of 59
– soft 62
– for thermal gradient-induced patterning 257
stamp copy 21
"step and flash" imprint lithography™ (S-FIL) 6, 27, 32
step-and-repeat processing 4, 6
substrate surface treatment 82f
superhydrophilic polymer surface 155f
superhydrophobic surface 313
surface capillary wave 218
surface energy 82f, 206, 237
surface fluctuation 222
surface instability
– capillary 248, 256
– commensuration between structural periodicity and length scale of 229, 235, 243, 248f, 252f, 255
– EHD 248ff
– external field-mediated 220, 246ff
– influence of substrate-polymer film interactions on 228
– length scale of 224
– nucleated 229
– of spinodally instable films 223
– spontaneous 219ff
– thermal gradient-induced 256
– in thermodynamically unstable ultrathin films 220
– thermodynamic origin of 220
– in thick metastable films 220, 247
– of thin polymer films 217ff
surface roughening 155, 218
surface roughness 18, 307
surface tension 250
surface wetting 184

t

template
– for assembling polymer particles 182
– for colloidal lithography 185f
– for dewetting experiments 233ff, 257
thermal expansion mismatch 4
thermal gradient-induced patterning 255ff
thermoplastic polymer 75
thin polymer film
– energetically classification of 221
– excess free energy per unit area 221
– instability-mediated disintegration of 218
– pattern formation in 217ff
– polygonal patterned 227
– preparation of, for dewetting experiments 225
– self-organized patterning of 218
– surface instability of 217ff
– 'switching' of surface instablities in
three-dimensional colloidal crystal 187
three-dimensional (3D) microperiodic array 94
three-dimensional (3D) microscale patterning 43, 46ff, 77, 79f
three-dimensional plotting 298f
three-layer polymer structure 81

tissue engineering 276ff, 294
– using silk scaffolds 99f
topographically directed self-assembly 199
TP, see transfer printing
TPP, see two-photon polymerization
transfer etching 3
transfer printing (TP)
– advantages and disadvantages of 85
– main characteristics of TP techniques 87
– polymers in 86
– principles of 71ff
two-dimensional colloidal crystal 186
two-dimensional nanoparticle array 185f
two-photon cross section 113f
two-photon excitation 110ff
two-photon polymerization (TPP)
– coating of polymer microstructures 130
– fabrication of photonic elements 130f
– fabrication of three-dimensional photonic crystals 128ff
– fundamentals of 110f
– imaging techniques for characterization of TPP fabricated microstructures 125ff
– intensity threshold in 121f
– monitoring of 119
– monomer-oligomer mixtures for 114f
– photo-initiator for 113ff
– process steps of 111f
– resolution in 121, 158
– shape of voxels created by 122
– strategies to decreased processing time 131f
– three-dimensional microstructures fabricated by 120f, 123f
two-photon stereolitography, see two-photon polymerization (TPP)

u
ultrashort pulsed laser 109, 144
ultraviolet excimer gas laser 109
UV curing of UV-NIL resists 12f
UV laser 147ff
UV laser ablation 147
UV molding 64f, see also UV-NIL
UV-NIL
– resist 12, 20
– stamp fabrication 16f
– step-and-repeat process 5ff
– tool 9ff
– using large stamps 7
UV-sensitive photo-initiator 12

v
vector-by-vector microstereolithography 158
vector-scan mode 118, 158
volume hologram 348f
voxel 119

w
waveguide 353f
waveguide grating 354
wavelength division multiplexing (WDM) 354
wavelength filtering 357
wave vector 338
"wet-and-drag" mold filling 72
wettability contrast 236
whole-layer transfer 75
wire-grid polarizer 35

x
xerography 247
X-ray lithography 57